For Kate,
who said yes

FALLEN DRAGON

CHAPTER ONE

TIME WAS WHEN THE BAR WOULD HAVE WELCOMED A MAN FROM Zantiu-Braun's strategic security division, given him his first beer on the house and listened with keen admiration to his stories of life as it was lived oh so differently out among the new colony planets. But then that could be said of anywhere on Earth halfway through the twenty-fourth century. In the public conscience, the glamour of interstellar expansion was fading like the enchantment of an aging actress.

As with most things in the universe, it was all the fault of money.

The bar lacked money. Lawrence Newton could see that as soon as he walked in. It hadn't been refurbished in decades. A long wooden room with thick rafters holding up the corrugated carbon-sheet roof, a counter running its length, dull neon adverts for extinct brands of beers and ice creams on the wall behind. Big rotary fans that had survived a couple of centuries past their warranty date turned above him, primitive electric motors buzzing as they stirred the muggy air.

This was the way of things in Kuranda. Sitting high in the rocky tablelands above Cairns, it had enjoyed long profitable years as one of Queensland's top tourist-trap towns. Sweating, sunburned Europeans and Japanese had made their way up over the rain forest on the skycable, marveling at the lush vegetation before traipsing round the curio shops and restaurant bars that made up the main street. Then they'd take the ancient railway down along Barron Valley Gorge to marvel once again, this time at the jagged rock cliffs and white foaming waterfalls along the route.

Although tourists did still come to admire northern Queensland's natural beauty, they were mostly corporate families that Z-B had rotated to its sprawling spaceport base that now dominated Cairns physically and economically. They didn't have much spare cash for authentic Aboriginal print T-shirts and didgeridoos and hand-carved charms representing the spirit of the land, so the shops along Kuranda's main street declined until only the hardiest and cheapest were left—themselves a strong disincentive to visit and stay awhile. Nowadays people got off the skycable terminus and walked straight across to the pretty 1920s-era train station a couple of hundred meters away, ignoring the town altogether.

It left the surviving bars free for the local men to use. They were good at that. There was nothing else for them. Z-B brought in its own technicians to run the base, skilled overseas staff with degrees and spaceware engineering experience. Statutory local employment initiatives were for the crappiest manual jobs. No Kuranda man would sign up. Wrong culture.

That made the bar just about perfect for Lawrence. He paused in the doorway to scan its interior as a formation of TVL88 tactical support helicopters thundered overhead on their way to the Port Douglas practice range away in the north. A dozen or so blokes were inside, sheltering from the evil midday sun. Big fellas, all of them, with fleshy faces red from the first round of the day's beers. A couple were playing pool, one solitary, dedicated drinker up at the bar, the rest huddled in small groups at tables along the rear wall. His brain in full tactical mode, Lawrence immediately picked out potential exit points.

The men watched silently as he walked over to the counter and took off his straw hat with its ridiculously wide brim. He ordered a tin of beer from the middle-aged barmaid. Even though he was in civilian clothes, a pair of blue knee-length shorts and a baggy Great Barrier Reef T-shirt, his straight back and rigid crew-cut marked him out as a Z-B squaddie. They knew it; he knew they knew.

He paid for the weak beer in cash, slapping the dirty Pacific Dollar notes down on the wood. If the barmaid noticed his right hand and forearm were larger than they should be, she kept quiet about it. He mumbled at her to keep the change.

The man Lawrence wanted was sitting by himself, only one table away from the back door. His hat, crumpled on the table next to his beer, had a rim as broad as Lawrence's.

"Couldn't you have chosen somewhere more out of the way?" Staff Lieutenant Colin Schmidt asked. The guttural Germanic tone made several of the local men look around, eyes narrowing with instinctive suspicion.

"This place suits," Lawrence told him. He'd known Colin for all of the full twenty years he'd spent in Z-B's strategic security division. The two of them had been in basic training together back in Toulouse. Green nineteen-year-old kids jumping the fence at nights to get to the town with its clubs and girls. Colin had applied for officer training several years later, after the Quation campaign: a careerist move that had never really worked out. He didn't have the kind of drive the company wanted, nor the level of share ownership that most other young officers had to put them ahead. In fifteen years he'd moved steadily sideward until he wound up in Strategic Planning, a glorified errand boy for artificial sentience programs running resource-allocation software.

"What the hell did you want to ask me you couldn't say it down at the base?"

"I want an assignment for my platoon," Lawrence said. "You can get it for me."

"What kind of assignment?"

"One on Thallspring."

Colin swigged from his beer tin. When he spoke his voice was low, guilty. "Who said anything about Thallspring?"

"It's where we're going for our next asset realization." On cue, another flight of TVL88s swept low over the town; with their rotors running out of stealth mode the noise was enough to rattle the corrugated roof. All eyes flicked upward as they drowned conversation. "Come on, Colin, you're not going to pull the bullshit need-to-know routine on me, are you? Who the hell can warn the poor bastards we're invading them? They're twenty-three light-years away. Everybody on the base knows where we're going—most of Cairns, too."

"Okay, okay. What do you want?"

"A posting to the Memu Bay task force."

"Never heard of it."

"Not surprised. Crappy little marine and bioindustry zone, about four and a half thousand kilometers from the capital. I was stationed there last time."

"Ah." Colin relaxed his grip on the beer tin as he started to work out angles. "What's there?"

"Z-B will take the biochemicals and engineering products; that's all that's on the asset list. Anything else . . . well, it leaves scope for some private realization. If you're an enterprising kind of guy."

"Shit, Lawrence, I thought you were a straighter arrow than me. What happened to getting a big enough stake to qualify for starship officer?"

"Nearly twenty years, and I've made sergeant. I got that because Ntoko never made it back from Santa Chico."

"Christ, Santa fucking Chico. I forgot you were on that

one." Colin shook his head at the memory. Modern historians were comparing Santa Chico to Napoleon's invasion of Russia. "Okay, I get you posted to Memu Bay. What do I see?"

"Ten percent."

"A good figure. Of what?"

"Of whatever's there."

"Don't tell me you've found the final episode of *Fleas on the Horizon*?"

"That's *Flight: Horizon.* But no; no such luck." Lawrence's face remained impassive.

"I got to trust you, huh?"

"You got to trust me."

"I think I can manage that."

"There's more. I need you at Durrell, the capital, in the Logistics Division. You'll have to arrange secure transport for us afterward, probably a medevac—but I'll leave that to you. Find a pilot who won't ask questions about lifting our cargo into orbit."

"Find one who would." Colin grinned. "Bent bastards."

"He has to be on the level with me. I will not be ripped off. Understand? Not with this."

Colin's humor faded as he saw how much dark anger there was in his old friend's expression. "Sure, Lawrence, you can rely on me. What sort of mass are we talking about?"

"I don't know for certain. But if I'm right, about a backpack per man. It'll be enough to buy a management stake for each of us."

"Hot damn! Easy meat."

They touched the rims of their tins and drank to that. Lawrence saw three of the locals nod in agreement, and stand up.

"You got a car?" he asked Colin.

"Sure: you said not to use the train."

"Get to it. Get clear. I'll take care of this."

Colin looked at the approaching men, making the calculation. He wasn't frontline, hadn't been for years. "See you on Thallspring." He jammed on his stupid hat and took the three steps to the back door.

Lawrence stood up and faced the men, sighing heavily. It was the wrong day for them to go around pissing on trees to mark their territory. This bar had been carefully chosen so the meeting would go unnoticed by anyone at Z-B. And Thallspring was going to be the last shot he'd ever get at any kind of a decent future. That didn't leave him with a lot of choice.

The one at the front, the biggest, naturally, had the tight smile of a man who knew he was about to score the winning goal. His two compadres were sidling up behind, one barely out of his teens, swigging from a tin, the other in a slim denim waistcoat that showed off glowmote tattoos distorted by old knife scars. An invincible trio.

It would start with one of them making some comment: *Thought you company people were too good to drink with us.* Not that it mattered what was actually said. The act of speaking was a way of ego pumping until one of them was hot enough to throw the first punch. Same dumb-ass ritual in every low-life bar on every human planet.

"Don't," Lawrence said flatly, before they got started. "Just shut up and go sit down. I'm leaving, okay."

The big fella gave his friends a knowing I-told-you-he-was-chickenshit grin and snorted contempt for Lawrence's bravado. "You ain't going nowhere, company boy." He drew his huge fist back.

Lawrence tilted from the waist, automatic and fast. His leg kicked out, boot heel smashing into the big fella's knee. The one in the denim waistcoat picked up a chair and swung it at Lawrence's head. Lawrence's thick right arm came up to block the unwieldy club. One leg of the chair hit full on, just above his elbow, and stopped dead. Its impact didn't even make Lawrence blink, let alone grunt in pain. The man staggered back as his balance was slung all to hell. It was like he'd hit solid stone. He stared at Lawrence's arm, eyes widening as realization hammered through the drink.

All around the bar, men were pushing back chairs and rising. Coming to help their mates.

"No!" the man in the waistcoat shouted. "He's in Skin!"

It made no difference. The youngster was going for the big bowie knife in his belt scabbard, and nobody was paying any attention to warnings as they closed in.

Lawrence raised his right arm high, punching the air. He could feel a gentle rippling against his wrists as peristaltic muscles brought the darts forward out of their magazine sacs into launch tubules. A ring of small dry slits peeled open above his carpals, black nozzles poking out. The dart swarm erupted.

As he left the bar, Lawrence turned the cardboard sign on the door so it said CLOSED and shut it behind him. He made sure his hat was on square, a fussy action, covering his anger. God damn the Armory Division. Those bastards never erred on the side of caution, always on the side of overkill. He'd seen two of the men lying on the floor start to convulse, the dart toxin levels set way too high for a simple incapacitation sting. The bar was going to get very noisy with police, very quickly.

A South American couple was sitting at one of the tables on the bar's veranda, studying the laminated menu. Lawrence smiled politely at them and walked off down the main street back to the skycable terminus.

Ambulances and police vehicles were parked down the length of Kuranda's main street when Simon Roderick's TVL77D executive liaison helicopter whispered over the town. They were at all sorts of angles to each other, completely blocking the road for thirty meters on either side of the bar. There were obviously no traffic regulator nodes to guide anybody through Kuranda's streets. Thoroughly in keeping with the town's doughty throwback nature. He shook his head in bemusement at the chaos. Emergency service drivers could never resist the dramatic slam-halt arrival. Tough luck if one of the injured needed a paramedic crew urgently; the closest vehicles were all police. Paramedics clad in green boilersuits were maneuvering stretchers around awkward angles, sweaty faces straining from the effort.

"God, what a bunch of no-brainers," Adul Quan complained from the seat behind Simon. The Third Fleet intelligence operative had pressed his face against the helicopter's side window so he could view the town directly. He never liked utilizing sensor feeds through his direct neural interface, claiming the viewpoint switch made him giddy. "We should bid to manage the state's civil operations. At least offer them AS coordination, bring them into this century."

"We have the urban area franchise," Simon replied. "And all our people have some kind of medical monitor fitted in case there's a problem. We can retrieve them wherever they are. That's what matters."

"Good PR, though. Devoting resources to helping civilians."

"If they want our help they should take a stake in us, contribute and participate."

"Yes, sir."

Simon heard the skepticism in the other's voice and made no comment. To get where he was, Adul had built up a large stake in Z-B, but even that couldn't make him understand what true belonging meant. In truth, Simon thought, no one except himself did. That would change eventually.

Simon used his DNI to feed a series of commands to the

autopilot, and the helicopter swung round over the little circular park at the top end of the main street. As he came back to the scrubland truck lot he'd identified as a landing zone, he saw that some kids had spray painted an open eye on the corrugated roof of a derelict shop. The fading green and blue symbol was big enough to stare up at all the strategic security division helicopters that zipped through the tropical skies above the town. Like a perfect portrait painting, its gaze followed Simon as the TVL77D extended its undercarriage and sank down on the baked-mud surface. Rotor downwash sent a flurry of crushed tins and junk-food wrappers tumbling away from them as the fuselage lost its gray sky-blur integument, reverting to ominous matte black.

He paused for a moment as the turbines wound down. His personal AS had extended trawlers to retrieve all the emergency service e-traffic within the local datapool. The relevant messages were relayed straight through his DNI. A display grid snapped up within his apparent field of vision, its indigo color, invisible to the human eye, ensuring it didn't obscure anything in his actual physical sight. But for all the torrent of information presented to him, he was still left lamentably short of hard facts. Nobody on the scene had yet established what had actually happened. So far they just had the one unconfirmed report of a suited Skin running amok.

His attention flicked to one of the medical grids. He called it up, and five high-resolution graphs expanded for him as he stepped from the helicopter cabin. The handheld blood analyzers that the paramedic teams were applying to the victims were establishing links to the Cairns General Hospital's databank, working through chemical profiles to identify the agent involved in the poisoning.

Simon put on a pair of old-fashioned wraparound sunglasses. "Interesting," he murmured. "Do you see this?" He had sent copies of the analyzer results to Z-B's bioweapons division AS, which gave him a positive match on the agent. His DNI relayed the secure package to Adul.

"Skin toxin," Adul observed. "An updosed incapacitation shot." He shook his head in disapproval before unfolding his own sunshade membranes across his nose. "One definite fatality. And those two with allergic reactions are going to wind up with nerve damage."

"If they're lucky," Simon said. "And only if these paramedics get them to the hospital fast enough." He ran a hand over his brow, dabbing at a thin layer of perspiration that had already accumulated in the intense heat.

"Shall I have the antidote dispatched to the emergency room?"

"Incapacitation toxins don't need an antidote; they clear automatically. It's what they're designed for."

"That dosage level will put a hell of a strain on their kidneys, though."

Simon stopped and looked at Adul. "My dear fellow, we're here to investigate how and why it was used, not to act as nursemaid to a bunch of retarded civilians who are too slow to duck in the first place."

"Yes, sir."

It was that tone again. Simon thought he might soon be reconsidering Adul's usefulness as a security operative. In his business, empathy was a valuable trait, but when it veered into sympathy . . .

The pair of them threaded their way through the maze of emergency vehicles parked along the main street. The few clear passages were clogged by people: locals, sullen and silent, and a few tourists, frightened and excited. Around the bar's veranda, police officers in their shorts and crisp white shirts milled about trying to look as if they had a reason and purpose. Their chief, a tall captain in her mid-forties, wearing full navy blue uniform, stood beside the rail, listening to a young constable making his excitable report.

Simon's personal AS informed him the officer in charge was Captain Jane Finemore. A script page containing her service record expanded out of the grid. He scanned it briefly and dismissed it.

All the police fell silent as Simon and Adul made their way forward. The captain turned; there was a flash of contempt as she took in Adul's mauve Z-B fleet tunic; then her face went protectively blank as she saw Simon in his conservative business suit, jacket slung casually over his shoulder.

"Can I help you fellas?" she asked.

"I rather fear it's the other way round, Captain . . . ah, Finemore," Simon said, smiling as he made a show of reading her discreet lapel name badge. "We intercepted a report that indicates someone in a Skin suit was engaged in hostile action here."

She was about to answer when the bar's doors slammed open and a paramedic team carrying a stretcher hurried out. Simon flattened himself against the veranda railing, allowing them past. Various medical bracelets had been applied to the patient's neck and arms, small indicator lights winking urgently. He was unconscious, but twitching strongly.

"I haven't confirmed that yet," an irritated Captain Finemore said when the paramedics were clear.

"But that was the initial report," Simon said. "I'd like to establish its validity as a matter of urgency. If someone in Skin is running loose, he needs to be dealt with immediately, before the situation deteriorates any further."

"I am aware of that," Captain Finemore said. "I've put our Armed Tactical Response Team on standby."

"With all respect, Captain, I feel this would be best dealt with by a counter-insurgency squad from our own internal security division. A Skin suit would give the wearer an enormous advantage over your ATR team."

"Are you saying you don't think we can handle this?"

"I'm offering every facility to ensure that you do."

"Well, gee, thanks. I don't know what we would do without you."

Simon's smile remained in place as various police officers snickered around him. "If I could ask, where did that original report come from?"

Captain Finemore jerked her head toward the bar. "The waitress. She was hiding behind the bar when your man opened fire. None of the darts hit her."

"I'd like to talk to her, please."

"She's still in a lot of shock. I've got some specially trained officers talking to her."

Simon used his DNI to route a message through his personal AS. The captain wouldn't have a DNI herself—Queensland State Police budget didn't run to that—but he could see her irises had a purple tint; she was fitted with standard commercial optronic membranes for fast data access. "Did nobody else witness this man in a Skin suit? He would hardly be unobtrusive."

"No." The captain stiffened as the script scrolled down across her membranes. "There was just the one sighting." She was talking slowly now, measuring every word. "That's why I haven't ordered a general containment area around the town yet."

"Then finding out is your first priority. The longer you wait, the wider the containment area, and the less likely it will succeed."

"I've already got cars patrolling along the main road to Cairns, and officers are covering the skycable terminus and the train station."

"Excellent. May I sit in on the waitress's interview now?"

Captain Finemore stared at him. His warning message had been very clear and backed by the state governor's office. But it had been private, enabling her to save face in front of her officers—unless she chose to make it public and destroy her career in a flare of glory. "Yeah, she'll probably be over the worst by now." Said as if she were granting a favor.

"Thank you. That's most kind." Simon pushed the bar's door open and went inside.

Over a dozen paramedics were in the bar, kneeling beside the toxin victims. Orders and queries were shouted among them. They rummaged desperately through their bags to try to find relevant counteragents; medical equipment was strewn about carelessly. Their optronic membranes were thick with script on possible treatments.

The victims shuddered and juddered, heels drumming on the floorboards. They sweated profusely, whimpering at painful nightmares. One was sealed in a black bodybag.

It was nothing Simon hadn't seen before during asset-realization campaigns. Usually on a much larger scale. A single Skin carried enough ammunition to stop an entire mob dead in the street. He stepped gingerly around the bodies, trying not to disturb the paramedics. Police officers and forensic crews were examining walls and tables, adding to the general melee.

The waitress was sitting up at the counter at the far end of the bar, one hand closed tightly round a tumbler of whiskey. She was a middle-aged woman with a fleshy face and permed hair in an out-of-date fashion. Not really seeing or hearing anything going on around her.

Clearly there wasn't a single viral-written chromosome in her DNA, Simon decided with considerable distaste. Given her background, the absence of such v-writing inevitably meant she had low intelligence, bad physiology and zero aspirations. She was one of life's perpetual underdogs.

A female police officer sat on a barstool beside the waitress, a sympathetic expression on her face. If she'd taken in any of her specialist training, Simon thought, the first thing she would have done was move the woman outside, away from the scene.

His AS was unable to find the waitress's name. Apparently, the bar didn't have any kind of accountancy and management programs. The AS couldn't even find a registered link to the datapool; all it had was a phone line.

Simon sat down on the empty barstool next to the waitress. "Hello there. How are you feeling now, er . . . ?"

Weepy eyes focused on him. "Sharlene," she whispered.

"Sharlene. A nasty thing to happen to anyone." He smiled at the police officer. "I'd like to talk to Sharlene alone for a moment, please."

She gave him a resentful look, but got up and walked off. No doubt going to complain to Finemore.

Adul stood behind Sharlene, surveying the bar. People tended to take a wide detour around him.

"I need to know what happened," Simon said. "And I do need to know rather quickly. I'm sorry."

"Jesus," Sharlene shivered. "I just want to forget about it, y'know." She tried to lift the whiskey to her lips. Blinked in surprise when she found Simon's hand on top of hers, preventing the tumbler from moving off the countertop.

"He frightened you, didn't he?"

"Too damn right."

"That's understandable. As you saw, he could cause you a great deal of physical discomfort. I, on the other hand, can destroy your entire life with a single call. But I won't stop there. I will obliterate your family as well. No jobs for any of them. Ever. Just welfare and junk for generations. And if you annoy me any more, I'll see you disqualified from welfare, too. Do you want you and your mother to be whores for Z-B squaddies, Sharlene? Because that's all I'll leave you with. The pair of you will be fucked into an early diseased death down on the Cairns Strip."

Sharlene's jaw dropped.

"Now, you tell me what I want to know. Focus that pathetic mush of flesh you call a brain, and I might even see you get a reward. Which way do you want to go, Sharlene? Annoyance or cooperation?"

"I want to help," she stammered fearfully.

Simon smiled wide. "Splendid. Now, was he wearing a Skin suit?"

"No. Not really. It was his arm. I saw it when he bought his beer. It was all fat, and a funny color."

"As if he had a suntan?"

"Yeah. That's it. Dark, but not as dark as an Aboriginal."

"Just his arm?"

"Yeah. But he had the valves on his neck, too. You know, like Frankenstein bolts, but made from flesh. I could see them just above his collar."

"You're sure about that?"

"Yes. I'm not making this up. He was a Zantiu-Braun squaddie."

"So what happened: he walked in and shot everyone?"

"No. He was talking to some bloke. Then Jack and a couple of the others went over. I guess they were looking for trouble. Jack's like that; a good bloke really, though. That's when it happened."

"The man fired darts that knocked everyone out?"

"Yes. I saw him hold his hand up high, and someone shouted that he was in Skin. I got down behind the counter. Then I heard everyone screaming and falling. When I got up, they were all just lying there. I thought . . . thought they were all dead."

"And you called the police."

"Yes."

"Had you ever seen this man before?"

"I don't think so. But he might have been in. We get a lot of people in here, you know."

Simon glanced round the bar, and just avoided wrinkling his nose in disgust. "I'm sure you do. What about the person he was talking to—have you seen him before?"

"No. But—"

"Yes?"

"He was Zantiu-Braun as well."

"Are you sure?"

"Yeah. I've worked in bars all around Cairns. You get to recognize the squaddies, not just from their valves."

"Very well. So the shooter came in and bought a beer, then went straight over to the other squaddie, is that right?"

"Yes. That's about it."

"Try to remember, did either of them seem surprised that the other was there?"

"No. The one who was here first was drinking by himself, like he was waiting for the other."

"Thank you. You've been most helpful."

Captain Finemore gave Simon a surprised look when he emerged from the bar. "What happened?"

"Nothing," he said. "It wasn't a Skin suit. He was using some kind of scatter pistol. I expect the dart toxin was produced in an underground lab. Shame the chemist wasn't a bit more attentive to the actual molecular structure when he attempted to retrosynth it."

"A shame?" The line of Captain Finemore's lips was set hard. "We've got one dead, and Christ alone knows if the rest of them will recover."

"Then you'll be glad we're getting out of your hair." Simon gestured along the clutter and confusion of Kuranda's main street. "It's all yours. But if you do need any help rounding the shooter up, then don't hesitate to ask. Our boys can always do with a bit of live training."

"I'll keep it in mind," Finemore said.

As before, police and civilians parted for him with sullen, silent resentment. He ran quickly through the TVL77D's start-up procedure and lifted from the baked mud. His personal AS reported there was no unauthorized removal of a Skin suit from the Cairns base armory.

"Check this out for me," he told Adul. "I want to know who it was walking round in Skin."

"Some squaddie got jumped in a bar. Do you really think it's that important?"

"The incident isn't. The fact that there's no reference to Skin missing is. And I'm curious why two of our people should choose to meet in such a godforsaken place."

"Yes, sir."

Zantiu-Braun's Third Fleet base centered on the old Cairns International Airport just to the north of the town. There were no commercial flights there anymore; the main transport link was the TranzAus magrail train, bringing cargo and people northward with smooth efficiency at five hundred kilometers an hour. Now the parking aprons held squadrons of Third Fleet helicopters along with scramjet-powered spaceplanes and a few dark, missilelike executive supersonic jets; eight old lumbering turboprop craft maintained by Z-B provided a civil coastwatch and rescue service all the way out to New Guinea. As a result, the airspace over Cairns was the busiest section in Australia apart from Sydney, where the remaining airlines had their hub. Synthetic hihydrogen fuels had replaced natural petroleum products, ecologically sounder but relatively expensive to produce, the cost pushing air travel right back where it started in the twentieth century, the preserve of governments, corporations and the rich.

With mass tourism dying and agriculture effectively eliminated by vat-grown food and worsening ultraviolet infall, Queensland was fast becoming an economic wasteland in 2265 when Zantiu-Braun was offered zero-tax start-up incentives to site a new wave of Earth-to-orbit operations there.

In those days, the operation was purely commercial. Freight spaceplanes boosted factory station modules to low-orbit stations and returned with valuable microgee products, while the passenger variants ferried colonists up to starships. After 2307 that all began to change. Asset realization became the new priority, and the nature of the cargoes that the spaceplanes hauled up to low orbit switched accordingly. The number of colonists flying from Cairns fell to zero inside of a decade, replaced by strategic security personnel. Third Fleet support systems took over from industrial shipments.

The base expanded, throwing up barracks and married quarters for the strategic security division squaddies. Engineering and technical support constructed themselves ranks of blank warehouselike buildings. New hangars and maintenance shops sprang up to house and service the helicopters. Huge swaths of government land were rented for training grounds. And, essentially, all of the new arrivals required administration. Glass and marble office towers rose up in the foothills, overlooking the base and the ocean beyond.

Simon Roderick had an office that occupied half of the top floor of the Quadrill block, the newest and plushest of Z-B's little managerial division enclaves. As soon as he landed the helicopter on the rooftop pad he was plunged into yet another round of planning committees and tactical meetings. Senior staff came in and went out of his office as if it were some kind of transit lounge, each with his own proposal or complaint or report. For an age that relied so heavily on artificial sentience, it always astounded Simon that so little could be achieved without human intervention and supervision. People, basically, needed a damn good kick up the ass to get them motivated and acting like adults. Something not even quantum-switch neurotronic pearls could provide.

After three years on-site Simon knew he was going to have to make a drastic recommendation to the Zantiu-Braun Board after the Thallspring campaign. Forty-five years' constant expansion had made the Third Fleet strategic security division so top-heavy with officers and management specialists it was in danger of grinding to a halt from datalock. Everybody's office generated reports and requests on a continual daily basis; coordinating them even with AS routing management was becoming progressively more difficult. Loop involvement, which was the preparatory-stage management strategy, was a grand forward-looking idea, but after four decades of accumulated optimization the Third Fleet software had become classic bloatware, total deadweight. The theory behind loop involvement was excellent. Experience from the last campaign was inserted at base level. Last campaign, these specific platoons ran out of Skin bloodpaks ten days before the usage programs projected; this time therefore they added a special requirements appendant to the logistics profile of those same platoons. Who could argue against providing first-rate support on the front line? But the additional bloodpaks had to be lifted into orbit, which meant more spaceplane flights, which needed maintenance and flight crew time allocation, and fuel, all of which had to be meshed with the existing schedule. A domino effect that triggered an avalanche every time. Simon was convinced the entire Third Fleet structure needed simplifying to such an extent it would actually have to be decommissioned and a new organization formed to replace it. One that had modern management procedures incorporated from the start.

For the last four months, since the Thallspring campaign planning had begun in earnest, he had concentrated on personally supervising the practical essentials, such as starship refit schedules, Skin numbers, helicopter availability, and basic equipment readiness. But then his total priority requests and orders had to be integrated into the already saturated command structure, creating another authority layer that the base management AS struggled to accommodate. He liked to think his intervention had speeded up the overall process, but there was no way of telling. Vanity of the ruling classes. *We make a difference.*

Adul Quan reappeared as the sun sank below the hills behind the base. Simon stood by the wall-window watching the thick gold sunbeams probe around the curved peaks as the last of the starship commanders filed out. In front of him the runway lights were becoming more pronounced, the street grid of some imaginary city, calling the helicopters home for the night. Away to the south, the neon corona of Cairns's Strip was already rising into the darkening sky. Down along the waterside, the clubs and casinos and bars were opening for trade, trashy games and girls smiling with bright come-on calculation at the squaddies.

There were times when Simon almost envied that kind of simple existence: fight, fuck and tune out, even though it was the antithesis of all he believed in. They didn't have to endure the same kind of pressure he experienced on a daily basis. It was one of the reasons he'd given the Kuranda shooter a higher priority than he probably should have. An excuse to get out of the office.

The office door swung shut after the last commander. "Have you got a name for me?" Simon asked.

"I'm afraid not, sir," Adul said. "It's a bit puzzling."

"Really?" Simon went back to his desk and sat down. He cleared the holographic panes of their script and graphs, giving the intelligence operative an expectant glance through the transparent glass. "Proceed."

"I checked the armory first. Skins that are being repaired seemed an obvious choice. Our man could have taken an arm while the computer log registered the suit as being worked on. I got every technician to report to me personally. They all swore their suits were fully integrated. No missing arms."

"One of them was the shooter?" Simon queried.

"Not a chance. You could slip out for maybe half an hour without anyone noticing, but not a trip to Kuranda. I had my personal AS review the internal security cameras, as well. They were all there."

"Okay. Go on."

"Next obvious source was a squaddie on training who slipped away. It can happen in the field easily enough. There were eighteen platoons undergoing training in Skin suits today. The nearest training ground to Kuranda was sixty-five kilometers. All the Skins arrived there this morning, and my AS queried every platoon leader to do an immediate head count this afternoon when I started investigating."

"Nobody missing?"

"Not one. I even got a list of squaddies who weren't actually on the training ground this afternoon. Three of them were injured; the hospital confirms where they were. Two had suit faults and got sent back to base; the armory confirms their location."

"Interesting."

"So I checked with skyscan." He nodded at the holographic panes. His DNI routed the file images for him.

Simon watched the picture form in front of him. Kuranda's main street from directly above, reproduced in a slightly washed-out color. He recognized the roof with the graffito open eye. From there it was easy to work out which building was the bar. A couple of pickup trucks were using the street; a few people were scattered about. A white cursor ring began flashing around one man.

"That's our man," Adul said. "And God knows what he looks like."

Simon ordered an image expansion and smiled, rather enjoying the way this was turning out. Worthy opponent, and all that. The image quality left a lot to be desired. The little spy sats that Z-B used to monitor the entire Earth's surface were intended to provide only a general review cover. Their designated function was real-time coverage, where they could be programmed for full-focus resolution. But even so, the memory capacity was adequate for this; he couldn't mistake what he was seeing. "A big hat."

"Yes, sir. I backtracked the time index and followed him from the moment he stepped off the train at Kuranda station. He's wearing it the whole time, and he never looks up."

"What about the man he was meeting?"

"Same problem." The picture changed, with a time index eight minutes earlier. It showed a snap-motion image of a four-wheel-drive jeep pull up at the back of the bar. Someone got out and walked inside.

"Shopkeepers are obviously doing a roaring trade in these hats," Simon muttered. He leaned forward, peering at the frozen picture. "Isn't that one of our jeeps?"

"Yes, sir," Adul said heavily. "The skyscan got its number: five-eight-six-seven-ADL-nine-six. According to the transport pool inventory, it was parked here all afternoon. I even used skyscan to track it leaving and arriving back at the base. It used gate twelve on both occasions, and I have the exact times. No record in the gate log."

"Is the gate log e-alpha guarded?" Simon asked sharply.

"No. Nor is the transport pool inventory. But it does use grade-three security encryption."

"They're good, then." Simon nodded approvingly at the holographic pane. "I'll bet you won't be able to backtrack the shooter getting on the train down at Cairns, nor off the sky-cable terminal, either."

"My AS is working on it."

Simon dismissed the image and swiveled his chair so he was facing the wall-window again. The impressive sunbeams had gone from the hills, leaving just stark silhouettes jutting against the fading sky. "They know how to avoid skyscan, and they can help themselves to equipment from the base without leaving any trace. That means they're either officers with high-level access codes, or very experienced squaddies who know the system from the inside. That waitress said she thought they were squaddies."

"That doesn't make any sense. Why would a couple of

squaddies go to all that trouble just to have a drink together? They bust over the wire every goddamn night to get down to the Strip."

"Good question. They obviously thought it was worthwhile."

"What do you want me to do?"

"Keep working on it. But if that last backtrack doesn't produce any results, don't bust a ball. Oh, and keep in touch with dear Captain Finemore. I doubt she'll come up with anything, but you never know; we might see a miracle yet."

"So they get away with it."

"Looks like it. Whatever 'it' was."

CHAPTER TWO

It had rained steadily overnight, leaving Memu Bay's stone-block streets slippery with water early in the morning when everyone was trying to get to work. Soon after, when the tropical sun rose above the ocean, the pale stone began to steam, boosting the humidity to an intolerable height. But by the afternoon everything had cleared, leaving a sweet cleanliness in the air.

Denise Ebourn took the children outside to enjoy what remained of the day. The playschool building was mostly open to the air, with a red clay tile roof standing on long brick pillars. Vigorous creepers swarmed up the pillars, crawling along the roof and clogging the gutters with diamond cascades of purple and scarlet flowers. Staying underneath the eaves wasn't exactly arduous, but like her little charges Denise wanted to be outside with the freedom it represented.

They raced out across the walled garden, cheering and skipping about, amazingly full of energy. Denise walked between the swings and slides, checking that they weren't overexerting themselves, or daring each other into anything dangerous. When she was happy they were behaving themselves as well as any five-year-olds could, she put both hands on top of the chest-high wall and took in a deep breath, gazing out across the little city.

The bulk of Memu Bay occupied a crescent of alluvial land at the end of a mountain range, a perfect natural sheltered harbor. Its more expensive homes clung to the lower slopes of the grassy hills: Roman villas and Californian-Spanish haciendas with the long steps of terraced gardens spilling down the slope in front of them. Sometimes a glimpse of shimmering turquoise betrayed a swimming pool lost amid palisades of tall poplars and elaborate rose-twined columns that surrounded broad sundecks. However, the majority of the urban zone sprawled out around the base of the mountains. As with all new human cities, it had broad tree-lined boulevards slicing clean through the center, fanning out into a network of smaller roads that made up the suburbs. Apartment blocks and commercial buildings alike were all painted plain white, dazzling in the bright afternoon sun, their smoked-glass windows inset like black spatial rifts. Balconies foamed over with trailing plants. Flat roofs sprouted sail-like solar panels that turned lazily to bake themselves in the intense light: they cast long shadows over the silver-rib heat dissipater fins of air conditioners that sprawled horizontally below them. Several parks broke up the city's aching glare, verdant green oases amid the whiteness; their lakes and fountains sparkled in the sun.

Denise always found the terrestrial vegetation a peculiar color, paradoxically unnatural. If she squinted inland, she could see the boundary just visible against the large mountains in the far distance. Terrestrial grass had pushed right up to the edge of the area sterilized by the gamma soak. Beyond that, Thallspring's indigenous vegetation swept away into the haze horizon. A more resolute color, reassuringly blue green; plants out there had bulbous, heavier leaves and glossy stems.

She'd grown up in the hinterlands—Arnoon Province, where human colonization had little impact on native life. Valleys of settlers escaping the restrictions of the majority civilization, as can be found on any human frontier. They lived amid alien beauty, where the vegetation could prove harmful to the unwary. Thallspring's botanical chemistry didn't produce the kind of proteins people or animals from Earth could digest. However, Arnoon's highland forests did cultivate the willow web, which the settlers harvested. When woven correctly it formed a silky waterproof wool that the city dwellers valued. It wasn't a fabulously profitable activity, but it allowed them to sustain their loose community. They were a quiet folk whose chosen life had given Denise a happy childhood, benefiting from the kind of rich education that only a starfaring species could provide while remaining firmly rooted in the nature of her adopted world. A life that was more secure than she ever realized because of their private cache of knowledge, subtly enforcing every core value of their lifestyle.

Her good fortune had lasted right up until the day the invaders arrived.

A burst of giggling broke her reverie. Several of the children were clustering behind her, urging Melanie forward. It was always Melanie: the boldest of them all, she didn't need any encouragement. A natural leader, not quite like her father the mayor, Denise thought. The little girl tugged at Denise's skirt, laughing wildly. "Please, miss," she implored. "A story. Tell us a story."

Denise put her hand to her throat, feigning surprise. "A story?"

"Yes, yes," the others chorused.

"Please," Melanie whined, her expression trembling into unbearable disappointment and the threat of tears.

"All right then." She patted Melanie's head as the others cheered. It was moments like this, when their smiles and adulation fell on her, that Denise knew everything was worthwhile.

At first, Mrs. Potchansky had been dubious about taking her on at the school. So young, barely in her twenties, and brought up in the hinterlands as well. Her youthcare certificates were all in order, but . . . Mrs. Potchansky had some very quaint old notions about propriety and the *right way* of doing things, ways probably unheard of in Arnoon Province. With a show of cool reluctance she'd agreed to Denise having a trial period; after all, a lot of *very important people* sent their children to the playschool.

That was a year ago now. And Denise had even been invited to Mrs. Potchansky's house for Sunday lunch with her own family. Social acceptance didn't come much higher in Memu Bay.

Denise sat herself down on one of the wooden swings, arms wrapping round the chains as she slipped her sandals off. The children settled on the grass in front of her, fidgety and expectant.

"I'm going to tell you the story of Mozark and Endoliyn, who lived a long time ago in the early days of the galaxy."

"Before the black heart started beating?" one of the boys shouted.

"Around the time it began to beat," she said. Many times she'd told the children of the galaxy's black heart, and how it ate up stars no matter what the Ring Empire did to try and stop it, which made them all squeal and gasp in fright. "This was when the Ring Empire was at the height of its power. It was made up from thousands of separate kingdoms, all of them united in peace and harmony. Its people lived on the stars that circled the core of the galaxy, trillions and trillions of them, happy and contented. They had machines that provided them with whatever they wanted, and most of them lived for thousands of years. It was a wonderful time to be alive, and Mozark was especially lucky because he was born a prince of one of the greatest kingdoms."

Jedzella stuck her hand up, fingers wriggling frantically. "Were they people just like us?"

"Their bodies were different," Denise said. "Some of the races who were members of the Empire had arms and legs similar to ours, some had wings, some had four legs, or six, or ten, some had tentacles, some were fish, and some were so big and scary that if you and I saw them we'd run away. But how do we judge people?"

"What they say and do," the children yelled happily, "never how they look."

"That's right. But Mozark did come from a race that looked a little like us. He had four arms, and eyes all the way round his head so he could see in every direction at once. His skin was bright green, and harder than ours, like leather. And he was smaller. Apart from that, he thought like we do, and went to school when he was growing up, and played games. He was nice, with all the qualities a prince should have, like kindness and wisdom and consideration. And all the people in the kingdom thought they were lucky to have a prince who was so obviously going to be a good ruler. When he was older he met Endoliyn, who was the most beautiful girl he'd ever seen. He fell in love with her the moment he saw her."

The children sighed and smiled.

"Was she a princess?"

"Was she poor?"

"Did they get married?"

"No," Denise said. "She wasn't a princess, but she was a

member of what we'd call the nobility. And he did ask her to marry him. That's where this story starts. Because when he asked her, she didn't say yes or no; instead she asked him a question right back. She wanted to know what he was going to do with the kingdom when he became king. You see, although she was very comfortable and had great wealth and friends, she worried about what would fill her life and how she would spend it. So Mozark answered that he would rule as best he could, and be just and listen to what his subjects wanted and endeavor not to let them down. Which is a very reasonable answer. But it wasn't enough for Endoliyn: she'd looked round at everything the kingdom had, all its fabulous treasures and knowledge, and it made her sad."

"Why?" they all gasped in surprise.

"Because everybody in the kingdom saw the same things, and did the same things, and was happy with the same things. There was never anything different in the kingdom. When you know everything and have everything, then nothing can be new. And that's what made her sad. She told Mozark she wanted a king who would be strong and bold, and lead his people. Not follow along and try to please everybody every time, because no person can really do that, you just wind up pleasing nobody. So she would only ever love and marry someone who inspired her."

"That's rude," Melanie declared. "If a prince asked me to marry him, I would."

"What prince?" Edmund sneered.

"Any prince. And that means when I'm a real princess you'll have to bow when I walk past."

"I won't!"

Denise clapped her hands, stopping them. "That's not what being a prince and princess in this kingdom was like. It wasn't a medieval kingdom on Earth, with barons and serfs. The Ring Empire nobility earned the respect they were given."

Edmund wound himself up. "But—"

"What about Mozark?" Jedzella asked plaintively. "Did he get to marry Endoliyn?"

"Well, he was terribly disappointed that she didn't say yes straightaway. But because he was wise and strong he resolved to meet her challenge. He would find something to inspire her, something he could dedicate his life to that would benefit everyone in the kingdom. He ordered a great starship to be built so that he might travel right around the Ring Empire and search out all of its wonders, in the hope that one of them might be different enough to make people change their lives. All of the kingdom marveled at his ship and his quest, for even in those days few people undertook such a journey. So he gathered his crew, the boldest and bravest of the kingdom's nobility, and said farewell to Endoliyn. They launched the amazing ship into a sky the like of which we'll never know. It was a sky to which night never truly came, for on one side was the core with a million giant stars shining bright, and on the other was the ring itself, a narrow band of golden light looping from horizon to horizon. Through all these stars they flew for hundreds of light-years, onward and onward until they were in a part of the Ring Empire where their own kingdom was nothing more than a fabled name. That's where they found the first wonder."

"What?" one of the boys squealed. He was quickly *shushed* by all his friends.

"The planet's true name had been forgotten centuries ago. It was just called The City now. A place as mythical to Mozark as his kingdom was to its inhabitants. The people who lived there had devoted themselves to creating the most beautiful buildings it was possible to make. All of them lived in a palace with its own parkland and lake and river, and their public buildings were as majestic as mountains. That's why their world was called The City, because every building was so big and grand and had acres and acres of its own grounds that they'd spread right over the entire surface, from the deserts to the polar caps. There was nowhere without a building. Now you might say this would be easy, given that the Ring Empire had machines that could build anything. But The City dwellers didn't want machines building their homes; they thought every person should build his own home; they believed that only if you build it yourself can you appreciate its true grandeur.

"Now, Mozark and his crew landed there and walked amid all these fantastic buildings. Even though they weren't the same species as The City's inhabitants, they could appreciate the splendor of what they were seeing. There were cathedral-like towers slicing kilometers into the sky. Crystal tubes that spiraled up entire mountains, which housed every kind of plant to be found on the planet in every environment. Starkly simple buildings, exquisitely ornate buildings, buildings that flowed into the landscape, they were so naturalistic. The City had them all, visual marvels everywhere you looked. Mozark spent many weeks there, he was so staggered by everything he saw. He thought it was the most superb accomplishment any race could make, for every citizen to live in luxury surrounded by beauty. But eventually he called his crew back to his ship and told them that for all its magnificence The City was not for the kingdom. They left, and continued their flight around the core."

"Why?" the children asked.

"Firstly, because The City had already been done," Denise said. "And secondly, because after a time Mozark began to see what a folly it was. All the inhabitants of The City did was maintain their buildings. Some families had lived in the same palace for twenty or thirty generations. They added to it, but never changed the nucleus, the essence that made them what they were. The only real interest in The City was shown by outsiders, different species from across the Ring Empire who flocked to marvel at its intricacy and debate its significance.

Mozark knew that people could be inspired to build beautiful or gigantic structures, but after that it is always time to move on. The City was magnificent, but decadent. It celebrated the past, not the future. It was everything Endoliyn so dearly wished to escape from. He had no choice but to continue."

"Where did he go?"

"What happened next?"

Denise glanced at her antique watch. A man's watch, bulky for her slim wrist; her grandfather had carefully adjusted its quartz innards to synchronize with Thallspring's twenty-five-and-a-half-hour day. "You'll have to wait until tomorrow for the next part," she said.

A huge barrage of groans and boos greeted the announcement.

"You knew that," she protested, acting astonished. "The Ring Empire is vast. Mozark had lots and lots of adventures on his voyage round it. It'll take me weeks to tell them all. Now make sure you put the games and toys back in the bins before you go. The right bins!"

Slightly mollified with the promise of more tales of the Ring Empire to come, they wandered back across the grass to pick up the discarded toys.

"You have such an imagination, my dear."

Denise turned to find Mrs. Potchansky standing a couple of meters away, giving her a slightly concerned look.

"Ring Empires and little green princes on a quest, indeed. Why not just give them the classics like Pratchett and Tolkien?"

"I don't think they're very relevant to today."

"That's such a shame. They might be archaic, but they're lovely stories. I really liked dear old Bilbo Baggins. I even have a hard copy book of *The Hobbit,* printed on Earth for Tolkien's bicentennial."

Denise hesitated. "The stories I make up do have a moral center."

"I noticed. Although I think I'm the only one who did. You are very subtle, my dear."

Denise grinned. "Was that a compliment?"

"More an observation, I feel."

"Do you want me to stop telling them about the Ring Empire?"

"Heavens no." Mrs. Potchansky was genuinely surprised. "Come along, Denise, you know how good you are with the children. You don't have to fish for compliments from me. I'm just worried you'll turn professional and put all these colorful thoughts of yours straight down into i-media. Who would I get to replace you?"

Denise touched the old lady on her arm. "I'm not going to leave you. I love it here. What could ever change in Memu Bay?" It came out before she could stop it.

Mrs. Potchansky glanced up at the clear turquoise sky, wrinkles around her eyes creasing into a burst of bitter resentment totally at odds with her air of gentility.

"Sorry," Denise said immediately. Mrs. Potchansky had lost her son during the last invasion. Denise knew few details other than the date of his death.

"That's all right, dear. I always look at how we live now. This is a good life we have here, the best of all the settled worlds. That's our revenge against them. They can't destroy our nature. They need us just as we are. I enjoy that irony, I think."

At moments like this, Denise just wanted to blurt out everything to the sweet old lady, all the anger and plans she and the others had brought with them to Memu Bay. Instead she gave Mrs. Potchansky a tight hug. "They won't beat us, not ever. I promise."

Mrs. Potchansky patted Denise's back. "Thank you, dear. I'm so glad you found this school."

* * *

As usual, some of the children were collected late. Old Mr. Anders, picking up his grandson. Francine Hazeldine, the mayor's fifteen-year-old daughter, scooping up her little sister, the pair of them laughing happily at the reunion. Peter Crowther eagerly beckoning his quiet son into a huge limousine. Denise gave them big media pads to finger sketch on while they waited in the classroom.

It took her nearly a quarter of an hour after the last one had left to get everything ready for tomorrow. She wiped the psychedelic patterns from the media pads, sorted the games and toys into the right bins, put the chairs back into line and reflated their one leaky jelfoam mattress. Mrs. Potchansky came in before she'd loaded the dishwasher with all the mugs and cutlery and told her to get off into town. It was a lovely day and she should enjoy herself. The old woman didn't quite ask if Denise had a boyfriend yet, but it wouldn't be long. The query came every three weeks or so, along with associated helpful observations on where nice boys were to be found. Denise always hated the embarrassment of having to deflect her from the topic. There were times when it was like spending the day with her mother.

The school was a couple of kilometers inland, so it was an easy downhill walk to the marina for her. On rainy days she would take the trams that ran through the major boulevards, but today the afternoon sun continued to shine through a clear sky. She strode easily along the sidewalk, making sure she kept under the broad shop awnings: she was wearing a light dress, and at half past four in the afternoon the sun was still strong enough to be avoided. The route was familiar enough, and she was on nodding terms with several people on the way. So very different to her first days in the city, when she jumped every time a car's brakes squealed, and more than five people gathered together made her claustrophobic. It had taken over a fortnight before she was comfortable just going into one of Memu Bay's plentiful cafés and sitting there with friends.

Even now she wasn't quite used to the triads she saw together out on the street, though she made a point of not staring. Memu Bay was proud of its liberalist tradition, dating right back to the founding in 2160. The city fathers, having left an Earth that they considered to have encroached upon personal freedoms, were determined to encourage a more relaxed and enlightened atmosphere on their new world. Communes were prevalent during the early days, along with cooperative industrial enterprises. Reality had gradually eroded this gentle radicalism; collective dormitory halls were slowly refurbished into smarter individual apartments and shares were floated and traded to raise capital for factories to expand. The most prominent leftover from all this early social experimentation was the trimarriages, whose popularity continued long after other hippie chic traditions lost their bloom. Though even that wasn't as popular as it used to be. Trendy liberalism and those first youthful hot randy nights of threesomes tended to deteriorate and sour when middle age approached, inevitably accompanied by mortgage payments and domestic demands with their three-way arguments. And trimarriage divorces were notoriously bitter, scarring a lot of children who swore blind they wouldn't repeat the mistake. Certainly less than a quarter of registrations at City Hall were now trimarriages, and most of those were of the one-male, two-female variety. Gay and lesbian trimarriages were an even smaller percentage.

Car traffic eased off as Denise entered the Livingstone District behind the waterfront; the streets here were narrower and clogged with bicycles and scooters. This was the city's primary retail zone, where quirky little shops alternated with clubs, bars and hotels. As this was where the tourists flocked, the city planners had re-created the look of an old-style Mediterranean town. Small windows and slim balconies overlooked squares full of café tables that were shaded by citrus trees. At first the streets had confounded her, as if they were deliberately laid out in the most confusing maze possible. Now, she slipped through like a native. The marina itself was packed with sailing yachts and pleasure craft. Farther along the shore, jetskis and windsurfers curved through the water, weaving round each other with curses and elaborately obscene fist gestures. Passenger boats were bringing divers and snorkelers home after a day exploring the reefs. Several of the archipelago islands were visible out toward the horizon, tiny cones of native coral clotted with a tangle of vivid terrestrial vegetation. They looked superb, motes of paradise scattered on an alien ocean. In fact, the gamma soak had killed the coral down to three meters below the surface. Civil engineering teams had gone out and capped the islands with concrete to stop them from breaking up. Sand was laboriously dredged up from their surrounding lagoons to form the exquisite white beaches, and desalination plants watered the vegetation through a buried irrigation network. It was all done for the benefit of the tourists. The living coral in deeper waters was spectacular enough to attract thousands of visitors each year, while the marina catered to watersports enthusiasts. Those physical activities combined with Memu Bay's reputation for an easygoing lifestyle made the city a premier lure for Thallspring's younger element, intent on having a fun time away from the capital and other sober cities.

The Junk Buoy was right on the waterfront, a popular tavern for tourists on the way back to their hotels and chalets. Not particularly smart, nor expensive, but it was the place where boat and dive instructors hung out as the day wound down, which gave it a big boost of kudos. Tourists could sit out under thatched awnings, watching the sun set behind Vanga peak as they drank cocktails with saucy names served up in long iced glasses.

Denise pushed her sunglasses up on her forehead as she walked in. Several young men watched her move across the room, smiling hopefully for her to join them. Denise ignored the come-ons as she made her way over to the far corner of the tavern, where she knew her colleagues would be. The nightly meat market had begun. The tourists were all in swimware or small tight T-shirts, casting eager curious glances at each other. Over half were wearing Preferred Sexual Acts bracelets. Some of the devices were flashy gold Aztec charms or mock–high-tech strips encrusted with blinking LEDs, while others preferred to use discreet black bands, or simple readouts incorporated into their watches. They would tickle the wrist unobtrusively when someone else with the same PSA loaded in their bracelet was within ten meters, and conversation would dip as directional displays were suddenly checked with avid enthusiasm.

She was aware of some bracelet wearers checking their directional graphics anyway, seeing if they matched her position. The anything-goes set, not unsurprisingly those with the most flamboyant bracelets.

One-night stands Denise could live with, not that she ever would for herself. But it was the coldness of the PSA system that she resented. It took everything that was human away from what should be the most enjoyable part of a relationship, the discovery of someone else.

Raymond Jang and Josep Raichura were sitting on their usual stools. Also as usual, they had a pair of girls with them, young and impressionable, wearing swimsuits and sarongs. Ray and Josep didn't need PSA bracelets. For them, this part of the mission was heaven-sent. When they arrived in Memu Bay, they immediately signed up as diving gill instructors with one of the big leisure companies, which brought them into daily contact with a parade of girls in their teens and early twenties. Diving instructors were universally slim and fit, but Ray and Josep now had perfect mesomorph physiques, tanned to a golden sheen. These days Denise had to think hard to remember the two awkward little boys she'd grown up with in Arnoon, one all gangly, the other who hardly ventured out of doors. Now the dweebs were babe magnets, and relishing every second of it. Even better for them—and worse for Denise—they were supposed to develop casual relationships. It would be essential for the next stage of the plan.

The four of them were having such a good time together that Denise almost felt guilty for butting in. She cleared her throat to attract their attention. The two girls instantly looked her up and down with hostile eyes, working out if she was competition or not. They decided not. Denise was the same age group as their catches, with the kind of slim healthy build that could mean she was a fellow gill instructor, and her impatient expression clearly marked her down as a no-fun person.

"Hello?" one of the girls said, her voice rising an octave with mild derision. "Were we friends in a former life?"

Any decent comeback escaped Denise. The girl's breasts were so large that for the first time Denise got an inkling of that most infuriating male reflex; she just couldn't help glancing at her cleavage. Surely she was too young to have undergone v-writing enlargement?

"Hi, Denise." Ray got up and gave her a demure peck on the cheek. "Girls, this is our housemate, Denise."

They consulted each other silently, and said a resentful, "Oh hi," to Denise.

"We just need a quick chat with Denise," Josep said. He gave his girl a quick pat on her bum. "Won't be a minute, and then we'll see where we can go to eat out tonight."

The girl licked some salt off the rim of her margarita glass. "I'd like that." She walked off with her friend, the pair of them whispering in sultry amusement. There were several coy glances thrown back at the boys.

"Working hard, I see," Denise said. Every time she found them with new girls she told herself it didn't bother her. Every time, her disapproval just spilled out.

Ray grinned. "Just following orders."

Denise steeled herself and sat on one of the vacated stools. There was nobody near them, and a melodic guitar tune was playing through the tavern's sound system. Not that Memu Bay's police were surveilling them—or even knew about them, but basic precautions now would save a lot of trouble later on. "We're clear today," she said quietly. "Prime didn't pick up any encrypted signals on the spacecom network."

"They'll come," Josep said.

His tone was understanding, more like the old Josep. He must have picked up on her frustration—he'd always been the more emotionally sensitive one. She flicked a modest grin of thanks at him. His face was broad, with high cheekbones and lovely wide brown eyes. A thick mop of floppy blond hair was held back from his forehead by a thin leather band—a gift from some girl ages ago. Raymond, by contrast, had round features and a narrow nose, his brown hair cut short. Other than that . . . She looked from one to the other. The only garment Raymond had on was a pair of old green shorts, while Josep's denim shirt was open down the front. Twin bodies. Did the girls they shared in bed ever comment on that? she wondered.

"I know." She got a grip on her free-flying thoughts. "Anything new from your side?"

"Actually yes," Ray said. He indicated the girls. "Sally lives in Durrell. She's at college there, a geology student."

"Okay, that's promising."

"And there's a possible contact we think should be checked out," Josep said. "His name's Gerard Parry. He started on my six-day diving proficiency course today. We got chatting. Turns out he's local. Works up at Teterton Synthetics, a distribution manager."

A cluster of neural cells in Denise's brain had undergone a d-written modification for direct communication with the local datasphere, an enhancement that human v-writing couldn't yet match. The cluster linked her directly to the pearl ring on her index finger. Her Prime program produced a brief summary of Teterton, scrolling an indigo script across her vision that detailed a small chemical processing company that supplied local food producers with specialist vitamin and protein concoctions. "Did he sound sympathetic?"

"That's for you to find out. But a contact there could be very useful. There's some compounds we still haven't acquired."

"Okay, sounds good. How do I meet him?"

"We promised him a blind date. Tonight."

"Oh God," she groaned. There would barely be time to go home and change.

"He's a nice bloke," Josep protested. "I like him. Sensitive, caring, all that bull chicks go for."

"Just as long as he's not like you," Denise snapped back.

"Ouch." He smiled. "Well, here's your chance to find out. Here he comes."

"What!"

Ray stood up and waved happily. Denise turned to see the man approaching. In his thirties, overweight, with thinning hair. The restrained smile of a professional bachelor, desperate to hide how desperate he was. A broad black-glass PSA bracelet was worn on his right wrist. Several girls around the tavern checked their directional displays, and hurriedly looked away.

Denise stood up to greet him, the heel of her right foot making solid contact on Josep's toes.

She didn't get home until well after eleven o'clock that night. By that time the weary anger had become a kind of numb indifference to life. All she wanted to do was go to bed and forget the entire evening.

Despite his appearance, Gerard Parry wasn't a bad man. He could hold a conversation, on local issues at least, and was willing to listen up to a point. He even had a few jokes, though he lacked the nonchalance to tell them properly. She could imagine him working hard to memorize them when he heard them around the office.

They had started off having a couple of drinks with Ray and Josep, much to the obvious disgust of the girls. Then dinner was mentioned, and Josep managed to split them up. Gerard took her to a fairly decent restaurant, which left her free to establish his political sympathies. That was when it all fell apart.

Denise never knew how much blame she should take for personal catastrophes like this. It was strange, considering how she almost unfailingly managed to befriend potential recruits who weren't single and male. She asked Gerard the questions she needed to, and tried to ask others, to show an interest in his personal life. But he figured out pretty early on that she wasn't interested in any kind of long-term relationship, or even a brief passionate affair. Men invariably figured that out about her at some time. Always, at the end of such evenings, it finished with her being told she was too intense, or cool, or aloof. Twice she'd been sneeringly accused of being a lesbian.

She didn't even mind the fact that she never made the connection. What she hated was that she could never tell them why. The fact that she'd committed herself to something more important than them, or her. It justified the way she was. But they'd never know. To all of them, she was just another wasted evening.

Gerard Parry got drunk very quickly, especially for a man of his bulk. His conversation turned into a bitter monologue; there were morbid complaints about how he missed out because girls never looked behind his size for the real him, and rhetorical questions about what did she and the rest of the female universe want from a bloke anyway. During his ramblings, he managed to spill half a glass of red wine over the table, which splashed across her skirt. She got up and didn't look back. The headwaiter called a cab for her.

She sat in the back of the AS-driven vehicle, refusing to cry as the lively town slid by beyond the windows. Inner strength was something that could never be installed, unlike her physical ability. That, she had to supply by herself.

The Prime program in her pearl had recorded the encrypted emissions from Gerard's PSA bracelet. A gross breech of etiquette; PSAs were supposed to be exchanged. As she reviewed the data she gained a degree of satisfaction knowing what a pig he was. It made her feel a hell of a lot more justified leaving him weeping into his wine.

The bungalow she shared with Ray and Josep was in a small, prim housing estate spread along the Nium River estuary, outside the center of town. It meant a twenty-minute commute to work in the morning on the tram, but the rent was relatively cheap. At night there was just enough of a breeze coming up the estuary to keep it cool once the big archway windows were folded back. Jasmine grew up the external walls, a mass of scarlet flowers giving off a sweet scent.

Denise came through the front door and dropped her little shoulder bag on the hall table. She pressed her back to the cool plaster, arching her spine and inhaling deeply. All in all, a really shitty day.

The lights were on in the lounge, turned down low. When she peered in, one of the girls from the Junk Buoy was lying facedown on the sofa, snoring with the erratic snorts of the comatose drunk. There were muffled voices and giggles coming from Josep's bedroom, along with familiar rhythmic sounds. Josep, Ray and the huge-breasted girl energetically straining seams on the jelfoam mattress together.

It would be all right, Denise thought, once she was in her own room with the door shut. From past experience she knew the soundproofing was good enough to give her complete silence to sleep in. When she looked down at her skirt, she could see it needed spraying right away to get the wine stain out. Once she'd put it in the washing cabinet and programmed the cycle, she remembered the pile of clean laundry hurriedly dumped in the linen basket this morning, including all her other work clothes. She'd intended to do them when she got back from playschool in the afternoon. So there she was at quarter past midnight, tired and utterly miserable, standing in the kitchen in her robe, ironing her blouse for tomorrow while the shrill whoops of other people's orgasms echoed along the hall.

If there was such a thing as karma, somebody somewhere in this universe was going to get hurt *bad* to level this out.

CHAPTER THREE

LAWRENCE NEWTON NEVER SAW A CLOUD UNTIL HE WAS TWELVE years old. Until then, Amethi's light-time skies had been an unblemished azure from horizon to horizon. When the planet's orbit around its gas-giant primary, Nizana, eventually propelled it into dark-time and the stars came out, they would burn with a steadiness unnatural for any atmosphere, so clear was the frigid air. And with Templeton, the capital where young Lawrence lived, on the hemisphere that permanently faced away from Nizana, he never realized it was possible for anything exciting to exist overhead. In terms of landscape and environment, Amethi was crushingly boring. Nothing moved above, nothing grew on the icy tundra.

To the McArthur Corporation, whose exploratory starship the *Renfrew* discovered it in 2098, such conditions were perfect. In the late twenty-first century, interstellar expansion was at its height, with the big companies and financial consortia funding dozens of colonies. Any planet with an oxygen/nitrogen atmosphere was being claimed and settled. But these ventures were expensive. The alien biospheres that had produced that precious blend of breathable gases were inevitably hostile and poisonous to terrestrial organisms, some immediately fatal. Establishing human communities amid such conditions was extremely costly. Not so Amethi.

For all it was technically a moon, Amethi's evolution had been fairly standard for a world of its size. It started normally, with a reducing atmosphere that slowly changed as life began to emerge from primordial seas. Primitive organisms that could photosynthesize released oxygen. Carbon was consumed by new lichens and amoebas. An unremarkable cycle that was repeated across the universe wherever such conditions occurred.

Evolution was progressing along standard lines until the asteroid was drawn in by Nizana's immense gravity field. Two hundred million years after the first primitive amoebas began fissioning, the seas were full of fish; plants had established themselves across the land. There were big insects with thistledown wings, and even small creatures not far removed from terrestrial amphibian genealogy. They all died in the aftermath.

The impact explosion threw up enough dust and steam to obscure the entire surface. In doing so, it triggered the ultimate ice age. The glaciers that thrust out from the polar caps in the aftermath encroached farther and farther through the temperate zone until they actually merged at the equator. Seas, oceans and lakes surrendered their water to the single megaglacier as it continued to expand. Temperature plummeted right across the planet, combining with the water loss and darkened atmosphere to eliminate all forms of life except the most resilient bacteria. Amethi returned to an almost primordial state. But now with a fifth of the surface covered in ice to a depth of several kilometers, and the remainder a desert that was Mars-like in its desolation, there was no potential catalyst left to precipitate change. It had become a world trapped in stasis. The isolock.

For the McArthur board members Amethi was perfection, an existing breathable atmosphere and no indigenous life. All that was needed was a slight rise in global temperature to end the isolock and restart a normal meteorological cycle.

Templeton was founded in 2115. At first it was nothing more than a collection of prefabricated igloos with a single track linking it to a runway bulldozed into the frozen dunes. The engineers and administrators who lived there were tasked with establishing a manufacturing base that would be self-sustaining, the idea being that once the initial investment was made, all you needed to do was shovel in local raw materials at one end and ultimately any product you wanted would pop out the other. After that, the only imports would be people and new designs to upgrade and expand the first few factories. Information cost nothing to transport between stars, while people would buy their own tickets to a new land with immense opportunities.

Over the first three years, spaceplanes ferried down their cargo from eight starship flights. At the end of it, industrial facilities in heavily insulated factories could supply most of the burgeoning colony's needs. But not all. There were always a few specialist systems or chemicals essential for the economy or special projects that only Earth with its abundant production facilities could provide. Time after time Templeton's governor sent back requests for additional units to be flown out, without which the whole project would stall.

The financial strain that Amethi placed on McArthur wasn't as bad as that for most other colony worlds, where human biochemists fought desperate battles against alien biospheres. Here there was just HeatSmash, the climate project, to initiate. Templeton's first indigenous industrial undertaking was to establish an orbital manufacturing station, Tarona. With that up and running in 2140—after nearly a third of its systems had been shipped in from Earth—they began local production of asteroid capture propulsion engines. Nizana had so much junk rock strewn around in orbit it could have provided HeatSmash with enough material to reheat a dozen worlds. The inaugural impact came in 2142, when a lump of stony iron rock measuring eighty meters across smacked right into the center of Barclay's glacier.

The explosion vaporized nearly a cubic kilometer of water and melted a considerably larger quantity. It had refrozen within a week. The steam clouds never even reached the edge of the glacier before they condensed into bullet-hard snowflakes and rained down.

Once the planetary engineers had correlated all the data from their sensors, they estimated that the atmosphere would have reheated sufficiently to induce and sustain glacial melt after 111 years of one impact per year, involving asteroids four times the mass of the test impact. With this mildly favorable prognosis, the colonists set about building their new world. By the time Lawrence Newton was born in 2310, economic and social changes on the old homeworld had modified the nature of the colony. Although the physical task of terraforming the world had progressed without interruption, it was no longer a destination for exultant pioneers searching for a little homestead amid a wilderness that was slowly being resurrected.

* * *

The big school bus rolled easily along Templeton's main north highway, fat tires clinging to the grubby concrete with its lacework of fine cracks. Twenty-five kids, aged nine to twelve, chattered excitably or threw crumpled biscuit wrappers at each other before ducking down behind their seats to avoid retaliation. Mr. Kaufman and Ms. Ridley, their teachers, sat up at the front, doing their best to ignore what was going on behind. They'd left the school dome only ten minutes ago; it was going to be a long day.

Lawrence was sitting midway along the bus. The seat next to him went unoccupied. It wasn't that he didn't have friends at school; he did, as well as several cousins and a tribe of more distant relatives. He just didn't have any close friends. Teachers described him as restless. He was clever enough, naturally, given he was a Newton, but that intelligence was never quite captured by any of his academic subjects. Report after report filed with his parents had the age-old comment: can do better. In the competitive environment of the school, where application and achievement received the highest accolades, he was too different to fit in comfortably. Not quite a rebel—he was still too young for that classification—but there were plenty of danger signs that he could fall into the dropout category if something wasn't done fairly soon. It was an almost unknown development among Amethi's well-ordered population. For a member of a Board family it was unthinkable.

So he sat by himself ignoring the antics of his peers, watching the city go by outside. On either side of the highway were drab curving walls of nullthene; huge sheets of the ultrathin translucent gray membrane from which the city domes were made. The standard size was four hundred meters across, produced in one piece by the McArthur factory, and wholly indigenous. Relatively cheap, and simple to establish, it was used by every town and city on the planet. All you needed was a flat patch of land over which to spread it.

The sheet had a built-in hexagonal web of slim tubing made from buckyfilament carbon (extruded up at Tarona) that was pumped full of epoxy. The resultant force was enough to lift the lightweight nullthene off the ground like some giant balloon that never quite managed to become airborne. The edges had to be buried hurriedly as the membrane's molecular structure had been designed to act as a near-perfect heat trap. Air inside quickly warmed to temperate and even tropical temperatures, exerting quite a lifting pressure from within. Large circulation and thermal exchange units (also built locally) were installed around the edge, helping to maintain the required climate inside. Once the dome was up and regulated, all that was needed to reinvigorate the soil was water and terrestrial bacteria, and it was ready for planting.

Right at the heart of the city, most of the domes were communal. Above average in size at six hundred meters in diameter, they had a single apartment block skyscraper in the center, acting as an additional support for the vaulting surface. Inside, rich parkland had been established around the skyscrapers, complete with artificial lakes and streams. Nobody outside top-level management used cars to get about within the city; the domes were all linked by a comprehensive rail transit network. The only vehicles on the road with the school bus were twenty-wheel juggernauts, agroform machinery, and civil engineering trucks, all of them cheerfully pumping hihydrogen fuel fumes out into the atmosphere.

Factories filled the gaps between the dome rims, squat bunkers built from glass and aluminum. Encrustations of dust streaked the big panes, built up over years as heat and moisture creeping out of the city structures loosened up the frozen ground. Even here, the air suffered as it did in every human city, a pollution of particles and vapor that hadn't known freedom for a hundred thousand years, churned up by the whirling zephyrs thrown off by the trains and road vehicles and dome circulation fans—for decades, the only wind on the whole planet. But it allowed plants to flourish. All along the side of the road, Lawrence could see tufts of dark green grass clogging the ruddy native soil. There were even little fissures where free water had on occasion run, fed by trickles of condensation along badly insulated panels or tattered slits in the nullthene.

Farther out from the city center, food refineries began to replace the domes—industrial sites the size of small towns where pressure tanks and enzyme breeder towers and protein convectors were woven together with a maze of thick, insulated pipes. Hot vapor shivered the air for hundreds of meters above the dulled metal surfaces as small fusion plants pumped their megawatts into the elaborate processes that kept Amethi's human population alive. Each refinery had its own quarry, huge vertical-walled craters gouged deep into the frozen soil by AS-driven bulldozers. Caravans of big utility trucks trundled up and down the pitside ramps all day long, bringing hundreds of tons of elusive, rare minerals to the catalytic furnaces.

The trans–Rackliff Basin pipe ended somewhere on this side of the city, too. It stretched a quarter of the way around the planet to the Barclay's glacier runoff, bringing that essential component of life: water. It was actually cheaper to pump it in than to melt it out of local soil. Both the domes and the refineries were greedy consumers.

Lawrence watched the various human enterprises that made up the city with detached interest, visualizing how Templeton and its peripherals must look from space. Some weird plastic flower seventy kilometers in diameter that had blossomed on this barren alien world as the atmosphere warmed. One day it would burst, the nullthene membranes ripping open in the wind so that the terrestrial spawn nurtured within could be flung out across the entire planet. Only with that kind of image did he ever begin to appreciate the enormity of the undertaking that was his homeworld. It was the endless statistics and enhanced images that he could never get his head around, everything the school felt impelled to provide and emphasize.

Out past the last of the refineries, the tundra extended away to the sharp horizon—dirty vermilion soil broken only by rocks and ancient crumbling gullies. Swaths of darkness cut through it at random. When Barclay's Glacier formed, sucking the moisture out of the air and sending the temperature plummeting, the forests were still standing. Their trees had long since died from the cold and lack of light, but the slumbering glacier calmed the air rather than enraging it. There were no winds or sandstorms to abrade the sturdy trunks. The scatterings of moisture left in the soil turned to ice, transforming the surface into a hard concrete mantle, keeping a possessive grip on sand and dust particles.

In the centuries after the glacier formed, Amethi's dead, blackened plants stayed resolutely upright in the still air. Time alone aged them, for there were no elements anymore. Over a hundred thousand years, even petrified wood lost its strength. They corroded slowly, snowing ebony flakes onto the surrounding soil until enough had been shed to make the whole unstable. Then the entire brittle pillar would crack, tumbling over to shatter as if made from antique black glass. More often than not, in the denser forests, they would bring down a couple of their neighbors, initiating cascades of devastation. Where the forests once stood were now areas where the soil was blanketed with low black dunes of congealed grit.

The children quietened at last as this new landscape unwound beyond the bus; here was where their future was birthing with pained deliberation. The first delicate effects of the HeatSmash were proudly visible. Crevices and tiny rills in the hard ground were host to tiny arctic plants. They were all heavily v-written for this world, to endure not only its coldness but also the long light-times and dark-times. Plants that grew above Earth's Arctic Circle through long wearisome days and equally oppressive nights had the closest environmental conditions to those on Amethi. This meant their genes needed the least amount of viral modification to withstand the hostility of this frigid wilderness.

Several of them boasted flowers, tiny dainty coral trumpets or golden starbursts. The most significant accomplishment of the geneticists had been modifying the pollination cycle so that the spores were expelled by ripening anthers into the quiescent air. A haze light enough to drift on Amethi's minimalist breezes, little more than a draft of perfume, yet eradicating the necessity of insects. None of these perennials had needed nurturing in greenhouses and planting out; they were self-set. The first naked terrestrial colonists.

While dark bottle-greens flourished in the earth's crannies, dry-rubber blotches of sulfur yellow and cinnamon brown encrusted exposed rock, from entire cliff faces down to pebbles scattered amid the carbon dunes of the old forests. The lichens that were first spread across Amethi's continents from high-flying robot aircraft in order to kick-start the new ecological cycle, expanded now as never before in the rush of warmth and rising humidity.

Lawrence liked the color invasion stampeding across the bleak tundra. It signaled an astonishing level of achievement. Fundamentally reassuring, that human beings were capable of such visionary endeavor. He began to smile, letting his daydreams build out of the landscape where the impossible was happening. It was easy out here; the demands of his family and school restrictions were falling behind as the bus raced on into the realm of possibilities.

His gaze drifted around and up. He squinted, suddenly alert. Hot urgent hands wiped the bus window where his breath was steaming it up, despite the insulative quality. There. In the sky, something very strange was moving. He knocked on the glass to try to show people where they should look. Then, realizing nobody would ever listen to him, he put his hand up above the window and found the red emergency handle. Without hesitating, he tugged it down hard.

Antiskid brakes engaged as the AS driver program brought the bus to a halt as fast as its engineering parameters would allow. A signal was flashed to the Templeton traffic authority, putting emergency services on immediate recovery standby. Sensors inside and outside the vehicle were reviewed for any sign of abnormality. Nothing was found, but the human/manual intervention was not one the AS could ignore. The bus continued its abrupt deceleration, engine and gearbox whining sharply in mechanical alarm. Kids were hauled back hard into their seats as the safety webbing contracted. Yells and screams ran the length of the aisle. Mr. Kaufman lost hold of his coffee cup and biscuit as he cried: "Fatesbloody-sake . . ."

A second later the bus was motionless and silent, a state almost as alarming as the sudden braking. Then the horn started a repetitive bleat, and amber hazard strobes on the front and back blazed away. Mr. Kaufman and Ms. Ridley gave each other a frantic uncomprehending look and slapped at their web release buttons. The red light above one of the emergency stop handles was flashing urgently. Mr. Kaufman never got a chance to ask whose seat it was before Lawrence was running past him to the front door, which had popped open automatically. The boy was zipping up the front of his baggy coat.

"What . . . ?" Ms. Ridley blurted.

"It's outside!" Lawrence yelled. "In the air. It's in the air!"

"Wait."

She was yelling at nothing. He had already jumped down the steps onto the highway. The other kids wanted a piece of the action; they were laughing wildly, shock already fading as they dashed after Lawrence. They formed a big group standing on the sandy verge. Coats were hurriedly zipped up and hands stuffed into gloves as the bitter air nipped exposed skin. Lawrence stood a little way ahead of them, searching around for the bizarre shape he'd seen. There were several titters behind him as the wait grew.

"There!" he shouted. His finger pointed westward. "There. Look."

The rebuke Mr. Kaufman had been forming died away. A patch of tufty white cloud was floating serenely through the air, the only blemish in a perfect bright azure sky. Silence fell over the kids as they watched the implausible miracle.

"Sir, why doesn't it fall?"

Mr. Kaufman stirred himself. "Because the density is equal to the air at that altitude."

"But it's *solid.*"

"No." He smiled. "It just looks like it is. Remember when we looked at Nizana through the telescope relay and you could see the clouds that made up the storm bands. They were flowing. This is the same, but a lot smaller."

"Does that mean there are going to be storms here, sir?"

"Eventually, yes. But don't worry, they'll be a lot smaller, too."

"Where did it come from?"

"Barclay's glacier, I suppose. You've all seen the pictures of the runoff. This is one of the results. You're going to be seeing a lot more as you grow up." He let them stare at the harbinger for a while longer, then shooed them back onto the bus.

Lawrence was last up the steps, reluctant to abandon his remarkable discovery. And there was also the inevitable censure to face . . .

The teachers were a lot more lenient than he expected. Ms. Ridley said she understood how strange the cloud was, but he must ask permission to ever do anything like that again. Mr. Kaufman gave a gruff nod, enforcing everything she said.

Lawrence went and sat down as the bus moved forward

again. The rest of the kids forgot their games to chatter in an animated fashion about what they'd just seen. Already, this was the best ecology field trip ever. Lawrence joined in occasionally with a few observations and speculations, his discovery giving him kudos previously not experienced. Mainly, though, he tried to keep tracking the cloud through the bus window.

He couldn't stop thinking about the journey it had made. Traveling halfway around the world, with so much unknown territory laid out below it. How ridiculous that a cloud had seen more of Amethi than he ever had. Lawrence wanted to be up there with it, soaring over the land and empty seabeds, swooping down to zoom along the crumbling edge of Barclay's glacier where he could see the runoff, a waterfall as long as a continental shoreline. How fabulous that would be. But here he was, stuck in a bus on his way to a poxy slowlife farm, learning about ecology when at some other school he could be learning how to fly. It just wasn't fair.

The slowlife farm was, like all Amethi's industrial facilities, an uninspiring glass and aluminum box. It was situated all by itself on the side of a gentle valley, with an empty river course meandering away below it. The arctic plants were particularly prolific along the low slopes, clustering thickly on the silt bed itself.

Several of the kids remarked on it when they scuttled from the bus to the warmth of the factory. Lawrence was still trying to see the cloud, which had disappeared off to the north some time ago. The lobby's big outer doors swung shut, and a gust of air washed over the group. They'd all been expecting that. The thermal trap lobby was standard across Amethi, a giant leaky airlock arrangement with thermal recyclers instead of vacuum pumps to prevent temperature drop in the domes. Here, there didn't seem much point. The factory was nothing like as warm as any of the city domes, barely a couple of degrees above freezing. They all kept their coats sealed up.

The supervisor came out to meet them, dressed in padded purple coveralls with a tight-fitting hood. Mrs. Segan, who with her three coworkers ran the whole operation. She tried hard not to show how annoyed she was with another bunch of kids touring around and screwing up her timetable.

"What you're going to see here today has no analogue in nature," she told them as they made their way into the building. This first zone seemed more factory than farm, with dark metal corridors lined with sealed glass windows that looked in on vats of some kind. "We grow fatworms here. I'd like to say breed, but the truth is, every one of these creatures is cloned." She stopped beside a window. The room beyond was filled with racks of trays, filled with a clotted jelly similar to frogspawn. "All slowlife is completely artificial; its DNA was designed for us by the Fell Institute in Oxford, back on Earth. As you know, the more complex an organism is, the more prone to illness and other problems it becomes. Therefore, fatworms are kept very simple indeed. The principal biological streamlining is their complete lack of reproductive ability. That's also very useful to us, as they are only needed for this stage of the terraforming process. They've got a lifetime of about ten years, so when we stop making them, they'll die out." She held up a jar of the jelly substance, handing it to the closest boy. "Pass that around, and please don't breathe on it. All slowlife is optimized to function at subzero temperature; your breath is like a flame to them."

When it came to Lawrence all he could see was a mass of translucent eggs with a pinhead of darkness at the center of each. They didn't quiver or shake about as if they were about to burst open—which would have been something. Boring.

Mrs. Segan took them through into the farm's main rearing arena, a long hall with rows of big rectangular plastic boxes separated by raised metal grid walkways. Pipes overhead sprayed gloopy fluid into each of the open boxes with short regular pulses. The air smelled of crushed grass and sugar.

"Each of the fatworms is essentially a miniature bacteria reactor," Mrs. Segan said as she led them along one of the walkways. "We place them on a new section of tundra and they burrow their way through the ground chewing the dead vegetable matter in the soil. When it comes out, it's suffused with the bacteria that lives in their gut. This prepares the ground for terrestrial plants, which all need the bacteria in the soil to live on."

The kids all leaned over the side of the box she indicated, their interest suddenly regained with the prospect of creatures that could poo out fungus and stuff. A glistening mass of oyster-gray fatworms covered the bottom of the box, squirming slowly: they were about fifteen centimeters long, a couple wide. Everyone *oohed* and *yucked* as they pulled rictus grimaces at the slimy minimonsters.

"Is that why they're called slowlife?" someone asked. "Coz they don't move fast?"

"Partly," Mrs. Segan said. "The temperature they encounter outside means they don't have a particularly fast metabolism, which makes their physical motion correspondingly slow. Their blood is based on glycerol so they can keep moving through the coldest ground without freezing solid."

Lawrence sighed impatiently as she droned out long statistics, then started to explain about other slowlife forms. Some were like fish, swimming in the snow-slush runoff rivers round Barclay's glacier; others were distant relatives of caterpillars, munching their way across the huge dunes of carbon granules left behind by Amethi's original forests. He glanced down into the big box again. It was a bunch of worms wriggling around sluggishly. So what? Who in Fate cared what grubbed around under the soil? Why didn't they come up with birds or something *interesting*? Dinosaurs maybe.

Mrs. Segan moved on, the group buzzing along behind her. Lawrence trailed at the rear. He craned his neck back, looking through the farm's grimy glass roof to see if the cloud had returned. The next thing he knew, he'd tripped on some ridge in the walkway, and went flailing onto his back. One scrabbling hand caught a shallow plastic bin, and when he landed painfully a whole bunch of fully grown fatworms were dropping on top of him.

He rolled away from them quickly, disgust overriding the pain along his jarred spine. These adults were some forty centimeters long, seven or eight in diameter. Their tips waved about blindly. Lawrence clambered to his feet, automatically checking the position of the teachers. Nobody had actually seen him fall. He looked down at the fatworms, the only evidence. Gingerly, telling himself they weren't in the slightest bit dangerous, he groped down and tried to pick one up. It was revoltingly cold and slimy, with a texture like sodden carpet, but he managed to grip it tight. As he lifted it up, the slight wavering motion began to speed up. Instead of putting it back in the bin, he held on and watched. After a while the fatworm was almost thrashing. He dropped it back down onto the floor, and it slithered off along the walkway. There was a claret-colored patch around its midsection where his hand had been. "All right," he murmured. "Not so slow after all." Which was logical. They were slow in the cold; therefore they'd be fast in the warmth.

He scuttled after the group. "Alan," he hissed. "Hey, Alan. Come and look at this."

Alan Cramley stopped munching on his Toby bar, curious about the furtive tone. "What?"

Lawrence took him back to the adult fatworms and showed him. They quickly turned the discovery into a challenge. Pick up the fatworms in tandem and hold them for thirty seconds, then drop them on the walkway and see which reaches the end of the grid first. In the end they were holding on to two each, turning it into a real race.

"What exactly is going on here?" Mr. Kaufman demanded.

Lawrence and Alan hadn't seen him approach from a walkway intersection. He was staring down at the four fat-worms twisting their way across the metal. Several of the other kids were behind him, and Mrs. Segan was scurrying up, anxious to see what the fuss was about.

"I knocked a bin over, sir, and we were trying to pick them up," Lawrence said, holding out his icy hands as proof. Slime dripped from white, cold-crinkled fingertips. "I'm really sorry."

Mr. Kaufman was frowning, not fully convinced.

"Don't touch them," Mrs. Segan called urgently. She slipped past Mr. Kaufman, pulling on a pair of thick gauntlets. "Remember what I told you about them being adapted to the cold."

Lawrence and Alan traded a look.

Mrs. Segan picked up the first fatworm. Her eyes narrowed as she took in the big red mark around its middle. She took it over to the nearest bin. "What have you done?" she yelled. All the fatworms inside had the same red mark. None of them were moving. She hurried to the next bin, and gasped. In the third bin there were some fatworms left undulating slowly; Lawrence and Alan hadn't raced all of them yet. She whirled around. Lawrence took a step back, afraid she'd strike him. Her face was rigid with fury. "You burned them all, you little . . ." She turned to Mr. Kaufman. "Tour's over. Get these brats out of here."

Lawrence had taken over the robot garage several years ago. The compact tracked machines that originally tended to the elaborate gardens of his family's domes had been replaced by newer, more efficient models when they upgraded their

AS groundskeeper program. He'd found the old concrete ramp in the middle of a clump of copper-flowering bushes that had been allowed to expand and merge into a shaggy wall now the entrance was no longer needed. At the base of the ramp was a swing-up door with stiff old lever arms. It took a commendable amount of effort and persistence for a nine-year-old to prize it open, but Lawrence did it, to be rewarded with a musty concrete cave stretching out ahead of him for a good ten meters. Its roof was less than two meters from the ground, and it had strange metal tracks bolted to the floor, walls and ceiling where waldo arms had once run. But there was still power, and a data node.

Since then it had become his den. He'd moved in life's essentials, cluttering it with a dilapidated magenta-colored leather settee, piles of cushions, a couple of tables, an old-model desktop pearl, a sound system with a decibel level that most hard-rock bands would envy, two active memory towers his father had salvaged from the office for him, an eclectic array of tools and boxes of toys he never played with. He'd tacked sheet screens over the walls and even part of the ceiling. A mosaic of images played as soon as the door was opened, some from the memory towers while others broadcast live camera feeds from the datapool.

It was a grand refuge from his family and the rest of Amethi. Even his four younger siblings knew to stay out unless he explicitly invited them in.

He'd gone there as soon as he got back after the ecology field trip. The sheet screens were showing several images of Templeton from cameras mounted on the apex of various domes. One of them showed Nizana's bright crescent, relayed from a near-side school's astronomy department telescope. Another was a telescope tracking Barric, the third-largest moon.

Lawrence told the desktop pearl to find a spaceport feed and switch it to the biggest sheet screen, the one hanging opposite the sofa, which took up half the wall. The camera must have been sited on the control tower: it showed the thick gray runway stabbing out across the bleak rusty-colored tundra. Nothing was landing or taking off.

"Get me a *Flight: Horizon* episode," he instructed the pearl.

"Which one?" its AS program asked.

"Doesn't matter. No. Wait. Series one, episode five: 'Creation-5.' I want third person with the edit I chose last time. Put it on the big screen, and close down the others." He flopped into the settee and stuck his feet up on the armrest. The remaining sheet screens blacked out. Right in front of him the credits started to roll and the soundtrack kicked in, making the thin screens tremble.

He'd found *Flight: Horizon* two years ago when he sent an askping through the catalogues of Amethi's multimedia companies; as far as he was concerned it was the greatest science fiction series ever made. Not fully i, but it allowed personae selection so the episode could be viewed from any of the principal characters' viewpoint. And it wasn't *educational* like all of Amethi's i-dramas aimed at the youth audience.

Set hundreds of years in the future, it featured the amazingly cool starship *Ultema,* which had been sent to explore a section of the spiral arm halfway round the galaxy from Earth: several of the crew were alien, and the weird planets they visited were superbly scary. They were also facing some awesome evil aliens, the Delexians, who wanted to prevent them from getting home. It had been imported from Earth thirty years ago, though the copyright was 2287. There were only thirty episodes in the multimedia company's library, and Lawrence knew them so well now he could almost recite the dialogue from memory. He couldn't believe that was all that had been made. The Earth datapool address of the show's fanclub was tagged in the expanded features menu of every episode, so he'd paid a starship carriage fee and sent them a text message asking for more information. Every time a starship arrived back at Amethi he checked its communication AS, but they'd never sent him a reply.

The *Ultema* was locked in a gigantic energy battle against a blue dwarf star that the Delexians had imprinted with a sentience matrix when a green priority script icon opened in the center of the sheet screen. The starship froze, and the script scrolled down.

Lawrence, please come to your father's study.

He checked the clock. Quarter to six. His father had been back home ten minutes. Mr. Kaufman hadn't wasted any time filing his report package. "Gimme the study pearl," he said to the den's AS.

"I have it online," the AS said.

"I'm busy right now," Lawrence said. He pushed some injured annoyance into the tone. The AS running in the study's desktop pearl was smart.

"Lawrence, please, I accessed the message from your school and prioritized it. Your father wants to see you now."

He kept silent.

"Do you want me to bring your father into this conversation on real time?"

"All right." Grudgingly. "I'm coming, I suppose. But you have to explain to the school AS why my homestudy is short tonight."

"You're not doing homestudy."

"I am. I just have *Flight: Horizon* on as background."

Lawrence swung the garage door shut behind him and wriggled through the bushes. The garage was near the rim of the main dome, which was approaching the end of summer. There were six of the big structures making up the Newton family estate: the large one in the center with its temperate climate, and five smaller ones ringing it, each with a different environment inside. It was one of the larger estates in the Reuiza District, which was where the capital's wealthiest citizens clustered.

He had a three-hundred-meter walk across the grounds to the house itself. The landscape designer had gone in for split levels, with a chessboard of English manor lawns walled in with near-vertical borders of flowering shrubs and perennials. Each lawn was themed with classical plants: one had roses, one with fuchsia, another had begonias, magnolias, hydrangeas, delphiniums; for variety several lawns were enclosed by rockeries sprouting dozens of alpines. Two serpentine pools led away into shallow rocky cascades with reeds and lilies sprouting from outcrops and shelves in the slope. Tall trees stood above the corner of every lawn, again selected for traditional appearance: willow, spruce, birch, horse chestnut, larch. Each of them had had boughs that drooped, either naturally or by judicious shaping, forming massive verdant skirts that swept the grass. Fabulous adventure caves for small children. Lawrence had enjoyed a lot of summers playing in the gardens, as his siblings did now.

A stream ran through the dome, a rough horseshoe shape around the outside of the formal lawns, where the grass was permitted to grow shaggy and daisies and forget-me-nots flourished. He crossed over a narrow moss-cloaked humpback bridge and walked the flagstone path to the house, going up or down steps at the end of each square lawn. Ahead of him, the Newton residence was a stately home built from a yellow stone, with big bay windows protruding from walls swarmed by honeysuckle. Several peacocks strutted around on the gravel path surrounding the house, long folded tails swishing the pebbles about. Their mad penetrating cackle-cry was virtually the only sound in the dome. They scattered as Lawrence crossed the path and made his way up the steps to the front door.

The entrance hall inside was cool. Heavy polished oak

doors opened into the formal ground-floor rooms. Their furnishings and decorations were all exquisite antique pieces. Lawrence hated them; he was frightened to go into any of the rooms for fear of breaking some priceless chunk of the family's precious heritage. What was the point of having a house like this? Nobody could use it properly, not like the real homes some of his schoolmates had. It cost a fortune to build. And it didn't belong on Amethi anyway. This was how people used to build. It was the past.

A wooden staircase curved up to the second-floor landing. He trotted up it, footfalls absorbed by the dark crimson carpet.

His mother was standing at the top, holding two-year-old Veronica on her hip. She gave him a worried look. But then that was Mother, always worried about something. His little sister smiled brightly and held out her hands to him. He grinned and kissed her.

"Oh, Lawrence," his mother said. Her voice carried a unique tone of despair and disapproval that always made him lower his head. It was awful, not being able to look at his own mother. And now he'd upset her again, which was a terrible thing, because she was six months' pregnant. It wasn't that he didn't want another brother or sister, but pregnancy always tired her so much. Whenever he said anything she smiled bravely and said it was why she had married his father, to continue the family line.

Family. Everything was for the family.

"Is he really cross?" Lawrence asked.

"We're both disappointed with you. It was a dreadful thing to do. Imagine treating Barrel like that."

Barrel was one of the family's dogs, a shaggy black Labrador. Lawrence's favorite out of the pack that roamed around the house. They'd grown up together. "It's not the same," he protested. "They're just worms."

"I'm not arguing with you. Go and see your father." With

that she turned her back on him and started down the stairs. Veronica gurgled happily, waving.

Lawrence waved back forlornly and walked slowly to the study. The door was open. He knocked on the wooden frame.

Kristina was just coming out. The new junior nanny for the Newton children. She gave Lawrence a sly wink, which lifted his spirits considerably. Kristina was twenty-one and utterly beautiful. He often wondered if he had a crush on her, but wasn't sure how you knew. He certainly thought about her a lot, if that's what qualified. Anyway, crushes were stupid. Beauty aside, it was great when she was on duty: she was fun, and she joined in the games his brothers and sisters played, and she didn't seem to mind what he got up to or how late to bed he was. All his siblings liked her as well, which was fortunate because she wasn't a whole load of use when it came to changing diapers and preparing food and things. Pity she wasn't on duty more often.

Like the rest of the house, the study wasn't for the use of children. There was a high marble fireplace that had never seen any flames other than the holographic variety. A couple of green leather reading chairs. You had to look hard to see any accommodation to modern technology: the two largest oil paintings were actually sheet screens, and the desk diary was a case for a pane. The walls were covered in bookcases holding leather-bound volumes. Lawrence would have loved to open up some of the classics (definitely not the poetry) and read what was inside. But they weren't books to be read, just to be looked at and assigned high dollar values.

"Shut the door," his father ordered.

Sighing, Lawrence did as he was told.

His father was sitting behind the walnut veneer desk, throwing a silver Dansk paperweight from hand to hand. He was Doug, to his friends—and a lot of people in Templeton fought over gaining that classification. In his mid-forties, though his extensive germline v-writing made it difficult to tell. With a lean build and a face to which smiling came easy, he could have passed for twenty-five without too much problem. Rivals on McArthur's Board had mistaken that smile for an easygoing nature, an assumption they never repeated.

"All right," he said. "I'm not going to shout at you, Lawrence. At your age it's just a waste of time. You just curl up into a sulk and let it all wash over you. If I didn't know better I'd say you were hitting puberty."

Lawrence blushed furiously. This wasn't what he was expecting, which was probably why his father was talking in such a fashion.

"Want to tell me what happened today?"

"I was just messing about," Lawrence said, making sure there was plenty of regret in his voice. "They were only worms. I didn't know getting them hot could kill them. I didn't mean to do it."

"Only worms. Humm." Doug Newton stopped throwing the paperweight and stared at the ceiling as if lost in deep thought. "That would be the same fatworms that are vital for preparing our ecology, would it?"

"Yes, but they clone millions of them in there every day."

The paperweight was tossed between his hands again. "That's not the point, son. This is just the latest episode in a very long line. You're twelve. I can put up with you misbehaving and slacking off at school; it comes naturally at your age. That's why teachers send us reports; so we can make you do your homestudy and ground you when you pee on the security cameras at the museum. What I don't like is the pattern that's developing here. Lawrence, you show a disturbing lack of respect for everything we're doing on this world. It's as if resurrecting the ecology doesn't matter to you. Don't you want to be able to walk outside the domes in just a T-shirt and shorts? Don't you want to see grass bloom on the deserts and watch forests grow?"

"Course I do." He was still smarting over the peeing remark. He hadn't known his father had heard about that.

"Then why don't you show that? Why don't your actions betray these thoughts? Why are you being such a total pain in the ass these days, and incidentally upsetting your mother who happens to be pregnant and is in no condition to be worried by your absurd antics?"

"I do think it. I saw a cloud today."

"And pulled the emergency stop on the bus. Yes, most impressive."

"It was fantastic. I really loved that part of the ecology."

"Well, that's a start, I suppose."

"It's just . . . I know how important HeatSmash is for Amethi, and I really admire everything McArthur is doing here. But it doesn't apply to me as much as it does you."

Doug Newton caught the paperweight in his left hand and stared at Lawrence, quirking an eyebrow. "As I recall, we had you v-written for an improved physique and intellect. I don't recall specifying traits that allow you to live naked and alone on an unmodified isolocked planet. In fact, I'm pretty sure about that."

"But, Dad, I don't want to live on Amethi. At least, not all the time," he added hurriedly. "I want to be a part of McArthur's starflight operations."

"Oh shit."

Lawrence's jaw dropped. He'd never heard his father swear before. He knew now he must be in some pretty deep . . . well, shit.

"Starflight operations?" Doug Newton said. "Has this got anything to do with that stupid show you're always watching?"

"No, Dad. I watch *Flight: Horizon* because it interests me. It's just a drama show. But that's the kind of thing I want to do. I know I can qualify. I'm doing well in all the subjects you need to get into flight training. I've accessed the application packages and the career structure."

"Lawrence, we're a Board family. Don't you understand that? I sit on McArthur's Board. Me. Your dear old father. I make the decisions when it comes to running this entire planet. That's your future, son. Maybe I haven't emphasized that enough. Maybe I shied away so that you'd grow up as normally as you can, without always having that prospect gnawing away at the back of your mind. But that's the way it is, and I think deep down you know it full well. Maybe that's what's upsetting you. Well, I'm sorry, son, but you're a crown prince in this bright new land of ours. It's not easy, but you gain a hell of a lot more than you lose."

"I can come back and be a Board member. Captaining a starship will be the best sort of training for that."

"Lawrence!" Doug Newton stopped himself, and groaned. "Why do I feel like I'm telling you Father Christmas doesn't exist? Listen to me. I can see how flying a starship looks like the greatest thing ever. But it's not, okay? You go from Amethi to Earth, and then Earth back to here. And that's it. Six weeks locked up in a pressurized module with other people's farts and no window. Even calling the staff crew members is a polite lie. People on starships either interface with an AS, or they're mechanics trained in freefall engineering maintenance techniques. You can interface with an AS here in safety and comfort from an office or a park seat. If you do it from a starship cabin for any length of time, your body will suffer. We've got good medicines to cope when your bones thin, your heart muscle decays and your head clogs with every body fluid in existence. They can just about get you through a flight without thinking of suicide—God knows enough of us have done it. I hated going to Earth and back. I was throwing up half the time, I bounced around so much I had more bruises than stepping into a boxing ring can ever give you, and it's impossible to sleep. But a trip back to Earth is a one-shot, people can endure that. If you stay up there ten or fifteen years, even with long planet leave periods, the effects are cumulative. That's just the ordinary damage. It's also a high-radiation-risk profession. Cosmic radiation will tear your DNA to shreds. And all this is the good job; I'm not even going to mention what'll happen to you if you become an engineer who has to go spacewalking. If you think I'm joking, or painting it blacker than it really is, just look up the death rates and life expectancy among crew members. I'll get you access to McArthur's classified personnel files if you want to do it."

"That's not the kind of starflight I'm interested in, Dad. I want to join a starship on a deep exploration mission."

"Really?"

Lawrence didn't like the look of his father's amused smile. It implied some kind of victory. "Yes."

"Find new planets to colonize, make first contact with a sentient alien race, that kind of thing?"

"Yes."

"When you trawled up McArthur's application form for starship crew, did you bother to look up which of our starships are dedicated to interstellar exploration? It's in the same information block."

"It doesn't say. That part of operations is all run from Earth." He watched his father's smile widen. "Isn't it?"

"Nothing is run from Earth, son, not since twenty-two eighty-five. In any case, McArthur canceled all interstellar exploration missions in twenty-two thirty. We haven't flown one since, not one. Know why?"

Lawrence didn't believe what he was hearing. It was all part of some fancy ploy to make him study harder at school, or something. "No."

"Too expensive. Starships cost a fortune to build, and a fortune to run. And I do mean fortune. We got nothing in return for scouting around this section of the galaxy. It's an investment black hole."

"We got Amethi!"

"Ah, at last, some pride in your home planet. Yes, we got Amethi; we also had Anyi, Adark and Alagon. That's what twenty-two eighty-five was all about. We had to get rid of them. Colonization costs money that shareholders on Earth will never see a return on. We're never going to make a commercial consumer product and ship it over interstellar distances and sell it for less than it costs to be produced locally. Investment must come from Earth. There was no way McArthur could fund four planets, so we sold three of them to Kyushu-RV and Heizark Interstellar Holdings in merger deals. That canceled a huge part of the debt we were running up, and in parallel with that we divested some other assets to holding companies and reassigned share ownership of the core company to Amethi residents. It was quite innovative really. Several other companies copied us later. The result is that fifty-eight percent of McArthur shares are owned by Amethi residents. The company on Earth, with all its factories and financial services, now exists for one thing, to fund Amethi. It also offers Earth-based shareholders the eventual dividend of emigration—it's like the ultimate benefits and pension scheme."

"But there's so much out there in space we need to see and understand."

"No, there isn't, son," his father said firmly. "Government space agencies sent ships to just about every kind of star there was to collect data right back at the start of the interstellar age. We've examined every stellar anomaly within range and found more planets than the human race can afford to exploit. We've been out there and done it all. That's over, now. This is the time when we benefit from all that knowledge and effort and expense. It's our golden age. Enjoy it."

"I'll go to another company then, join their starship program."

"Hello? This universe calling Lawrence. Did you not hear everything I just said? Son, *nobody* is exploring anything anymore. There is nothing left to explore. That's why your school concentrates on the courses you'll need to manage Amethi. You have to know what's required to complete the terraforming project. Your future is here, and I want you to start focusing on that, right now. So far I've been tolerant of all this misbehavior, but it ends today. It's time you started measuring up to this family's expectations."

CHAPTER FOUR

"THE WORLD HAD BEEN CHOSEN BY THE LAST CHURCH TO SITE its Supreme Temple because it was close to the Ulodan Nebula, which was remarkable for its darkness. Normally nebulas can be the most glorious of all stellar objects. They're bunched-up, twisted cyclones of gas and dust that measure light-years across, so big they often have several stars inside. The light from those stars makes them glow, fluorescing the dust and vapor into a blaze of scarlet or violet or emerald. But not the Ulodan. The Ulodan was mostly made up from carbon dust, as black as the gulf between galaxies. There were stars inside, including one very famous one that was home to the Mordiff; but they were all invisible from outside. There was no glow, not even a glimmer. The Ring Empire called it the cloud of the dead, especially after their explorer ship found the Mordiff planet. For the Last Church, it was perfect. Standing on their planet and looking up into the sky, the Ulodan eclipsed half of the core suns. It was as if they were being eaten away.

"Mozark's ship landed there on the fifth year of his journey. I suppose it was inevitable he would go to the Last Church at some point during his voyage. Everybody at some time in their life at least considers religion, and Mozark was no different. He left his ship at the spaceport and went to the city of the Supreme Temple. Over the next few weeks he had many meetings with the priests who ran it. They were pleased to receive him, as they were all people. But of course in Mozark's case they made a special effort. He was a prince from a kingdom in a part of the Ring Empire where they had few churches, and he was looking to enlighten his whole people. With his patronage they could convert many new worlds to the true cause."

"What cause, miss?" Edmund asked. "Did they have Buddha and Jesus and Allah?"

"No." Denise laughed, running a hand through her newly shortened hair. "Nothing like that. You have to remember the Ring Empire was a very old civilization. They were long past believing people who claimed to have spoken to God, or to be related to Him, or to have been sent on a divine mission to enlighten this universe. I'm not even sure 'Church' is a very good translation for what the Last Church represented. It was a kind of evangelical physics, really. Unlike all our religions, there was nothing in their doctrine that was contrary to scientific fact, no way their teachings could be weakened as people learned and understood more about the universe. Instead they were a product of that same learning that had given the Ring Empire all of its fabulous technology. They worshiped—again if that's the right word—the black heart of the galaxy."

The children drew in breaths of astonishment. There were a few nervous titters.

"How could they worship nothing, miss? You said the heart of the galaxy is a black hole."

"I did," Denise agreed. "And that's what it is. A huge great hole into which everything falls and from which nothing ever returns. It's already eaten millions of stars, and one day it will finish devouring the whole galaxy. But not for billions and billions of years. And that's why the Last Church revered it and studied it. Because finally, all that will be left of the universe is black holes. They will consume galaxies and superclusters alike. Every atom that ever was will be locked inside them, and then they'll merge, eventually into one. And after that . . ." she teased.

"What?" It was a frantic cry from over a dozen small mouths.

"That's why there was a Last Church, because of the uncertainty. Some of the Ring Empire's astrophysicists said that at the moment when the black holes unite and become one, then a new universe will be born, while others claimed that it's the end of everything forever. The Last Church wanted people who believed that after the unification would come a new universe. You see, as everything in this universe would be absorbed by the black holes, they thought they might be able to influence the outcome. Matter is crushed to destruction inside a black hole, but the Last Church believed it is possible for energy to maintain its pattern inside, either by inscribing it on the crushed matter, or as some independent form. They wanted that pattern to be thought. Souls, if you like. They wanted to send souls into the black heart so that when the end of time came, and the neat order of physics and time fell into chaos, there would be purpose.

"Now as you can imagine, this appealed to Mozark. The sheer worthiness of the concept dazzled him: making sure that existence itself continued. It was something to which the kingdom could devote itself with vigor and enthusiasm. It would also appeal to Endoliyn, he thought. But then he began to have doubts, the same kind of doubts that always threaten religion, no matter how rational its basis. Life is a natural product of the universe; to believe its purpose is to artificially impinge upon the end of time is a huge article of faith. The more he thought about it, the closer the Last Church's gospel seemed to be taken from divine intervention. Their very first physicist-priest had made a choice, and in his vanity wanted everyone else to agree with it. Mozark wasn't sure he could do that for himself, let alone his whole kingdom. For all its grandeur, life is small. To expend it all on a mission that may or may not be necessary in hundreds of billions of years' time was to demand just too much faith. A life in the service of the Last Church wouldn't be spent wisely, it would be wasted. That wasn't what Endoliyn wanted.

"Once again, Mozark returned to his ship, and left the First Church planet to continue his voyage. He rejected the Last Church's abstract spirituality just as firmly as he rejected The City's devotion to materialism."

Denise looked round at her little audience. They weren't quite as enthusiastic as they had been when she'd told them about the wonders to be found in The City. Hardly surprising, she chided herself; they're too young to be preached at.

"Sometime soon," she said in a low, awed voice that immediately gained their attention, "I'll tell you about the Mordiff planet and all of its terrible tragic history."

The Mordiff planet was another of those legends of the Ring Empire that made the children shiver with chilly delight every time she mentioned it. Thanks to the vague hints she'd dropped, it had taken on the form of a particularly aggressive hell populated by well-armed monsters. Which wasn't quite a fair description, she thought, but for using as a bogeyman threat to get the garden tidied at the end of the day it was just about peerless.

* * *

After work Denise took a tram up to the Newmarket District of town. A twenty-minute ride, moving slowly away from the substantial buildings clustered around the marina and docks, out into the suburbs where the roads were broad, and the shops and apartment blocks had flat unembellished fronts. Long advertisement boards hugged the street corner buildings, no longer screen sheets but simple paper posters. Side roads showed long rows of nearly identical houses, whitewashed concrete walls scabbing and crumbling in the humid salty air, small gardens overflowing with ferns and palms.

She got off a stop from the enclosed mart she wanted, and walked. There were no tourists here, just locals. She strolled casually, taking her time to look in shop windows. The bars that were open all had tables and chairs on the pavement outside; their patrons preferred the inside where the lighting was low and the music loud. A scent of marijuana and redshift lingered around the darkened doorways, thick and sweet enough for her to imagine it spilling over the step like a tide of dry ice.

As she approached one, a triad stumbled out into the bright sunlight, blinking and shielding their eyes while their slim wingshades unfurled from gold nosebridges. They giggled with the profound scattiness that only the truly stoned can manage. Two men in their late twenties, large, manual workers of some kind judging by their overalls, and a woman. She was in the middle, with her arms slung around both of them. Not much of a figure, not terribly pretty, either. Her tongue glistened in the sunlight as she licked one man's ear, shrieking with delight. His hand closed on her rump, squeezing hotly.

Denise stopped abruptly and turned away. Despite the sunlight and humidity, her skin was suddenly chilly. She cursed herself, her weakness. It was just the combination that had caught her off guard. From her angle: two men dragging the woman off. Incipient sex. Laughter indistinguishable from cries.

Idiot, she raged against herself. There was a wild impulse to slap herself hard across the cheek. *Knock some sense into you, girl.* Would have done it, too, if this wasn't so public.

It was crazy that her body could be so strong, while her mind was so feeble. Not for the first time, she wondered if Raymond and Josep had asked for subtle neurochemical alterations to be incorporated into their modifications. Human psychology was highly susceptible to chemical manipulation. Could you get a drug for cool?

The triad wobbled away around a corner, and Denise started walking again. A couple of deep breaths and squaring her shoulders tautly returned her traitor body to equilibrium.

A curving glass roof ran the length of the mart, branching out in a cruciform shape a third of the way down from the entrance. Inside, the air was conditioned, scrubbed of moisture and dust. Open-fronted shops had speakers blaring out music and amplifying the spiel that the owners shouted without pause. At the front, the majority of shops were protein knitting; taking raw protein cells from the city's food refineries and blending them with various hydrocarbons and baseline compounds to produce textual approximations of original terrestrial food. There were greengrocers with colored globes purporting to be fruits and vegetables, butchers with burger-steak approximations of every animal from sheep to ostriches, fishmongers with glistening white slivers of flesh on crushed ice; shops with fresh pasta, new-baked bread, rice, curry, cheeses, chocolate, speciality teas and coffees. The smells were enticing as she walked past. Plenty of people loitered, haggling over portions, testing their consistency.

Denise made her way to the back of the mart where Likeside Bikes had their shop. Like every bicycle shop in the universe it had a small front area cluttered with bikes still in their wrappings, while a counter partitioned off a workshop full of tools and small boxes of spares. There were three main work areas, centered on elaborate clamps that held the bikes at chest height. All of them were occupied with machines in various states of assembly with mechanics working on them. Cycling was a popular mode of transport in Memu Bay, and business was brisk.

The assistant manager, Mihir Sansome, looked up and immediately abandoned the child's bike he was working on.

"Hi." Denise flashed him a bright smile. "Has my order come in yet?"

"I believe so." Mihir glanced at his two colleagues and gave Denise a twitchy grin.

She kept her gaze level: it was almost a rebuke.

Mihir cleared his throat. "I'll check." He went back into the workshop and picked up a box from his work bench. "Here we go. Front suspension pins, five sets."

"Thank you." She put cash down on the counter, separating the notes into two piles. Mihir made a show of swiping the box's strip into the till. Five notes went into the cashbox; the larger pile was deftly folded and slipped into his pocket without his colleagues' seeing. He put the box into a carrier bag and handed it to Denise.

As she walked back down the mart, she allowed herself a small smile. Mihir wasn't the greatest actor, but the bicycle shop with its autoclaves and catalytic bonders was incredibly useful. The risk of his activities being noticed was tiny. And even if he was queried by his colleagues or the manager, they'd just assume it was some kind of illegal scam he'd got himself involved in. That was the beauty of every cell-structured underground group: outside of the command group, nobody knew anybody else.

Even if the worst-case scenario came about and the authorities became aware of a cell, they'd only be able to close down that one unit. Taken by itself, the items Mihir had produced for them would mean nothing to the police. He'd probably be able to describe Denise, but as far as he knew she was just a courier. He'd been recruited by members of another cell, who had been given the information that his cousin had died during the last invasion. After skirting around his sympathies, they'd asked him if he could help out making life difficult for the next occupation force. It wouldn't even cost him anything; the movement would be happy to pay for his trouble. Once he'd agreed, the only contact he had was through encrypted packages containing the specifications of components. And Denise.

Had it been a normal radical movement, then they would have used a low-level courier to collect the box. This was a little different.

Indigo data scrolled across her sight as the Prime in her pearl ring trawled the datapool for real-time police messages. There were hundreds of them, the majority simple routine contacts and location monitors. There were even some special investigation branch operations. None of it related to her.

Even so, she kept an eye on her fellow pedestrians, noted the few cars and vans parked along the street, watched the cyclists. None of them seemed interested in her, except for a couple of lads. But then surveillance operatives wouldn't show an interest; it was recurring faces that she was hunting for.

Only two people got on the tram with her. She switched trams twice before she finally arrived at the workshop, confident no one was following her. It was one of twelve identical workshops in a two-story block designed to accommodate light industry. The whole place had a fairly dilapidated appearance, with windows covered up with reflective shields or wood panels. A faint whine of air-conditioning sounded along the narrow deserted street that led to the rear loading bays. Piles of discarded packaging were accumulating by several of the roll-up doors. She'd never seen anyone put rubbish out, or a city council crew collect any. But the size and position of the piles changed on a weekly basis, so someone else used the workshops.

Denise asked her neural pearl to check the workshop's security network, which reported that the perimeter was secure. She waved her left hand over the lock sensor and pushed the door open. It was a large concrete-walled room inside, empty apart from a long wooden carpentry bench they'd set up in the center and a metal storage rack that took up half of the loading bay wall. Both windows and the roll-up door had been bricked up and reinforced with carbon webbing.

Josep was already sitting at the bench, milling cylinders of stainless steel on a programmable electron beam lathe. "Did you get them?" he asked.

"Hope so." She dropped the box on the bench and broke the seal. Two dozen black cylinders spilled out. They both started examining them.

Mihir had produced slightly conical tubes of boron beryllium composite ten centimeters long. The narrower end was open, while the base was sealed with a small hole in the center and an outer ridge. Denise wondered if he knew he was producing bullet casings. The shape was obvious enough, though the high-strength composition could be misleading.

"Not bad," Josep said. He was measuring his casing with calipers, the liquid crystal display blurring as they closed around the base. "Not bad at all. He's got the dimensions within spec."

"I'll start filling them," she said. The casings were the last component. They already had the bullets, the caps and enhanced explosive. Combined with the rifle they'd assembled, a single shot would be able to punch clean through Skin from over two kilometers away.

The rifle was just one of the weapons they planned on using. Other weapons and booby traps were being put together by cells across Memu Bay. Innocuous little components locking together into lethal combinations. This time when the invaders arrived, the resistance movement would be there and ready to make life hell for them.

Platoon 435NK9 had to wait in the base's transit lounge for five hours. Lawrence didn't mind that for himself—the lounge was air-conditioned, he had a memory chip loaded with a good multimedia library, the drinks machine was free, missiontime pay had begun that morning. Squaddie heaven. He stretched his legs out over three chairs and relaxed while the big departure sheet screen kept repeating the same messages about their scheduling delay and mechanical service requirements. Somewhere out across the hot runway, teams of mechanics were peering quizzically into the inspection hatches of their assigned spaceplane, trying to find which one of the fifty thousand subcomponents the AS pilot was bitching about. AS pilots monitored every component parameter constantly, running the results against International Civil Aerospace Agency performance requirements. Lawrence had heard that operating companies often rebooted their vehicle electronics with AS programs downrated from the manufacturer's primary installation, allowing a degree more flexibility when it came to determining flightworthiness. The letter of ICAA's law equaled huge maintenance costs.

If a Z-B AS pilot wanted repairs before it would fly, Lawrence was very happy to have the procedure carried out. The spaceplane would definitely need it.

The enforced hiatus didn't sit so well with the rest of his platoon. Worst hit was Hal Grabowski, the youngest member, just past nineteen years old. Hal's flight experience was limited to one subsonic transocean flight to Australia and five short helicopter trips during the last phase of their training. He'd never been on a spaceplane, let alone experienced freefall. Spaceflight was a novelty he was hungry for, prowling around the lounge in search of some sign that they could embark. A sure giveaway that he'd never seen active service before, either. The ancient armed forces maxim—never volunteer—had streaked over Hal's head at near-orbital altitude.

"It's been three hours!" the kid complained. "Fuck this. Hey, Corp, if they don't fix it, will they give us another spaceplane soon?"

"Yeah, I expect so," Corporal Amersy muttered. He didn't even glance up from the screen on his media player card.

Hal's arms flapped about in disgust. He stomped off to annoy someone else. Amersy looked up, watching the kid's back, then turned and smiled at Lawrence. The two of them shook their heads in unison. Amersy was a good ten years older than Lawrence, though his thinning hair was the only outward sign of aging. He was very careful to keep in shape, spending hours each week in the base gym. Good physical condition was a non-negotiable requirement Z-B placed on all its strategic security division squaddies. Amersy was never going to rise above corporal; he had neither the stakeholding nor the connections. It didn't bother him; the position meant he could take good care of his family, so he worked hard at maintaining it. That worked to Lawrence's advantage; Amersy was the most reliable corporal in the Third Fleet.

Only his face betrayed the time he'd devoted to the front line of Z-B's asset-realization policy. A wide patch of skin at the rear of his left cheek was slightly chewed up where a Molotov cocktail had burned through his helmet fifteen years earlier in the Shuna campaign, before Skin reached anything like its current level of ability. Even that shouldn't have been too visible, not with the dark ebony color of Amersy's skin. But that day the Third Fleet field hospital had been inundated with casualties; at the end of a twenty-two-hour shift, the trauma doctor was too fast applying dermal regeneration virals. They'd done the job they were designed for, infiltrating the corium layer to implant new genetic material that would build his epidermal layer back over deep char ridges. Unfortunately the genes that the virals carried were tailored for a Caucasian. Half of Amersy's cheek was white, resembling some kind of flat tumor.

Amersy allowed rookie squaddies to have one joke about it. Hal, naturally, had made a second. The kid was taller even than Lawrence, topping out over two meters, with muscles that could match a Skin suit's strength. It didn't make any difference; he'd limped for a week after landing badly. The kid had shown the corporal plenty of respect since then; it was about the only lesson he had ever learned properly in the whole nine weeks since he'd joined the platoon.

"Are there going to be stewardesses?" Hal asked Edmond Orlov. "You know, some decent-looking pussy."

"It's a fucking military flight, you dipshit," Edmond sneered at him. "Officers and management get freefall blow jobs. You get to fuck Karl."

Karl Sheahan lifted his head, blinking his eyes open. Tiny colored silhouettes shivering over his optronic membranes shrank to nothing. He gave the pair of them the finger.

"What about the starship?" Hal persisted. "Any chicks in the crew?"

"I haven't got a fucking clue. And even if they were all female, it wouldn't make any goddamn difference to you. Crew only ever get the best, that means their fucking coffee machine is smarter and more attractive than you."

"Aww man, that is such a waste. I mean, how many times does a guy have this kind of opportunity? The way I figure, I'll see six, maybe seven campaigns. That'll give me a total of fourteen spaceflights. I don't wanna waste none; that's criminal."

"Waste them how?"

"Boomeranging the padding, man. The big freefall free-for-all. A midair rodeo." He clenched his fists and held them up, pleading. "I wanna have sex in zero-gee, man! Every unnatural position you're not built for. Holy shit. I get hard just thinking about it."

"Shut up, you arrested pervert. There's no such thing. The whole idea's a myth dreamed up by corporate publicity back when they started flying orbital sight-seeing tours. Get it? You even twist your head around fast in freefall and you throw up. You start tumbling around the way you're thinking of, and every orifice lets fly. And I mean every. Now forget about it and give the rest of us a break."

Hal backed off, looking wounded. Edmond was the closest he had to a genuine buddy in the platoon. The two of them had broken base curfew enough nights to go cruising the Cairns Strip together.

Lawrence waited silently, hoping the kid would finally shut up. There were ten other platoons waiting in the lounge with them, all of them hyped with the prospect of the flight. It wouldn't take much to start a fight. He didn't want to start ordering the kid about before the mission had even taken off. None of the others were such a pain, but then they were older, half of them had families, too, which acted like a damping rod on wilder aspects of their behavior. And all of them had seen duty together.

Hal walked over to one of the big picture windows, pressing his face against it to look eagerly out at the huge spaceplanes that were managing to take off. He took a swig from a Coke can.

"Hal, stop drinking now," Amersy said. "You don't want any fluid in your stomach when we go into orbit. You'll throw up even if you don't twist your head."

Hal glared at the can. He dropped it and kicked it in the direction of the nearest wastebasket. There was no other form of protest.

The kid would do all right, Lawrence decided. He just

needed guiding through the first few crowd encounters and he'd start to learn caution. Pity he didn't have a steady girlfriend; that was always a calming influence. But at nineteen he was only interested in screwing as many girls as he could impress by his muscles and his credit card.

Four and a half hours into the wait, and the departure sheet screen changed their flight status to boarding. Hal let out a loud whoop and snatched up his small bag. The rest of Platoon 435NK9 lumbered up out of their chairs and made their way over to the designated gate. Their spaceplane was rolling slowly into the departure bay as they assembled at the clearance desk.

The Xianti 5005h3 spaceplane was a well-proven commercial ground-to-orbit vehicle; the Beijing Astronautics Company had first flown the original 5005a mark in 2290. Since then there had been over forty variants produced as the manufacturer gradually expanded capacity and smoothed out early bugs. The 5005h3 was a stretched delta planform 120 meters long, with a wingspan of a hundred meters. Eighty percent of its volume was taken up by fuel tanks. Its carbon-lithium composite fuselage had a broad center section with graceful curves blending it cleanly into the wing section, a softness in sharp contrast to the knife-blade leading edges. A third of the way down the belly was a single oval scoop intake with an airspike protruding several meters from the rim.

Several gantry service arms rose out of the departure bay's concrete, carrying pipes and utility cables that were plugged into sockets along the Xianti's belly. Technicians in silver fire suits were walking about underneath, inspecting the huge wheel bogies and keeping an eye on the fueling process. A tall girder tower at one side of the bay had clean white vapor flowing silently out of a nozzle on the top, dissipating fast in the warm breeze. That was the only sign that the spaceplane used cryogenic fuel. Its fuselage remained remarkably free from condensation as the on-board tanks were chilled down and filled.

A pair of Z-B spaceflight division staff stood behind the clearance desk, handing out protective black plastic helmets, similar to the kind cyclists used. They made sure everyone put theirs on before embarking. At the end of the sealed walkway a small grimy window looked back along the huge vehicle. That was the last sight Lawrence was given of the spaceplane—a vast expanse of silver-blue wing surface, its size the only indication of the raw power to be unleashed in the flight. As he walked past he felt that familiar small twist of envy, wishing that he were the pilot who hauled this superb monster up through the atmosphere into space and freedom. Except, as all the years since Amethi had shown him, it wasn't true freedom. At some time, you always had to come back down to earth. That wish was the wonderful deceit that had so far cost him twenty years of his life.

The Xianti's passenger cabin was remarkably similar to that of a standard aircraft. Same worn-down blue-gray carpet, not just on the floor but the walls and ceiling, too; pale gray plastic lockers above the seats, harsh lighting, small vent nozzles hissing out dry air a couple of degrees too cool for real comfort. There was plenty of headroom, though, and the chairs had deep jelfoam padding as well as being spaced well apart. All that was missing were windows.

Lawrence made sure the platoon stowed their bags and strapped themselves in securely before he fastened his own buckles. Seatback screens ran through a few brief safety procedures. Lawrence ignored them. Not that he was blasé about spaceflight, more like pragmatic. At takeoff, the spaceplane carried nearly five hundred metric tons of cryogenic hydrogen. No major emergency was survivable.

The Xianti taxied to the end of the runway, and the human pilot cleared the AS for launch. Four Rolls-Royce RBS8200 turbojets throttled up, producing seventy-five tons of thrust.

They began to race down the runway. Seatback screens showed Lawrence the scenery flashing by; the green blur transferred smoothly into a pale blue as they lifted from the tarmac. Then the huge bogies retracted with a noise more like sections of fuselage tearing off. The blue slowly began to darken.

With full afterburn the turbojets pushed the Xianti up to Mach 2.6 somewhere above the Willis Islands. The scramjet ignited then, liquid hydrogen vaporizing in carefully designed supersonic plume patterns within the hot compressed airflow before combusting in long, lean azure flames. It produced 250 tons of thrust, shaking the cabin with a gullet-rattling roar as it pushed the spaceplane ever higher through the stratosphere.

Lawrence clamped his teeth together as the G-force crept upward and the scramjet's fierce vibration blurred his vision. The pressure on his lungs increased toward the verge of pain. He tried to concentrate on breathing regularly—not easy through the building anxiety. The enormity of their power-dive up into orbit made him understand just how insignificant he was in relation to the energies driving them, how hopelessly dependent they were that the obsolete design programs had been used properly fifty years ago, calculating theoretical parameters of aerothermaldynamic flow; that everything was going to work and keep on working under obscene stresses.

Stars began to appear in the seatback screen as the velvet blue panorama drained away into midnight black. The AS pilot began to throttle back the scramjet as they reached Mach 20. They were at the top of the atmosphere now, still soaring upward from the impetus of the burn. Even at that speed, the oxygen density was falling below sustainable combustion levels. Two small rocket motors in the tail fired up, producing a mere fifteen tons of thrust each, which gently eased the spaceplane up to orbital velocity. They created the illusion that the spaceplane was standing vertical on a low-gravity moon. Lawrence's chair creaked as its struts adjusted to the new loading. At least the pounding roar was over.

The glaring blue-white crescent of Earth slid into the bottom of the seatback screen as the rockets cut off, taking with them the last percentage of G-force. Every nerve in Lawrence's body was screaming at him that they were now falling back to the ground, ninety kilometers below. He took some quick shallow breaths, trying to convince himself that the sensation was perfectly natural. It didn't work particularly well, but he was soon distracted by the sounds of worse suffering from his fellow passengers.

For forty minutes the Xianti glided along its course, passing over Central America and out across the Atlantic. Seatback screens flashed a quick warning, and the small rockets fired again, circularizing their orbit at four hundred kilometers' altitude. After that Lawrence heard a whole new series of mechanical whines and thuds. The spaceplane was opening small hatches on its upper fuselage, extending silver radiator panels to shed heat generated by the life support systems and power cells. Its radar began tracking the *Moray*. The orbital transfer ship was twenty kilometers ahead, in a slightly higher orbit. Reaction control thrusters adjusted their trajectory in minute increments, closing the gap.

Lawrence watched the screen as the *Moray* grew from a silver speck to a fully defined ship. It was three hundred meters long, and about as basic as any space vehicle could get. Habitation cabins were five cylinders clustered together, thirty-five meters long, eight wide. They'd been sprayed with a half-meter layer of carbon-based foam, which was supposed to act as a thermal shield as well as providing protection from cosmic radiation. Lawrence had checked the solarwatch bulletin before they took off. Sunspot activity was moderate with several new disturbances forming, one of them quite large. He hadn't told the rest of the platoon, but he was quietly relieved the transfer would take only thirty hours. He didn't trust the foam to protect him from anything serious. Its original white coloring had darkened to pewter gray as the years of vacuum exposure boiled the surface, and even with the spaceplane camera's poor resolution he could see pocks and scars from micrometeorite impacts.

Behind the cylinders was a life support deck, a clump of tanks, filter mechanisms and heat exchangers. A broad collar of silver heat-radiator panels stretched out from the circumference, each segment angled to keep the flat surface away from direct sunlight.

Next was the freight section, a fat trellis of girders sprouting a multitude of loading pins, clamps and environmental maintenance sockets. For the last three weeks, the Cairns base spaceplanes had been boosting cargo pods up to the *Moray* and its sister ships. They'd been ferrying them up to Centralis at the Lagrange four orbital point where the starships waited, then returning to low Earth orbit for more. Even now there wasn't a single unoccupied clamp. They contained the helicopters, jeeps, equipment, armaments and supplies for the Third Fleet's ground forces; everything they'd need to mount a successful mission.

The final section of the ship was given over to propulsion. It housed two small tokamak fusion reactors and their associated support machinery, a tightly packed three-dimensional maze of tanks, cryostats, superconductor magnets, plasma inductors, pumps, electron injectors and high-voltage cabling. The fifteen heat radiator panels necessary to cope with the system were over a hundred meters long, sticking out from the ship like giant propeller blades. The tokamaks fed their power into a high-thrust ion drive, eight grid nozzles buried in a simple box structure that was fixed to the base of the ship almost as an afterthought.

Moray's length drifted past the camera as the Xianti gently maneuvered itself to the docking tower at the front. Reaction control engines drummed incessantly, turning the spaceplane along its axis as it was nudged ever closer. Then the airlock rings were aligned, snapping together with a clang.

Lawrence took a look around the cabin. Several squaddies had thrown up, and the larger air grilles along the ceiling and floor were splattered with the residue. Checking his own platoon he could see several of them showing signs of queasiness. Hal, of course, had an expression of utter delight on his face. Zero-G didn't seem to have any adverse effect on him at all. Typical, Lawrence thought; he could already feel his own face puffing up as fluids began to pool in his flesh.

The airlock hatch swung open, and the cabin PA hissed on. "Okay, we're docked and secure," the human pilot said. "You can egress the transorbital now."

Lawrence waited until the platoon sitting in front of him had gone through the airlock before releasing his own straps. "Remember to move slowly," he reminded his people. "You've got a lot of inertia to contend with."

They did as they were told, unbuckling the restraints and gingerly easing themselves out of the deep seats. It had been over eighteen months since any of them had been in freefall, and it showed—sluggish movements suddenly becoming wild spins. Desperate grabs. Elbows thudding painfully into lockers and seat corners. Lawrence Velcroed his bag to his chest, and used the inset handles along the ceiling to make his way forward. In his mind, he tried to match the process with climbing a ladder. A good grounding psychology: always try for a solid visual reference. Except here his legs wanted to slide out to the side and twist him around. His abdominal muscles tensed, trying to keep his body straight. Someone knocked into his feet. When he glanced around to glare, Odel Cureton was grimacing in apology, his own body levering around his tenuous handhold, putting a lot of strain on his wrists.

"Sorry, Sarge." It was a fast grunt. Odel was trying not to puke.

Lawrence moved a little faster, remembering to kill his motion just before he reached the airlock. He slithered easily around the corner and through the hatch, pleased with the way the old reflexes were coming back.

The *Moray* was as crude inside as it was outside, stark aluminum bulkheads threaded with dozens of pipes and conduits, hand loops bristling everywhere. The air reeked of urine and chlorine. It must have been strong: Lawrence's sense of smell was fast diminishing beneath clogged sinuses. One of the crew was waiting for him on the other side of the airlock. Lawrence gave him their platoon number, and in return was told what berth they'd been assigned. Each of the big habitation cylinders was color coded. Lawrence led the cursing, clumsy platoon down into the yellow one. Voices echoed about him, coming from open hatchways—other platoons bitching about the conditions and how ill some of their buddies were and why didn't someone do something about bastard freefall. Twice Lawrence banged himself on the walls as they scrambled their way along the central tubular passageway; elbow and knee. By the time he slipped into their compartment he could feel the bruises rising.

The others crawled in after him, moaning and wincing, looking round sullenly. Their compartment was a simple wedge shape. It had three rows of couch chairs with simple hold-you-down straps, a pair of freefall toilets, a locker full of packaged no-crumble meals, a microwave slot and a water fountain with four long hoses ending in stainless-steel valves. Someone had written: *Don't even think about it* on the aluminum concertina door of one toilet. What would become the ceiling when the ion drive was running was covered with a sheet screen. It was orientated so that anyone in the chair couches would have a reasonable view. A Z-B strategic security logo glowed faintly in the center: purple omega symbol bracketing the earth, crowned by five stars.

Hal stowed his bag and flew across the compartment, turning a fast somersault on the way. "This is fucking amazing. Hey, what kind of i-media are they going to give us, anyone know?"

"You don't get i-media in a crate like this," Odel said in exasperation. "This isn't a pleasure cruise, kid. You worked that out yet? You'll be lucky if they've got black-and-white films." He put his glasses on, leaving the display lenses clear, but settling the audio plugs in his ears. Thin vertical scarlet lines appeared on the lenses as he called up a menu. He worked down a playlist of rock tracks from centuries-old classics right up to Beefbat and Tojo Wall, then settled back contentedly as they began to play.

Lawrence sighed, fastening himself loosely into one of the couch chairs. It could have been worse. Some platoons had boosted out of Cairns ten days ago. At least he only had another four days until the Third Fleet departed from Centralis. Perhaps they could put something in the kid's food.

Simon Roderick went down to the observation gallery half an hour before the portal opening. There were over a hundred people crammed into the small chamber protruding from the surface of Centralis. Somehow they contrived to give him a little patch of free space in front of the thick glass where he could stand by himself. They were silent, though Simon could sense their minds spiking with resentment and disquiet. As always, he ignored their pusillanimous nature with his usual contempt. The physical discomfort of Centralis itself couldn't be dismissed quite so easily, however.

Centrifugal force didn't make him giddy, although he often caught himself wishing for a full one-gravity field.

Centralis was too small for that, its rotation producing slightly less than two-thirds of a gee around its outermost level.

Back in the mid-1970s when Gerard K. O'Neill was putting his High Frontier concept together he produced several designs for space "islands." Starting with the simple Bernal sphere at four hundred meters in diameter, the concepts progressed up to the paradise garden of "Island Three": linked twin cylinders twenty kilometers long. All of them were admittedly achievable with relatively simple engineering procedures. The problem came in gathering together that much material with the requisite construction crews and their assembly equipment in an era when it cost upward of two hundred million U.S. dollars to launch a single space shuttle.

Scramjet spaceplanes eased the problem of cheaper access to space. But as they helped to build low-orbit stations and their associated industrial modules, so they reduced the need for vast habitats. Even in a rampant consumer society, the quantity of ultraspecialist crystals and chemicals that could be produced only in microgee facilities was limited to a few hundred metric tons a year—a figure easily supervised by small tough crews paid exorbitant salaries to endure the generally unpleasant conditions to be found in Earth-orbit space stations.

It was only in 2070, when a method of faster-than-light travel was developed, that there was any need for large high-orbit dormitory towns. Starships were neither compact nor cheap: they needed thousands of people to construct the massive superstructure and integrate hundreds of thousands of components into a functional whole. As they were too big to take off or land on a planet, they had to be built in space. O'Neill's old ideas were pulled out of university libraries and studied afresh.

One critical development since O'Neill's day was synthetic food production. The old island designs were driven by the need to provide vast amounts of farmland to feed the indigenous population. His cylinders had been given multiple-wheel crystal palace geometries and elegant necklaces of agricultural modules so they would be self-sufficient. The new designs discarded all that baggage. All they needed was a couple of refinery modules to process excrement back into protein cells. The starship companies still kept the idea of a central garden park; that kind of open space was acknowledged as a primary psychological requirement for the crews that would have to spend months if not years up in the islands. And that much biota was a reasonably cost-effective fail-safe air regeneration system. But the rest of it, the luxurious landscapes, meandering freshwater lakes, the giant windows with their fans of mechanically swiveled mirrors, Caribbean climates and nuclear family neighborhoods of open-plan Italian villas—that was all condensed and modernized.

Centralis, which was Z-B's primary Lagrange point facility, adopted a plain cylinder geometry, five hundred meters in diameter and one kilometer long. Its apartment complexes were as cramped as any low-rent city skyscraper, only these were ring-shaped, occupying the lower fifty meters of each circular endwall. The garden between them, like any urban park, was overutilized and overmaintained. Shrubs and trees grew tall and spindly on the thin layer of crushed rock sand that passed for soil, the fusion-powered plasma tube running along the axis never providing the right level of UV. But it had ponds with fountains and expensive koi carp, and picnic tables, and a jogging track, and baseball diamonds, and tennis courts. Although it took months for the Johnny-come-latelies to pick up on the Coriolis-driven double curve flight of balls.

Radiation shielding, too, was a constant worry beyond Earth's protective atmosphere. The only true defense against gamma rays and high-energy particles was mass, big fat solid barriers of it. Centralis was given a rocky external shield two meters thick, which was encrusted with black radiator fins. There were a few gaps through, for the axial corridors connecting the main cylinder to the nonrotating docking net at each end, shafts for the pipes carrying fluid to and from the fins, and the observation gallery.

Constellations traced a slow curve outside, although the portal was always in view. It hung fifty kilometers off Centralis's rotation axis like a blue pole star. Simon knew where it was only from the cluster of colony trains keeping station around it, slim silver bars agleam with reflected sunlight, forming their own tight little cluster.

He used his DNI to receive sensor feeds from the trains, closing his eyes to be greeted with the sight of the portal directly ahead. A simple ring five hundred meters in diameter, it looked like a toroid of black hexagonal netting encasing a weak-glowing neon tube. The sight of one never failed to inspire him, reaffirming his faith in Z-B and all its endeavors. A portal was the most sophisticated, and expensive, technological artifact that the human race was capable of constructing. Only Zantiu-Braun had the facilities, money and determination to build them. Portals were one-shot wormhole gateways. Rather than the continual spatial compression that ordinary starships generated for themselves to fly through, these opened a wormhole down which any vehicle could travel. It took a phenomenal amount of energy, correctly applied against the fabric of space-time, to create the rift. The combined fusion generators of twenty starships wouldn't be able to produce a fraction of the power necessary. So Z-B manufactured an isomer based around hafnium 178 and boosted it up to its K-mixing state; the subsequent decay to its ground state produced the amount of energy required to distort space-time once it was correctly channeled and focused. But the isomer was incorporated into the solid-state wormhole generation mechanism itself, which meant that once the decay was complete the whole edifice was not only useless but highly radioactive as well. You couldn't take it apart and refill it with a fresh isomer. A new one had to be built each time.

That one-shot limitation meant Z-B had to extract the maximum amount of use from every portal. As a result, they'd designed the colony train, a spacecraft every bit as crude as an orbital-transfer vehicle, but on a much larger scale. They had the same type of engine system, fusion generators powering high-thrust ion rockets, sitting at the base of a kilometer-long girder tower. Simple enough to build, they were assembled in the redundant starship yards floating in attendance around Centralis.

The fleet of colony trains that Simon could view through his sensor feed were fully laden with descent capsules, as if the tower had been swamped by metallic barnacles. Each of the 840 capsules was an identical cone, six meters wide at the base, coated with a silvery ablative foam that would allow them a single airbrake into a planetary atmosphere. They carried four people and all the basic equipment they'd need to start life on a new planet from scratch.

To qualify for a berth, the colonists had to be thirty or younger (that is, of childbearing age) and have an appropriate stakeholding in Z-B. It was a qualification that millions of people still strove for, even though the portal worlds were nothing like the original interstellar colonies. There would never be a follow-up, no second portal delivering supplies to expand the colony, no regular starship flights from Centralis. After they arrived the colonists were on their own. If they wanted to get back in contact with Earth and the other developed worlds, they had to build their own FTL vehicles. Best estimates put that level of financial and industrial ability being achieved at least a century after the founding flight, and probably a lot longer.

As financial analysts never tired of pointing out, Z-B's portal colonies were extremely risky ventures. In an age in which most interstellar flight was now concerned with asset-realization missions, Z-B's attitude seemed almost anachronistic, especially as it was itself heavily involved with asset realization.

Simon observed the colony trains patiently, waiting as the small digital timer on the rim of his vision counted down the minutes left until activation. When it came, there was little physical evidence. The blue light of the portal ring brightened by two orders of magnitude as the isomer cascade was initiated. The aperture turned blank, obscuring the stars that had been visible in the center. Slowly the blue luminescence contained in the ring began to seep inward, forming a solid sheet of photons. It twisted in an instant, distorting backward and opening into a seemingly infinite tunnel. The intensity of the light withdrew until the wormhole walls were defined only by a faint violet haze that neither camera nor eye could quite bring into perfect focus.

A whole grid of script tables displayed by his DNI provided Simon with more information than he wanted about the wormhole's stability and its endpoint coordinate. The target was Algieba, a yellow giant binary system 126 light-years distant. Easily the farthest out that any kind of colony venture had been attempted; it was just approaching the range of the current portal capacity.

The portal operations AS picked up the beacon signal left behind by the explorer starship, allowing it to confirm the endpoint coordinate was within ten million kilometers of the target planet, an Earth-equivalent world orbiting the smaller of the two stars. It flashed a go signal to the colony trains.

Ion rockets burned with a painful near-ultraviolet gleam, moving the massive craft out of their holding pattern. The first five to slide into the wormhole were carrying descent capsules full of industrial equipment and civil engineering machinery, a basic infrastructure that would support the entire colony throughout its early years. Twenty trains of descent capsules followed, each one traversing the twenty-five-kilometer internal length of the wormhole in a little over two minutes, emerging into the double glare of the yellow stars.

Data traffic within the wormhole reached a crescendo as the trains sent back their status and location. Then the isomer cascade was exhausted and the transdimensional fissure collapsed.

Cold blue light faded from within the portal's external netting, revealing a complex braid of flat gold, ebony and jade filaments. The luster had gone from the materials; they now possessed the tarnished and brittle look of an antique, as if the wormhole had aged them centuries.

People started to make their way out of the observation gallery. Simon waited until he was alone. He canceled his DNI link with the Centralis datapool to stare at the patch of space where the dead portal floated. It was as if the extinguished circuitry still exerted some kind of weak gravitational pull on his mind. He felt an almost childish burst of jealousy against those who had gone through. They were free of Earth's myriad problems, its contamination and sullying of all human events. Even their passing made it harder for those left behind. Zantiu-Braun had weakened itself still further by giving them their fresh chance. The company could barely afford to fund a portal colony every eighteen months now; not even asset realization was plugging the financial gap anymore. Every time Simon stood in the observation gallery watching colleagues and family depart, his resolution to stay and hold back the barbarian horde decayed a fraction more. He often wondered what his mind's half-life was, at what point he would give in to pessimism and abandon Earth for his own new beginning. It could happen. The sheer inertia of humanity's stupidity would see to that.

The *Moray* arrived at Centralis thirty hours after leaving low-Earth orbit. Lawrence had resisted the urge to eat as best he could, limiting himself to one meal. That way he only had to use the toilet once; even with the ship's tiny acceleration helping directionwise, it wasn't an experience he wanted to repeat if he could help it. At the same time he forced himself to drink every hour or so: with only a minute G-force when the ion engines were on and several hours' coasting in freefall between burns, it was easy for his body to become dehydrated. Without gravity pulling fluids down to his feet, his deep instincts were confused and unreliable. At least pissing in space was simple—assuming you were male.

Hal had to be ordered to drink from the water fountain hose on more than one occasion. He wasn't the major problem. Lewis Ward got a bad dose of space sickness, throwing up every time he even tried to drink water. After a couple of hours dodging revolting yellow stomach juices, Lawrence called for the doctor. When she did arrive twenty minutes later, she simply used a hypospray to give him a mild sedative and told the platoon to try to get him to drink in an hour or so.

"Don't let him eat for the rest of the trip," she warned. "*Koribu* has a one-eighth gravity field. He can last until then." With that she zipped out of the door with the agility of a shark.

Hal gave a disgusted snort as she left; she'd been in her fifties, and a decade of service in low gravity had seen her body gently swell out as her legs and arms became more spindly. He'd brightened at the prospect of her house call. Since she'd arrived, he hadn't said a word.

"Sorry, guys," Lewis whispered. There was a single strap over his legs, allowing him to adopt a semicurled position on the couch chair. His face was gleaming with sweat. When it came to training and maneuvers, Lewis could move quicker than just about anyone else in the platoon. He had a ratlike agility, allowing him to vanish into some crack or corner that would give him cover whatever the terrain was. His thin body had the kind of stringy standout muscles and tendons Lawrence associated with marathon runners. But he could dash along a ten-meter suspended pole without even having to hold his arms out for balance. Funny how space sickness had hit him worse than any of them.

"No problem, my man," Odel Cureton said. "Statistically, one and a half of us will suffer some kind of aggravated motion sickness per twenty-five hours of flight. You coming down with it means the rest of us are in the clear." Odel was what passed for the platoon's electronic specialist. At thirty-two he didn't have any degrees or qualifications from colleges or even Z-B, at least none that he produced for the personnel department. But as Odel admitted to four ex-wives, and those in just the last six years, Lawrence could appreciate the man's need for blurring his background. Who knew how many other women could legally lay claim to part of his salary packet? Odel was what Lawrence's old teachers had called bookish; the voice was distinctly upper-class English, too. Normally, Lawrence would have deep misgivings about anyone with those characteristics; they were too much like officer material. But Odel couldn't be faulted for his frontline performance, which was all everyone really cared about. The platoon entrusted a lot of its equipment field maintenance to him, knowing he'd do a good job.

"Thanks, cretin," Dennis Eason said. He turned back and applied a medicsensor to Lewis's damp forehead, checking the readout that popped up on his field-aid kit.

"Do you even know what you're doing?" Karl Sheahan asked.

Dennis tapped the Red Cross symbol on his tunic shoulder. "You'd better hope so, pal. I'm your best hope for survival."

"You couldn't even give him a fucking aspirin without checking with that whale they call a doctor."

"I'm not authorized to administer anything," Dennis said tightly. "Not when there's a qualified ship's doctor on call. It's a jurisdictional thing."

"Yeah? Is that what you told Ntoko? Huh? Too much blood, it's a jurisdictional thing."

"Fuck you!"

"Enough," Amersy announced quietly. "Karl, watch the fucking movie and stop causing me grief."

Karl grinned as he performed a neat midair spin and landed gently on a chair couch. It turned out the sheet screen didn't have an interactive driver; all it showed were third-person fixed-view dramas. Up on the ceiling the young actress was tooling up to slaughter the vampires that were taking over Brussels. It meant she had to wear a lot of tight black leather.

Hal was in the chair couch next to Karl, staring up at the movie. He hadn't even heard the goading. Karl held on to a strap and leaned over to slap Hal on the arm. "You could give her one, couldn't you, dickbrain. Huh?"

Hal's leer grew broader.

Some of them managed to sleep for small periods during the rest of the flight. It wasn't easy. There was always noise, if not in their compartment, then from others, drifting incessantly through the ship like an audio-only poltergeist. People flailed about the cabin, sucking noisily on the water hoses, and microwaving snacks, then bitching about how they tasted of nothing. The toilets were used, which always ended in exclamations of misery, and the small cubicles let out a smell each time the doors were folded back. Who had the worst-smelling farts was an ongoing topic of conversation. Nic Fuccio kept score. Those who couldn't sleep kept their eyes shut from exhaustion, shifting around fitfully in the tiny gravity field. At some stage, almost everyone shouted at Dennis to give them tranks. He refused.

Lawrence nearly cheered out loud when the crap movies finally ended and the *Moray*'s captain showed them an external camera view. The *Koribu* was five kilometers distant, surrounded by a shoal of smaller support and service ships.

Even now, after twenty years in strategic security and flying to eleven different star systems, Lawrence still got an adrenaline buzz from seeing the huge starships. Like every other starship still flying, the *Koribu* was designed as a colonist carrier. Not that there were any other designs—even the explorer craft that had once probed interstellar space around the Sol system had the same layout. Only the size varied.

Their shape, and to some degree, form, was constrained by the nature of the compression drive. Although FTL capability was a scientific and technological breakthrough of the highest order, it didn't have the kind of commercial viability Earth's corporations and financiers would have liked. The development team had originally talked about starflights taking the same time as intercontinental aircraft journeys. A more honest analogy would have been with sailing ships. Like a portal, the FTL drive generated a wormhole by compressing the fabric of space-time with a negative energy density effect. As such an energy inverter consumed a colossal amount of power simply to open a wormhole, and the only practical source was a fusion generator, the subsequent wormhole was extremely short in comparison to the distance between stars. That wasn't a technological problem, as the starship flew down the wormhole it was creating, so the drive would simply redefine the endpoint, moving it ever forward. Although a valid solution, it also stretched out the flight time.

Modern starships could make the Centauri run in a week, giving them a speed of just over half a light-year per day.

The *Koribu* was such a ship. She had been intended as a colonist carrier forty-two years ago when she was being assembled in one of Centralis's freeflying shipyards. Cost structuring by Z-B's accountancy AS had given her an effective range of forty light-years. With that as their mandate, the designers had housed her energy inverter in a drum-shaped superstructure two hundred meters in diameter that made up the entire forward third. Seventy percent of its volume was given over to the eight fusion generators required to provide the massive quantities of power that the drive consumed, an engineering reality that explained why the outer surface was a mosaic of thermal radiators, mirror-bright silver rectangles five meters long, throwing off the phenomenal heat-loading produced by the generators' support systems during flight.

Because of the debilitating effect of freefall on human physiology, especially on twenty-seven thousand untrained colonists over ten weeks, some kind of gravity field had to be provided for them and the crew. It came in the simplest fashion there was: six life support wheels, fat doughnuts thirty meters wide and two hundred in diameter. They were arranged in pairs, counter-rotating around an axial shaft to balance precession. Their hulls, in common with all spacecraft operating outside the protection of Earth's magnetic field, were blank, without ports or markings; just the standard coating of light gray foam rucked from particle impacts and bleached from the light of different stars.

Adapting them for strategic security transport was an easy refit. Z-B turned common rooms and lounges into gyms and sim-tac theaters; some dormitories were taken out of commission and used as Skin suit depots, while the remaining dormitories were unchanged. Between them, they could billet twenty thousand squaddies.

Behind the life support wheels came the cargo section, a

broad open cylinder section built up from a honeycomb lattice of girders, which formed deep hexagonal silos. They had once carried modules of industrial machinery and essential supplies that the colonists needed to maintain their settlements. Modifying the silos for asset-realization missions was a simple matter of changing the hold-down clamp designs.

Now, seven orbital transfer ships held station three hundred meters away, encircling the *Koribu.* One-man engineering shuttles glided backward and forward between them and the starship with halo bursts of green and blue flame, carrying the Third Fleet's lander pods, which they slotted down into waiting silos. At one end of the honeycomb, silos had been merged to form long, deep alcoves. They contained spaceplanes, the familiar sleek profile of Xianti nose cones just visible rising above the shadows.

Koribu's final stage was its main reaction drive, five direct fusion rockets in the shape of elongated cones over three hundred meters long, ribbed with a filigree of pipes and cables. Big spherical deuterium tanks were plugged into the stress structure at the head of the cones, along with ancillary support equipment and ten small tokamaks that provided power for the main engine ignition sequence.

The *Moray* docked just ahead of the life support section, nuzzling up to a tunnel that had extended out beyond the starship's body. Lawrence had to wait for another twenty minutes listening to the clamor of other platoons banging their way through the orbital-transfer ship's habitation cabins and into the tunnel. Finally, they were given clearance to disembark.

It was a long trek through the starship's freefall corridors to the rotating transfer toroid of their wheel. Inside the top of the wheel spoke was an elevator that was barely high enough to take an adult. They all aligned themselves, tucking the boots into the floor hoops. The G-force built as they descended, much to Lewis's relief. They stopped on the middle of the three decks occupying the wheel itself, where the gravity was an eighth Earth-standard. Enough to settle their stomachs and restore normal circulation patterns. But with it came a disconcerting spinning sensation, as if the decking were about to heave over. They emerged from the elevator, reaching out to steady themselves against the wall.

Every time he came down into one of the wheels, Lawrence swore he wouldn't let the effect trick him again. Every time his body promised him he was about to flip over. He gingerly took his hand away from the wall. "Okay, I know it feels like we're washing about. Ignore it. You're all down and stable. Let's go find our quarters."

He set off down the corridor. After ten paces he had to move to one side to avoid Simon Roderick and his retinue of senior managerial staff. The Third Fleet's board representative was so busy snapping out instructions to a harried aide he never even noticed the platoon. Lawrence kept his own face impassive. He'd followed the investigation Roderick and Adul Quan had launched in the wake of the bar fight in Kuranda. His Prime program had loaded unobtrusively into the base's datapool, passively observing the surge of traffic shunting between AS programs, the information requests to skyscan. Their inquiries had withered away after a couple of days, and the police had never turned up anything. Even so, it was a shock coming face-to-face with a board representative who'd taken such a keen interest in his off-base activities.

Roderick and his entourage disappeared up the curve of the corridor, and Lawrence walked on without breaking stride.

The dormitory that they'd been given was probably only double the size of the compartment on the *Moray*. It had two ranks of bunk beds each with its own locker containing a standard clothes package for everyone, a couple of aluminum tables with chairs and a sheet screen. There was a small washroom next door.

Hal looked around, his face screwed up in dismay. "Oh man, what is this shit?" he exclaimed.

Amersy laughed. "Best quarters in the fleet, welfare boy. Lie back and enjoy. You get fed, you get paid and nobody shoots at you. Now find a bunk and make the most of it."

"I'll go fucking stir crazy." He made to climb onto a top bunk, only to find his way blocked by Karl's forearm.

"Bottom rung, kid," Karl said, grinning a challenge.

"Jesus fucking wept." Hal threw his small bag onto a lower bunk and hopped on after it. "I can't take these closed-up rooms."

"You'll put up with it," Lawrence said. He dropped his own bag on a top bunk, momentarily fascinated by the weird curve of its fall. "Settle down, all of you; you know the on-board drill. I'll find out what our canteen schedule is, and then we work training and fitness around that. Lewis, how are you feeling?"

"Not too bad, Sarge. Guess the doc was right."

Lawrence made his way over to the small keyboard set into the wall beside a sheet screen. Platoon dormitories didn't rate an AS program, but the operating system was sophisticated and easy enough to operate. He called up their basic shipboard data: where they ate and when, what the local time was, when departure was scheduled.

"Hey, you guys want to know where we're headed?" he asked.

"Thallspring," Karl shouted back. "Didn't they tell you, Sarge?"

Hal gave him a puzzled look. "How did you know that? It's like top secret."

Karl shook his head. "Fuck, you are a big waste of space, kid."

They were due to depart in twenty-two hours. Lawrence

read the Third Fleet data from the screen and muttered, "Jesus."

"Problem?" Amersy asked quietly.

Lawrence took a quick glance round the dormitory. Nobody was paying attention to them. "Seven ships. Is that what the Third Fleet is these days?"

"More than a match for Thallspring. Their population is small, barely seventeen million."

"Projected," Lawrence said. "That's no true guide. But it's not what I'm worried about."

"The ships?"

"Yeah. Fate! My first mission, to Kinabica, that took seven weeks of spaceplane flights just to lift us and our equipment offplanet. There must have been thirty-five starships on that mission."

"We don't have that many starships anymore. Not since Santa Chico."

"Not just there. Second Fleet lost two ships on approach to Oland's Hope. No one projected they'd have exo-orbit defenses. But they did."

"You want to eject?"

"Hell no. I'm just saying this one could be tough. We're going in too small."

"They'll cope." Amersy clapped him on the shoulder. "Hell, even the kid will pull through."

"Yeah, right." Lawrence began pulling menus from the starship's computer, seeing what he could throw up on the screen. He read one schedule and smiled, hurriedly calling up supplements. "You might want to see this," he told the platoon. "You'll probably never have another chance for ringside seats this good."

The screen brightened with an image from one of *Koribu*'s external cameras. It was centered on the portal, glowing a hazy blue against the void. Colony trains were clustered around like a shoal of eager technological fish.

"Two minutes to the starting gun," Lawrence announced happily. Despite all he hated about Z-B, he had to admit, they got this absolutely right.

His mood was broken by Hal's petulant voice asking, "What the fuck is that thing, a radioactive doughnut? Order me a couple of coffees to go with it, Sarge." He trailed off fast at Lawrence's look.

Lawrence just managed to stop himself from bawling out the kid. He couldn't believe anyone was that ignorant about the most important endeavor the human race was undertaking. But then Hal was just some teenager from a welfare block in some godforsaken city. Lawrence himself had been a teenager with the best education his home planet could provide, as well as apparently unlimited data resource access, and he hadn't known that portals existed. It had been Roselyn who told him.

CHAPTER FIVE

IN FIVE YEARS, AMETHI'S CLIMATE HAD UNDERGONE A PROfound degree of alteration. The changes wrought by Heat-Smash had become self-sustaining and were now accelerating on a scale that allowed human senses to register them. Locals were calling it the Wakening. Instead of surprise and delight at seeing a single cloud, they now welcomed the sight of a small patch of sky through the sullen cloud mantle.

Now that the overall air temperature had risen several degrees above freezing, the Barclay's Glacier meltdown exhaled water vapor into the atmosphere at a phenomenal rate. Giant cloud banks surged out from the thawing ice sheet, reaching almost up to the tropopause where they powered their way around the globe. In their wake, warmer arid air was sucked in, gusting over the ice where it helped transpiration still further, keeping the planet-sized convection cycle turning.

When the clouds rolled over the tundra they began to darken, condensing to fall as snow. By the time the flakes reached the ground they were miserable gray smears of sleet. Great swaths of slush mounted up over the entire planetary surface, taking an age to drain away in stubborn trickles that were often refrozen by fresh falls. On the continental shelves, muddy rivers slowly began to flow again, while across the dead ocean beds, the deep trenches and basins were gradually filling with water. The thin viscous sheets of dirty liquid that rolled sluggishly downslope across the sands carried along the crusting of salt that had lain there undisturbed since the glacier had formed. It was all dragged down into the deepening cores of the returning oceans, dissolving to produce a saturated solution every bit as dense and bitter as Earth's Dead Sea.

Above it, meanwhile, the air was so clogged with hail and snow that flying had become hazardous. Spaceplanes were large enough to power their way up through the weather, but smaller aircraft remained sheltered in their hangars for the duration. Driving also was difficult, with trucks newly converted into snowplows running constantly up and down the main roads to keep them clear. Windshield wipers were hurried additions to every vehicle. Major sections of the Amethi ecology renewal project had been suspended until the atmospheric turbulence returned to more reasonable levels. The insects already scheduled for first release were as yet uncloned; silos holding the seed banks were sealed up. Only the slow-life organisms remained relatively unaffected, carrying on as normal under the snow until they were unlucky enough to be caught by a fast flush of water. Lacking even rudimentary animal survival instinct, they never had the sense to wriggle or crawl away from the new torrents raking across the land.

This particular phase of Amethi's turbulent environmental modification was proceeding as expected, claimed the climatologists, it was just more vigorous than most of their AS predictions. Some quick revisions incorporating new data estimated the current turmoil wouldn't last more than a few years. Specific dates were not offered.

Lawrence rather enjoyed the Wakening, secretly laughing at all the chaos it had brought to McArthur's meticulously laid plans and the amount of disturbance it caused his father. This was nature as it existed on proper planets, playing havoc with human arrogance, exactly what he wanted to witness firsthand in star systems across the galaxy where alien planets produced still stranger meteorology. However, after the first nine months or so of Amethi's whiteouts and oppressive obscured skies, even he grew bored with the new phenomena.

That boredom was just one of the contributory factors suggested to his parents for his continuing behavioral problems. By the time he reached sixteen his thoroughly exasperated father was already sending him on weekly trips to Dr. Melinda Johnson, a behavioral psychologist. Lawrence treated the sessions as a complete joke, either exaggerating grossly or simply answering every question with a sullen yes or no depending on how pissed off he felt at the time. It probably helped disguise just how alienated he was from the rest of Amethi's society, which was why she never made any progress with him. Lawrence knew he was growing up in the wrong place at the wrong time. He should have been an American astronaut in the 1960s or a deepspace astrophysics officer in the last decades of the twenty-first century when starships first set out to explore the new worlds around Sol. Yet telling that to the professionally sympathetic Dr. Johnson would have been a huge admission of weakness on his part. No way was he giving in to her. She, and everything she stood for, the normality of Amethi, was the problem, not the solution. So the lies and moods just kept on swinging a little further each time, picking curiously at the envelope of acceptability as if it were an interesting scab. All the while he built a defensive shell of stubborn silence around himself, which grew progressively thicker each time his father raged and his mother showed her quiet disapproval. Nothing apart from i-media interested him, nothing apart from gaining more i-media time motivated him. He had few friends, his teachers virtually gave up, and sibling rivalry at home began to resemble a full-blown war zone. With his hate-the-world attitude and his rampaging hormones, he was the basic teenager from hell.

That was why his father had totally surprised him one morning at the breakfast table when he said: "I have to go to Ulphgarth tomorrow for a conference, fancy coming with me?"

Lawrence glanced round his siblings, waiting for them to answer, then realized everyone was staring at him, including his father. "What, me?"

"Yes, you, Lawrence." Doug Newton's lips twitched with his usual lofty amusement.

"Why?" Lawrence grunted suspiciously.

"Oh dear." Doug Newton rubbed his fingertips against his temple. "Well, quite. Why indeed? To reward your exemplary behavior, perhaps? Or your grades? Or just for keeping your data access costs below the K-pound mark this month? Which do you think, Lawrence? Why should I be nice to my eldest son?"

"Why do you always do that? Why are you always so

damned sarcastic? Why can't you just ask me like a normal person?"

"As opposed to the way I put the question?"

Lawrence turned bright red as Janice and Ray started sniggering at his expense. He glared around at everyone, angry with himself for being caught out. But it was such an unusual thing for Dad to ask . . . "Well, what's there, anyway?" He managed to sound as if nothing in the universe could ever interest him in Ulphgarth. Not that he'd actually heard of it before.

"A first-rate conference center, where we're discussing the final stage bidding with contractors for the new Blea River bridge."

"Oh yeah, thanks, like I'm really gonna want to be a part of that."

"Which is what I shall be attending, while you can just stay in the five-star resort hotel next door. One of my aides has pulled out, leaving a room already paid for. You can sleep in as late as you like, or even for the whole five days if you want. You can have room service meals on a twenty-four-hour-a-day basis. There's a fully equipped sports center and pool free to guests. The dome lighting is rigged for tropical climate if you want to lounge around getting a tan. Your room includes unlimited datapool access. There's live music every night. And you don't have to see me or even have a meal with me the entire time. So . . . do you want to give your mother a break for a few days before term starts?"

Lawrence looked across at his mother, who was smiling gamely. Her stress lines had become permanent since his last brother had been born. He knew she was taking prescription antidepressants, washed down with vodka, and hated her for being so weak. He hated himself even more for being so harsh on her. It was this whole fucking stupid world that was rotten. "I . . . Yeah. Great. Sounds cool. Thanks."

"Thanks. Good Fate, wonders never cease on this planet, do they?"

Lawrence scowled again.

Three days later, he wasn't actually enjoying himself, but he was relaxing. The hotel building was in a dome all by itself, a fifteen-story triangle of broad glass-fronted balconies right in the center where guests could look out over humid, verdant parkland. It seemed as if every bush and tree was sprouting some kind of brightly colored flower. Branches and leaves had been infused with a vitality lacking to ordinary plants—you could virtually watch the glossy shoots growing. The tough Bermuda grass was mown every night by the gardening robots, but it was still like walking over a layer of thick sponge in the morning.

Lawrence lay back on the sun lounger, shifting his shoulders around on the cushioning until he was completely comfortable. The big lights overhead were warmer than the ones in the tropical dome of his family's estate, sending out rays that soaked right through him. He'd found a spot on the broad curve of paving that surrounded the big circular swimming pool, away from everyone else, but close enough to the open bar to signal to the waiter. Amazingly, nobody bugged him about how old he was when he ordered drinks! He'd started out on beers yesterday before moving on to the list of cocktails. Some of them were pretty disgusting despite the intriguing colors and foliage, and he'd almost gone back to beer. Then he found margaritas.

The girl was in the pool again. Lawrence moved up the backrest slightly so he could see the whole area without having to turn his head. He was wearing mirrorshades with a built-in audio interface to his bracelet pearl, while optronic membranes covered his eyes underneath. So he could either play some i's or sneak a look at the people in the pool or even doze off, and nobody would be able to tell. Yesterday he'd been playing *Halo Stars* and guzzling down his beers before he noticed her.

She was, he guessed, about sixteen, blond, her thick straight hair cut off level with her shoulders, and tall with legs that were fabulously athletic. In fact her whole body was lithe and trim. He could see that easily enough thanks to the small black bikini she'd worn.

Lawrence had spent the rest of the afternoon watching her and sipping his margarita. There was a whole gang of kids messing about around the pool, from his own age down to about seven or eight. Conference kids, he guessed, left to themselves while the adults discussed the intricacies of bridge building. He didn't join in. For one thing he wasn't so hot on socializing. Never knew what to say to a complete stranger. And then there was his body. He wasn't self-conscious, of course. But out here in the open wearing just his swim trunks he was keenly aware how much heavier he was than the other seventeen-year-old boys. Despite his height and general size, which the school's coaches were convinced would be advantageous for football and field events, he had no interest in joining any of the teams and wasting valuable i-hours by training. That lack of exercise meant that unlike the rest of his year his puppy fat hadn't burned off. It was unusual in a world where most children had been given some degree of germline v-writing to improve their general physiology, as he could see around him. It wasn't just the girl who glowed with health. Even so, she stood out: the other girls having fun in the pool were attractive, but she was stunning. He couldn't say why he found her so irresistible, exactly. She had a narrow face, with wide lips and prominent cheekbones, features that were attractive, but not outstanding. And her gray eyes were never still, always taking in the world around her with wonder. In the end he decided that was her magic—she was so full of life. Others obviously agreed with him; she had a harem of boys longer than a comet's tail following her around.

He watched silently as she splashed about in the pool. Then the group were diving and jumping in. Chasing about around the side, throwing each other in. Lobbing a ball about. Rushing over to their sun lounger to grab a quick gulp of Coke before jumping back in. All the while she was laughing and shouting.

She levered herself up out of the water directly ahead of Lawrence, lean muscles taut, water glistening over her skin. His breath grew hot as he pictured that incredible body shivering in delight while he ran his hands over her, taking as much time as he wanted. Sweet Fate, he wanted to fuck her badly. Really badly. His cock was growing hard inside his trunks. He had to hurriedly activate the bracelet pearl, optronic membranes wiping out the sight of her behind a deluge of astronomical data.

Running away would have looked odd. And he'd seen Naomi Karamann using one of the sun loungers on the other side of the bar. She was—allegedly—his father's executive assistant. Lawrence didn't have to be told she was the same as all the assistant nannies who came and went on a near-monthly basis. A beautiful girl in her early twenties, with dark ebony skin and a very full figure. She walked about the side of the pool in a scarlet swimsuit designed for provocation rather than swimming. At no time had she shown any interest in the conference. The night before, Lawrence had seen his father and her join a big group of businessmen for dinner in the hotel's restaurant. She'd been dressed in some silver backless gown, her hair glittering with embedded gold.

No doubt if she saw him acting strangely his father would hear about it. So he stayed immersed in *Halo Stars,* gliding over the astonishingly detailed cityscapes of alien cultures. The i-media game was the new market leader. It was an import from Earth, where teams of designers and AS extrapolators must surely have spent years generating the concept. It featured a large band of inhabited stars wrapped around the center of the galaxy, where hundreds of alien races coexisted in a peaceful commonwealth. The first-person player was the pilot of a trade and exploration ship, the *Ebris*. Whatever settled world the ship landed on, there was some problem or requirement that could be solved by tracking down a resource that another world in the Halo possessed, be it technological, artistic, raw material, medical, or even spiritual. Lawrence was in the middle of a sequence where he was making his way toward a domain that had bred the methane-grazing botanical organisms that a species of sentient octopeds needed to complete their colonization of a new planet. But he could only get the botanicals by trading them for a specific mineral that formed on low-gravity planets with an argon atmosphere. To do that he first had to put together a survey and mining team. Once that was done he would fly scouting missions through a dozen likely star systems, hunting for the right class of planet. And this particular segment had already opened up several further opportunities for his ship.

The sheer wealth of detail, both economic and physical, was astounding. The stars, planets, stellar phenomena and species of the Halo were so *real*. They'd even got the quasar locations right. The whole thing interlocked perfectly; in the three months since he uploaded the base chapter he hadn't found a single continuity flaw. Flying his ship around the arc of the magnificent glow thrown off by the galactic heart he felt as if he were on a genuine training mission at McArthur's starship officer academy—as it should have been if the company wasn't so stupid. Small wonder the import company with the license was making a fortune.

After scanning three star systems with swarms of microsatellites he finally found one that had the kind of planet he was looking for. He landed the *Ebris* at the end of a valley cloaked in a turquoise grass, where a binary of yellow and green dwarf stars were setting in the saddle of the hills. Tomorrow he would supervise the mineral extraction. He noted several potentially dangerous-looking animals slinking through the long grass, loaded their profile into the ship's computer, then saved and exited.

On the opposite side of the pool, the girl was lying on her sun lounger, big gold-orange glasses over her eyes. Several of the younger kids were clustered around, laughing and giggling together. Three of the more persistent boys were sitting on the edge of the sun lounger next to hers, squashed together uncomfortably. Each was doing his best to be charming, witty, knowledgeable and casual. She occasionally laughed at their jokes and joshing. From where Lawrence was it looked as if she was just being polite rather than genuinely amused.

His margarita ice had melted in the bottom of the glass, producing an undrinkable slush. Naomi Karamann had disappeared. Several adults were in the pool and more were walking across the lawn from the hotel. The day's conference had obviously finished. Lawrence picked up his towel and went back inside to order another room-service meal.

That was yesterday. Today, he'd come down early, by his standards, before ten o'clock. His reward was the well-positioned sun lounger and the girl's prompt appearance. This morning she was in a white bikini, but she was just as lively as she had been before. He found himself smiling at the way she enjoyed herself so effortlessly. Two of the smaller girls arrived with her, chattering excitedly, one no more than eleven while the younger was about six or seven. He realized the three of them were all sisters, sharing roughly the same facial features. That explained why the older boys of the wishful harem had been so tolerant of them yesterday.

It wasn't long before the whole group was gathered together again. Laughs and shrieks carried across the humid landscape as they began pushing each other into the water.

Lawrence tensed when one of the older boys, around his own age, shoved the girl in with too much force. But she broke the surface smiling. He let out a sigh, wishing there were some way he could go over and introduce himself and ask if he could join in. It would seem weird now, though, after he'd spent a day slobbing out by himself, mark him out as a creepy freak. What could he say, anyway? Does anyone want to link in to *Halo Stars*? He didn't think this physically active bunch would have much interest in i's. And she certainly wouldn't.

He told the bracelet pearl to return to the game, and the shadowed valley materialized around him. A small convoy of hoverjeeps roared out of the *Ebris*'s lower cargo hold, with him navigating in the lead vehicle. A satellite survey map was projected onto the windshield, showing him the direction he needed to take. And some distant animals were growling aggressively, hidden by the blue grass.

"Hi there, can you help us out?"

Lawrence told the bracelet pearl to suspend the game. His membranes cleared and he was looking up at the girl. She was standing at the side of his sun lounger, dripping wet and glorious. He pulled his mirrorshades off in a hurried awkward motion, twisting the earpieces out.

"Sorry, what?" Was he staring too hard? The dome lights were directly above her, forcing him to squint. *Damn it, I must look a total idiot.*

"Can you help us?" She held out a ball. "We need one more to make the teams even."

"Teams?" He could have smacked himself one. He sounded so dumb.

"Yes. We're playing water polo. We're one short."

She had a lovely accent, her voice all blurring and soft. Where had that come from? "Er, yeah, sure." He pushed himself up, standing beside her, holding in his belly. She was only a couple of centimeters shorter than him. For some reason that made no sense, he liked that. But then he liked everything about her. She was utter perfection. "I haven't played for a while. I'm probably a bit rusty." He'd never played before.

"That's okay. Myself, I've never had a game in my life. And I don't think too many of us know the rules anyway."

"Oh, great. Probably best if I'm goalie. Do less damage there." Ask her what her name is, you asshole. Ask!

She smiled brightly. "I fancied that gig myself."

"Sure. Fine. Whatever."

She lobbed the ball at him, which he just managed to catch. "Were we interrupting anything?" She gestured at the mirrorshades and bracelet.

"No. Not at all. I was just going through an i-media, that's all. It's stored."

"Fine." She turned and started back to the pool. "Got him!" she yelled at her friends. The harem of boys greeted the news with unwelcoming smiles.

"Uh, I'm, er, Lawrence."

"Roselyn." She dived cleanly into the water.

It was almost the last he saw of her for the next twenty minutes. Water polo was every bit as bad as he imagined it would be. Twenty minutes in water five centimeters too deep to stand comfortably, while people powerslammed the heavy, wet ball at him. Chlorine spray got in his eyes. He swallowed liters. His breath was hauled down painfully, feeling wretchedly exhausted.

The game finally dissolved into some kind of ending, which was mainly an argument about the score. Twenty, thirty, probably. A lot of shots had got past him. He wheezed up out of the chrome steps with a shaky hold on the rails.

"Are you all right?"

Roselyn was in front of him, squeezing water from her hair.

"Yeah, I'm good." He was too puffed to pull his belly in anymore.

"I fancy a drink." Her expression was mildly expectant.

Lawrence couldn't believe this was happening. "Me too," he blurted.

He received a barrage of evil-eye stares from the harem as he walked with her over to the open-air bar. Several of the boys called out at her to join in with their latest game. She just waved and told them maybe later.

"I need a break," she told Lawrence. "Jeez, where do they get their energy from?"

"I know what you mean. I'm here to chill out."

She sat on the stool right at the end of the wooden bar, which meant nobody but Lawrence could sit next to her. He held back on a smirk as he sat down.

"You here by yourself?" she asked.

"No, with my father. He's at the conference."

"Right." She asked the waiter for a Coke.

"Me too," Lawrence said. It would look like he was showing off if he went for a margarita. "Where's your accent from? I haven't heard anything quite like it before. It's very nice," he added hurriedly. It didn't look like she'd taken offense, and he couldn't think of anything else to say.

"Dublin."

"Where's that?"

She burst out laughing.

He grinned bravely, knowing he'd been stupid again.

"I'm sorry," she said. "Dublin's in Ireland, on Earth. We arrived three days ago."

"Earth?" he said, amazed. "You came from Earth? What was the flight like? What did you see?" It seemed wholly unreasonable that girls as young as her two sisters had experienced a real live starflight while here he was, forever trapped in protective domes under an opaque sky.

Her small nose wrinkled up. "I didn't see anything.

There's no window. And I had motion sickness the whole way. Not as bad as Mary, mind. Urrrgh, we must have used up the whole ship's supply of paper towels."

"Mary?"

"My sister." She pointed at the elder of the two sporting in the water. "The other one's Jenny, there."

"They look like they're okay kids."

"Really?"

"Oh yeah. I've got five younger brothers and sisters myself. I know what it's like."

"Five. Wow. Your parents must be pretty devout Catholics."

"Ah. I know that's a religion, right? There's not much religion on Amethi. People here all tend to know the universe is natural."

"Do you now?"

"Yeah." He got the feeling he was being teased, somehow. "So why did you come here?"

"My father died."

"Oh shit, I'm sorry. I didn't mean to, well . . ."

"That's all right. It was over a year ago now. It was a car accident. Very quick. All the people at the hospital said he wouldn't have felt anything. I've got used to it. Still miss him tons, though. But we were stakeholders in McArthur, and there was a lot of insurance, so Mother decided to cash it all in and make the proverbial new start. I'm glad she did. Leaving Dublin took me away from the bad memories, and Earth's pretty crappy these days. This place is just fabulous."

"Er, yeah."

"What's the matter with it?"

"Nothing. You're right. It's just that nowhere you live can be exotic. That's only ever somewhere else."

Her smile lingered for a long time. "Very profound, there, Lawrence. I'd never thought of that before. So do you think I'm going to be bored with Amethi in a while?"

"Actually, no. It's starting to liven up a bit right now."

"Come on, let's go and see it." She picked up her glass of Coke and stood.

"What?"

"Amethi. Let's go see it."

"Sure. Okay." He smiled at how impulsive she was.

Roselyn set off across the lawns with Lawrence hurrying to keep up. She kept asking what various plants and bushes were. Some of them were similar to those planted within the family's estate, but for the life of him he couldn't give them a name. She didn't seem to mind.

They arrived at the rim of the dome, where the nullthene was anchored in a band of concrete. Thick moss had swamped the crumbling gray surface, though it couldn't get a grip on the slippery nullthene itself. Roselyn pressed herself up against it.

"How can you not find that incredible?" she asked. "I've only got this bikini on, and I'm a millimeter away from an arctic blizzard."

"That's technology, not geography. But you're right. It's pretty spectacular." He was looking at her back, the way she'd arched herself slightly to rest her hands against the thin nullthene. Her skin was smooth and mildly tanned; intriguing bands of muscles slid around just below the surface. "Of course, the technology isn't perfect. And in some cases it's too good."

"What do you mean?"

"McArthur worked out the general effects HeatSmash would have on the environment, but they didn't always follow it through to its conclusion. When the snow started falling, it landed on the domes just like every solid surface. Trouble is, nullthene is a perfect insulator. The cold doesn't get in, but neither does the heat leak out. So the snow stuck, especially up on the top of the domes where it's flatter. When the original designers came up with the particular domes we employ, they made allowances for the next stage of Heat-Smash, when it will rain. The nullthene can take the weight of water running down the outside, but nobody thought about the piles of snow that would accumulate up there. There were splits and miniavalanches in every city. It was damned dangerous. A ton of snow can kill you just as easy as a ton of steel if it falls on you. Over a dozen people were killed, and plenty of buildings were damaged. We had to shore up the support grids in every dome. All the civil engineering robots on the planet were switched over to reinforcement work. It took months, cost a fortune, and everyone's still arguing about who's to blame and what sort of compensation there should be."

She gave him a quick, incredulous glance, then gazed out at the flurry of tiny hailstones drumming against the null-thene. The tundra outside was completely white, even the rugged tufts of grass were no more than spiky white mounds. "It's still impressive to me. All this is the result of human ingenuity."

"Amethi wasn't like this when I was younger. All I ever saw was a frozen desert."

"But to change a whole planet. And not through ecocide."

"Ecocide?" He was beginning to think he should start paying a bit more attention in school. She knew so much more about the universe than he did.

"On most planets that people have colonized there's an existing biosphere," she said. "And none of them are compatible with terrestrial biology. So we come along and kill it off with gamma blasts or toxins and replace it with our own plants and animals. Ecocide, the worst kind of imperialism there is."

"It's only the area around settlements that's cleared, not entire planets."

"Spoken like a true galactic overlord. Each habitable planet had its own indigenous species. They're unique and evolved to live in a reasonable balance. Then we come along and introduce competitive species, our own. At first terrestrial biology zones are enclaves wrapped around our settlements, but then the population rises and the zones expand until they're in full-blown conflict with the natives. And we always back ours up with technology, giving ourselves the edge. Eventually, every planet we've ever landed on will have its indigenous life swamped by ours, and become a poor copy of Earth. That's what some projections say, anyway."

"That's all a long way off."

"Yes. But we've set it in motion." She gave the icy landscape a sad smile. "At least we're not guilty of it here. Do you fancy some lunch?"

Lawrence would have liked to be able to think back to the last time he'd been alone with a beautiful girl, strolling through a lush parkland setting. It had never happened, of course. There'd been no girlfriends, just blue-i's, and fantasies over girls at school. Now here it was, the real thing, and it was so easy he kept wondering if he'd fallen through a wormhole into some alternative universe. Roselyn was gorgeous, she seemed to like him, or at least accept him, and she was easy to talk to. Chatter, actually, which he'd never done with anyone, let alone a girl. But when they got back to the pool they sat together in the restaurant—at a small table with only the two seats—and carried on talking. Lawrence ordered a cheeseburger with extra bacon and a large portion of fries; Roselyn asked for a tuna salad.

She was, she explained, only staying at the hotel for a few days. "It's a sort of treat for us, Mother said; we're here to recover from the starflight just until our apartment is ready. Then we move straight into Templeton and start school. What a bore."

"I live in Templeton," Lawrence blurted.

"Great, maybe we can meet up some time. I'll have to get

settled in first, though. I'm going to Hilary Eyre High; it's supposed to be very good."

He swallowed some of the burger before he'd even chewed, clogging his throat. "My school."

"Pardon?"

"I go there!"

His yell drew another round of glowers from those members of the harem who had wandered in for lunch, hoping Roselyn would be sitting at one of the large tables.

She smiled delightedly. "That's fabulous, Lawrence. You'll be able to show me around and introduce me to everyone. There's nothing so horrible as starting somewhere new when you don't know a single person, don't you think?"

"Uh, yeah, I'd hate it."

"Thanks, Lawrence, that's really sweet of you."

"No problem." He was desperately trying to think who the hell he could introduce her to. Alan Cramley might play along, and one or two others. *Worry about it when the time comes,* he told himself. *All that matters is managing to stay with her right now. Don't blow it. Just don't say anything dumb or pathetic. Please!*

After lunch they went back to the poolside. Roselyn put on a white blouse and settled back on her sun lounger. Lawrence took the one beside her, bringing his bracelet and towel over. It turned out she'd never heard of *Halo Stars*. He found that puzzling; it must surely be one of Earth's major i-media games. But he spent a while explaining and showing her the game before some instinct told him to shut the hell up and move on to another topic.

When she asked him what he was doing that evening he said: "Dunno. Nothing yet."

"I'm going to listen to the band in the hotel bar. They're very good. I heard them last night, too. I didn't think I saw you there."

"No. I was . . . out. But, er, I'd like to go with you. If you're free tonight."

She appeared satisfied with that. He'd noticed slight dimples appear when she was pleased with something. It wasn't a smile, more like demure approval. "Date then."

Lawrence smiled wide, covering his urge to howl out in victory. A date! But . . . had he asked for a date and been accepted? Or even more unlikely, was she the one wanting a date with him? It didn't matter. He had a date!

"I just love dancing," Roselyn said contentedly.

Lawrence nearly groaned out loud.

How could it be so easy to get a date with the loveliest girl on Amethi for the one thing he was completely useless at? He spent ninety minutes getting ready in his room. That was seven minutes in the shower using up most of the hotel's stupid poxy-sized complimentary soaps and deodorants. Three minutes getting dressed in his pale green trousers and gray-blue shirt, with a black-and-gold waistcoat; just about the smartest set of clothes he owned. Mother had insisted he bring them in case his father wanted him to go to dinner—thanks, Mum! And eighty minutes with his optronic membranes presenting him with a phantom dance instructor; he had to access the hotel's i-tutorial class for that, because he certainly didn't have anything like it in his own memory chips. Thankfully he did at least know a few of the basics; his family had two or three formal parties each year when he was expected to partner obnoxious great-aunts and revolting ten-year-old nieces on the floor. It was just a question of brushing up.

Only when he checked himself in the mirror when he was on the way out did he realize he didn't know what sort of band it was, nor the type of music they played.

It was Lucy O'Keef, Roselyn's mother, who answered the door when he knocked. She was younger than his own mother and possessed a lot more energy. Lawrence was reminded of an aunt on his father's side of the family, one of those independent women who spent a couple of months each year doing consultancy or software design work, and the rest of the time partying and playing tennis. Clever, active, healthy, pragmatic and good fun. He could also see where Roselyn inherited her beauty from: they shared the same small nose and pronounced cheeks.

"So you're Lawrence." Her voice was husky with amusement.

"Yes, ma'am."

"Come in. She's almost ready."

The O'Keefs had a suite with three bedrooms. This meant the younger sisters were in the lounge, giggling. He'd met them that afternoon, and the three of them had spent a short, sparky time establishing boundaries. True, like all younger children, they were irritating, but they were too wrapped up in the wonder of a new world to be completely odious. He took their teasing in good humor, reminding himself that Roselyn would have to endure his own siblings one day. That is . . . he hoped she would.

When she came out of her bedroom she was wearing a simple navy-blue dress with a skirt less than halfway down to her knees. It made her even more alluring than her bikinis.

"Have fun," Lucy said.

The bar was of a type indigenous to five-star hotels the universe over. Straight ahead of the door was a small semicircular marble counter with dozens of liquor bottles displayed on mirrored shelves. Deep settees and plush chairs were arranged around small tables. A high ceiling was cloaked by low lighting. And inevitably, a grand piano stood on a central podium where a tuxedoed crooner would sit and entertain elderly guests for the evening with tunes never less than a century old.

Tonight a less respectable culture had taken over. The band up on the podium was all electric, playing power ballads.

Bottles of beer were cooling in tubs of ice on the bar, and a buffet had been laid out along one wall. Half the floor was given over to dancing, where holoprojection rigs sent iridescent seaswell waves crashing across the energetic boppers in showers of dazzling kaleidoscopic spray.

Lawrence recoiled slightly as the elevator doors opened on the lobby. He wasn't used to quite so many people packed together. There were a number of the teenagers from the water polo game in there, throwing themselves about enthusiastically. Roselyn grinned wolfishly at the sight and grabbed his hand, pulling him through the doors.

In the end it didn't matter that he didn't know how to dance like the others. There were too many hot bodies pressing in against him to allow any vigorous moves. He just shuffled about, watching Roselyn. She danced a dream, swaying in lithe slow motion, her arms flexing in time to the rhythm.

They grabbed food from the buffet and talked by shouting over the music. She drank her beer straight from the bottle. They danced some more. Drank some more.

With blood pounding, his skin sticky with sweat and alcohol humming sweetly in his head, Lawrence folded his arms around her in the middle of all the swaying people. She flowed up against him, resting her head on his shoulder for a slow number. Golden light broke over her, shimmering into deep violet. They smiled in lazy unison. Lawrence tilted his head forward, and they were kissing.

The band called it a day at two o'clock in the morning. Lawrence and Roselyn were among the five couples left standing.

"That was lovely," she murmured. "Thanks, Lawrence."

When the elevator doors closed, they kissed some more. There was an urgency about it this time. Lawrence pushed his tongue deep into her mouth. Then the elevator door opened. They kept on making out all along the corridor. He slid his hands all over her back before finally clutching at her buttocks. Somehow, he didn't have the courage to grasp her breasts or slide his fingers up inside her skirt.

"I can't," she said breathlessly in his ear. Her tongue licked at him, making him quake. "Mother will wonder where I've been." The door to her suite opened.

"Tomorrow?" he gasped.

"Yes. See you at the pool. Nine o'clock."

His head was spinning so hard it was a miracle he even made it back to the elevator, never mind his own room.

I can't. That's what she said.

Lawrence dropped on his bed, still fully clothed, as the room wobbled about dangerously. *She was talking about sex. With me. We were kissing all night.* When he closed his eyes and breathed in deeply he could still feel where she'd rested against him. The skin she'd touched seemed to glow hotly.

But what had she meant when she said yes? All he'd asked was: tomorrow? Nothing else, it had been completely open. And she'd said Yes. *Yes.*

The sleep that should have arrived instantaneously thanks to all that beer he'd guzzled eluded him for hours.

Lawrence was sitting on a sun lounger by the pool by twenty to eight. He was the first guest to get there. Several gardening robots scuttled out of his way as he walked across the lawn. A faint mist from the irrigation system hovered over the grass, making the blades glisten under the coral light. Visually, it was an inspiring start to the day.

Roselyn arrived at ten to nine wearing a loose midnight-black toweling robe and carrying a shoulder bag. They grinned at each other; Lawrence tried not to make it too uncertain and sheepish.

"You're early," she said.

"Didn't want to miss any of the day."

"Are you all right? You look tired."

"I'm fine. Didn't sleep much. My feet ached after all that dancing."

"Poor thing." She kissed the top of his head and plonked herself down on the sun lounger opposite. "Have you had breakfast?"

"Not really." He'd rushed out as soon as the alarm woke him. Hadn't even cleaned his teeth—probably a tactical error if he hoped to kiss her again.

"I know just what you need." She went over to the bar, which was still closed up, and started talking into the phone handset. A few minutes later two waiters arrived carrying trays.

They sat up at the bar, peering under the silver lids covering a profusion of plates and dishes. Roselyn made him swallow a couple of pills first: headache and stomach settler. He was only allowed to sip his iced orange juice for a few minutes until they took effect.

She'd ordered popped rice, yogurt with fruit slices, scrambled egg with hash browns, sausages, bacon, black pudding, button mushrooms and tomato, then finishing up with crepes in honey. There was toast and blood-orange marmalade if he wanted it, too. And a pot of Assam tea.

"This is good," he said loyally. Normally he got up at about half past ten and breakfasted on hot chocolate and chocolate cookies. Actually, although the yogurt and fruit was a bit crummy, the rest of it was pretty tasty.

Roselyn spread some of the marmalade on her toast. Apart from the yogurt and fruit it was all she had. "Most important meal of the day."

His mother always said that, but coming from Roselyn he could understand and appreciate the meaning. "Any plans for today?"

"Just going to hang," she said lightly.

"Me too."

She rested her elbow on the bar and put her chin in her palm to give him a quizzical look. "You're funny, Lawrence. I've never met a boy quite like you before."

"What do you mean?"

"Half the time you act like you're terrified of me."

"I'm not!" he protested indignantly.

"Good to know. You've got lovely eyes, halfway between gray and green."

"Oh. Um, thanks."

She broke off a corner of toast and popped it in her mouth. "Which is your cue to give me a compliment. Any part of me you like?"

A strength of will that he never knew he had stopped him from looking directly at her chest. Instead he gazed right back into her shining gray eyes. "I wouldn't know where to begin," he said softly, and blushed.

For a moment she held still; then a wide smile spread across her lips. "That sounded like a pretty good beginning to me. For someone who comes over all reticent, you've got the moves, Lawrence."

"That wasn't a move. That's what I really think."

Her hand touched his knee, squeezing gently. "You're really sweet. I didn't understand that. I thought you were just being Mr. Chill, sitting back while the rest of us were charging around the pool like crazed kangaroos. Like some big wolf eyeing up the flock to decide which one to eat."

"Sorry, but you're a really terrible judge of character. I was sitting there because I didn't know what to say to anyone. Stupid really."

"No. Not stupid. There's never anything wrong in being yourself. I think I was hoping you weren't a phony like those lads who've been trying to chat me up these last few days."

He grinned. "It's like you're a boy magnet. I was watching, you know, when I was sitting being Mr. Chill; their tongues were rubbing against the floor when they trailed after you."

"You should have heard some of the lines. 'I'd love to show you around.' 'Dublin sounds just like my dome, you must visit.' As if some polythene greenhouse could possibly be like a thousand-year-old city. Jeez! I came off a starship, not the ark. It's like they're all country cousins from Einstein County."

"Right," he said cautiously.

"Einstein County." She raised an eyebrow at him. "Where everyone's a relative, to be sure."

Lawrence burst out laughing. "God, you are just so amazing."

She pulled a modest face and did some make-work tidying of their trays. All the while they just smiled at each other. He'd never been so perfectly comfortable with anybody before.

"Did you have a boyfriend back in Dublin?"

"Not really. I was quite keen on one. We went out a couple of times. Nothing happened. Well . . . nothing too serious, anyhow. Thank Mary. We both knew I was leaving, see. I figured out in the end he thought that meant whatever he wanted would be for free. I wouldn't be there afterward, so he wouldn't have to go through all that emotional crap to dump me for the next girl. Can you believe it? What an asshole."

"He's bonkers. If I'd been him, I would have found a way to follow you here. Stowed away or something."

"Dear Mary, what have I gone and found?" She stroked his cheek, almost as though she was testing to see if he was real. "So what about you, Lawrence, have you got a girlfriend? You can be truthful with me, now. I won't mind."

"Nothing for you to mind. I don't have anybody."

"Now I know I'm on an alien planet. Let me tell you, in Dublin you'd have been triple-dating at least."

"Any chance the two of us can go back there together?"

"There now, just when I think you're smart you go and say something like that. Dublin's the same as the rest of Earth; it's old and tired. And we're both here now. On a planet that's got a future without any of the problems others have. Are you still so sure there's not a big fella up there rolling the dice for us? Seems to me I couldn't be this lucky naturally."

"I'm the lucky one." He leaned forward quite deliberately and kissed her. Her hands went around his head, mussing up his hair, holding him closer as they grew more passionate.

People were talking noisily as they walked over from the hotel to the pool. Lawrence and Roselyn ended the kiss and stared at each other. He didn't feel a trace of embarrassment. Quite the opposite, he felt certainty without arrogance. Both of them knew what they'd started, and knew that the other knew. It was almost relaxing.

"Won't be long before my sisters get here," she muttered.

"Oh, great."

They both laughed, and made their way back to the sun loungers. The newcomers were mostly the younger kids. None of them paid much attention to Lawrence and Roselyn.

"We'll have to wait half an hour for our food to go down before we swim," she told him.

"Right." He watched eagerly as she slipped out of her robe. Today it was a scarlet bikini, and he stared without shame. She blew him a mock-coquettish kiss and settled back on the sun lounger.

Her sisters arrived soon after. Lawrence greeted them with a cheery hello. The four of them chattered away, with the young girls giggling every time the band and dancing of last night was mentioned.

When they all jumped and dived into the pool later on, he endured the girls' attempts to push him under and bounce the big beach ball off his head, retaliating by diving and grabbing their ankles underwater. They laughed and shrieked happily.

He was quite surprised when Roselyn eventually said: "That's it for me." He threw the beach ball as far as he could, laughing as Mary and Jenny raced off in pursuit.

Roselyn was squeezing her hair dry when he got back to

the sun lounger. He held out a hand, which she took hold of. "I need a fresh towel," he said. There was a moment of horrendous vertigo while she gave him a level gaze. Then she nodded. "All right," she murmured. "It had better be your room, though."

He regressed to his original self for a while. All he could do on the walk back to the hotel was give her sheepish, nervous looks. She was equally timid, almost as if she were puzzled by who she was with and where they were going. In the elevator, they kissed again, but it was awkward this time. When he closed the door of his room, anxiety was making his fingers tremble.

Roselyn gestured at the broad balcony with its glass wall. "Can you shut the curtains? I know it's silly, but . . ."

"No." He almost ran across the room to pull the heavy fabric along the rail. When he finished the room was suffused with a warm golden glimmer, and Roselyn's superb body was cloaked in alluring shadow. She was looking at the big double bed, a slightly forlorn expression on her face. That wasn't what he wanted at all. He wanted her to be smiling and begging him to hurry.

"Look," he said in despair, "we can really just collect some towels if you want."

She turned from the bed and held out her arms for him. "No," she said when they were touching. "I don't want towels." She kissed him again, and this time the old heat was back. "And I know exactly what you want."

"You."

She slipped free and took a step away. Her hands reached behind her back, flicking the bikini top's clasp. The scrap of cloth fell from her, exposing wonderfully pert breasts.

"You're beautiful, Roselyn," he said, so quietly it was as though he was speaking to himself. Cursing his clumsiness, he closed his fingers around her nipples, tweaking the dark erect buds of flesh. He heard her inhale, a hiss of pain. She frowned in protest.

"Sorry. Sorry." He eased his grip slightly, but never let go. He couldn't do that; he'd never believed she would be so firm, so smooth, warm.

She took his hands gently and slid them up to her shoulders so she could kneel before him. Lawrence whimpered as she pulled his trunks down. She looked at his rock-hard erection with a blank curiousness, then tilted her head back to smile up at him. When she stood up he pulled hurriedly at her bikini bottom, tugging it down her legs. One hand kneaded her breast while the other ran down her belly, feeling the soft pubic hair, the wetness and the heat.

He half-pushed, half-carried her onto the bed. Their hands clutched at each other, mouths open, licking, sucking, devouring and tasting flesh. Breathing came hard and harsh. The sensations she left across his skin were driving him crazy.

Lawrence knew from all the i-blue shows he'd accessed how you were supposed to go slow, to caress and stroke a woman, to arouse her, to consider her feelings. But in the heat and semidarkness he could barely remember the facts he'd been shown. In the here and now he'd got the most beautiful and randiest girl in the universe panting and twisting underneath him. Her delectable legs were flung wide. There was a quick flinch of apprehension scarring her face as he penetrated her; it changed to a kind of dismayed delight. "Oh bloody hell," she grunted. "Just go easy, all right?"

"Of course," he promised. "Of course." As if he would ever do anything else. He began to move in a slow rhythm, as gently as he possibly could. He couldn't believe it was possible for anything to be this exquisite. Her incredible body squirmed beneath him, because of him. The grip she had on his cock was raw ecstasy. Little moans and surprised gasps of excited joy kept bursting from her clenched teeth.

Gentle and slow became impossible. He thrust into her fast and furious, fucking hard just like that vision the very first time he laid eyes on her. He came in great shudders while she cried out.

They rolled apart, him gasping for breath amid the wonder and glory. His head lolled over to see her chest heaving, and he just about came again. He was in love, smitten, besotted, obsessed. He would kill for her. Die for her.

He smiled in simple-minded happiness. "I'm yours, Roselyn. I mean it. You own me now."

The corner of her mouth lifted up, the nearest to a smile she could manage. Her expression was troubled, reluctant.

"What?" he cried.

"Lawrence. Please. Don't be so rough."

He wanted to throw up. He was the worst shit in the world. He'd hurt Roselyn, the only person who'd ever loved him. Hurt her! "Oh shit. I'm sorry." His fingers shook as they hovered above her. He was too afraid to touch her now. "I didn't mean to. Please, oh please."

"Shush. It's all right." She turned onto her side, and stroked his brow. "I'm all right. Just a bit sore, that's all."

"We won't do that again, ever. I promise."

"Yes, we will, Lawrence."

"But it hurt you," he protested.

"Lawrence, it was our first time. You'll . . . We'll learn to make it different." She grinned wryly. "The rest of the human race doesn't give up so fast, now, does it?"

"No."

She licked around his ear. "If I get as much pleasure out of it as you did just now, would you want to stop?"

"Oh, Fate no. No way."

"Well, then?"

"You want to try again?" Astonishingly, his cock was growing hard again at the mere thought.

"Not that, exactly. Not for a while. Can we try something else instead?"

"Sure!"

That was it for the rest of his holiday. Three days spent up in his room, the pair of them naked on his bed. Bodies locked together and writhing heatedly as they experimented with each other. They rested when they were too tired or sore to carry on, going back down to the pool for a swim, or eating in the outside restaurant. After walking the dome's perimeter they'd go back upstairs for another few hours of total physical excess. Lawrence accessed an i-sutra file, and they worked their way enthusiastically through the positions and different acts. The furniture was sturdy enough to be useful, and the big marble bathtub with its powerful spar nozzles was simply glorious fun.

Their lovemaking was only possible during the day. Roselyn insisted she still had to go back to her suite for the night. He didn't mind. He didn't mind anything she said or did. She was his for the day, and the definition of night was pushed back later and later every time. On the last day she didn't leave him until three o'clock in the morning.

"Our apartment is in the Leith dome," she told him as they clung to each other on top of the rumpled sheet in those last few hours. "Is that very far from you?"

"No. I got a trike for my last birthday. I can ride round in less than ten minutes. Or if we cut through the public 'tween-dome tunnels and walk it's about twenty-five minutes. Probably best while we're in the Wakening." In his mind he was working out the best route, which domes to go through.

"So it will be easy for us to see each other?" she asked anxiously.

"Very." He stroked his fingertips along the curve of her hips, the way he'd found excited her most.

She snuggled up against him, bestowing a multitude of quick playful kisses along his neck. They tickled.

"And you've got my dp-code?"

"Yes." He moved on top of her, pinning her arms down. "I'll call you as soon as I get home. I'll call you an hour later. I'll call you an hour after that."

"I'm sorry. I don't want to be a possessive bitch. I just want you."

"You'll be in Templeton a day after me. We'll see each other first thing in the morning at school."

"All right." She nodded slowly, as if they'd been discussing a legally binding contract. "I'll wait till then."

The limousine that picked up Lawrence and his father early the next morning took five hours to drive back to their home. Lawrence sat back in the leather seat and stared out moodily at the thick dancing snowflakes. The only thing he saw was Roselyn, curled up in his arms, smiling fondly as they soaked in each other's warmth.

"Is your bracelet pearl broken?" Doug Newton asked.

"Huh?" Lawrence shifted his attention back inside the limo. "No, Dad, it's fine."

"But you're not using it."

"Don't feel like it."

"Hell, we'd better go direct to the hospital emergency department."

"Dad?"

Doug caught the tone, and suddenly focused hard on his son. Indigo script faded from his optronic membranes. "Yes?"

"We've got house rules for everything."

"Look, Lawrence, I don't invent them specifically to annoy you. They exist so that we can all live under the same roof in a vaguely civilized fashion."

"Yeah. I know all that. But you've never said what the rules are about girlfriends."

"Girlfriends?"

"Yeah."

“But you haven’t . . . oh. You kept that very quiet, son. Do we get to meet her?”

“I don’t know, Dad, what are the house rules about that? Is she even allowed to visit?”

Doug Newton eased himself back into the seat and gave Lawrence a long look. “All right, son, you’re virtually old enough to use your voting shares, so I’m not going to treat you like a total child. In return I expect the same courtesy. Okay?”

“Yeah, right.”

“There are two sets of house rules. Your girlfriend will be very welcome to visit. In fact, as you damn well know, your mother will insist on it the instant she finds out you have one. When the young lady comes around, the pair of you can do what you want. Play tennis, soccer, go swimming, study together; all that jazz. She will also be welcome to join us for meals when she’s here. What she cannot do is stay the night, not in your room. Understand?”

“Yes, Dad.”

“The other set of rules are very simple, and they are the same as in real life. You do not get caught. Neither myself, your mother, and especially not your brothers and sisters, are ever to be put in the position of walking into a room and finding you screwing her ass off. Do you understand *that*?”

Lawrence knew his cheeks were bright red; he could feel them burning. This was turning into a hell of a week for fundamental life changes. “I get it, Dad. Don’t worry, that won’t happen.”

“Glad to hear it. Just make sure the lock on that cave of yours works properly.”

“It does.”

Doug Newton shook his head in bemusement. “I’ll say one thing, son, you never fail to amaze me. I take it she is real, not an i-program.”

“Of course she’s real!”

"Thank Fate for that. Does she have a name?"

"Roselyn O'Keef."

"Not sure I know an O'Keef family."

"They're not an Amethi family, Dad. They just got here."

"Really? Well, that means they have a decent stake then."

"Is that all you care about, that they're rich or players?"

"As it happens, yes, it does matter to me. But as we both know by now, what matters to me doesn't even register with you."

"It does. It's just . . ." Lawrence didn't want to say the wrong thing right now. He'd never talked with his father like this before. All this honesty was almost making him feel guilty for earlier behavior. He supposed he had been slightly inconsiderate to his parents recently. But life here wasn't easy. They always seemed to want so much for him and from him.

"I know." Doug held his hands up. "I'm an ogre. You think you're different to me? If you ever find the time to talk to your grandparents, ask them about the fun they had bringing me up."

"Really?"

"Like I said: if you ever talk to them."

"Yes, Dad."

"That's my son."

As soon as he got home, Lawrence loaded her dp-code into his den's desktop pearl and asked the AS to connect him. Her face filled the sheet screen, smiling down at him. The faint freckles dusting her cheeks were the size of his palm. They talked for an hour. He called her another three times that day before finally going to bed to sleep. During the night, he woke up twice, reaching for her. In those blurred moments before he was fully awake he was unsure if she wasn't just a dream. It was a terrifying experience.

Hilary Eyre High was in the center of its own dome, a three-story H-shape structure, big enough to provide first-class educational facilities for fifteen hundred pupils. The ground around it was mostly sports fields, with a constant all-year-round climate, approximating the start of a temperate zone autumn. It was an unusual sight for kids who'd grown up in a city where each dome took pride in its horticultural layout. There were no trees at all, just a flat expanse of verdant grass, interrupted by various styles of slim white goalposts.

Not quite as unusual, though, as the sight of Lawrence Newton standing on the steps ninety minutes before the new term officially started. Despite the weather, he'd driven his trike to school to make certain he wasn't late. Now he was shuffling his feet about impatiently as he tried to look at all nine 'tweendome tunnel arches simultaneously. Pupils were emerging from the twisting caverns to walk toward the school's glass entrance hall. Already, several groups were forming on the plaza outside, friends catching up with each other, sports teams bonding before the term's action, pupils behind on their coursework (usually Lawrence) desperately searching for a crib to download, in-crowds being cool together.

He saw her easily enough even when she was a hundred meters away. Shouted and stuck his hand up, ignoring the curious glances. She saw him and smiled. Waved back. He ran over and they embraced in the middle of amused onlookers. That kind of public kissing was against school regulations. Lawrence didn't care.

"You're here," he said dumbly.

"Yes." She grinned around nervously. "I didn't have anything else to do today."

They were attracting just too much attention for Lawrence to pretend to be blasé. He put his arm around her, and they walked to the side of the steps.

Roselyn said the trip from the hotel had been fine. The apartment in the Leith dome was okay, except for some problem with the building's network cables. They only had a few pieces of basic furniture, so her mother wanted to go around all the stores that weekend.

"Are these clothes all right?" she asked, fingering her sleeve. She was wearing a long dark skirt, with a white blouse and jade-green sweater. With her hair held back in an enameled butterfly clasp it made her look very prim.

Lawrence found the style arousing. "You look perfect." True, some of the other girls wore clothes that cost a lot more, but it sure as hell didn't make them more attractive.

He saw Alan Cramley giving them a sideways look, focused more on Roselyn than himself. They shared a lot of the same low-grade classes, although Alan had recently turned into a soccer maniac and was actually quite good at the game, which gave him considerably more kudos than Lawrence in their year's food chain.

Alan leered behind Roselyn's back and gave Lawrence a quick thumbs-up. Lawrence's immediate annoyance that anyone should disrespect his beautiful girlfriend in such a fashion was more or less canceled out by the gender bond approval. He'd never had that before.

"So what do I do now?" Roselyn asked.

Lawrence spent the rest of the morning taking her through registration, then showing her the layout of the building. He introduced her to as many people as he could—just about everyone he knew, actually. It didn't take him long to notice that with Roselyn by his side their greetings were warmer than they used to be, girls as well as the boys.

After lunch in the canteen they went back to the entrance hall, which was housing the sign-up session for that term's sports and activities. Roselyn put her name down for badminton, track training, girls' soccer, piano and accountancy.

"What are you after?" she asked brightly after they'd done a complete round of the tables.

"Not sure," he mumbled. He'd never even been to a sign-up session before. They did another slow circuit of the hall. Software development was the first choice for extra studies: he reasoned that whatever he wound up doing in adult life, that would come in useful, and it would help supplement his coursework. There was a flight club, which almost made him say: "I didn't know this was here." Flying would be cool; he'd played enough i-simulations (normally involving alien fighters and dogfights) to know he'd enjoy the real thing, and the whole concept was still a powerful totem left over from his old ambition to pilot starships. He put his name down for it, which won a smile of approval from Roselyn. It was games that gave him a real headache. In the end he went for cricket, mainly because the training was the same afternoon as her soccer, so they'd stay behind together, but also because it was about the most nonenergetic game he could find in the syllabus.

They had to part for the afternoon when classes started, but he waited for her in the entrance hall afterward and asked her home.

"You should know," he said apologetically, "Mum's been badgering me to bring you back. I can put her off for a couple of days, but it's like trying to stop Barclay's Glacier from melting. It's got to happen sometime."

"That's okay. I'd like to meet her."

"You would?" he asked cautiously.

"Yes."

"Oh. Okay, good. Uh, I brought my trike. We can get home on that."

"A trike? Lawrence! I've only got these clothes. I can't go outside."

"I know. I'm not totally stupid."

He led her down to the garage at the edge of the dome. His trike stood almost by itself in the rack, a small machine with two rear wheels powered by a hihydrogen combustion engine that was encased in metallic purple bodywork. A sleek elongated bubble of plastic gave the driver and passenger a degree of cover from the elements, although it was open along both sides. The three broad tires had deep snow treads, but even so he could never open it up beyond fifty kilometers per hour without risking a skid. Ten years ago every teenager in Templeton either had one or wanted one, but the Wakening had severely curtailed their use—yet another sign that Lawrence had been born into the wrong age.

He dived into the bin beneath the seat and pulled out two pairs of thermal overalls. "See?"

"Oh yes." Roselyn rolled her eyes. "Really useful when you're wearing a skirt."

"Er . . ." Lawrence knew his face was coloring.

"It's all right. I'll manage." She started to hitch the fabric up.

When she was riding pillion, with her arms tight around him, Lawrence steered them through the thermal cycle lock and out onto Templeton's roads. There had been a light drizzle of hail that lunchtime, which the snowplows had brushed away. The road surface was slick with antifreeze fluid that curdled with melted water, producing the dull shimmer of oil-rainbow patterns. Despite his thermals and helmet, he was glad of the bubble's protection. The wind chill was ferocious.

Templeton's domes glowed with a steady opalescence under the low, forlorn gray sky. The cityscape had acquired a blunter, more industrial-looking architecture these days, appearing less complete than it had during his childhood. The delicate fringe of grass and raoulia plants scrabbling for life along the side of the roads had virtually disappeared. Concrete drainage ditches had been dug in the icy mud along every major route, with excavation mounds piled carelessly alongside. The only remaining signs of botanical life to be found were the rancid green streamers of algae that clotted the deep thaw channels slicing through the scree.

Dome air intake vents were now all fitted with new filters to keep the powdery snow and sticky sleet out of the fans and heat exchange mechanisms, great boxy affairs of galvanized metal held together with crude rivets, standing on legs of steel I-beams. Similar ugly encrustations adorned the factories, additional shielding hastily erected over inlets and grilles.

Worst of all, for Lawrence, was the rust. He'd never realized there was so much metal involved with the city's construction, blithely assuming its component parts were all sophisticated modern composite, held together with intricate molecular bonds. But they weren't: metal remained the cheapest and easiest method of fabrication. Templeton had been screwed, riveted, nailed, reinforced and bolted into a cohesive whole like every other human conurbation since the Iron Age. And now it was paying the price of that cheapness in Amethi's Wakening climate. Rust oozed from every structure. It was the city's sweat, exuded from a million filthy pores. Grubby red-brown stains dribbled and wept along each surface, sapping its strength in an eternal drip of oxidation.

Lawrence was actually relieved when they turned onto the ramp down to the small underground garage that served his family's estate. There was nothing outside for him now. Amethi was squeezing the humans back into their ghettos of technology, veiling the landscape they aspired to conquer. One time at school, the teacher had told them how Scandinavian countries suffered the worst suicide rates on Earth during their long nights; Lawrence understood why, now. It wasn't just coincidence that the hours he spent with i-dramas and games had increased steadily since the Wakening started.

The steps up from the garage opened out into the semiarid dome. Roselyn looked round at a desert of rugged rocks and white sand. Glochidiate and tomentose cacti of every shape flourished amid the wiry scrub grass, their umbellate flowers forming vividly colored crowns. Palms and fig trees encircled quiescent oasis pools where lizards baked on flat rocks around the edges. After the drive from school, the air was wonderfully warm and dry.

"Doesn't anyone live here?" she asked.

"No, the house is in the main dome. This is like an environment park. We've got six." He caught her troubled expression. "What's the matter?"

She wouldn't meet his gaze, and if anything the question just upset her further.

"Roselyn, please."

She was suddenly in his arms, and crying. It was heartbreaking to see her so distraught. He felt as if he was about to cry himself. All he wanted was for her to stop. Every feeling he ever had for her was suddenly intensified. And even through the tears she was beautiful.

"I promised myself I wouldn't do this," she sobbed.

"Do what? What is it? Is it me?"

"No. Yes. Sort of. What you are."

"What do you mean?"

"I'm being so weak. But nothing's stayed the same after Dad died. Everything's different every month. Sometimes it seems like I have to face something new every day. I hate it. I just want to stay in the same place and have a dull boring routine each day, just so it'll give me some stability."

"Hey, it's all right." He stroked her back gently. "You're here to stay on Amethi, and believe me there's nothing more boring and routine than Hilary Eyre High."

She still wouldn't look at him. "I checked up on you."

"You did?"

"Yes. Your family's got a seat on McArthur's Board, Lawrence."

"Yeah. So?"

"You didn't tell me."

"Because it never came up. What's that got to do with anything?"

"I thought . . . You're rich, and you'll have a million connections and friends here. I know how much society and position means to this world. And I just got here, and we're not rich. I thought I was your little bit of holiday fun. You've had me now. I thought that was it, I wouldn't see you again, and you'd be laughing about how easy I was to all your friends. And then you were waiting for me this morning, and . . ." Her tears had returned.

He cupped her cheeks with his palms and gently tilted her head so she had to look at him. "I never thought that. I can't believe you thought it. Roselyn, you're going to have to put up with me for the rest of your life, because I'm never going to find anybody as wonderful as you. Never. And if anybody should be worried, it's me. You're going to take one look at all the jocks at school and realize what a mistake you've made."

"No!" Her hand found the back of his head, and pulled him down for a kiss. "No, Lawrence. I don't want some brain-dead jock. I want you."

They stood still for some time, arms wrapped around each other while the geckos and salamanders filled the dome with strange calls. Eventually, Roselyn smiled meekly and wiped her hand across her face, smearing the tear trails. "I must look a mess."

"You look beautiful."

"That's very sweet, but I'm not going to meet your mother like this."

"Er . . . we can stop off at my den first, I suppose."

Lawrence experienced a mild tingle of doubt as he opened the garage door. Looking at his den with Roselyn standing beside him, he was uncomfortably aware of how . . . well, nerdy it must seem. His own private empire. As such it revealed a little too much about his real self.

Roselyn walked into the middle, and turned a slow circle, taking it all in. "It's very—"

"Sad? Egomaniacal? Tasteless?"

"No. Just that it could only belong to a boy."

Roselyn ran her hand along the back of the battered leather settee. She looked at Lawrence. He stared back.

The bottom of the door hadn't reached the ground before they were tugging frantically at each other's clothes.

"What do you do in here?" Roselyn asked afterward. She was lying along the settee, her head resting comfortably in his lap.

Lawrence was still having trouble with the concept of a naked girl in his den. The two factors simply didn't compute. Although, now he thought about it, having sex in here had been severely exciting. The forbidden fruit syndrome. "I don't do a lot. It's just somewhere that I can come and relax, be myself."

"Okay, I can understand that. There's times I wish my dearly beloved sisters never existed, and I was cooped up on a starship with them for a month. No escape. But what do you do, when you're being yourself?"

"Nothing really interesting, I guess. I used to be quite into electronics and stuff. That's what most of the junk is, I just haven't got round to fixing it all. I do a lot of homestudy in here. Play a lot of i-games."

"Like the *Halo Stars*?"

"That's a new one, actually." He stopped, slightly abashed. But then he did have a nude girl half sprawled over him. You couldn't get more personal than that. "When I was younger, I'd spend hours watching my favorite show up on the big sheet screen."

"What was it?"

"I doubt you've heard of it. *Flight: Horizon.*"

Her nose wrinkled up. "I think I know the name. It's an old sci-fi show, isn't it?"

"Yeah. About a starship exploring the other side of the galaxy. Amethi only imported one series, though. I'll never know what happened to them, and if they made it home."

"Why didn't you send a message to the distribution company back on Earth? It can't cost that much to get the other series sent."

"I tried that a thousand times, but I never get any answer. I guess the company's folded."

"Nothing is ever lost from the datapool, that's why it's expanded beyond its homogeneity globe. It's not that the original network design was faulty, people just kept adding so much memory capacity that the interconnectivity broke down. There are whole sections that are almost autonomous, other sections don't know what's in them, or even that they exist. If you need anything slightly quirky these days, you've got to load in a dozen different askpings and hope one of them finds a metalink for you. When I was looking up Amethi, some of the data took days to get back to me. Nothing mainstream, just the peripherals, early survey reports, start-up finances, that kind of thing. Specialist stuff. There are even rumors about closedpools existing, sections that only have internal metalinks, and their AS controllers don't know they're no longer linked to the outside."

"That sounds crazy. You can't lose information in Amethi's datapool. One askping will find you anything."

"That's because it's still small. Earth's datapool breakdown was inevitable. There's too much data to be indexed in a single source, and the more the index is distributed the weaker the metalinks become. They're talking about giving it official subdivisions. Except, if you don't know where all the original data is stored, how are you going to rearrange it?"

"No wonder I couldn't get an answer."

"If you like, I can send a message to a friend I know. She can load an askping for the show."

Lawrence tumbled off the settee. He wound up kneeling in front of Roselyn, who was regarding him with intrigued amusement. "You can get the rest of the episodes for me?"

"We can find out if they exist, yes. Entertainment is still mainstream. Unless it's over a century old, of course. Even then, it's pretty easy."

"Please." He clamped his hands on her knees. "I would be eternally grateful, and I will sign that in blood."

"Humm." She pondered the notion for a moment, eyes unfocused on the ceiling. "There is one thing I'd like."

"It's yours."

She took hold of his hand and licked his fingers one by one, ending with a kiss at each tip. Then she began to move him slowly across her body until the place he touched made her gasp. "That," she murmured huskily. "I like that."

Every day for a week Roselyn went back to the Newton family estate after school. Sometimes they drove on the trike, but often they'd walk through the 'tweendome tunnels. It wasn't until the third day she was introduced to Lawrence's mother and brothers and sisters. He worried about the meeting a lot more than she did, wincing every time his mother was "nice" or asked a personal question; glaring at his siblings when they shouted a crass comment. Roselyn sailed through it with a grace he envied as much as he admired.

After that initial encounter was over, he wasn't obliged to bring her into the house every time, although it was made very clear she was to come to a meal whenever she was visiting. And it would be lovely to meet her mother for lunch one day. Soon.

"Parents," she sighed when Lawrence glumly relayed this latest development. "They never book themselves into the nursing home. They just stay home and embarrass their children."

He glanced up from licking her navel. "You know what'll

happen, don't you? My mum will start introducing your mum to eligible men."

Roselyn shifted around. They'd put a blanket on the settee now; the leather used to stick to her bare skin. "I doubt it."

He heard the tension in her voice. "Sorry. You don't talk much about him."

"No." She let out a long breath. "I don't. There's not much to say. He was a great father, I loved him lots. Then one day he was gone, and everything I thought was my world went with him. And just when I thought my life was going to be completely shitty from then on, I came here." She pinched a roll of flesh around his waist, which made him squirm. "And there you were waiting for me."

"Something else we've got in common. My life was pretty shitty, too, until I met you. I don't mean it was as bad as you losing your father, no way. Mine was all self-inflicted, most of it, I guess. Easier to bring that to an end."

"Well, I'm going to inflict some more suffering on you."

"What?"

"Lawrence, I can't keep coming back here after school."

"Why not?" he asked, shocked. "Don't you like this?"

Her grin was dangerous as she clambered on top of him. "Oh yes, to be sure, I like this. Way too much, in case you hadn't noticed. Two weeks with you, and I've turned into a complete slut." She pushed her breasts toward his face.

"Me too." He licked her nipples, urgent for the taste of them. Even after all this time he was still amazed at what she let him do. His own bravery in suggesting things was surprising, too. It was as if neither of them owned a single inhibition between them.

Roselyn lifted herself out of his immediate reach. "I've got to start doing some serious homestudy. Amethi's schools are ultra-fast-tracked compared to dear old Ireland's. If I'm not careful, I'm going to wind up the biggest dunce this planet has ever seen."

"You won't."

"Lawrence! I will. I'm serious now, I have got to get my homestudy done."

"Do it here," he said simply. "There's datapool access. You've got your bracelet pearl with you. No problem." His hand went up ready to fondle her breast.

Roselyn sat back, hands on hips, to stare down at him. "You know what'll happen if I come here to do it. You'll start cuddling up and then we'll wind up fucking, and I'll never get anything done. Do you want me to be a total idiot?"

"Of course not. But—" He couldn't bear the idea of not seeing her outside school. "I won't get fresh with you until you've finished your coursework. Promise. Just, please, come back here in the afternoons. Please?"

"Cross your heart and hope to die?"

His finger drew a cross on his chest. "Absolutely."

"Okay then."

"Great!"

"*But,* we go to the house first. Do our homestudy there."

"Ow, what?"

"That's the deal. We work together in the lounge or somewhere. That way neither of us can lapse."

"Oh hell. All right."

"And afterward"—she leaned down again, taunting—"afterward, we can come back down here, and I'll show you how grateful I am."

"Will you?"

Her tongue licked around the outside of his lips at the same time he could feel her nipples brushing against his chest. The provocation was a beautiful torment.

"Oh yes," she whispered.

"How grateful is that, exactly?"

"So grateful, I won't be able to talk, my mouth will be too busy."

Lawrence's moan was almost a whimper, his eyes were

half shut, pleasure blurring his vision with tears. Trepidation made him tremble as he felt her hand curl lovingly around his balls. Then—*bastardFate*—her other hand pinched the fat at the side of his belly, and he juddered free.

Her beautiful face was pouting with disappointment. "What's the matter?"

"I don't like that," he grunted shamefully.

"You mean this?" Her hand reached for the band of fat again.

"Yes!" He shifted sharply out of her way. "There's no need to remind me I'm overweight."

Roselyn frowned. "You are your body, Lawrence. Just like me."

But your body is fantastic, he avoided saying. *Whereas mine . . .* "I know. I keep meaning to get into better shape." He shut his mouth quickly, before anything else stupid could escape.

"Really?" Her face lit up, and she kissed him enthusiastically. "That would be such a turn-on for me."

CHAPTER SIX

CAN-TIME—THE PERIOD THAT GROUND FORCES CAN SPEND IN transit before their combat performance will start to deteriorate—was a factor that military commanders had known about for centuries, building it in to all their tactical planning. According to Z-B's manual, their strategic security forces could endure a fifty-day trip in a starship without any noticeable decay in efficiency.

At forty days into the flight, which put them still three light-years from Thallspring, Lawrence was already wondering if any of Platoon 435NK9 would even get into the drop glider when the time came to go planetside. Whatever office-lurker expert had come up with the fifty-day rule had clearly never been in low-Earth orbit, let alone a starship.

Day forty-one, at 09:30 shiptime, the platoon were in the gym. With the rest of the day given over to nonphysical training and mission revision, it was the wrong time to be doing anything strenuous. The high they'd come out with would take hours to fade, leaving them hyped and edgy. But every platoon was scheduled for ninety minutes a day in one of the life support wheel's gym compartments, keeping their muscle and bone structure up to scratch. There was no getting out of it.

Even knowing it would screw with the rest of his day, Lawrence concentrated hard on his exercise regimen, pushing rhythmically against the stiff resistance of the handlebars. He was prone on one of the starship's standard apparatus benches, which used only springs or pistons to provide resistance. He tightened the resistance settings a couple of notches and carried on. Sweat began to build up on his forehead. His heart was pumping fast. That was the response he wanted, keeping every organ at its peak. He'd emphasized that enough times to the rest of them, and then led by example. Their Skin suits placed a lot of strain on a body, especially one that had been rotting away in an eighth of a gee for five weeks—something the can-time charts tended to overlook.

Glancing round the gym he could see Amersy and Hal Grabowski putting in a decent amount of effort; sweat was staining their scarlet T-shirts. Odel and Karl were getting away with the minimum, as always. Jones Johnson was barely moving his leg restraint, treating the session as some kind of personal rest period.

Typical, Lawrence thought: Jones was their platoon's mechanic, and damn good with just about any sort of machinery, including projectile weapons. Naturally, he assumed that ability compensated for his lack in others. Despite being a member for three campaigns, he never seemed to grasp that the platoon survived by teamwork, which started at the most basic level: physical adequacy.

Lawrence got up and casually threw a towel round his neck. He loped over to Jones and grasped the frame around the man's bench apparatus to give himself some leverage. His free hand slammed down on the leg restraint bars, forcing them to hinge round, and bending Jones's legs almost double.

"Fuck!" Jones yelled.

"You've just been ambushed. A mine has blown a wall down, pinning your legs under a shitload of stone, and three rebels with machetes are coming toward you. If you want to live, you've got to lift yourself free."

"Jesus fuck."

"Come on, you idle bastard, lift."

Jones's face was compressed into a rubber mask as he strained to bring his legs back level. Blood vessels stood out on his neck, pulsing fast.

When it was clear he'd never get the bars back, Lawrence let go. "You're fucking useless, Jones. I don't mind that it will get you killed; we might get a halfway decent replacement. But if you're immobilized it leaves the rest of us covering our asses. Keep up with us, or drop out now. I'm not carrying a liability."

"This is a fucking gym, Sarge. If we're out on patrol I'll be in Skin. This fitness crap we're supposed to stick to is total bullshit."

"The only thing you can ever truly rely on is yourself."

Lawrence caught Hal grinning at the scene. He turned to the kid. "And you can stop smirking. In six days we're going to be on the planet. Every welcoming smile you get means they hate you; the bigger the smile the more they want you dead. We've only got each other down there. Nobody else is going to look out for you. So I want you in the best shape it's possible to be. Not just your body, I want your attitude to be right, too, because Fate help me, I've got to depend on you."

He walked back to his own apparatus bench. Hal resumed his pumping, seemingly proud of how high he'd got the resistance turned up and how easily he handled it. Amersy, who hadn't stopped his forward presses, gave Lawrence a look of mild rebuke as he passed. The corporal wasn't out of line, Lawrence admitted; he had overreacted to Jones slacking off. But this time he wanted a lot more from the platoon than on any other mission. If he was ever going to achieve his personal objective when they reached Thallspring he needed to have complete loyalty, and to do that he had to take care of them. Good care. They might not appreciate it up here, but on the ground these missions quickly became very warped. Society's assholes they might be, but they were streetsmart enough to know who they could trust when the shit hit the fan. And Z-B, in the form of Captain Douglas Bryant, didn't get a look-in.

Lawrence began working his apparatus again. He could see Jones pedaling wildly and let out a quiet snort of satisfaction. He was lucky the squaddie hadn't tried to smack him one. The frustration of can-time was twisting them up. At least back in Cairns they could sneak out of the base at night and screw the tension away with a girl on the Strip.

After gym the platoon was due two hours' equipment readiness training. Lawrence left Amersy to supervise them by himself. He'd got another meeting with the captain; this close to the end of the voyage, they were averaging out at nearly one per day.

Their briefing room was a rectangular compartment with bare aluminum walls, apart from one large, high-resolution sheet screen. The three other platoon sergeants, Wagner, Ciaran and Oakley, were already sitting at the composite table. Lawrence gave them a quick nod and took his own seat. Captain Douglas Bryant walked in a moment later, accompanied by Lieutenant Motluk. The sergeants rose to their feet, all of them with one hand gripping the edge of the table to keep their feet on the ground, while the other was used to salute.

"At ease, people," Douglas Bryant said cordially. He was twenty-eight, a product of Z-B's officer academy in Tunisia. A smart man, with a solid family stake in the company to propel him along the promotion path. When Lawrence accessed his record he found the only active duty the captain had seen was counterinsurgency missions in East Africa. Punishment raids on camps deep in the jungle, where the native tribes still fought the imperialist company mines stripping the minerals from their land. It was a qualification of sorts for asset realization, but Lawrence would have preferred someone with genuine experience.

If he was honest, his contempt for Douglas Bryant originated from knowing the young man was more or less what he would have probably turned out to be himself: genuinely concerned about the condition and morale of the men serving under his command, full of information and knowing shit about what really mattered.

"Ciaran, have you got your platoon's supply inventory sorted out?" the captain asked.

"Sir," the sergeant of Platoon 836BK5 answered. "It was a glitch. The supplies were in the correct lander pod."

The captain smiled around at his sergeants. "It's always software, isn't it? Have we had anything other than virtual problems since we left Centralis?"

They smiled back, tolerantly polite.

"Okay. Final suit tailoring, how are we doing? Newton, your platoon hasn't started yet, why is that?"

"I keep them going in for function tests, sir. I want to leave final tailoring until as late as possible. Even with the gym sessions, five weeks in this gravity is messing with their size."

"I can appreciate the reasoning behind that, but unfortunately it's not quite the procedure we're following. Your platoon is to report for final tailoring oh-eight-hundred tomorrow."

"Sir."

"I can't risk them not being ready when we emerge from compression. We must not be caught unprepared."

Right, Lawrence thought, *like Thallspring has moved and we're going to finish this flight early.* Final tailoring took a couple of hours per suit, at max. "I understand, sir."

And so it went. Bryant was obsessed with details; everything any experienced commander would leave to his sergeants to sort out he wanted a say in. He had to have the operation running perfectly along the standard track, a dead giveaway that he was concerned more with the impression he generated within the company than with the practicalities of the situation they'd be facing. He even wanted Oakley to cancel a request he'd made for more remote sensors when they went groundside. His platoon had been assigned to sweep through an urban area that was all narrow roads in a maze of cheap housing—and that was from a ten-year-old map; it could have decayed a lot since then. In other words, a perfect ambush territory for the local badboys. And they'd have a lot of bravado before Z-B established themselves and obtained their good behavior collateral. Lawrence would have wanted the same security those remote sensors could provide. But despite Z-B's vaunted policy of loop involvement, the beachhead plan already contained the number of sensors considered relevant. Bryant did not want anything to alter at this stage.

Oakley said yes, and got his bracelet pearl to rescind the request. They moved on to the landing operation's timing and how Bryant didn't want them to suffer undue delay on their way out of the drop gliders.

A gentle warm rain had been falling on Memu Bay for most of the day, the second unseasonable downpour in a fortnight. It meant Denise had to keep the children out of the garden and at the tables and benches sheltered by the roof. In the morning she'd handed out the big media pads and got them to paint the shapes they saw in the clouds, which resulted in a splendid collage of strange creatures in glowing blues, reds and greens. By the afternoon, when it was obvious the clouds weren't going to blow away any time soon, she settled them in a broad semicircle and sat on one of the tables in the center.

"I think it's time I told you about the planet of the Mordiff," she said. "Even though Mozark never actually visited it himself."

There were several sharp intakes of breath. The children gave each other excited looks. The dark history of the Mordiff planet had only ever been hinted at before whenever she talked of the Ring Empire.

Jedzella stuck her hand up. "Please, miss, it's not too horrid, is it?"

"Horrid?" Denise pursed her lips and gave the question some theatrical consideration. "No, not horrid, although they fought terrible wars, which are always evil. I suppose from where we are today, looking back, it's really quite sad. I always say you can learn the most from mistakes, and the Mordiff made some really big mistakes. If you remember what they did, then, I hope, you'll be able to avoid those same mistakes when you grow up. Do you want me to go on?"

"Yes!" they yelled. Several of them gave Jedzella cross glances.

"All right then. Let's see: Mozark never went there, although he did fly close to the Ulodan Nebula where the planet and its star were hiding. There wasn't a lot of point to him going. Even in those times the Mordiff were long gone, and nothing they'd left behind could have helped him in his quest for a grand purpose in life. Although, in a way, a very warped and twisted way, the Mordiff had an overriding purpose. They wanted to live. In that they were no different to all the rest of us: humans today and the sentient species of the Ring Empire all want to live. But by fate, or accident, or chance, or even luck, the Mordiff evolved on a planet in the middle of the darkest, densest nebula in the galaxy at that time. They had daylight, just as we do. The nebula wasn't thick enough to blank out their sun. But their night was absolute. The night sky on that planet was perfectly black. They couldn't see the stars. As far as they knew, they were completely alone; their planet and its sun were the entire universe."

"Didn't they send ships out to find other stars?" Edmund asked.

"No. Because they had no reason to explore. They didn't know anything else existed, and observation backed up the whole idea, so they didn't even know they could go looking. That was their downfall, and it's the lesson we must learn from them. You see, like most sentient species, they thought in the same fashion we do, even though their bodies were very different. They were big, almost as big as dinosaurs, and they had very clever limbs that could change shape. It meant they could slide their bodies along the ground, the way a snake does, or they could swim like fish by turning the limbs into fins, and some Ring Empire historians and archaeologists even thought they could fly, or at least glide. But that didn't stop them from having an ordinary civilization. They had a Stone Age, and an Iron Age, just like us; then they went on and had a Steam Age, and an Industrial Age, and an Atomic Age, and a Data Age. And that was where their troubles started. By then, they had developed their whole world, and they had good medicine that gave them a long healthy life. Their population was expanding all the time and consuming more and more resources. Whole continents became giant cities. They built islands miles across that were just floating buildings. All of their polar continents were settled. There was no room left, and all the surface was being exploited. It meant they had wars, horrible, terrible wars that killed tens of millions of them every time. But they were always pointless, as all wars are. After entire nations were destroyed, the victors would just move into the ruins, and within a generation the land would be full again. All the while their technology, especially their weapons technology, grew more powerful and more deadly. The wars they fought became worse, and more dangerous to the rest of the planet.

"Then one day, the biggest nation, which was ruled by the greatest Mordiff overlord, discovered how to create a wormhole."

The children let out a fearful *Ooooh.*

"Did they invade the Ring Empire, miss?"

"No, they didn't invade the Ring Empire. Have you forgotten? They didn't know it was there. They made their wormhole go in a very different direction. You see, wormholes are formed from a distortion of space-time. We use ours to create a tunnel through space so we can fly to the stars. The Mordiff traveled in time. Because the Ulodan Nebula denied them a vision of space, time was all they knew. The overlord ordered a single giant wormhole terminus to be built, standing at the center of his nation. It was the greatest device the Mordiff had ever constructed, for not only did the terminus generate the wormhole, its own structure was self-sustaining. As long as it had power, it would never decay or fail. And it got its power from the way it distorted space-time. In other words, it was eternal, almost like perpetual motion."

"My daddy says that's impossible," Melanie said with haughty self-confidence. "He says only fools believe in it."

"It is impossible," Denise said. "But that's the best way to describe how the Mordiff's terminus worked."

Edmund sneered at the girl, then turned to Denise. "Why did the overlord build it, miss?"

"Ah. Well, that's where the terror and the tragedy of the Mordiff begins. When it was finished, the overlord ordered an exodus of his whole nation. An armada of flying craft carried them all into the terminus, millions and millions of Mordiff. And when they were all safe inside, the overlord's personal guards set off the most terrible weapons the nation possessed. All of them, all at once. They were so bad, and so powerful, that they killed every living thing on the whole planet, and turned all the cities to rubble, even those of the overlord's nation."

The children stared at her, awed and troubled.

"Every Mordiff nation had the same awful doomsday weapons; some spread deadly diseases while some simply exploded hard enough to open up cracks into the magma below the continents," Denise said. "The overlord knew it would just be a matter of time until somebody used them. By then, each of the nations was so desperate for new land and resources that not using the doomsday bombs would mean they'd collapse from within.

"So now the overlord's nation was inside the wormhole, traveling further and further into the future, away from the time of the planet's death. Some of them, a scouting party, emerged a hundred thousand years later, flying out of the terminus—which had survived the explosions and radiation, of course. To the scouting party it was only minutes since they'd entered the wormhole, but as soon as they came out they found a sterile planet, with the ruins of the megacities crumbling into dust. By then, the radiation had decayed, and the plagues had died away. These Mordiff scouts dumped bacteria and algae over the surface and flew back into the terminus. Then they came out another thousand years later, when the bacteria had spread everywhere, bringing the soil back to life. This time they scattered seeds before they went back into the terminus. The third time they emerged, they left breeding pairs of animals and fish. A thousand years after that, the world had returned to the state it was in before their Industrial Age, with huge grassy plains and forests and jungles. That's when the whole of the overlord's nation came out of the wormhole. They'd only been flying inside the wormhole for a couple of hours, while outside a hundred and twenty thousand years had rushed past.

"They looked around at this beautiful, clean new world, and they rejoiced and thanked the overlord for delivering them to this wonderful place. Many of them forgot the crime that had been done to give them this chance at a fresh life and settled down to rebuild their original society. So once more they mined the land for metals and minerals, and their cities began to grow again, always expanding over the wilderness. After a few generations, some of the Mordiff forgot the debt they owed to the overlord family, which still ruled the original nation, and began to break away and form new nations of their own. Two and a half thousand years later, the planet was once again covered with cities. Once again, wars were being fought. So, the overlord of that time did what his ancestor had done. He gathered his nation into flying craft and sent them through the terminus. Behind them, the doomsday weapons exploded yet again.

"This wretched cycle turned another three times. Whenever the world grew too crowded to support the billions of Mordiff who filled the cities, the overlord's nation would escape through time and kill everyone left behind. But after the last time they fled into the terminus, the scouts came out a hundred thousand years later to find something unexpected had happened. Their sun had changed. When they looked at it, they could see dark sunspots swelling and bursting all across its surface. It was reaching the end of its main cycle and growing colder. Of course, as they'd never seen the other stars in the galaxy, they didn't know what was happening. They never knew that stars change and die; they'd assumed that their little universe was static and eternal. The physicists among them began to speculate and produce theories at once, and they probably worked out what was happening, because they were smart, don't forget. But knowing what's happening and being able to do anything are very different things.

"So the scout group took measurements and recordings of how cold the air was becoming, and how frigid the land had turned, and went back into the wormhole to report to the overlord. At first, he didn't want to believe what they told him, but, eventually, he came out and witnessed the star's winter for himself. By now, the ground and ruins were covered in a thick layer of frost, which glittered in the dimming sunlight, and the seas were frozen solid. For a long time the overlord raged against what he thought of as supreme injustice before he regained his senses. Scouts were dispatched far into the future: two hundred thousand years, five hundred thousand, a million, two million, even ten million. They all came back with worsening reports of how the sun grew colder and colder, swelling into a huge red monster that covered a fifth of the sky. At no time did it ever show signs of returning to its original state."

"Can stars do that?" Melanie asked quietly. "Get better, I mean?"

"No, dear, they can't. Not by themselves. There are stories

that some kingdoms in the Ring Empire tinkered with the interior of stars when the Empire was at the height of its powers, but they're only stories. And for all their knowledge and technology, the Mordiff were never as strong and wise as the Ring Empire. So the overlord had no choice, he had to order his people out of the wormhole as soon as the effect of the doomsday weapons had faded away, and while the sun still had some warmth. In that respect, he was a good leader, doing the best he could. He ordered that the new cities were to be built under protective domes. Their technology, he said, was enough to turn back the tide of night. Which, in truth, it was. They could still live on their planet, protected from the cold under skies of crystal. Fusion power would provide them with all the light and heat they could ever want. But these enclaves were harder to build, and took even more resources to maintain. It was a difficult life, and by now, the Mordiff had evolved for war and conquest. They knew nothing else. After so many generations devoted to endless conflict the outcome was inevitable. Once their population began to expand again, the ordeals and depravation hit them harder than ever before. The domed cities fought each other. It was insane, because they were so much more fragile than the open cities of old. And this time there was nowhere to flee if anyone let off the doomsday weapons. The only thing in their future now was cold and darkness.

"According to the Ring Empire archaeologists, the last of the Mordiff died out less than fifteen hundred years after they emerged from the terminus for the final time. The Ring Empire explored the Ulodan Nebula twenty-five million years later and found a few fractured remains amid the ice that shrouded the whole world, all that was left of a species that had covered their planet with cities and marvels."

The children sighed and shivered. Many of them glanced out at the sky for reassurance that their own sun was still there, as bright and warm as ever. The clouds were clearing above Memu Bay now, shredded by the offshore wind into gravid streamers. Broad white-gold sunbeams prized their way through the ragged gaps to chase over the land. Denise smiled with them in reassurance at the water that glistened so refreshingly on the plants in the garden and the trees outside.

"That was scary," Jedzella announced. "Why did they all have to die?"

"Because of their circumstances. The nebula meant they could only ever look inward. We're luckier than that. We know the stars exist. It should help us develop a more enlightened attitude toward the way we live and behave." Denise tried hard to keep the sarcasm out of her voice.

One of the girls waved urgently. "What's enlightened?"

"It means being nice and sensible, instead of being stupid and violent." She paused and smiled round. "Now, who wants to go out on the swings?" It was still too wet, and she'd get a telling off from Mrs. Potchansky for letting them get their clothes damp. But they were at their happiest when they were gallivanting around outside: she couldn't bear to take a moment of that away from them.

They pelted out from under the roof, cheering and racing each other to the swings. Denise followed at a slower pace. Running the Mordiff tale through her mind always conjured up a melancholic mood. The story of their tragedy had too many resonances with humanity. *There but for the grace of God . . .* Not that she believed in gods, human or alien.

Her Prime alerted her to a priority spacecom alert diving through the datapool. Two fusion plasma plumes had been detected eight million kilometers out from Thallspring. Spacecom was scanning for more. Data traffic rates between their offices and tracking satellites doubled inside fifteen seconds, then doubled again, increasing almost exponentially.

Denise's hand flew to her mouth as she looked round at the children. Their carefree shrieks, giggles and smiles pounded into her consciousness, and she was suddenly fearful for them. She tilted her head back, searching the section of sky that spacecom's coordinates indicated. In relation to herself, it was a nine-degree window just above the western horizon. There were too many clouds in the way to permit any sighting of the tiny blue-white sparks she knew were there. But their presence acted like an eclipse on her heart, making her world colder and darker.

It had begun.

Captain Marquis Krojen sat back in what he liked to think of as the command chair on the *Koribu*'s bridge. In practice, it was just another black office chair equipped with freefall restraint straps and bolted to the decking behind a computer station. There were eleven other identical stations in the square compartment, arranged in two rows of six, facing each other. Nine of them were currently occupied in readiness for exodus.

When he was a junior officer on his first couple of starflights, he'd managed to get a place in one of the observation blisters in the forward drive section for exodus, his captain of the time agreeing he wasn't essential for the operation. He'd waited spellbound with his fellow young officers as the moment approached, putting up with cramped limbs and stuffy air just for the chance to witness the transition. In the end it was as uneventful as most events aboard a starship. The wormhole wall, a blankness that wasn't quite black, slowly faded away, allowing the stars to shine through, almost like a lusterless twilight creeping up on a misty evening.

That had been thirty years ago. He hadn't bothered with visual acquisition since, preferring the more precise story of the display screen graphics and his DNI grid. Five of his own junior officers were currently crammed into the observation blister, a reward for reasonable performance of duties during the flight. They'd learn.

"Stand by for exodus," Colin Jeffries, his executive officer, announced. "Ten seconds."

There were so many displays counting down that the verbal warning was completely unnecessary. Tradition, though, like so many things on board, orchestrated the crew's behavior, helping to define the chain of command.

His DNI showed him the ship's AS powering down the energy inverter. The plasma temperature in the tokamaks began to cool as the magnetic pinch was reduced. Power levels fell toward break-even, producing just enough electricity to keep the ancillary support systems up and running.

All around the *Koribu,* the drab monotony of the wormhole faded away to be replaced by normal space. Holographic panes on top of the bridge computer stations turned black, showing the steady gleam of stars relayed from external cameras. The AS activated various sensors, aligning them on Thallspring. Several of the bridge officers cheered as the bright blue-and-white orb materialized on their panes.

Let's face it, Marquis thought, *we have little else to do.* Bridge officers were simply a last fail-safe mechanism, nothing more. The AS ran the ship, while humans made small decisions based on the minute fraction of tabulated information it provided them through holographic panes and DNIs. Summaries of summaries: there was so much data generated by the millions of onboard systems that it would take a human lifetime just to review a single frozen moment.

"Eight million kilometers, as near as you can squint," Marquis said, after analyzing his DNI information. "Radar active. We're searching for the rest of the ships."

Simon Roderick leaned on the back of the captain's chair, inspecting his displays. "Very good. I expect that as we tracked their compression distortion while we were in the wormhole, they won't be far behind."

Marquis didn't reply. Everything Roderick said, the way he said it, was an assertion of his assumed superiority. A captain should be master of his own ship; as indeed the other captains of the Third Fleet were. But with *Koribu* acting as the flagship on this campaign, Marquis had endured Roderick's presence for the whole flight. He'd been subject to a stream of advice and requests the entire time. Every night, Roderick had dined with the senior officers, making it a miserable meal. The man's conversation was rarefied, discussing culture and economics and history and company policy. Never a joke or a lighthearted comment, which put everyone on edge. And he'd occupied five cabins. Five! Although Marquis no longer begrudged him that. The Board member spent most of the ship's day cosseted away there in meetings with his ground force commanders and the creepy intelligence operatives Quan and Raines.

"What's the reaction drive status, Captain?" Roderick asked.

"Engineering crew are priming us for ignition." Marquis kept his voice level and polite. Roderick could access as much data as he could, probably even more, given the access codes he had. The question was just a reminder of the strategy he'd insisted on.

Normally, a fleet would hold its drift positions at exodus, waiting for every starship to arrive before maneuvering into formation and heading in to the target planet. This time, Mr. Roderick had decided that there would be no formation; each starship would start its Thallspring approach flight at once. With the starships strung out, the planet's hypothetical exo-orbit defenses would be more exposed when they deployed. The lead starship would take the brunt of the attack but provide the remainder with first-class targeting information.

Marquis had pointed out during this discussion at their nightly meal that a formation of starships multiplied the available firepower to generate an excellent shield, and provided a much greater all-round coverage than a singleton.

"Remember Santa Chico, Captain," Roderick had replied. "We should examine history and move on from our failings in an appropriate fashion. *Tempora muntantur.* Tactics evolve in association with technology."

Marquis hadn't been on the Santa Chico campaign, *thank God,* but that planet was always a one-shot. Thallspring wouldn't have anything like their level of technology. If by some miracle they had built exo-orbital systems, they'd be the old-fashioned kind.

"Course to six-hundred-kilometer orbit plotted, sir," Colin Jeffries said.

Marquis reviewed the fusion drive schematics that his DNI was scrolling. Overall failsoft was 96 percent, which was good. They'd spent three months before the mission in dock at Centralis having a C-list refit. Only if failsoft dropped below 70 percent would he cancel ignition.

"Cleared for ignition, Mr. Jeffries. Alert the life support wheels to secure for gravity shift."

"Yes, sir."

"Anyone know what's happening on the planet?" Roderick inquired lightly.

Adul Quan looked up from the bridge station he'd appropriated. He'd routed a lot of sensor readings to his holographic panes, where analysis routines were reinterpreting the raw data. "Standard microwave and radio emissions. I'm also seeing hotspots corresponding to known settlement sites. They're still there, and effective."

"Ah, some good news. Very well, they'll attempt to contact us soon enough. There's to be no response. I'll talk to the president once we're in orbit."

"Understood."

Amber lights began to flash, warning them of the fusion drive ignition.

"Sir, the *Norvelle* has come out of exodus," Colin Jeffries reported.

"Excellent," Roderick said. "I'm going to my cabin. I doubt you need me breathing down your neck right at this moment, eh, Captain? You have my every confidence to deliver us into orbit unharmed."

Marquis didn't look round. "I'll inform you of any status change."

One thing Denise, Ray and Josep had never properly taken into account was how little lead time they'd have. Their Prime software might have trawled the spacecom alert from the datapool with a minimum delay, but that didn't mean there weren't others who were just as fast. Leaks were also a factor. The verified sighting was automatically distributed to over a hundred government personnel; most of them had family, all of them had friends and media contacts.

Fifteen minutes after spacecom's internal verification of starship exodus, the general media knew of the alert and started bombarding the president's office for official confirmation and a public statement. It was just after midnight in Durrell, the capital, but the president's praetorian aides responded swiftly. Their first cautious release that anomalous spacecom data was being reviewed hardly satisfied the howling mob, but it did give them enough justification to start breaking the story across the datapool and on the news shows. It was a story that fed on its own hysteria, expanding with each retelling. Recordings of the last invasion were snatched from their libraries and broadcast in extreme detail, reminding everyone of the oppression and brutality they'd suffered, as if they needed such cues. Thirty minutes later, just about all of Thallspring knew the starships had come again.

In their single act of public responsibility that day, the media announcers did keep repeating there was no need to panic—the starships were eight million kilometers away. Given how many people were desperate to hear the entire message, it amazed psychologists just how many managed to blank that part out.

Human nature being what it is, people's overriding instinct in times of danger is simply to head for home. It's a baseline refuge, seeking comfort and security from contact with your own family. In every city people walked out of work to hail the nearest taxi or jump on their tram; bikes and cars poured onto the roads. There hadn't been traffic snarl-ups and grid-lock like it for over a decade; in fact, not since the last time the starships arrived.

Denise's usual twenty-minute trip back to the bungalow on the Nium River estuary took nearly an hour and a half. She hadn't realized so many people even lived in Memu Bay, let alone had cars or bicycles or scooters. So much time had been wasted just sitting in trams expecting them to move any minute. Nobody ever drove down the tram lines in the center of the road—until today, when they blocked it solid. Eventually, she hopped out of the stationary vehicle and started walking.

Fortunately the local datapool retained its integrity through the chaos, though even its connection time had slowed appreciably as half the town used it to contact the other half and ask them where they were. She sent a stack of preformatted messages through her own ring pearl, using Prime to route the heavily encrypted packages to various resistance cell members so they would be untraceable. Acknowledgments returned sporadically, scrolling across her vision as she dodged traffic and slithered around lumbering pedestrians.

Outside the town's heart, the traffic wasn't so clogged, which allowed the vehicles to drive a hell of a lot faster.

They'd all had their AS programs taken offline, with the human driver ignoring the speed limits. Denise jogged along the suburb pavements, sprinting across intersections. Not even being young and female saved her from vigorous hand signals as cars swerved.

When she did trot up the gravel path to their front door she was sweating enough to make her blouse and skirt cling annoyingly. Ray and Josep still weren't home: they'd been out on a boat when spacecom's alert was given. Their last message said they were less than ten minutes behind her now. She wondered how they'd managed that with the melee constricting the center of town.

The bags they'd need were permanently packed. Denise disengaged the bungalow's active alarm and tugged them from the hall cupboard where they were stowed—a couple of sports shoulder bags, the kind anybody would take for a week's holiday. Inside were clothes—some needing washing—toiletries, coral souvenirs, several bracelet pearls with the supplements any student would own. All the items would pass a spot inspection. It would take a detailed lab analysis to detect any sort of subterfuge. Her ring pearl interrogated the hidden systems, running final function and power checks. Once their validity was confirmed, she dumped them beside the door, then ran back to her own room, stripping off her blouse. Her blood still seemed to be hot and fizzing, even though her heart had slowed right down. Now the starships had arrived, she felt invigorated. A simple faded-copper T-shirt and black shorts gave her a great deal more freedom of movement. She twisted the plain gold band on her index finger that contained the pearl, reassured by the contact. A strange preparation ritual for a warrior about to go into combat, but then this was not an arena the gladiators, knights and ninjas of old would recognize.

Denise was lacing up her sneakers on the doorstep when the boys arrived. They'd acquired an open-top jeep from the diving school, which Josep was driving. He braked it to a sharp halt at the end of the drive. Ray jumped out and slung the cases in the rear. Denise took the backseat. She was still slipping on her safety harness when Josep accelerated away, sending gravel pelting into the jasmine.

"Which way are you going?" she asked.

"We figured the outer loop road," Ray called over his shoulder. "It's longer, but the traffic regulation AS says it's still relatively clear."

Denise visualized a layout of the town. Their bungalow was just about on the opposite side from the airport. Perhaps they should have planned that better, as well. But once they got onto the loop road, it would take them directly there.

"How long?" she asked Josep. She had to shout; the wind was whipping her short hair about as he sped them along the concrete road with its broad, neatly mown verges.

"Forty-five minutes," he said.

"You're kidding."

His smile was grim. "I can do it!"

"Okay." Denise started to instruct her Prime, and indigo timetables slid across her vision. Scheduled planes were still flying out of the airport. According to the bookings program, just about every tourist in Memu Bay was trying to bring his or her departure flight forward to today. Prime accessed the Pan-Skyways reservation system and searched through the passenger list on a flight to Durrell that was due to leave in an hour and ten minutes. Only a quarter of them had checked in so far. Several had contacted the airline to say they'd been caught up in the traffic snarl and were running late. Sensible people, she thought. She erased two of them and substituted Ray and Josep, under their ghost identities.

"We're in," she said gleefully.

The loop road was a big improvement. At first. Traffic on their side of town was light. It began to build, increasing in proportion to the distance to the airport. Even Josep had to slow down as both lanes began to fill.

"Where've they all come from?" she asked, looking round in dismay. Family cars, sedans with darkened windshields, jeeps like theirs, vans and trucks; every one had a driver gripping the wheel with an intent don't-mess-with-me expression on his face.

"I don't know," Josep muttered. "But I know where they're going." He swung the wheel sharply, sending the jeep around a big pickup and onto the hard shoulder. Free of the jammed-up main lanes he accelerated again. Tires bounced frantically through the potholes, the suspension juddering loudly.

Ray grinned happily. "There goes your license."

"It's a stolen jeep, and I'm not licensed to drive it anyway. Now smile for the traffic cameras."

Denise rolled her eyes and pulled a floppy old fishing hat on her head as other drivers shouted at them. To the side of them, traffic was grinding to a complete halt. She could see the kind of luggage people were carrying. Cars just had suitcases thrown into the backseats, but several vans and pickups were piled high with furniture, some even had pets, mainly dogs barking in confusion. A small pony peered nervously out of one rig. She couldn't understand where they thought they were going. It wasn't as if this continent had a big rural community that could absorb them. There was only the Great Loop Highway with its scattered settlements around the Mitchell Plateau mountains. And she knew what their inhabitants would think of refugees from the city.

"Damn," Josep grunted. Other people had started to swerve onto the hard shoulder. Vehicles stuck on the inside lane were tooting their horns furiously at the lawbreakers trundling past. It didn't take more than another five hundred meters before the hard shoulder was reduced to a parking lot. They were still a good twelve kilometers from the airport.

"Go around them," Ray said.

Sighing, Josep engaged the high-traction mode for the hub motors and urged the jeep off the hard shoulder and onto the verge. They bounced along the grass, tilted at quite an angle. Tires left long spin tracks in soil still wet from the morning's downpour. Cars on the hard shoulder tooted angrily as they bumped and fishtailed past the completely stationary lines.

That ride ended three kilometers short of the airport, when the verge turned into a cutting. The banks were too steep for the jeep to use even on high traction.

Josep braked and they slowly slithered down the slope until the tire rims were resting on the curbstones lining the hard shoulder. Nothing was moving on the dual roadway. People had climbed out of their vehicles, talking to each other in exasperated voices. Denise could hardly believe it, but the trams on the high-speed link between the roadways were also stationary. Maniac drivers had actually tried to use the rails as a road, ramming through the crash-barrier that guarded the outer lane. There was a long zigzag line of cars and vans bumper to bumper along the tramway, looking as if several dozen of them had all collided in slow motion. Those drivers were screaming at each other. She could see several fistfights had broken out.

"Out," Josep said. "Come on, we're close enough now."

A big DB898 passenger plane thundered overhead, its undercarriage bogies folding into its fuselage. Hihydrogen turbofans whined loudly as it rose in a steep climb. Everyone standing about on the road stopped what they were doing to watch it pass. The majority then started walking, as if the aircraft had been some kind of religious summons.

Denise, Josep and Ray started a fast, easy jog, drawing jealous stares from families and older people tramping along the concrete with moody desperation. Thanks to the d-written enhancements throughout their bodies, the weight of the bags and the intense midafternoon sun had no effect on them, so they were able to maintain a steady pace for the entire three kilometers. Denise had a mild sweat when they reached the arrivals hall, but that was all.

The crowds around the various airline gateways were thicker than fans going into a stadium turnstile on a league finals day, and a lot more restless. They pushed and shoved their way toward the front, either ignoring or giving out aggressive nose-to-nose stares to anyone who objected. Up on the walls, giant sheet screens were showing man-in-the-street interviews, with just about everyone the reporters found asking the same thing: when are our exo-orbit defenses going to blast these invader bastards into radio-active gas? Surely some clandestine top-secret government project had built them ready for this moment? Why are we defenseless?

They arrived at the Pan-Skyways gate three with five minutes left until boarding ended. There, in the middle of five hundred noisy, straining, angry people, Denise gave both of them a kiss and a hug. If they were surprised by the uncharacteristic display of affection they didn't show it. She'd often been exasperated by them during the last year; now she realized how much she cared for them, almost as much as for their mission. "Look after yourselves," she mumbled. It wasn't a wish; it was a command.

They returned the hug, promising her they would. When they showed their ghost identity cards to the gate it opened smoothly to let them through.

Denise wormed her way out through the crush of people and went up to the observation deck on the roof. She was the only person there. A humid offshore breeze plucked at her T-shirt as she stood pressed up against the railing. Twenty minutes later, the big Pan-Skyways jet taxied out onto the runway and raced into the hot sky. Denise watched it vanish into the hazy horizon, then lifted her gaze to the sky's zenith. Seven tiny, bright stars were visible through the azure veil.

Her arms were spread wide, hands gripping the smooth,

worn metal of the railing. When she took a deep breath, she could feel the oxygen flowing through her arteries, fortifying her enhanced cells. Her physical strength brought a cool self-confidence with it, a state of mind she relished.

Welcome back, she told the sparkling interlopers. *Things are going to be a little different this time around.*

Simon Roderick sat at the desk in his appropriated cabin, surrounded by data. Some of it came from holographic panes, the rest was provided by DNI. All of it flowed and flashed at his whim. Organization, the key to success in any field, even one with as many intangibles as this. He knew how Captain Krojen considered himself at the mercy of the *Koribu*'s AS, how isolated that made him from the physical running of the starship, a situation Simon never placed himself in, no matter what his supervisory assignment. The captain's trouble was his insistence on routing commands through his officers, keeping them *involved.* If he kept humans out of the equation he would find himself a lot closer to achieving true authority over his machinery.

The stream of information enveloping Simon shifted as the last of the Third Fleet starships reached its six-hundred-kilometer orbit. Its new pattern was close to the optimum he had envisaged. Needless to say, Thallspring had deployed no exo-orbit weapons against the starships during their approach. They had, though, endured a constant bombardment of communication traffic during the flight into orbit. Several tapevirals had been hidden in the packages, some of them quite sophisticated—for an isolated world. The *Koribu*'s AS had recognized and isolated them immediately. None of them had come even close to the Barbarian Sentience subversives that the antiglobalizers had used back on Earth.

Simon shifted his attention to the images building up from

the small squadron of observation satellites that the Third Fleet had released into low-Thallspring orbit. It was a world that had moved ahead in a steady pedestrian fashion since Z-B's last asset realization. Infrared mapping showed the settlements had expanded roughly as predicted, although Durrell was certainly larger than expected. Worst case, it gave them a hundred thousand more people, which the ground forces could certainly handle. Fortunately, that corresponded to an increase in industrial output. After all, those extra people had to be housed, clothed, fed and provided with jobs.

Several blank zones on the planetary simulation caused him a flicker of dissatisfaction. His personal AS noted the direction of his ire and informed him that three observation satellites and one geostationary communications relay had failed. The successfully deployed systems were being reprogrammed to fill the gaps.

He sent the planetary data into peripheral mode and established a link to Captain Krojen. The officer's sullen face appeared on a hologram pane. "I'd like you to begin the gamma soak, please," Simon said.

"I wasn't aware our reviews were complete," Krojen said. "There could be people down there."

"The primary scans haven't found any artificial structures in the location we selected. That's good enough for me. Begin the soak." He canceled the link before there was an argument, and expanded the *Koribu*'s schematics out of the grid.

Just behind the starship's compression drive section, their gamma projector began to unfurl. The mechanism had been included on all of Z-B's colonist carrier starships as fundamental to establishing a settlement. Basically a vast gamma ray generator and focusing array, it was a cylinder fifteen meters in diameter, and twenty long, riding on the end of a telescopic robot arm. Once it was clear of the drive section, the cylinder's outer segments peeled open like a mechanical flower. On the inside, the petals were studded with hundreds of black-and-silver hexagonal irradiator nozzles. A second set of segments hinged open around the first, followed by a third. At full extension, it formed a circular disk sixty meters across.

Thallspring's second-largest ocean was rolling past underneath the *Koribu,* with the coastline sliding into view over the horizon. Durrell was directly ahead of the starship, a gray smear amid the emerald crescent of land that was the settlement's enclave of terrestrial vegetation. Outside that, Thallspring's native aquamarine plants embraced the rest of the land.

Koribu's gamma projection array swung around on the end of its arm until it was pointing toward the settlement. Small azimuth actuators tweaked its alignment and began tracking. Tokamaks inside the starship's compression drive section started to power up, feeding their colossal energy output straight into the projector array. The amount of energy demanded by a starship to fly faster than light sliced down through the atmosphere in a beam that was no more than a hundred meters wide when it struck the surface.

The impact was centered on a patch of ground at the western perimeter of the settlement, just overlapping the border of the earth plants. No living cell of any type could survive such a concentration of radiation. Thallspring's plants, animals, insects and bacteria died instantly beneath the beam, a huge zone of vegetation that immediately turned bruise-brown and began to wither. Branches and leaves bowed down and curled up beneath the relentless invisible onslaught; fissures split open along tree trunks, hissing out steam from ruptured osmotic capillaries. Animals thudded to the ground, skins shriveled to black parchment and innards cooked, spitting out little wisps of smoke as they ossified in seconds. Even below ground, nothing was spared. The gamma rays penetrated deep into the soil, eradicating bacteria and burrowing insects.

Then the beam began to move, scanning back and forth across the ground in slow kilometer-wide swaths.

Simon shifted the soak data into peripheral. He used the Third Fleet geostationary relays to open a connection into Thallspring's datapool and requested a link to the president.

The man whose image appeared on his holographic pane was in his late fifties, heavy features roughened by lack of sleep. But there was enough anger burning in his eyes to compensate for any insomnia lethargy.

"Stop your bombardment," President Edgar Strauss growled. "For fuck's sake we're not any kind of military threat."

Simon's eyebrow twitched at the obscenity. If only Earth's politicians were as forthright. "Good day, Mr. President. I thought it best if I introduced myself first. I'm Simon Roderick, representing the Zantiu-Braun Board."

"Switch your goddamn death ray off."

"I'm not aware of any bombardment, Mr. President."

"Your starship is firing on us."

Simon tented his fingers, giving the pane and its reply camera a thoughtful look. "No, Mr. President; Zantiu-Braun is continuing to upgrade its investment. We are preparing a fresh section of land for the Durrell settlement to expand into. Surely that's beneficial to you."

"Take your investment and stuff it where the sun doesn't shine, you little tit."

"Is there an election coming up, Mr. President? Is that why you're talking tough?"

"What would the likes of you know about democracy?"

"Please, Mr. President, it's best not to annoy me. I do have to monitor our beam guidance program very closely. Neither of us would want it to move out of alignment at this crucial moment, now, would we?"

The president glanced at someone out of camera range, listening for a moment as his expression soured further. "All right, Roderick, what do you bastards want this time?"

"We're here to collect our dividend, Mr. President. As I'm sure you know."

"Why the hell can't you just say it? Too frightened of what we'll do? You're pirates who'll slaughter all of us if we don't comply."

"Nobody is going to slaughter people, Mr. President. As well as being a crime against humanity with a mandatory death penalty in the World Justice Court, it would be stupidly counterproductive. Zantiu-Braun has a great deal of money tied up in Thallspring. We don't want to jeopardize that."

Edgar Strauss became even more angry. "We're an independent world, not some part of your corporate empire. Our funding was raised by the Navarro house."

"Who sold their interest in Thallspring to us."

"Some goddamn tax-avoidance bullshit on a planet twenty-three light-years away. That doesn't entitle you to come here and terrorize us."

"We're not terrorizing you. We're simply here to collect what rightfully belongs to us. Your middle-class daydream existence was bought with our money. You cannot run away from your fiscal responsibilities. We need a return on that money."

"And if we choose not to?"

"You do not have that choice, Mr. President. As the lawfully elected head of state, it is your obligation to provide us with assets that we can liquidate back on Earth. If you personally fail to meet that requirement, you will be removed and replaced by a successor who isn't so foolish."

"What if all of us refuse? Think you can intimidate all eighteen million of us into handing over our possessions to you thugs?"

"That isn't going to happen, and you know it."

"No, because you'll fucking kill us if we try."

"Mr. President, as the officially designated retriever of your planetary dividend, I am serving you formal notice that it is due. You will now tell me if you will comply with its collection."

"Well, now, Mr. Board Representative, as president of the independent planet of Thallspring I am telling you that we do not recognize the jurisdiction of Earth or any of its courts out here. However, I will surrender to a military invasion fleet that threatens our well-being, and allow your soldiers to loot our cities."

"Good enough." Simon smiled brightly. "I will post lists of the assets we require. My subordinates will transfer down to the planet's surface to supervise their shipment. We'll also help reinforce your police force in case of any civil disturbance. I'm sure both of us want this to go as smoothly as possible. The quicker it's done, the quicker we leave." He canceled his link to Edgar Strauss and issued the general landing order.

"We have a go authorization," Captain Bryant informed Lawrence. "Get your platoon suited up. We'll embark the drop gliders in two hours' time."

"Yes, sir. Have we got the updated ground cartography yet?"

"Tactical support is processing the surveillance satellite data at this moment. Don't worry, Sergeant, you'll have it before you fly down. Now carry on."

"Sir." He turned to face the platoon. They were all hanging on the edges of their bunks, facing him expectantly. "Okay, we're on."

Hal let out a loud whoop of satisfaction and jacked out of his bunk. The rest followed, keen for any end to the voyage, even one that pitched them into a hostile environment.

Lawrence was first into their suit armory. One of the reasons *Koribu*'s life support wheels were so cramped was the amount of space the Skin suits took up in transit. Each one was stored in a bulky glass-fronted sustainer cabinet, which fed it a regulated supply of nutrients and oxygen. He walked down to the cabinet with his own suit inside and opened the small drawer on the bottom. It was empty apart from a plastic capsule containing a pair of full-spectrum optronic membranes. He slipped them onto his eyes and began to undress.

There was plenty of joshing and derisory comments as the platoon put in their own membranes and stripped off their one-piece tunics. Lawrence didn't join in; the banter had an edge to it as the reality of Thallspring crept toward them—their way of riding over the jitters.

He stripped naked except for a slim necklace with a cheap hologram crystal pendant. His thumb stroked the scuffed surface, and a seventeen-year-old Roselyn smiled brightly at him. Technically, even the necklace was against regulations, but Lawrence hadn't taken it off in twenty years. He pumped the small dispenser button next to the sustainer cabinet's drawer. The metal nozzle squirted out globs of pale blue dermalez gel, which he began to smear over his body. It took a good five minutes to cover himself completely, slicking down his short-cropped hair, rubbing it into his armpits and crotch. He and Amersy did each other's backs and shoulders. Only then was he ready to put his suit on.

His cabinet door opened with a quiet wheeze of cool air. He put his palm on the scan panel inside for a bone and blood review. The suit AS compared them to the patterns contained within its e-alpha section and agreed he was Lawrence Newton, the designated wearer. He waited for the disengage sequence to run, cycling the sustainer fluids out of the suit before disconnecting the umbilicals. Indigo script from the suit's AS scrolled down his optronic membranes, showing him its status. Bracing himself on the floor, he lifted the flaccid suit out. In the *Koribu*'s low gravity it didn't weigh much, but it had roughly the same inertia as his own body.

From the outside, it looked no different from any of the other Skin suits his platoon was struggling to remove from their cabinets. The flexible carapace was a dark gray color, without any visible seams or ridges. Its fingers had hardened, slightly pointed tips; while the feet were boots with toughened soles. To touch, it had a texture similar to human skin, although the outer layer was the one part that wasn't biological. A smart polycarbon with an external sheet of chameleon molecules, and woven with thermal fibers capable of redirecting its infrared signature. Even if a hostile did manage to locate it, the carapace was tough enough to protect him from all handheld projectile weapons, and a fair percentage of small artillery pieces.

Lawrence gave it the order to egress him, and it split open smoothly across the chest from crotch to neck. Inside the carapace was a stratum of synthetic muscle up to five centimeters thick. He pushed his foot into the right leg, feeling the gel ooze against his skin as the limb slithered deeper into the suit. Like squeezing into whale blubber, he always thought. The left leg followed; then the arms were inserted into their sleeves. He tilted his head back and reached round for the helmet, which was hanging loosely. Moving his arms through even a small arc was hard, as if he were trying to shove a gym bar that was on maximum resistance. Slowly, though, the helmet section came up, and he pushed his head up inside. The grille was open and inactive, allowing him to suck down some air. As always, he felt a quick chill of claustrophobia: it was difficult to move, he could see nothing and hear nothing through the helmet.

Indigo script blinked up as the AS reported it was ready for full integration. Lawrence gave it permission. The carapace sealed up. A ripple moved along the suit as the synthetic muscle adjusted itself to grip him correctly. The optronic membranes flashed elaborate visual test patterns at him, then began feeding him the picture from sensors mounted around the helmet. He swiveled his eyeballs from side to side, a motion picked up by the suit, which altered the sensor angle accordingly. Audio plugs wriggled into his ears, and he heard the grumbles and complaints of the platoon as they clambered into their own suits.

"Phase two," he told the suit AS.

With his legs held tight by the synthetic muscle, small nozzles extended into the valves on the top of his thighs, which had been surgically spliced into his femoral arteries and veins. A second set of nozzles coupled with the subclavian valves on his wrists; the last set were on his neck, plugging his carotid artery and jugular vein into the suit's own circulatory system. With the connections physically secure, the suit AS interfaced with the integral e-alpha guards governing the valves, enabling them. They opened, and his blood began to circulate round the suit muscle, blending with the artificial blood that the suit had been hibernating on in its support cabinet.

A checklist scrolled down, confirming the suit muscles' integrity. Internal blood bladders held a large reserve of the oxygenated, nutrient-rich fluid capable of being fed into the circulation system when any bursts of strength were required. Other than that his own organs would have to support the suit muscle entirely by themselves.

"Phase three."

The suit AS began to bring a multitude of peripheral electronic systems online: he'd enhanced the original program with his Prime, which he felt gave him a better response and interface. Nobody else knew about the addition. He still wasn't sure about Prime's legal status, and the Z-B armory technicians disapproved of such customizations.

Phase three started by providing him with several sensor options, all of which he could link to targeting grids. Communication links ran through their interfaces and encryption codes. Air filters slipped across the helmet grille, giving him immunity from chemical and bioviral attack. Integrated weapons systems ran through test sequences. He selected neutral carapace coloring, shifting it from the original dark gray to a bluer shading that the human eye had difficulty distinguishing. That was coupled with full thermal radiance, allowing him to jettison the heat generated by his body and the Skin suit muscle through the thermal fiber weave. His penis sheath confirmed it was secure and capable of allowing him to take a leak any time he needed.

Lawrence stood up and examined the range of articulation his new Skin gave him, moving his limbs in every direction, bending his body, flexing fingers. Sensors on the inside of the synthetic muscle picked up the initial movement, and in conjunction with the AS shifted the suit in a corresponding motion. As he worked methodically through the various positions and actions the yammer of claustrophobia vanished as it always did. Worming up from his subconscious to replace it was a mildly narcotic sense of invulnerability. Even on Santa Chico his Skin had never let him down. Anything that made him less reliant on Captain Bryant was a good thing indeed.

Lawrence looked around the compartment. Most of the platoon were already in their Skin and running preparation checks. He saw Hal, who only had his helmet left to fit. The kid was sitting on the bench, frozen with worry. Lawrence went over and stood directly in front of him. He flashed the kid a quick thumbs-up, unseen by the rest of the platoon. "You need a hand?" his amplified voice bounced round the aluminum walls.

"No, Sarge," Hal said gratefully. "I can cope, thanks." His suited hands scrabbled round slowly and awkwardly behind his head, finding the helmet. Then he was pushing himself into the dark covering.

The platoon trooped out of the suit armory and lumbered down the corridor to the munitions store. Each Skin's AS linked directly with the quartermaster AS to issue the weapons authorization. When he received his allocation, Lawrence's Skin split along the top of his arms, revealing various mechanical components that were melded with muscle bands to form hybrid guns and microsilos. He slotted his magazines into their receiver casings and watched as the thin muscle bands undulated, moving missiles and darts into their sacs and chambers. The punch pistol he'd been given was clipped to his belt, ironically the largest weapon and the least lethal.

For some unfathomable bureaucratic reason, the Cairns base AS had decided that the munitions store should also distribute Skin bloodpaks. Lawrence collected his four and secreted them in the abdominal pouches. They'd give him another few hours' endurance should they hit physically demanding conditions. Nice to have. Although, frankly, if the Memu Bay ground forces hadn't established their headquarters and barracks at the end of the first day, it wouldn't matter anyway.

Now that the squad was active, they took a lift up to the life support wheel's axis, then transferred down the wide axial corridor to the cargo section. The radial corridor that led out to their drop glider was even narrower, making life difficult for the bulky Skin suits. Not that the interior of their little landing craft—a short cylinder filled with two rows of crude plastic chairs—was much of an improvement. They strapped themselves in amid curses about lack of space and bumped elbows. Lawrence took the single chair at the front. It put his head level with a narrow windshield. A small console with two holographic panes was provided in case anything glitched the AS pilot and he needed manual control. For a vehicle intended to deorbit and deliver them to a specific ground coordinate with only a fifty-meter margin of error, the whole arrangement seemed totally inadequate.

Amersy closed the hatch and strapped himself in. Short trembles running through the fuselage indicated the other drop gliders were leaving their silos. Eight minutes to go.

"Hey, Sarge," Jones called out in their general channel. "I think Karl's testing out his vomit tube. Aren't you, Karl?"

"Fuck the hell off."

"Knock it off back there," Lawrence said.

His optronic membranes alerted him to a call from Captain Bryant, which he admitted.

"Tactical have completed the cartography of Memu Bay," Bryant said. "It's accessible now. Get your platoon to install it."

"Yes, sir. Any major changes?"

"None at all. Don't worry, Sergeant, we're on top of this one. I'll see you down there. Meteorology says it's a beautiful day; we might even have a barbecue on the beach this evening."

"Look forward to it, sir." He canceled the link. *Asshole.* The suit's AS gave him the platoon's general channel. "Okay, we've got the current map. Get it installed and integrated with your inertial navigation. I don't want anyone getting lost."

"Has it got any decent bars marked on it?" Nic asked.

"Hey, Sarge, can we have access to the Durrell guys?" Lewis asked. "Like to know how it's going."

"Sure. Odel, set it up."

"Absolutely, Sergeant."

Five minutes until their flight.

Lawrence began installing the new cartography into his Skin's neurotronic pearls. Out of curiosity, he accessed the traffic Odel was pulling out of the Durrell force's datapool. His membranes displayed a small five-by-five grid, with thumbnail videos from different drop gliders. He expanded one, seeing a shaky picture from the nose camera. A splinter of dark land rocked from side to side in an ultramarine void. Terse voices barked short comments and orders.

"No groundfire," Amersy observed. "That's good."

"Have you ever seen any?" Hal asked.

"Not yet. But there's always a first time."

Three minutes.

Lawrence dismissed the video grid and requested the new map of Memu Bay. It looked very similar to the settlement he remembered from the last time he was here: big features like the stadium and harbor were still there. Smaller, somehow. He superimposed the old map and let out a shallow breath of aggravation as he took in the new sprawl of outlying districts. Memu Bay had grown beyond Z-B's projections. A larger population would be harder to keep in line. *Oh, great.* No battle plan ever survived engagement with the enemy, but it would be nice to have one that was vaguely relevant when they hit the beach.

He opened a link to Captain Bryant. "Sir, the settlement's a lot bigger than we thought."

"Not really, Sergeant. A few percent at most. And physically there's been no change to the center since last time. Our deployment strategy remains effective."

"Are we getting any additional platoons?"

"From where? It's Durrell that's really grown over the last decade. If anything we should be supporting our forces there."

"Are we?" he asked in alarm. He'd never dreamed that the platoon might be switched. That would screw up everything.

"No, Sergeant," Bryant said wearily. "Please monitor your status display. And stop worrying. A bigger population just means more behavior collateral. We're carrying enough units down with us for that."

"Sir."

One minute.

The intermittent vibrations he could feel through the fuselage suddenly grew more pronounced. When he did check his status display, he saw the captain's drop glider had left the silo beneath them. Icons flashed an alert. Then Platoon 435NK9's drop glider was shaking as it slid down the silo's rails.

"Hang on to your hats, ladies," Edmond sang out. "We're going bungee jumping with angels, and someone just cut the cord."

Light burst in through the windshield. Lawrence saw the edge of the silo falling away from them, a dark hexagon framed in lusterless silver-white metal that shrank into the middle of a honeycomb of identical silos. Their retreat brought the rest of the starship into view. Once again, he could only smile at its functional beauty. Drop gliders and pods were being spat out of the silos at a furious rate. They retreated from the *Koribu* in an expanding cloud, dropping ass-first toward the planet below. Pods were just squat, rounded cones, with a collar of small rocket motors secured around their peaks. Drop gliders were also cones, but flattened into a standard lifting body shape and fitted with swept-back fins. They'd been coated in a thick pale gray foam of thermal ablative to get them through atmospheric entry. A rocket motor pack had been attached to their rear. Those he could see falling beside them were puffing out streamers of grubby yellow gas from the reaction control nozzles, turning as they fell.

The AS began to fire their own reaction control thrusters, orientating them so the rocket pack was aligned along their orbital track. Thallspring slipped into view through the windshield, a dusky ocean smeared with hoary clouds, its outer atmosphere a phantom silver corona caressing the water. Memu Bay was hiding over the horizon, a third of the planet away.

Orange sparks bloomed around the drop glider as the

squadron began to retro-burn, hundreds of solid rocket motor plumes flaring wide in the vacuum, blowing out a cascade of glimmering particles as though some iridescent fluid was part of their chemical formula.

Flight profile displays began a countdown for their own drop glider. The solid rocket at the center of the pack ignited, giving them a four-gee kick. It was little more than a mild discomfort for the platoon, encased in their protective Skin. Thirty seconds later it ended as abruptly as it began. Small thrusters fired again, turning them through 180 degrees. Now the nose was pointing along the line of flight. With their speed below orbital velocity, they began the long curve down into the atmosphere.

The rocket pack stayed attached for another fifteen minutes, maintaining their attitude with steady nudges from the reaction control thrusters. Up ahead of them, a multitude of sparks began to burn once more as the pods and gliders hit the uppermost fringes of gas. They were longer this time, a darker cherry red, and they continued to elongate as the ablative foam vaporized under the vigorous impact of gaseous friction. Soon space around them was drenched with inferno contrails, arching down toward the planet like the chariots of vengeful gods.

Lawrence felt the fuselage start to tremble as they sank deeper into the chemosphere. His communication links to the starship and relay satellites diminished, then dropped out altogether as ionization built up around the fuselage. The AS began to move the fin flaps, testing the vehicle's maneuverability. Once the air surfaces were providing a predetermined level of control, it fired the explosive bolts securing the rocket pack. The jolt flung Lawrence and the others forward into their straps, a motion cushioned by their Skin. There was nothing for him to see now; crimson flames from the slowly disintegrating ablative were playing across the windshield, lighting up the cabin.

They were flying blind at Mach 18 inside the crown of a three-kilometer-long fireball; gravity began to take hold, pulling them eagerly toward the ground. All he could do was wait and sweat and pray as the AS flicked the lean air surfaces with a dolphin's precision, maintaining stability within the hypersonic glidepath. This was the moment he hated and feared above all else. It forced him to invest trust in the cheapest craft Z-B could build to accomplish the job, with nothing he could do other than ride it out.

He reviewed the platoon, calling up a grid of video and telemetry windows. As expected, Amersy's heart rate was over a hundred while he quietly murmured his way through a gospel chant. Hal was asking a host of questions, which Edmond and Dennis took in turns to answer, argue about, or just tell him to shut up. Karl and Nic were talking quietly together. Jones had brought up maintenance profiles for the jeeps that the lander pods were bringing down for them. Whereas Odel . . . Lawrence enlarged the man's grid, scanning his suit function telemetry. Odel's head was rocking from side to side, while his hands palm-drummed rhythmically on his knees. He'd accessed a personal file block in his Skin's memory. As they were streaking through a planetary atmosphere with the savage brilliance of a dying comet, Odel was happily bopping away to a Slippy Martin track.

At Mach 8 the external flames began to die away. Clean blue daylight embraced the drop glider. Lawrence could see the residue of ablative covering their blunt nose, black bubbling tar that sprinkled droplets from the peak of seething ripples. The craft's antenna found the relay satellite's beacon and established a link.

Mission tactical data scrolled across his membranes. The other drop gliders bringing down the Memu Bay force had made it through aerobrake. One of them, Oakley's platoon, was going to undershoot, coming down fifty kilometers from shore. Their AS was already modifying the descent profile so they'd land at one of the larger archipelago islands. A helicopter could recover them later.

Captain Bryant had already begun shifting deployment patterns to cover the loss. Platoon 435NK9 was given an extra two streets to sweep.

"Always a pleasure," Amersy grunted as the fresh data installed into their mission orders.

"We'll assess on the ground," Lawrence told him. They both knew the extra streets would be left alone—privilege of having field autonomy, it gave him some leeway. Lawrence's priority was getting the platoon through the town without incident.

According to the tactical data, the landing pods were descending nominally. They'd taken a different profile from the gliders, using a longer, higher aerobrake path, then dropping steeply. They were scheduled to hard-land on the ground behind Memu Bay. Watching their tracking data, Lawrence could see they were already spreading too wide, and that was before chute deployment left them vulnerable to wind. From experience, he knew nearly half of them would scatter outside the designated area. Rounding them up would take a long time.

The coastline was visible ahead, growing rapidly. Just how fast they were losing altitude had become apparent with the way the horizon's curvature was flattening out. When he moved forward in his seat, he could see the archipelago spread out below him. It was as if the dark ocean had been stained with droplets of cream. Hundreds of isles and atolls had been created by the crests of coral mountains that had risen up from the ocean floor over a kilometer down, emerging on the surface to accumulate cloaks of white sand. Waves broke against the reefs in gentle sprays of surf. The larger spreads of coral were hosting tufts of vegetation. Dark meandering mounds were visible in the water between the atolls where the submerged reefs lurked. It reminded him of

Queensland's coastline, where Z-B's ecological restoration teams had worked their quiet miracles on the ailing Great Barrier Reef. Only the blue tinge of the vegetation was evidence that they were on an alien world.

Closer to the mainland the islands were larger, homes to thick forests. Then the plant leaves were a verdant green, and the beaches protected by long curving wave walls of broken coral. They all had wooden jetties extending out into the ocean. Huts were visible beneath the palm trees; sailing boats and canoes drawn up on the sand.

"Too good to be true," Dennis said. "Maybe we should just stay here when the starships leave."

"Nice idea," Nic said. "But the residents would slice you up into fishbait if they found you."

The drop glider shook enthusiastically for a few seconds as their speed fell below Mach 1. The nose dropped, and the familiar sight of Memu Bay was directly ahead, huddled in the folds of unnervingly tall mountains. The speed of their approach made Lawrence's natural skin crawl. Drop gliders had the aerodynamic characteristics of a brick; the only thing that kept them stable was their forward momentum. And they were shedding that rapidly.

The harbor drifted off to starboard, leaving them pointing at a shallow bay of gingery sand. A marble-walled promenade ran its entire length, separating beach from buildings. What looked like a line of police cars was parked along the top, with blue strobes flashing enthusiastically. Their AS tipped the nose up again, shedding more speed. They lost altitude at a dramatic rate once they leveled out. The beach was less than a kilometer away now, and the waves were only a couple of hundred meters below.

"Stand by," Lawrence called. "Brace yourselves."

Myles Hazeldine stood on the balcony that ran around the fourth floor of City Hall, watching the sky over the ocean.

His two senior aides hovered behind him. Don and Jennifer had been with him since he was a first-time councilman, twenty years ago now, one of the youngest ever to be elected in Memu Bay. They'd stayed loyal ever since, throughout all the wearying backbiting and dirt slinging of democratic politics; even the dubious deals with the business community that helped his campaign funding hadn't put them off. All of them had lost their naive idealism—probably back in that first term when he used to make hothead speeches condemning the then mayor. Now, they made a practical levelheaded team who ran the city with a decent level of efficiency, well equipped to deal with the new generation of young hotheads in the council who constantly criticized him. Goddamn, he was proud of the way he'd overseen Memu Bay's development in recent years. This was a prosperous settlement, high economic index, lowish crime rates . . . *Shit!* Social problems, unions, bureaucrats, finance, scandals—he could handle any of that. But this kind of crisis was beyond anybody's ability to survive.

If he took a heroic stand and resisted Zantiu-Braun, he'd aggravate the situation and the invasion force governor would sling him out anyway. He'd achieve nothing. While if he cooperated and worked alongside the governor to ensure the bastards stole everything they wanted, he'd be a collaborator, a traitor to his electorate. They'd never forgive that.

A swarm of black dots materialized high in the clean azure sky, moving with incredible speed as they sank toward the beach on the east of town. Myles hung his head in shameful fury. Edgar Strauss himself had called yesterday, urging him to cooperate. "None of us want a bloodbath, Myles. Don't let it happen, please. Don't let them take our dignity as well." Another good politician lost to events out of their control. Myles had almost asked: *In God's name why didn't you fund exo-orbit defenses? Why have you left us helpless against this?* But that would have been too much like kicking a man when he was down. The best missiles Thallspring could have come up with would have been a pathetic token gesture. God alone knew how advanced Earth's weapons technology was these days. And the Z-B starships would have retaliated, made an example. Myles shuddered as he remembered the last invasion: his son dead, the meager ration of food for months afterward as they struggled to get back on their feet. And everyone had accessed the pictures of the new blasted land on the edge of Durrell, that highly unsubtle and very effective demonstration of their capability.

He knew what he would have to do, the public example he would have to set. It would ruin him. He might even have to leave Memu Bay after Z-B withdrew. But then he'd known that when he ordered the police to seal off the beach and clamp down on any physical bravado as the drop gliders arrived. Cooperation would mean keeping a lid on any stupid acts of defiance by the population. Lives would be saved. Although he'd never be thanked. Maybe he did owe Memu Bay's population for all those crabby back-room deals he'd put together down the years. It was a view that helped ease some of the numb depression.

A barrage of sonic booms made him jump. They were so like explosions. Glass rattled in just about every window. He could see flocks of birds taking to the air above the city, wings flapping in wild shock.

Out in the bay, the first of the drop gliders were splash-landing. Dumpy cones streaking down through the air at nearly forty-five degrees to smack into the lazy waves a couple of hundred meters offshore. Huge plumes of spray shot out from the impact point, then followed them as they skidded along the top of the water, gradually dying away as they slowed. Several of the craft careered into the sand with a drawn-out crunching sound, twisting around sharply. One almost made it to the promenade wall, its nose finishing only a couple of meters short.

"Pity," Don grunted.

The majority of drop gliders finished up bobbing in the shallows. Their hatches blew off. Burly dark figures jumped out and began wading ashore, kicking effortlessly through the water. Myles recognized that color, size and strength all too well.

A big banner suddenly unrolled down the promenade wall.

Die Screaming
Nazi Fuckheads

Kids raced away from it. The police officers leaning over the rail to watch the drop gliders made no effort to catch them.

"Oh, very original," Myles muttered under his breath. He could only hope that would be the worst the local hooligans would do.

He turned to Don and Jennifer. "Let's go."

The invaders were already running up the promenade steps and spreading out along the top. They seemed to be ignoring the police.

Myles took the elevator down to the mayor's private apartment at the back of City Hall. He didn't really like the place, the ceilings were too high and the rooms too big. It was no place for a family to live. But his own house was away on the other side of town, forty minutes away, so during the week they had to stay here.

His office had wide patio doors that opened onto a small central garden. He saw Francine out there, lying on one of the benches under the shade of a Japanese pine. She was wearing a simple black dress with white piping. The skirt was shorter than he approved of, well above the knee. But he hadn't won that kind of argument with her since she was thirteen. Cindy would have known how to cope with her, he thought. *Damn, I should have married again. Never finding the time is such a pathetic excuse.*

Francine adjusted her sunglasses. Myles could see a frown on her brow and realized she must be accessing the news channels. He wanted to go out to her and put his arm around her, and offer her some comfort, and promise that it would be over soon, and that she wouldn't be harmed. The sort of thing real fathers would be doing all over Thallspring right now.

But the senior staff and the party leadership were waiting for him, and they had family, too. He sat behind the desk with one last reluctant look at the patio door.

"I'd just like to say that if anyone wants to resign effective immediately, then I will accept it. It won't affect your pensions or benefits." There was a moment of awkward silence, but no one came forward. "Okay, then. Thank you for your support. I do appreciate it. As you know, I've decided to follow Strauss's lead with a policy of cooperation. They're a hell of a lot more powerful than us, and God knows, more evil. Trying to sabotage the chemical plants or throw rocks at their soldiers is just going to lead to retaliation on a scale I cannot accept. So we just grin and bear it, and hope their starships all hit a black hole on the voyage back. If we do that, I think we can come through this relatively unscathed, at least as far as infrastructure is concerned. Margret?"

Margret Reece, the chief of police, gave a reluctant nod. She was looking at the reports scrolling down her membranes rather than at anything in the room. "I studied the files from last time. They really are only interested in pillaging our industrial output. That's where their enforcement comes in. We can do what the hell we like in the rest of town, riot and burn it to the ground—they simply won't care. As long as the factories remain intact, they're supplied with raw material, and the staff turn up for each shift, they'll leave us alone."

"Then that's what we ensure happens," Myles said. "The rest of our civic business carries on as normal. To keep the factories operational, we keep the town functioning. That's the service we provide, no matter what."

"Do they steal our food as well?" Jennifer asked. "I remember there wasn't much to go around last time."

"They'll only take what they need to eat themselves," Margret said. "Given that thirty percent of the tourists managed to make it out before flights were grounded this morning, the food refineries we've got will give us a large overcapacity for the remaining population. The reason food was short last time is some rebel moron went and firebombed two of the production lines."

"Which we can't allow to happen again," Myles said swiftly. "I'm not having some heroic resistance movement putting innocent lives in danger."

"I doubt we'll get an organized resistance," Margret said. "Z-B always makes sure the punishment for any action against them outweighs the propaganda gains. But we're keeping a close eye on the people we know can make trouble."

"What about the tourists?" Don asked. "There's a lot of them didn't make it home; the airport looks like a refugee camp."

"Not my decision," Myles said. He had to squash his anger so he could speak in a clear voice. "The governor will say how much civil transport will be allowed. Given why they're here, I expect they'll want everyone at home being as productive as possible."

"One of their platoons has reached the main square," Margret announced loudly. "They'll be here any minute."

So quickly? Myles took a breath. So much would depend on what kind of working relationship he could establish with the governor. "Okay, let's go greet the bastards with a smile."

Denise milled with the crowd on the edges of the Livingstone District. Human curiosity had won out over trepidation, allowing hundreds of people to come watch the spectacle firsthand. Few children had been allowed out, though. This was mainly adults and older teenagers, staring grimly at the streets that led down to the waterfront where the police had established a no-go zone. Conversation was dark mutters of resentment, folklore of what Skin suits were capable of and the atrocities committed last time.

Bars were still open and well frequented. Most of the men were clutching cans of beer, drinking steadily as they watched on their glasses and membranes the drop gliders bursting out of the sky. The attitude reminded Denise of prematch anxiety, home-team fans barely tolerating the provocative antics of their rivals. Animal territorialism was still a strong component of the human psyche. That was going to work to her advantage. This was a very volatile situation, and most of the police were covering the waterfront and promenade. The mayor had been worried about his good citizens rampaging down onto the sand as the drop gliders beached. Idiot. An open beach was no place for urban conflict, not against well-organized troops.

Her sunglasses were showing datapool video relays of the gliders arriving. The discordant voice of the crowd rose around her. She dispatched a series of coded messages to cell members scattered along the street. Acknowledgments came back. Everyone was ready.

The first Z-B troopers appeared at the end of the street. Five of them, striding along confidently. There wasn't even a pause when they saw the crowd.

Denise raised her sunglasses and stared at the first one. Her irises focus-shifted for detailed close-up. The Skin was very similar to what she remembered, as if a bodybuilder were wearing a dark gray leotard. They all had very fat fingers and strange bulges along the arm. Their helmet design had altered; the Skin's pliability ended around the jaw, turning into a protective shell covering the upper face and skull.

There was a tiara band of sensors at eye level, and two gill-vents on the cheeks. The only visible weapon was a cumbersome pistol clipped to a belt along with some pouches (must be for effect, she thought). Heat profile was surprisingly uniform, with only a couple of degrees' difference across the whole suit surface.

Her view pulled back. There were nine Skins walking up the street. A chorus of obscene taunting chants rose from the crowd who were moving back and forth restlessly along the pavement. Nobody ventured closer than four or five meters. Then a young man walked out into the middle of the road directly ahead of the Skins. He was carrying a can of beer, which he drained in a couple of big gulps. The Skins ignored him as they got closer. So he turned his back to them, bent down and dropped his shorts.

"Kiss my ass!"

The crowd laughed and jeered. Several cans clattered onto the road around the Skins, spinning around as foaming beer sprayed out of the open tabs. Still the Skins kept going, silent and seemingly unstoppable. Denise had to admit, their discipline was good. Her ring pearl was picking up short databursts from individual suits. Her Prime started to break down the heavy encryption.

A rock sailed over the heads of the crowd to smack against a Skin's chest. Denise's enhanced vision captured the sequence as the outer layer hardened around the impact point. The Skin's stride halted momentarily as the rock bounced off him. Still none of them retaliated. Emboldened by their apparently passive attitude a couple of tough lads ran out and tried to rugby tackle the invaders.

One Skin stopped as the first lad charged toward him, turning so they were facing. The lad was yelling at the top of his voice as he spread his arms wide ready for the collision. A second before they hit, the Skin darted swiftly to one side, bending slightly, one arm coming round. It was a perfectly timed throw. The Skin's arm caught the lad in his chest and lifted with tectonic strength. He left the ground, momentum flipping him until he was upside down above the Skin. Then the powered push ended. His boozy battle-cry had turned to pure terror as he found himself inverted, three meters in the air, and hurtling toward a shop wall. His arms and legs flailed wildly as the now-silent crowd watched. There was a wet thud and the sudden loud *crack* of snapping bone as he hit the bottom of the wall. His cry cut off dead.

The other Skin simply extended his arm, fingers flat and pointing at his assailant. He never moved as the second lad cannoned into him, the extended fingers striking the middle of his chest. There was a bright flash of electricity, and the lad jerked backward, limbs thrashing madly from the discharge. He crumpled onto the pavement, twitching.

The crowd growled its resentment. They began to close in on the Skins. A swarm of beer cans and stones started to fly.

Lawrence had known it was a bad situation as soon as they got off the promenade and he saw the crowd lining the street ahead. He would have preferred the police to let the town's population through on the beach. The street pushed everyone together. It could cause serious casualties.

"Keep calm," he told the platoon, mainly for Hal's benefit. "They have to find out what we're capable of sometime. Might as well be now. A quickshock demonstration will make them think twice in the future."

The shouts and insults were nothing. Beer sprayed around their feet, and they splashed through. A very well aimed rock caught Odel on his chest.

"Ignore it," Lawrence ordered.

"Shouldn't we tell them to keep back?" Hal asked. There was a hint of unease in his voice. "They're just getting worse."

"This is nothing," Edmond said. "One Skin could take these pimps out. Stop sweating it, kid."

Lawrence expanded Hal's telemetry out of the grid, checking the kid's heart rate. Which was high, but acceptable.

"To these people we must appear invincible," Amersy said. "Half of that trick is making them believe it. So just swagger along nice and easy. Come on, remember your training."

Two fury-driven young men charged out of the crowd, heading straight for the platoon.

"No weapons!" Lawrence commanded. "Lewis, shock yours." The other was heading straight for Hal. Lawrence said nothing, wanting to see how the kid would handle it. As it turned out, the throw was perfect, sending the youth crashing against the bottom of a wall.

"Way to go, kid!" Nic whooped.

"Nice one," Jones said admiringly. "You could have turned faster, though."

"You couldn't," Hal said cheerfully. "Too old. Your reflexes are shot."

"Shit on you."

"Pull in formation," Lawrence said. He didn't like the mood of the crowd. "Hal, well done. Everyone, let's not get excited here."

The crowd was moving in, winding themselves up for a head-on clash. Cans and stones were coming at them from all directions.

"You going to dart them?" Dennis asked.

"Not yet." Lawrence switched on his external speaker and cranked the volume up. "Stand back!" He could see the people closest to him wince, putting their hands over their ears. "You are causing a civil disturbance, and I have the authority to disperse you with appropriate force. Now calm down and go home. The governor and mayor will address you shortly."

His amplified voice was lost under a howl of obscenities. Looking out at the raw hatred facing him he imagined what it would be like standing here without Skin. The lapse made him shiver. "All right, grab your punch pistols, I want . . ." His suit's AS flashed a warning at the center of the tactical display grid. Sensors had picked up a thermal point approaching fast.

The Molotov arched through the sky, trailing a streamer of bright blue flame from the hihydrogen fuel. It was spinning as it went, curving down toward Karl.

"Let it hit," Lawrence ordered.

Karl's arm was already extended, the nine-millimeter muzzle poking through the carapace. Targeting lasers had found the Molotov. "Oh, man," Karl grunted. "I hate this, Sarge."

The Molotov crashed down on his helmet. The glass burst, flinging out a sheet of dense flame that enveloped the whole suit. People nearby yelped, scrambling back out of the way as the flames grew hotter, gorging on the fuel. The rest of the platoon calmly took their punch pistols up and flicked the safeties off.

"Give them the talk, Karl," Lawrence said.

The flames died away, revealing the Skin suit standing unharmed. "The person who threw that is under arrest," Karl said through his speakers. "Step forward, please. Now." He took his own punch pistol from his belt. "I said, now."

The crowd began shouting and chanting again. More stones were flung. Then another three Molotovs appeared in the air. Again, they were all aimed at Karl.

Someone's organized, Lawrence realized suddenly. The Molotovs were aimed at the same place, and came from different directions at the same time. "Take them out," he ordered.

Karl and Amersy shot the bottles in midflight. Giant fireballs ruptured the air and poured down. Flame splashed over a dozen people, who ran screeching in agony. The crowd went berserk, and charged forward en masse.

"Disperse!" Lawrence yelled at them above the bedlam. He aimed his punch pistol and fired. The plastic bullet caught a man in the middle of his chest, slamming him back into the three behind him. They tumbled like human bowling pins. Rushing feet trampled them.

The platoon had formed up in a circle. The punch pistols began firing. Psychologically, they should have acted as a much greater deterrent than darts. A mean-looking weapon, a loud gunshot, and a man goes flying. It was obvious and physical, you could see it happening. You should run away lest it happen to you.

Lawrence's AS alerted him to the sound of gunshots, simultaneously running an analysis program. Someone in the crowd was firing a pump-action shotgun. He saw Dennis stagger backward, his Skin carapace totally solid.

"Where the hell did that come from?"

Three Skin AS programs coordinated their audio triangulation and indicated the line of fire. Lawrence's visual sensors showed him a man running through the crowd—something (long, dark) in his hand. He gave the image to Lewis and Nic. "Snatch. I want him."

They charged forward into the mob, ruthlessly thrusting people aside.

Someone jumped on Odel's back, an arm around his neck, trying to strangle him. He reached around and picked off the attacker effortlessly. Two men lunged at Lawrence. He hit one, going for the arm. Kicked at the other, hearing the leg splinter. Each time, the Skin's AS moderated the strength of the blow. A full strike from a Skin fist could smash clean through a human rib cage. Unless you wanted to kill somebody, always go for the limbs.

They were too close now for the punch pistol. He dodged one madman who was swinging a chair at his head. Another broke a bottle across his shoulder; ragged glass spikes slithered uselessly over the Skin carapace.

Jones screamed. Lawrence saw his grid turn red. Graphics swirled madly as the AS tried to make sense of the data. Visual sensors locked on. Jones was falling, arms waving slowly. He hit the pavement, and his fists cracked the stone slabs.

"Jones!" Lawrence yelled. "Status?"

"Okay," Jones gurgled. "Electric. Electric shock. I'm okay. Motherfuck. They zapped me with a charge. Goddamn, it was a brute."

"Amersy," Lawrence ordered. "Dart them."

Amersy held his arm up high. Nozzles slid out through the carapace around his wrist. Fifty darts puffed out.

It was as if God had reached down and switched people off. The front ranks of the mob crumpled with startled expressions that swiftly faded to the neutral face of the deep sleeper. Within seconds, a fifteen-meter logjam of inert bodies surrounded Lawrence and the platoon. Beyond that, the remainder of the crowd stared down at their comatose compatriots in numb horror.

Amersy fired another salvo.

Screams broke out as more people fell. The remainder began running, vanishing down side streets at an incredible rate.

"One for the good guys," Edmond said.

"They're crazy," Hal whined. "Totally fucking crazy. Is it going to be like this the whole time?"

"One sincerely hopes not," Odel said.

"Jones?" Lawrence walked over to the trooper, who was now sitting up. "You okay?"

"Shit. I guess so. The insulation blocked most of it. Bloody thing scrambled half of my electronics. Systems are coming back online. E-alpha fortress is rebooting the full AS."

Lawrence didn't like the sound of that at all. The suit should have shielded him from just about any kind of current, and the electronics were EMP-hardened. He looked round the deserted street. A lot of the unconscious bodies were bleeding, and he could see several who'd been caught by the Molotovs. The burns looked bad.

Rocks. Molotovs. Shotguns. Electric shock.

We were being tested, he thought. *Someone wanted to know our Skin capability.*

"Dennis, check Jones over, please."

"Yes, Sarge."

"Did anyone see who hit Jones with the shock?"

"I was busy," Karl said. "Sorry."

"That's okay, we can run the sensor memories."

"Newton?" Captain Bryant said. "What the hell's happened?"

"Crowd got out of control, sir. I don't think . . ." The display grid with Nic Fuccio's video and telemetry flickered and turned black. A medical alarm began to shrill in Lawrence's ears.

"Sarge!" Lewis cried. "Sarge, they shot him. Oh Jesus. Oh fuck. They shot him."

"Dennis!" Lawrence yelled. "With me." He was sprinting, moving at incredible speed over the sprawled bodies, then powering down a narrow side street. Bright indigo navigation displays scrolled down, guiding his feet. Left turn. Right turn. Curve. Right turn. Clump of people across the narrow road, standing staring. He slammed them aside, ignoring the pained protests.

A Skin was lying spread-eagle on the cobbled road. Dark red blood was spreading out from it in a thick glistening puddle. A fist-sized hole had ripped into the carapace between Nic's shoulders. It was bad, but his Skin could have sustained him. The suit's circulatory system was still plugged into the jugular and carotid splices; in such extreme damage situations the AS would keep the brain supplied with blood until the field medics arrived. Whoever the sniper was, he must have known that. The second shot had been fired when Nic was down. It had taken off the top half of his head, leaving nothing from the nose upward.

Lewis was kneeling on the road beside him. Emergency disposal valves had opened on his lower helmet, allowing a stream of vomit to splash down his chest.

"He's dead," Lewis wailed. "Dead. Never had a chance."

Lawrence glanced around. The civilians were backing off fast. Heads vanished into windows, which were slammed shut.

"Where did it come from?" Lawrence asked.

"Oh God. Oh God." Lewis was rocking back and forth.

"Lewis! Where did the shots come from?"

"I don't fucking know!"

Lawrence looked up and down the nearly empty street, reviewing the last of Nic's telemetry. He was running eastward, so judging from the impact he had been shot from behind. There was no obvious window or balcony for the shooter. When Lawrence raised his view, he saw a church tower standing above the roofs. The whole street was exposed to it. But it must have been over a kilometer away.

Myles Hazeldine's single quiet hope that the governor would be a shrewd political operator open to compromise vanished into the air before they even met. He stood outside the main doors of City Hall, watching the Skin-suited invaders march across the main square. The few locals who stubbornly stood their ground were shoved violently out of the way. Z-B's goons never bothered to modify their suits' strength, so the victims really did tumble backward to land awkwardly on the hard slabs.

The three leading the column trotted up the broad stone steps to the doors. At the last minute Myles realized they weren't going to stop. He hurriedly stepped aside as they barged in, nearly breaking the heavy glass-and-wood doors.

It wasn't their strength that made Myles's heart sink, but the deliberate arrogance. "Hey!" he began.

"You are the mayor?"

It was an unnecessarily loud voice booming from one of the Skins that had stopped in front of Myles and his people.

"I am the democratically elected leader of Memu Bay Council, yes."

"Come with us."

"Very well. I'd like to—"

"Now."

Myles shrugged to his aides and went back into City Hall. The Z-B goons were spreading out through the large entrance hall. Their tough heels made a clattering noise like hooves on the marble tile flooring. Nervous staff peering through open doorways moved aside briskly as the big, impassive suits started to check out all the offices. Several of them were jogging up the twin looped stairs to the first floor.

The main group made their way directly to the mayor's apartment. Myles had to take fast steps to keep up with them. Nobody asked him directions. The layout would be in their suit memories, of course.

I should have changed the architecture around inside, he thought. *That would have pissed them off and spoiled the know-it-all effect.*

The doors to his inner study were flung open. Seven of the Skins walked in. Myles saw Francine jump up from the bench out in the garden. She grabbed hold of Melanie and lifted the little girl up so she was cradling her. Melanie's face was sulky with resentment, but not fearful, Myles saw proudly. He made a brief calming gesture at his daughters.

One of the Z-B goons stood by the door and pointed at Myles's aides. "You," the voice reverberated. "Wait out here." A chubby finger beckoned Myles. "You, inside."

Myles found himself standing in front of his own desk as the doors were slammed shut behind him. One of the suited figures sat down in his own chair. Myles winced as the antique pine creaked under the immense weight.

"You should learn to control your suits more carefully," he said calmly. "There won't be a door left in Memu Bay by the time you leave."

There was silence for a moment; then the figure's suit split open down the chest. That was where the impressive routine of invincibility fell apart slightly. He had to struggle to pull his head out of the helmet, and when he did his face was covered in a sticky blue goo.

Myles grinned. "Did you sneeze in there?"

"I am Ebrey Zhang, commander of Z-B forces in Memu Bay and the surrounding settlement regions, which makes me governor of the civil population. I'm now going to give you the only piece of advice you'll get for the whole occupation: don't play the smartass with me. Understand?"

He was about what Myles had expected: somewhere in his forties with dark Asian skin and slightly narrowed eyes; black hair that was receding. His eyeballs were covered in an unusually thick optronic membrane, similar to lizard scales. It didn't make his scowl any more effective. Just a standard-issue military bureaucrat trying to appear uncompromising and totally in control.

"Straight talk, huh?" Myles asked.

"Yes. I don't like politicians. You twist words too much."

"I don't like occupying armies. You kill people."

"Good. Then we have a deal. You're the mayor, Myles Hazeldine, yes?"

"Yes."

"I want the access codes for your civil administration network."

They didn't need them, of course. With their software they could probably establish total control over the network in seconds. That wasn't the point. This was the defeated barbarian chief kneeling before Caesar, acknowledging Rome's authority and glory.

"Certainly," Myles said. He told his desktop pearl to display the codes.

Ebrey turned to one of the faceless suits. "I want us interfaced and supervising the local datapool in ninety minutes. Get me a full industrial capacity review and a police file interrogation. I want to know what they've got, and who's likely to resist."

"Sir," the suited figure replied.

"Mr. Mayor, I'm officially appointing you as my civil deputy. It's now your job to make sure that civil services in this town carry on working smoothly, so you'll be doing essentially the same thing as before but with some exceptions. We keep an eye on your work. The council is suspended for the duration—I'm not putting up with a herd of blabbermouths whining away to me night and day. Second, you can't resign. Third, in public you will give me your full and utmost cooperation as an example to everyone else. Fourth, my second in command will now assume control of your police force. Laws will remain the same, with one principal addition. Interfering with our activities is a capital crime. And we're going to start with the little shit who just went and shot one of my men."

"Shot?"

"Killed, actually. I take it you deny all knowledge."

Myles looked round the suited figures, wishing desperately that he could see their faces. "I didn't know that . . ."

"I'll accept your avowal for now. But believe me when I say we'll find whatever resistance movement you people have cobbled together and exterminate it. I will not tolerate interference with our operation, and certainly not at that level."

"Somebody shot one of you?"

"Yes. And the platoon leader seems to think it was a deliberate trap."

"But . . . wasn't your man in a Skin suit?"

"He was. That's what I really don't like about this."

"Jesus."

"Quite. Now, I take it you've heard about our good behavior collateral policy?"

The news about the death had made Myles's heart jump in panic. Z-B hadn't been in Memu Bay thirty minutes, and already their commander was being forced to consider reprisals. Now the mention of collateral made the muscles across his chest tighten up. "I've heard."

"Of course you have." Ebrey Zhang reached into one of the pouches on his belt, and produced a loop of what looked like white plastic string. "We are going to select a thousand or so honest and true citizens of Memu Bay and put these necklaces on them. Each necklace contains a small discharge mechanism filled with nerve toxin. It's quite painless—after all, we are not savages—but it will kill the recipient within five seconds, and needless to say, there is no cure or antidote. Every mechanism has a specific number, and for every act of violence committed against Zantiu-Braun one or more of those numbers will be selected at random. They will be transmitted by our satellite. The mechanism will discharge, and the wearer will die. If anyone attempts to tamper with or remove their necklace, the mechanism will discharge. The mechanism also has an inbuilt twenty-four-hour timer, which the satellites have to reset every day, again by broadcasting a code. So if anyone thinks he can escape by hiding away underground or in a shielded room, he will only be able to do so for twenty-four hours. Any questions?"

"I think you've made yourself clear."

"Very well. Let us hope that it works, and we don't have a repeat of today's murder." The plastic was rubbed absently in his thick Skin fingers.

Myles couldn't shift his gaze away from the awful thing. "Are you going to put that on me now?"

"Good heavens, no, Mr. Mayor. What would be the point in that? They are supposed to guarantee good behavior in others. If your political opponents saw you'd been fitted with one, I imagine they'd go straight outside and start hitting my people over the head with rocks. You see, I don't want to make you a martyr, Mr. Mayor, I simply want you to back up all those fine words of conciliation and submission with some positive action. Let me show you how that's achieved." He twisted around in the chair and smiled at Francine, who was still standing in the middle of the little garden.

"No!" Myles shouted. He began to lunge forward, but a heavy Skin hand clamped down hard on his shoulder. It was impossible to shift. His vision blurred with tears as the hand gripped tighter; he was sure his collarbone was about to snap.

Ebrey Zhang beckoned. Francine gave him a sullen, rebellious look, then gently put her sister down and whispered a few words in her ear. Melanie ran away across the garden, disappearing through a door on the other side. Francine straightened her back and walked into the study.

"I have a gift for you, my dear," Ebrey Zhang said. The loop of plastic came open.

"For fuck's sake," Myles shouted. "She's only fifteen."

Francine gave her father a brave little smile. "It's all right, Daddy." She knelt in front of the governor, who put the length of plastic round her neck. The two ends melded together, and it contracted until it was tight against her skin.

"I know," Ebrey Zhang said sympathetically. "You want to kill me."

Francine ran across the room and threw her arms around Myles. He clung to her, stroking her chestnut hair. "If anything happens to her, you will die," he told the governor. "And it will be neither quick nor painless."

* * *

It was one of Memu Bay's attractive wide boulevards in the center of town, the pavements lined with tall sturdy trees whose canopy of leaves created a pleasant dappled shade for pedestrians. Karl Sheahan walked along the center of the tram lines, praying that some shithead civilian would try to trip him up or just look at him funny. Anything that would give him a legitimate excuse to smash some local bastard's skull open. He wanted revenge for Nic, no matter what the price.

They'd left Amersy and the kid standing guard over the body to continue their deployment pattern. Karl had argued against that. They should all stay: it was respect if nothing else. But the goddamn sarge had insisted they carry on. So they'd taken their assigned streets, and now he was supposed to be checking for signs of organized resistance.

At least the anger was helping to cover his nerves. Some of them. Goddamn, this bunch of fish fuckers had guns that could shoot through Skin as if it weren't there. That was bad, real bad. It meant they'd all be vulnerable right up until the moment the guys from intelligence tracked down the cache. They'd do that, though. They *would* find it. He had to believe that. Intelligence division was creepy, but effective. In the meantime, he had to walk about in the open with his ass hanging out ready for someone to kick. Bad. Bad. Bad.

He kept a keen lookout as he walked along, scanning anything that looked remotely like a rifle barrel. His punch pistol was held high and prominent; so far it looked like it was intimidating people like it was supposed to. They were all staying indoors, glancing out at him through windows. There'd been a few catcalls, but that was all. News about the shooting had flooded the local datapool. That and the mass darting had cleared people off the streets pretty fast.

Some old geezer shuffled out of a side road, a walking stick waving about aggressively in front of him. Acting like he owned the place. Karl kept walking.

"Hey, you, sonny," the old man called.

"What?"

The old man had stopped at the edge of the pavement. "Come here."

Karl swore inside his helmet and angled his walk so he'd pass close. "What do you want?"

"I'm looking for your mother."

Karl's sensors zoomed in for a closer look. The old man really was ancient. Probably caught too much sun over the years. "My mother?"

"Yes. She pimps your sister, doesn't she? I want to know how much she charges. I'd like to give you people a good fucking."

Karl's fists clenched. The Skin AS had to modify his grip on the punch pistol to prevent him from crushing the casing. "Get back to the nuthouse, you old fart." He turned away and started walking. Goddamn parasite colony bastards. He never did understand why Z-B didn't just gamma soak the whole lot of them and send down its own people to run the factories.

The walking stick whistled through the air to crack across Karl's back. His carapace didn't even have to harden to protect him.

"Goddamn! Stop that. Crazy old bastard."

"They're going to bury him here, sonny."

The stick had a pointed end, which the old man was now using to try to gouge out one of the helmet sensors.

"Stop that!" Karl gave him a light shove. He nearly fell backward, but quickly regained his balance to make another stab with the stick.

"You can't take the bodies home, they weigh too much, and Z-B's too cheap. Your friend will have to be buried here. I'm going to dig him up again when you're gone."

"Fuck off." Karl swatted the walking stick away.

"We'll piss on him and use what's left of his skull as a trophy. And we'll laugh about how he died, with shit dribbling out of his ass and pain blowing his brain apart."

"Bastard!" Karl grabbed the insane old jerk, and drew his fist back. The old man started a cackling chuckle.

"Karl?" Lawrence asked. "Karl, what's going on?"

That goddamn suit telemetry circuit! Karl had lost count of the number of times he'd wanted to rip it out. He took a breath, his fist still cocked back. "Caught a ringleader, Sarge. He knows about the gun they used."

"Karl, he's about two thousand years old. Put him down."

"He knows!"

"Karl. Don't let them get to you like this. It's what they want."

"Yes, *sir.*" Karl let go of the old man, then realized there was a form of revenge available to him. "Hey, fuckface, you're my trophy now. How do you like that, huh?" He opened the pouch on his belt and pulled out a collateral necklace. The deranged old fool just kept laughing at him the whole time he fitted the thing around his neck, as if it were the best thing that could ever happen.

Michelle Rake had spent the whole morning sitting on her bed hugging her legs. She was fully dressed, but couldn't bring herself to venture out from the little apartment. Some of the other students in the residence house had gone out to see the invaders march through Durrell. Michelle knew what that meant. They'd end up throwing stones at the terrifying Earth-army troops, who would shoot them with agonizing stun bullets and drag them away to have explosive collars fastened around their necks.

So she had kept indoors and accessed the datapool news services. That way she'd been given a close-up view of the drop gliders landing on the edge of town and disgorging thousands of the big Skin-clad troops, who had promptly swarmed along the streets. And she was right. People had lobbed rocks, and bottles, and even some kind of firebomb. Barricades were thrown up across streets, then set on fire. The troops just walked through as if it were rain, not flame. Nothing affected them or slowed them down.

There had been other forms of resistance. The news reported that one of the spaceport's hydrogen storage tanks had exploded. A few civic buildings had been set on fire, sending up thick columns of smoke over the capital city. The datapool was slow, and sometimes her connection dropped out for minutes at a time as strange software battles were fought within the city's electronic shadow.

A quarter of an hour after the gliders arrived, small pods full of equipment fell out of the sky, dangling beneath big gaudy yellow parachutes. They were all drifting into the parks and meadows to the west of Durrell. Cameras followed several whose chutes had tangled, hurtling down to smash apart in a cascade of metal and plastic fragments.

To start with, she'd kept a line open to her parents over at Colmore, a settlement two thousand kilometers to the south. It might have been weak of her, but they understood how frightened she was by the invasion. This was her first year at the university, and she didn't make friends too well. All she wanted was to go home, but the commercial flights had all stopped within half a day of the starships being detected. She was stuck here for the duration.

Every time she thought about it, she told herself that she was an adult and should be able to cope. Then she started crying. Durrell was the capital, there would be more of the invaders here than anywhere else. Everything was bigger in Durrell, including the potential for trouble.

An hour after the drop gliders landed, her link to Colmore was cut. Nothing she could do would bring it back; the datapool management AS kept saying that the satellite links were down—nothing about how or why they were down.

She'd hugged herself tighter, flinching at every tiny sound

in the building. Her imagination filled the stairs and corridors with Skin suits as the invaders dragged students out of their rooms and snapped the explosive collars around their necks. They'd do it because everyone knew students always caused trouble, and rioted and demonstrated, and campus was a perpetual hotbed of revolutionaries.

There was a knock at the door. Michelle squealed in shock. The knock came again. She stared across the room at the door. There was nowhere for her to hide, no way she could escape.

She uncurled and stood up. The knock came again. It didn't sound authoritative or impatient. Hating herself for being so fearful, she padded across the threadbare carpet and turned the lock. "It's open," she whispered. She was trembling as if the world were in winter while the door slowly swung back. Somebody was standing there, giving her a curious look. He was so totally out of context that she thought her feverish brain was producing hallucinations.

"Josep?" she muttered.

"Hi, babe."

"Ohmygod, it's you!" She jumped at him, clutching him so hard she would surely squeeze him to death. But . . . Josep!

They'd met that summer when she was on vacation, celebrating her entrance exams—the first vacation she'd ever had by herself. It had been the most incredible time. Before then she'd always laughed at the clichéd stupidity of a vacation romance. But this had been different, she really had fallen in love. And at night she'd almost been frightened by her body's passion, the things they did with each other in her hotel bed. Almost. Leaving Memu Bay had torn her very soul in half.

She sobbed helplessly as he held on to her. "I thought you were one of them," she babbled. "I thought I was going to be made a hostage."

"No, no." His hands stroked her back. "It's only me."

"How did you get here? Why are you here? Oh, Josep, I've been so frightened."

"I caught the last flight out of Memu Bay. I told you, I wanted to come with you and enroll at the university here. I'd just decided to leave the diving school when these Z-B bastards arrived."

"You came here . . . for me?"

He took both of her hands, pressing them together inside his own until they stopped shaking. "Of course I did. I couldn't forget you, not ever."

She started crying again.

He kissed her gently on the brow, then moved down her cheek. Each touch of his lips was like a blessing. He was here, wonderful Josep with his strong, exciting body. And all the badness that had fallen upon their world wouldn't, couldn't touch her anymore.

Steve Anders made his way carefully down the concrete steps into the basement underneath the bar. The concrete steps had worn and crumbled in the coastal humidity, making them treacherous. He hadn't even known the bar had such a room underneath, but then it was a long time since he'd been in one of the tourist traps along the marina waterfront. His walking stick tapped its way gingerly across each curved surface before he put his feet down. At his age he didn't want to risk a broken bone.

He chuckled at that. It was his age that had brought him here. By God, it was good to be helping fight back against the swine who'd killed his son last time around. Good that he could do something, that his age was finally an asset.

It was a typical bar's storage room. Crates of empty and full bottles stacked against the walls. A trapdoor with a power platform to bring the beer barrels in and out. Broken chairs, advertising placards from years ago, boxes of old tankards, torn sheet screens rolled up and stuffed behind a pile of elaborate clay pots that still held desiccated plants.

He reached the floor and peered around the gloomy shapes. The place was lit by a single green-tinged light cone.

"Hello, Mr. Anders."

He squinted at the girl who came out of the shadows. Pretty, young thing. "I know you," he said. "You're the schoolteacher."

"Best not to label people," Denise said.

"Yes. Yes, of course. I'm sorry."

"That's all right. I thank you for what you've done. It was very brave."

"Pha." His free hand came up automatically to stroke the plastic collateral necklace. "It was easy enough to get. And I had fun annoying that young shit who put it on me."

Denise smiled and indicated a chair. Steve nodded gruffly, covering his rising nerves, and sat down. He watched with interest as she took a standard desktop pearl from her canvas shoulder bag. The unit was a rectangle of black plastic, fractionally larger than her hand, with its pane furled up along one edge. Nothing special.

She put it on her open palm, as if she were holding an injured bird. Her eyes closed and the slightest frown creased her forehead.

Steve Anders wished he were sixty years younger. She was enchanting. Some young lad didn't know how lucky he was.

The desktop pearl changed shape, stiff plastic flowing into a crescent with needle-sharp tips.

"That's unusual," Steve said, trying to keep his voice light. Before he'd retired, he'd been a protein cell technician. Nothing fancy, just a time server at Memu Bay's food refinery. But he knew Thallspring's level of technology.

Denise's eyes fluttered open. "Yes. Are you ready?"

Steve suddenly had a lot more confidence he was going to live through this. "Go ahead."

Denise brought the device up and touched its tips to the collateral necklace. Steve tried to look down at what was happening.

"It is melding with their systems," she said, understanding his apprehension. "By echoing them we can understand their function. Once that state has been reached, they lie open to us."

"It sounds more like philosophy than hacking." Did she mean duplicating their software, or hardware? Either way, he'd never heard of a gadget that acted the way this one did. It excited and disturbed him at the same time.

"There we are," she said contentedly.

The necklace loosened its grip. Denise took it from his neck. Steve let out a whoosh of breath. He saw that the tips of her gadget had sprouted a kind of root network, fibers as thin as human hair that dipped into the necklace plastic.

No, nothing native to Thallspring could do that.

"That's it?" he asked.

"That's it."

CHAPTER SEVEN

THE SCRUM DOWN-FORMED WITH A HEFTY BONE-CRUNCHER thud as the heads of the prop forward locked together. Each of the boys tensed, gritting their teeth, breathing hard as they waited for the scrum-half to slip the ball in.

From his flanker's position, Lawrence could just see through the tangle of mud-smeared legs. The ball was a blur of darkness as it entered the narrow gap. He yelled with the effort as he helped his teammates push. The hookers went after the ball like a pair of human jackhammers.

Lawrence's boots began to skid backward. The Lairfold team's prop forwards were the biggest (supposed) eighteen-year-olds Lawrence had ever seen. The Hilary Eyre High first fifteen were losing almost every scrum, and it was costing them in points.

This time Nigel, the Eyres hooker, managed to snag the ball for his team. It went sneaking back through the second row. The Lairfold team saw what was happening and started to wheel the scrum. Rob snatched the ball out of the second row and gave it a flying pass out to the Eyres wing just before he vanished below the painful slam-down of the enraged Lairfold scrum-half.

The scrum broke apart with jostling aggravation, and the heavy boys began to lumber out toward the wingers who were running with the ball. It was passed three times before Alan caught it just short of the halfway line. He was smaller than most of the team, but his stocky frame carried a lot of strength. He sprinted downfield faster than the opposition expected. The twenty boys converging on him had to alter direction, gaining him a few extra seconds before one of Lairfold's flankers crashed into him. It was a tumbling impact, both boys leaving the ground, legs akimbo. The ball flew straight and purposeful out of the melee with Alan screaming, "Go, you fucker!" and Lawrence caught it without even stopping. He pounded toward the Lairfold goal line.

The cheering from the touchline rose to a bombardment of yells, catcalls and chants. Out of the corner of his eye he just saw the scarlet and turquoise pompoms sashaying about as the Eyres cheerleaders gave it their raucous all. Couldn't make out which one was Roselyn. Then he saw the Lairfold fullback coming straight at him, and the lanky bastard was faster. He wasn't going to make the touchdown. On the other side of the pitch Vinnie Carlton was keeping pace with Lawrence's dash, making sure he didn't get in front.

Two seconds before the fullback tackled him, Lawrence turned and flung the ball. The fullback's arms wrapped around his legs and he crashed to the sodden grass with a bruising impact. The ball arced across the field, turning slowly end over end. Everyone watched its silent flight; even the supporters on the sideline abandoned their clamor. Vinnie carried on running. And the Lairfold team noticed him. Their gorilla-men prop forward bellowed a furious war cry. But nobody was even close.

Vinnie caught the ball beautifully, ten paces from the line. He sailed over with a joyful whoop, holding it aloft as he pelted in toward the big goalposts, slamming it down onto the grass.

The crowd was jubilant. Lawrence laughed madly as he clambered out from under the angry fullback. His ribs and shoulder hurt like a bastard, and the tackle had left him partly winded, but he was still clapping and hollering in elation. The Eyres team swooped on Vinnie, who hugged Lawrence.

"Great pass, man!"

"Better try."

"One point down," Alan said, always eager to spread gloom.

Lawrence shook his head. "Two up, you mean. No sweat. Richard'll get it."

They walked back toward their own half as Richard hacked into the ground with his heel, then carefully stood the ball upright. Lairfold lined up between the goal, facing him. But for Richard, Eyre's prize kicker, the three-point goal was a simple jog forward and a swift boot. The ball flew sedately between the tall white posts.

There was another three minutes left to the game. Eyres played it tactical. Not giving ground. Kicking it into touch. Holding the ball in the scrum.

The referee blew the whistle. Both captains did the gentlemanly thing and shook hands in the middle of the pitch. Lawrence stood with his teammates and gave their opponents three hearty cheers as they left the field.

Alan was laughing cruelly. "Look at them. Bunch of jerkoffs. Go home and kill yourselves, guys!"

Nigel's hand clamped over his mouth. "Show some dignity, man."

"I am." Alan smirked. "I'm fucking enjoying myself. I love it when people that arrogant take a dive."

"Hey, man of the match!" John wrapped an arm around Vinnie's shoulder, and pulled his hair down over his face. "What a run!"

Vinnie grinned happily. "Wouldn't have meant a thing without Lawrence."

Lawrence put on his most humble tone. "I do what I can."

"Yeah," Alan grunted. "Only if Roselyn lets you."

Several of the cheerleaders were running across the field to greet their heroes. They were dressed in short scarlet skirts and cornflower-blue sports halters.

"Now that's what I call a welcome home," Alan said. His laugh was like a bad case of hiccups. He put his arms out wide and ran toward them. They scattered.

Roselyn swatted him with a pompom and danced around to reach Lawrence. "You won!" she squeaked as she kissed Lawrence.

"It was a team effort."

"No, it wasn't. It was your brilliant throw that clinched it. I saw it all. You were magnificent. Kiss me."

"Oh, for fuck's sake," Alan grumbled. He ambled off toward the changing room.

Lawrence and Roselyn laughed at his departing back.

"Ugh, you're filthy," she complained suddenly. Streaks of cold, wet mud from his shirt had soaked into her halter. "Go and wash."

"Yes, ma'am."

"Be quick. It's freezing out here." She rubbed her arms and gave the dome's conditioning fans a suspicious glance. The school always lowered the temperature for rugby and soccer so the players wouldn't get too hot, but this felt as if the atmosphere had circumvented the inlet grids to blow straight in.

"Are you going to the party tonight?" Nadia asked. She was leaning against Vinnie, with his arm casually possessive around her waist. But it was Lawrence who was receiving her intent stare.

"Yeah, sure," he said, very careful to keep the tone neutral. Roselyn seemed to have some kind of telepathic ability when it came to detecting his thoughts on other girls. Not that he did have thoughts on other girls, of course. Funny thing was, for years not a single girl at Hilary Eyre High had shown any interest in him whatsoever; but now he had Roselyn he'd started to get definite signals. Not just from Nadia, either.

"See you later," Roselyn said. She turned, then bounded back. "One more kiss."

He obliged.

"So is she pregnant yet?" Alan asked in the locker room.

"What? Who?" Lawrence had showered, managing to grab someone else's shampoo. Now he was toweling his hair dry beside his locker.

"Roselyn."

"No!"

"So what's all the practice for?" Alan's question trailed off into his hiccup laugh.

"God, you're such a pervert."

"God? Ah, this would be Roselyn's God you've borrowed, would it?"

"Fuck off."

"Listen." Alan's voice rose in volume so he could appeal

to the rest of the locker room. "Three times I asked if he was coming out for an evening last week. Every time," his voice became all whiny, *"I can't, we have to study together."*

"Which bit of her were you studying?" Rob shouted.

"Yeah." Nigel laughed. "Don't you know all the working parts yet?"

"Fuck *off,*" Lawrence yelled at them, hoping he wasn't grinning too much. It was quite a prestige thing, having a girlfriend for so long that everyone knew for sure that the relationship was solidly physical.

"They're just jealous," Vinnie said. "Freaks without chicks."

Lawrence gave him a small bow. "Thank you." He liked Vinnie Carlton. The boy had arrived on Amethi only eighteen months ago, just after Roselyn's family. But already it was as if he'd been there forever. Lawrence had started getting pally with him around the same time he was reintegrating himself with his own peers. Vinnie didn't have any family in Templeton. His father was still back on Earth wrapping up contracts for his software business before flying out to live permanently on Amethi. As Vinnie was seventeen when he disembarked the starship, he was legally able to live without any guardian supervision. He had his own apartment, and some legal firm took care of his finances and other official stuff, such as getting him a place at school. Lawrence had been incredibly jealous of that apartment at first. But they had a lot in common—shared academic classes, both in the flight club (Vinnie had actually flown an aircraft back on Earth—he claimed), got roped into the same team games, enjoyed duking it out in the i's together. They even looked similar, though Lawrence's hair was a couple of shades lighter, and Vinnie's eyes were deep brown instead of gray green. "I think you're cousins," Roselyn had said once.

Lawrence laughed at that and said: "No way." Although a couple of months after they'd been hanging out together he did ask Vinnie about his family. That was when he discovered the Carltons were the ones who'd imported *Halo Stars* to Amethi. Which made Vinnie a seriously good person to know—he got the upgrades before anyone else. Not that Lawrence was playing the i's anything like as much as he used to. He simply didn't have the time these days.

"Alan, we've got to find you a girl before your mind goes into meltdown from hormone overload," Vinnie said. "You're getting worse every day. You are coming tonight, aren't you?"

"Course I am, this party was my goddamn idea, remember?"

Lawrence could remember Roselyn and Nadia saying the team should all go out together after the game to either celebrate or commiserate. He chose not to mention it at that point.

"We should ask a few extra girls along," Richard said.

The idea of Richard even knowing a few extra girls was also something Lawrence kept quiet about. Richard had been going steady with Barbara for ages. *One* extra girl, and she'd kill him.

"Don't you worry about me, mate," Alan said in his most annoyingly cocky voice. "I've got a foolproof system to get laid."

"What?" Nigel snorted. It was supposed to be contemptuous, but a small note of interest had crept in.

The changing room magically quietened down as the other guys in the team just happened to overhear Alan's brag. Not that any of them needed a system, but it never hurt to know.

"Simple," Alan said, delighted by his audience. "My mate, Steve, you remember him, the bright one that went to university last year? Yeah. Well, he swears this works; he does it all the time. You go into the party and look around to find the most beautiful girl there. Then you walk straight up to her and say: will you sleep with me tonight?"

There was a moment of silence as the rugby team absorbed this news.

"Crap."

"You asshole."

"That's such a bunch of shit."

A shoe thrown by a disbeliever hit Alan's leg. He yelped and searched around for the offender. "Hey, look, I'm not kidding around here," he exclaimed. "Steve says it works. He gets laid every weekend. Seriously."

"Oh yeah," John jeered. "And the most beautiful girl in the room takes one look at a toxic midget like you and just says yes."

"Well, maybe," Alan said. "If you get really lucky."

"I think I'll stick to the traditional method of giving her too much to drink," Lawrence muttered.

The noise level rose. People started getting dressed again.

"Hey, *listen,*" Alan protested. "This is statistics. That's solid mathematics. It can't fail."

"But you just said this mythical supermodel was likely to turn you down," Nigel complained.

"So? Doesn't matter. You find the second-most-beautiful girl, and ask her the same thing. If she says no, you just keep moving along down the beauty scale until one of them says yes."

John's expression was pitying. "Alan, *none* of them are going to say yes. Not to that."

"Yes, they will. They're at the party for exactly the same reason we are. It's just that they're not as honest about it as we are."

"You're lecturing on honesty," Lawrence said. "Oh, my sweet Fate. We're doomed."

"Girls like you being honest," Alan insisted.

"They like politeness and flattery a lot more," Richard said.

"Most of them most of the time, yeah. But this is a party,

right? They've been drinking, the evening's moving on and they haven't scored yet. One of them's bound to say yes. It's statistics. I told you."

Vinnie's despair had caused his head to sink into his hands. "Alan," he asked, "do you ever wonder why you haven't got a girlfriend yet?"

"Hey, I've had hundreds of girls, okay."

"When?" Lawrence demanded. "Tell us when this system ever got you a girl."

"Tonight."

"I knew it. You're talking bullshit."

"Durr! No! This is completely for real. Steve's screwed half the babes on campus. It's amazing. You've just got to have the balls to use it."

"Your balls have got to be where your brain is before you'll use it, more like," John grunted dourly.

Alan jabbed his thumb proudly against his chest. "Listen, mate, I'm the one that's going to get laid tonight. It's you sad joes who'll be left propping up the bar and going home all by yourselves. I'm telling you, it works."

The party, like all parties, started out with good intentions. At seven-thirty, the first fifteen team and friends headed over to Hillier's, which was in a dome they could all walk to. It was a big old club buried under a residential tower, with three main oval-shaped sections comprising lounge, dance floor, and brasserie, that joined together at a central circular bar. In its heyday, Hillier's had been the center for younger members of Board families, a place where the jazzy hung out and the pool sharks lay in wait. But time and fashion had moved on.

Now it was the even younger members of second-echelon families who congregated there in the evening. They, of course, thought it was superb, a real nightclub that didn't kick up a fuss and ask for proof of age at the door. Hillier's couldn't afford to get that choosy about its paying customers anymore. And these kids did seem to have access to large amounts of money.

The plan was to start with a meal, then move on to a drinking and dancing session. When Lawrence arrived, the boys were all in the lounge, having a drink before hitting the brasserie for something to eat.

"You're late," Vinnie said. He was already on his second beer.

"I had some news," Lawrence said modestly. He'd thought he was in for another lecture when he got home after the match. His father had called him up into the study, and he was never summoned there for any other reason. But when he arrived, his father was smiling as he held out a sheet of hard copy. "Thought you might want to see this," Doug Newton said blithely.

Lawrence took the sheet from his father with some trepidation and began to read. It was a provisional acceptance from Templeton University, offering him a place to study general science and managerial strategy.

Doug clapped his son on the back. "You did it, my boy. Congratulations. I didn't even have to pull any strings."

Lawrence had just stared at the sheet, elated and frightened by what it meant. Everybody applied to Templeton University: the candidate rejection rate was 80 percent. "Only if I get the qualifying grades in my final exams," he said cautiously.

"Lawrence, Lawrence, what are we going to do with you? You'll get them. We both know that. The way you've turned your schoolwork around these last couple of years, you'll probably get a distinction." He gripped his son's shoulders. "I'm proud of you. Genuinely proud."

"Thanks, Dad."

"You off to celebrate tonight? I heard you won the game."

"Some of us are thinking of going down to Hillier's, yeah."

"That old place still going, huh? Ah well, good for you. But I think you deserve something a bit more tangible for this result. I've booked you in for ten days at Orchy. You can go skiing on Barclay's. How does that sound?"

"Pretty amazing!" His enthusiasm faded. "Uh . . ."

"It's for two," Doug had said gently. "If you have a friend you'd like to take."

Lawrence looked around Hillier's lounge. "Where's Roselyn?"

"Haven't seen her yet." Nigel signaled the barmaid for two beers. She was in her mid-twenties, and immune from his hopeful boyish smiles.

"Oh." Lawrence kept looking. "What about Alan?"

"Am I your personal news trawler? He's around somewhere, talking to a girl."

"What?" Lawrence gaped at Nigel. "You don't mean his system worked?"

"Oh, get fucking real," Nigel exclaimed. The barmaid frowned at his language and put the beers down in front of him without saying a word. Nigel winced at her departing back, then glared at Lawrence. "Thanks."

"You're as bad as Alan. A girl like that and you is never going to happen."

"Maybe if I left a big tip . . ."

"Don't even think it." Lawrence picked up his glass and took a sip. The beer was so cold it disguised any taste. "So how is Alan doing?"

"One slap on the face, two cocktails thrown at him, and he's been told to piss off a few times as well," Vinnie said happily. "We're thinking of running a book on it."

"Put me down for a day five years hence." Lawrence saw Roselyn moving across the lounge and waved. She was in a green dress that had a big oval patch open at the front to show off her navel. Whatever she wore, she always looked sensational. It was just a knack she had. But as usual it made Lawrence terribly self-conscious about his own clothes. He worried that his bronze-shimmer jacket would look awfully crass beside her.

Roselyn arrived at the bar at the same time Alan staggered in from the other side. A long strip of pink toilet paper was tucked into the back of his trousers. Half of the lounge clientele were mesmerized by this flimsy tail sliding along the floor behind him.

"Damnit," Alan whined. "They're all playing hard to get."

"Who are?" Roselyn asked.

"All the babes." Alan glanced around accusingly at his friends. "Did you guys warn them?"

Nigel bent over, his face radiating martyred dismay, and tugged the toilet paper free. "We didn't have to."

"What?" Alan did a double take at the paper. "Oh, thanks. It must have got stuck in my cleft. My round." He clicked his fingers loudly at the barmaid. "Oi, how about some service?"

"I have some news," Lawrence told Roselyn.

She grinned. "Me too."

"You first."

"No, you."

They both laughed.

"Ladies first," Lawrence said.

"I'm going to throw up," Alan muttered.

"Okay." Roselyn fished round in her small handbag and produced a memory chip. "I'm late because I was downloading this from the *Eilean*'s communication AS; it's just arrived in orbit. Judith sent me another series."

Lawrence gagged in wonder. He took the chip from her hands with a great deal of reverence. "Series six?" he asked.

"Uh-huh." She accepted a margarita from John and carefully wiped the salt from a section of the rim. "The last one."

"Hellfire. The final episode. I wonder if they get home."

Roselyn cocked an eyebrow demurely. "Only one way to find out. Oh, and there was some stuff from the fan site, too. Half a dozen series-related i-games, I think, and a whole load of generated graphic follow-ons."

"Fantastic."

"Damn." Alan grinned at Roselyn. "This is a moment like that stunt your God does. What is it? Oh yeah, he turns up again or something."

"The Second Coming of The Christ. A time of revelation throughout the universe."

"That's the one." Alan raised his beer glass. "Here's to Lawrence finally finding out what happened to a bunch of jerkoff actors when they asked for a pay raise in series seven."

"There was a proper story arc," Lawrence protested. Too late he realized the fatal mistake of letting Alan know you cared about something.

"Whoo ho! I was right, it's a revelation! Please, Lawrence, do us all a big favor and get a life."

"Alan?" Roselyn asked in a voice tinged with curiosity. "Do you know that girl?"

"Which one?"

"Over there, in the blue top."

"Her?" His glass slopped about in the girl's general direction as he laughed his short dirty laugh. "Damn, see what you mean, two puppies wrestling in a blue sack."

Roselyn's face remained serene. "Yes. Her."

"Never seen her before in my life, Your Honor. And I would definitely remember." He drained the last of his beer and burped. Fortunately, he'd ordered too many, so there was a fresh glass he could lift straight off the bar.

Over Alan's head, Lawrence gave Vinnie a frantic grimace and mouthed: "When did he start?"

Vinnie shrugged helplessly.

"She's been looking at you," Roselyn said.

"Fuck! Really?" Alan laughed again and poked Richard in the chest. "I told you. It's statistics." He straightened himself up and walked over to the girl. There was a momentary flash of panic on her face when she saw him approach.

"Remind me never to annoy you," Nigel told Roselyn.

Lawrence was wincing as he followed Alan's progress. "I'm not sure I can watch this. The pain level's too high."

"So what did you want to tell me?" Roselyn asked.

"Oh, yes." The joy returned to Lawrence's life. He pocketed the memory chip. "I got a letter from Templeton University today."

Roselyn's gaze was one of pure admiration as he explained about his preliminary acceptance and the skiing trip. "I knew you could do it, Lawrence," she murmured quietly. "Well done." She kissed him just below his ear.

"What about your mother?" he asked apprehensively. "Do you think she'll let you come to Orchy with me?"

"You leave her to me."

His hands went around her, pressing into the small of her back. "Sounds good to me." They kissed. He could taste the sharp tang of the margarita on her lips.

"Er, guys, I think we should get over there," Vinnie said.

Alan was so engrossed with making obscene small talk to the girl in the blue top that he hadn't noticed her boyfriend standing behind him.

"No way." John was shaking his head. "Look at the effing size of him!"

"Bigger they are, the harder they fall," Rob declared. He was almost as drunk as Alan.

"As long as he falls on you, not me," Nigel said.

"He's our friend," Lawrence said. Somehow he couldn't summon up much conviction. The boyfriend had a couple of friends with him, too.

"Just tell the bar staff," Roselyn said urgently. "The bouncers will sort it out."

"Too late," Vinnie groaned.

Alan had finally noticed the boyfriend.

They looked on incredulously as their friend employed his own never-fail method of getting out of sticky situations by telling the one about the parrot and the starship stewardess.

". . . the airlock slammed shut, and as they were tumbling through interstellar space the bloke turned to the parrot and said, Pretty ballsy for a guy with no spacesuit." Alan giggled hysterically at the punch line.

The boyfriend, it turned out, didn't have much of a sense of humor.

Lawrence finally got home at half past three in the morning, after his father and the family lawyer bailed him out from the police station.

Amethi's turbulent climate was changing again, emerging from its snowfall phase. Over the last few years, billions of tons of water had been liberated from Barclay's Glacier as the meltoff accelerated. The contribution it made to atmospheric pressure and density was small, but effective. Thicker and heavier, the planet's envelope of gas now retained more heat than before. Overall temperature was up by a couple of degrees. On the side of the planet away from the glacier, the snow was giving way to rain. Templeton even had weeks of broken cloud cover as the winds slowly strengthened.

A lot of people saw that as a bad omen, predicting the Wakening would end in hurricanes ripping the domes apart. The official line was that increased air speed was a natural and inevitable part of acquiring a normal weather pattern. There might be a few peaks on the graph along the way, but it would level out in the end.

Whether you believed that or not, the clearer skies did mean that passenger jets were returning to commercial service after their near-hiatus of the preceding years. Lawrence and Roselyn caught the morning flight out from Templeton, taking fifteen hours to reach Oxendale. One day, Oxendale would be the major city on a long chain of islands in the middle of the ocean. For the moment, it was sitting on the top of a massive, flat-topped mountain, the largest in a ridge of similar mountains rising out from a slushy saltwater quagmire.

On this side of the planet, facing Nizana, the glacier still dominated the environment. The air was a lot colder, and clouds still sprinkled snow as they migrated out to the warmer tropics. Their jet touched down on a runway that was coated in white, powdery ice. They glimpsed it only a few seconds before the wheels hit. For the last hour, they'd been flying blind through thick fog. Oxendale's altitude a kilometer above the salty marsh meant that it was almost permanently in the clouds.

They had a half-hour wait in the airport lounge while their luggage was transferred; then they trooped on board a thirty-seater STL plane, built for arctic conditions. Orchy was another two hours' flying time away. Forty minutes after takeoff, they cleared the base of the cloud layer to see Barclay's Glacier in the distance.

With Amethi a quarter of the way around its orbit from superior conjunction, the sun was shining almost directly onto the vertical cliff face of the glacier. It split the land from the sky with a silver-white glare stretching from north to south, as if a crack had appeared in the landscape to allow another, closer sun to shine through from behind the planet. Lawrence had to put his sunglasses on to look at it directly. Colors here were all monotone. The surface of the glacier was pure white; even the clouds didn't seem to cast a shadow. Features, at least from this distance, were nonexistent. The most that could be said was that the ice was rumpled, with long, gentle curves overlapping all the way to the boundary. Overhead, the sky shone with an astonishingly bright metallic-blue sheen. Nizana's dominant ocher crescent appeared intrusively alien, its darkness in some way negative. Squashed streamers of cloud swirled about, almost as bright as the glacier itself. All of them were sliding in the same direction, out from the ice shelf and away over the ocean floor.

When Lawrence looked straight down, he could see nothing but dunes of slick auburn mud, their crests dusted white. Slivers of grubby water shimmered in the cirques amid the dunes, forming an infinite plexus of connected rills. Every few kilometers there would be a deep river cutting its way through the mud. Here the water was fast-flowing and filthy, clawing at the gully sides to loosen great swaths of mud. Lumps of ice bobbed along, colliding against each other with enough violence to produce small explosions of splinters, or even split apart.

For all the physical activity, the vista got to Lawrence. He used to think the tundra desert outside Templeton was bleak, but this was pure desolation. There was no sign here that any of the terraforming algae had ever bloomed in the slushy puddles, no meandering tracks of slowlife organisms as they impregnated the mud with their spores and bacteria. This was impassive, ancient geology at its most aloof, untouched by life's Machiavellian tendrils. It made him feel small, irrelevant.

After a while, the little aircraft curved around and headed in over the glacier. A lot of the edge was still sheer cliff, but a quantity had crumbled into giant talus falls extending for kilometers out into the mud. The top of the glacier was bisected by deep rifts that carried the rivers out from the interior. Some of these fractured canyons were over a kilometer deep and still expanding as the water gnawed away at their floor, but that still left them terminating high over the ocean floor. The edge of Barclay's Glacier was host to the most spectacular array of waterfalls on any known world. Over a thousand prodigious rivers ended abruptly hundreds of meters above the ground, projecting their waters in monumental arcs to thunder into ragged craters gouged out by their own relentless torrent.

The town of Orchy was situated on the top of one of these rifts, Coniston's Flaw, a long jagged gully extending well over a thousand kilometers toward the east. In some places it was over three kilometers wide, its steep angled sides resembling the Alpine valleys of France and Switzerland. Orchy was currently sitting on top of a broad, curving section, with the river churning along the rift floor six hundred meters below. The curve meant that the water constantly chewed into the ice, an erosion that pulled down vast avalanches from the sides. Once they'd settled, they were excellent skiing slopes, although the flow of water that created them would ultimately undermine them, changing the valley's profile once again. The entire length of Coniston's Flaw was a variable geometry, flexing in month-long undulations, with only its terminal waterfall holding reasonably steady. Even the tributaries would forsake it after abrupt and violent shifts, defecting to other rivers.

Orchy moved to accommodate these whims, a truly mobile town, made up from oblong building modules that could be carried by large flatbed trucks. Whenever the slopes decayed or quaked or collapsed, the silvery modules would be unbolted and hauled along the top of the Flaw to the next suitable site.

The STL plane extended its ski blade undercarriage and skidded along a length of flat ice marked out by flare strobes. Fans howled as the AS pilot reversed pitch and brought them to a halt at the center of a microblizzard. A bus took them into town, dropping them off at the Hepatcia Hotel. It was identical to every other cluster of metal modules that made up the town. They were laid out in a fat fishbone pattern, standing on legs that left a seventy-centimeter gap between the floor base and the ice. Reception was at one end of the spine, with the bar, lounge and dining room at the other. The interior was smart without being ostentatious. It reminded Lawrence of aircraft furnishing.

Their room was made up of three modules, which gave them a bedroom, a small bathroom and what the bellboy insisted on calling a veranda room. It was essentially an alcove with lounger chairs and a wide floor-to-ceiling triple-glazed window giving them a view out across Coniston's Flaw.

"Wonder what old Barclay would make of this?" Lawrence mused. Thick clouds were boiling overhead, but they were pure white, fluoresced by the sun. Ice and snow gleamed underneath, making it difficult to know where the horizon was. Orchy was at the center of its own little closed radiant universe. With his new sunglasses, Lawrence could just make out tiny, dark figures zipping down the slopes below the hotel.

"I think he'd be impressed," Roselyn said. Her dimples had returned as she took in the view. "I am."

He glanced around the room. "Not quite up to the same standard as Ulphgarth."

"We'll have to make do." She offered him a small jeweler's box.

"What's this?"

"Open it."

There was a slim silver necklace inside, with a hologram pendant. When he held it up to the light, a small Roselyn in a blue dress smiled at him from inside the plastic.

"So I can be with you all the time," she said, suddenly bashful.

"Thanks." He slipped the chain round his neck and fastened the clasp. "I'll never take it off."

Her hand turned his head to face her, and they kissed passionately. He began tugging at her blouse.

"Wait," she murmured. "I'll just be a moment."

Lawrence did his best not to show his frustration as she

picked up a bag and went into the bathroom. "You could get ready, too," she said as she slid the door closed. "And I like the lights low, remember."

He stared after her for a second, then raced over to the door and locked it. Over to the big veranda windows and opaqued them. Swept the hand luggage off the bed. Pulled the cover onto the floor. Struggled to push his trousers down, dancing on one foot when his shoe became stuck. Got a shirtsleeve caught as he pulled it off over his head. Set the communication panel to call guard. Landed hard on the bed, and let out a small whoop of delight when the mattress rippled underneath him. Plumped up the pillows and flopped back onto them, hands behind his head, grinning oafishly at the ceiling.

Ten days!

Roselyn walked out of the bathroom. She was wearing a white silk negligee that couldn't have weighed more than ten grams. He'd never been so scared of her sexuality before.

"You're magnificent," he whispered.

She sat on the side of the bed. When he rose up to embrace her, she held up a finger, shaking her head softly. He let himself down again, not sure how long his self-control would last.

"I so hoped you would enjoy me like this," she said quietly.

"Fat chance I wouldn't—" He broke off at the slight frown on her face.

She reached out with one hand to touch the pendant, then gently traced the shape of his pectoral muscles. "I wore this because I wanted to please you. I need you to know how much tonight means to me."

"It means a lot to me, too."

"Does it, Lawrence?" Her hand stroked down his abdomen.

The eroticism of the motion was an insanely beautiful torture. It almost brought tears to his eyes. All he could do was draw breath in sharp little gasps as her gray eyes searched his face, divining everything he felt. He'd never been so naked before.

"We're going to spend the night together," she said. "Do you understand that?"

"Of course I do."

"Do you now? Well, I'll tell you anyway. It means that we can make love for as long as our bodies can last. That there will be nothing else to consider; no timetable, no having to go home, no caution about someone coming in. Just you and me alone with as much joy as we can create. And then when we're done with each other, we're going to fall asleep in each other's arms. We've never known that before, Lawrence. And it's going to be the most exquisite moment of all for me, because I'll do it knowing I'm going to wake up with you beside me. You don't know how long I've wanted that to be."

Even in the dusky light he could see the admiration on her face, and the hope. "I want that just as much as you do," he said. "I wish you'd said something before. We could have worked out a way to make it happen before now."

"Would you have done that? For me?"

"Yes."

"I love you, Lawrence." Her expression became rueful. "And you know all of me now, everything I am, however foolish that is." She swung her legs around and straddled him just above his hips.

"You're not foolish," he told her earnestly.

The grin that dawned on her mouth was wicked and knowing. Fingers slid back up his chest. "You're so fit now," she said huskily. "It's indecent."

"You're the one who wanted me in this condition."

"I did. And I'm a grateful girl." She arched her back, then slowly, tauntingly, began to undo the lace bows running down the front of her negligee.

* * *

They missed their first scheduled skiing lesson, staying in their room together for over a day. Not that it particularly mattered. Amethi wasn't going to move into Nizana's penumbra for another sixty hours. It would remain light for all of that time.

After they did finally get out of bed to have breakfast, Lawrence called the school and arranged another lesson. The AS receptionist told them another slot wasn't available for five hours.

They took a walk through the town, looking at the restaurants and cafés and bars. Pavements were slatted aluminum walkways set up between the buildings, standing on the same kind of legs. Lawrence loved it. The first open-air town he'd ever been in; the sensation of freedom was invigorating. Temperature was at least fifteen to twenty degrees below freezing. Not that he cared about that; they both wore their brand new ski-suits: colorful one-piece garments with a lacework of active thermal strips whose conductivity could be set by an integral thermostat, allowing you to choose whatever temperature you wanted to be at. The hoods were close-fitting and had extra flaps, which could be pulled across the face. They were essential to stop windburn when you were skiing, but in town most people let them hang free.

"It's like you can feel the ice pulling heat from your skin," Lawrence exclaimed. He was leaning over a walkway's rail, looking down what passed for Orchy's main street. Buses and ice bikes roared about, carrying vacationers between the hotels and the runs.

"Nice to know," Roselyn said. Every flap on her hood was closed tight, leaving just her goggles poking through. Even so, she stood slightly hunched, as if fighting the cold.

Lawrence laughed and kept walking. They stopped off in a couple of stores. The only difference they could find between them were the names of the owners. Both were franchises to the company that ran Orchy. And both of them sold the same ski equipment; there weren't many manufacturers on Amethi yet.

"Business opportunity," Roselyn observed. She giggled at Lawrence, who was trying on a different hood: its style was awful, all pink and orange stripes. "Two business opportunities," she corrected.

"I want to be seen on the slopes," he said with pained dignity.

"What as?"

They moved on. The trouble with a town made out of identical modules, they decided, was that you didn't know what kind of businesses they contained until you were inside. The names flashing over the doors didn't offer much of a clue. Accessing the datapool for a local directory was a pain, and too functional. They just wanted to stroll and take in the sights. Orchy wasn't really built for that. There was no civic identity; its purpose was simply to house and feed people in between skiing jaunts.

They did find a reasonable café eventually. The Flood Heights was positioned as close to the edge of the rift as safety would allow. So Lawrence and Roselyn sat at one of the window tables and ordered hot chocolate and a plate of Danish pastries.

He sat sipping at his mug, looking up into the sky with a kind of wistful admiration. He'd never seen Nizana like this, not with his own eyes. Here on the near-side it hung directly overhead, a massive circle sliced by a thousand compacted cloud bands, clearly defined lines of rust red and grubby white grating and tearing at each other with hooked curlicues. Hundreds of runaway cyclone storms the size of moonlets were constantly on the prowl amid the upper layers. They distorted the neat arrangement of bands, chaos engines churning the usual colors into freakish shades with oceanic-sized upwells of weird chemicals from the unseen depths. Sheets of electricity surged outward from their eyes, too vast to be called mere lightning bolts: continents of electrons birthed and extinguished in microseconds. Their ephemeral illumination ensured that Nizana's nightside was never dark; a jade aural phosphorescence writhed permanently within the cage of the ionosphere, while the discharges themselves fluoresced ragged patches of cloud thousands of kilometers across.

"They're going so fast," Roselyn said, gazing down at the skiers sliding along the snow. "Do you think we'll learn to go that fast this time?"

"Huh?" Lawrence brought his attention back to the ground, looking where she was. "Wrong question. You've got strips of polished composite strapped to your feet, and you're standing at the top of a mountain of ice. The trick is learning to go down slowly."

She stopped dropping sugar lumps into her chocolate and flicked one at him. "Prat. You know what I mean."

"Yeah. I don't suppose it's that difficult, not on the nursery slopes. They claim they can get you up to moderate grade by the end of a week."

"It looks scary, but I think I'm going to like it." She watched several skiers as they reached the bottom of the main slope, curving to a halt in a graceful spray of snow. The cable lift began tugging them up to the top again. On the other side of the rift, slim-line fissures extended deep into the ice cliff, intersecting each other and twisting around in convoluted geometries. Sunlight shone into them to be refracted in glorious iridescent rainbows, forever encased below the translucent surface.

Roselyn sighed contentedly. "I'm so happy. I've got you, I've got a life. It's funny, I never thought leaving Earth would allow me to be happy. You know the only thing I miss?"

"What's that?"

"Boats." She gestured around extravagantly. "I mean, Amethi's leisure industry is starting to lift off. There's this, and all those hotel domes in the middle of nowhere, and that ridiculous five-city motor rally race they've got planned for next year. But there are no boats."

"Give it time. Our oceans are filling up, and there are lakes forming on the continents."

"Ha! It'll take another thousand years to melt this glacier. So I'll see none of that till I'm either dead or too old to care. Such a shame. It would have been nice to stand on the prow with the sails creaking away, and feeling the wind on my face."

"When did you ever do that?"

"Dublin has a port, I'll thank you. Although it's mainly for the big cargo ships that come in from England and Europe. But there are sailing clubs along the coast. I know how to crew a dinghy. I was even getting quite good at windsurfing." Her gray eyes stared off beyond the horizon. "But I've done it once. Better that, than never."

Lawrence slouched down in his seat. "And I never will."

"You poor old boy." She pouted. "I fell off a lot. The water was freezing, and didn't taste so good either. Heaven alone knows what pollution was in that sea. That's the thing with memories, you only ever dwell on the good parts."

The lesson went the way of all first skiing lessons. Lawrence and Roselyn spent a lot of time slipping about and falling over. But they did make a kind of progress, enough to slide down the nursery slope several times without landing in a tumble of limbs and poles, enough to get an idea of how much thrill there would be from descending the main slope, enough to promise faithfully they'd be back on time tomorrow.

It wasn't until they got back to the hotel room that their muscles began to protest at the way they'd been abused. Ankles and calves ached as they stiffened up. Lawrence's shoulders throbbed as if they'd been bruised, which he could only put down to the way he'd pushed himself along with the poles. With laughter tinged by winces they stripped off and got into the bath together. Soaping each other down was an erotic foreplay that quickly evolved into full sex, sending water all over the floor. Drying each other in the big soft towels had the same effect. Then they moved out into the main room, where the bed waited invitingly.

After their third bout of lovemaking they ordered a huge room-service dinner, complete with iced champagne. The mattress was too unstable for them to eat in bed, so they sat in front of the veranda window wearing big toweling robes and tucked in.

"Those slopes are going to look beautiful after sunset," Roselyn said.

The instructor had told them that when Amethi moved into the umbra the runs were all illuminated by orange and green lamps. Skiers themselves wore red and white torches on their helmets. It was as if the whole valley side was invaded by swarms of dancing starlight.

Lawrence took her hand and gave it a squeeze. "We'll see it. Our last days here are in the conjunction night. We'll be good enough to be using the main slope ourselves by then. They say that when we're in the heart of the umbra, Nizana is like a flaming halo, as if the sun's set the edge of the atmosphere on fire."

"I can't wait."

They took the half-empty bottle of champagne and a box of chocolates back to the bed. Lawrence lay on the mattress, a flute of champagne in one hand, the box of chocolates in reach of the other, and Roselyn curled up beside him.

She squirmed around for a moment until she was perfectly comfortable, then said, "Go on then."

"Thanks." He kissed her brow, and told the room AS, "Access my personal file, entertainment section, and play *Flight:*

Horizon, series six, episode five. Give me the standard third-person view edit."

"Happy now?" Roselyn asked.

She always watched *Flight: Horizon* with him, though he was pretty sure she was humoring him rather than developing any deep interest in the crew of the *Ultema*. "I am, thank you," he said with dignity. She snuggled in a fraction closer and took a sip from her own flute as the credits rolled and the signature tune began its fanfare.

Eighty minutes later the *Ultema* had managed to prevent a planetary collision that would have wiped out three sentient alien species. One of the species was furious with this interference in their glorious destiny as angels of the apocalypse and came gunning for the starship with some very nasty weapons. Three of the crew had been killed before the end, two of whom had just got engaged.

"Seven crew in three episodes," Lawrence said in dismay. "That's as many as in the whole of series four."

"Oh dear." Roselyn's lips were pressed together to hold back her giggles. She attempted to put on a grave expression. "That's not good, then?"

"It doesn't help their chances, no."

"Oh, poor baby." She wriggled around until she was on top of him and gave him a wet kiss while she giggled.

Lawrence played stubborn.

Roselyn laughed outright. "Oh, I'm sorry. It's just that you take it so seriously."

"I used to take it very seriously. They were good role models when I was younger. It meant a lot to me then. Now it's like having old friends around; I can appreciate it without adulating it. You showed me there's more to life than the i's. But I still claim it's a pretty good show."

"Oh, Lawrence." She turned back to give the big sheet screen a remorseful look. "That was nasty of me. I sometimes forget how different our backgrounds are."

"Hey." He stroked her back gently. "You couldn't be nasty if you tried."

"Except to Alan."

Lawrence sniggered. "That wasn't nasty; that was funny."

"True." She lay down beside him, their faces a couple of centimeters apart. "And you were right, *Flight: Horizon* isn't a bad influence for a growing boy."

"Well, I'm growing out of it now. Damn, taking an administration class at university. That's about as far away as possible from what I used to want."

"No, it's not. Command qualities are the same no matter what fancy name you stick on them. And it will be a damn good basic if you ever change your mind and go in for officer training."

"Ha! Training for what? Dad said it: we just run a passenger service. You should know, you've been on it. I wanted to be a part of exploring the galaxy, pushing back the frontiers. That's all over, now."

Roselyn propped herself up on an elbow to look at him. "This is what I can never understand about you, Lawrence. You always tell me how much you hate McArthur for shutting down its exploration program. Yet you never talk about anything else but staying here and making your contribution to Amethi, to the company. That's dichotomous to the point of schizoid, especially for you."

"What the hell are you talking about?"

"If you can't do what you want here, then leave and do it somewhere else."

"There is nowhere else," he said in exasperation.

Her perplexed look was equally impatient. "Well, not apart from Earth with its half-dozen exploration fleets, no."

Despite the warmth of the room and her body, the lazy fizzing of the champagne in his blood, Lawrence was abruptly cold and terribly alert. What she'd said simply wasn't true. Because it contradicted his whole world, everything he'd known and done since that hot-tempered day when he'd ruined the fatworms. "What did you say?"

"That you should go to Earth and sign on with another company if you feel so strongly about all this."

His hands closed about her upper arms, squeezing hard. "What other fucking companies?"

"Lawrence!" She looked from his hands up to his face.

"Sorry." He let go. Tried to haul in his temper. It seemed to be as intense as his fright. "What companies? Are you telling me that someone is still running explorer fleets?"

"Of course they are. Zantiu-Braun is the biggest space-active company of all, but Alphaston, Richards-Montanna, Quatomo are all still funding missions. None of the fleets are as big as they used to be before everyone started their asset-realization atrocities, but they still send out starships to survey fresh stars. And Zantiu-Braun has its portal colonies as well."

"Somebody's still founding colonies?" His voice had dropped to an aghast whisper.

"Yes. Lawrence, didn't you know any of this?"

"No."

"Shit." She was giving him a very troubled look. "Lawrence, I . . ."

"I want a full datapool trawl," he told the room AS in a flat voice. "Get an askping to pull all the information you can on current interstellar exploration. Specifically, the activities of the Alphaston, Richards-Montanna, Quatomo and Zantiu-Braun companies."

"There are no files on current interstellar exploration," the AS reported. "All information pertaining to current human starflight activities concern commercial flights and asset-realization missions."

Lawrence emitted a punch-drunk snort, astonishment momentarily overcoming his anger. "He lied to me. He fucking lied. My father lied to me. That *bastard.*"

"Lawrence?" Roselyn reached out tentatively, her hand touching his shoulder.

"This whole world is a lie. Everything I'm doing is a lie. Nothing is true." He jumped off the bed as if it had burned him, standing with every muscle tense. "I could be doing it right now. I could be on Earth at an officer academy. And what am I doing? I'm taking fucking *administration.* That's what I'm fucking doing. And I was so pleased about qualifying I celebrated. Celebrated! Sweet Fate . . ." His fists rose up, searching for something to strike. Something to punish. The rage felt superb, making everything so clear.

"Lawrence, calm down."

"Why?" he shouted. "I've been calm for four years. Which is what *he* wanted. That piece of shit. That's what McArthur's rigged this whole world to be—nice, quiet, obedient little drones doing as they're told to boost share prices."

"Lawrence, please." Roselyn was close to tears. "Stop it."

The hurt in her voice tripped every defensive reflex he had. Roselyn should never be upset; that was his reason for being alive. "Okay." He held his hands up, a conciliatory gesture. "Okay, you're right. This isn't you, you're not to blame." He hunted around the room, not knowing what he was looking for. Nothing here, that was for sure. "We're leaving. Get your stuff together."

"Lawrence, we can't leave."

"I have to." He lowered his voice, almost pleading. "Roselyn, he lied to me. He lied so big he warped the whole world around me. He trashed everything I wanted, everything I was. Can you understand that?"

She nodded slowly. "What are you going to do?"

"Ask him—no, make him—tell me the truth. I want to know if Amethi university degrees qualify me for another company's starship officer academy. I want to know how to get there. I want to know how much it costs. I want to *know.*"

* * *

They caught a taxi from Templeton airport. Lawrence told it to drop Roselyn at her dome first, then take him on to the Newton estate. It was midafternoon Templeton time when he finally got home, and he'd been traveling for nearly twenty hours. Changing his flights around had been relatively easy. The airline was used to people leaving Orchy early with injuries that had put an end to their skiing, and passenger manifests were drawn up to accommodate last-minute additions.

Full-spectrum lights were shining above him as he walked into the estate's main temperate dome, filling the vast enclosure with a harsh glare. The sun had fallen below Templeton's horizon days ago as Amethi's orbit carried it toward inferior conjunction. Somehow, the artificial lighting always seemed wrong to him, as if the engineers were using the spectrum of a different star altogether.

Faint multiple shadows fanned out around him as he walked along the stone path. The red-and-gold climbing roses that swarmed up the pillars on either side were beginning to fade, shedding their petals across the ground. As he walked along, he heard the shouts and whoops of his siblings playing in one of the sunken lawns, so he made a right-angle turn at the end of the rose walk, taking a longer route to the house, making sure he avoided them. He didn't want anyone to know he was back. It was strange, but he still felt protective toward his siblings. They were too young to know what kind of person their father really was. That childhood innocence should be preserved: it was too precious for him to ruin in the flare of temper and reckoning.

When he got to the landing he heard the soft murmur of voices coming from the study. He knew his father would be in there at this time, although it was unusual for someone else to be with him.

The door was partially open. Lawrence edged closer, careful not to make a sound. His father was one of the people in the room; he knew that cheerfully confident voice anywhere. The other was female. He thought it was Miranda, the latest junior nanny, another awesome beauty in her early twenties.

"... not even make it to the ski slopes," his father was saying in amusement. "The two of them away together for a week. Hell, he'll come back screwed senseless. I'll probably need to send an ambulance helicopter for him."

Miranda giggled. "That's what you wanted. You said."

"Yeah, I know. Damn, she's good at her job. Cheap at the price. And those legs of hers; have you seen them?"

Job. The word echoed silently around Lawrence's brain. Job?

"Yes, I've seen," Miranda said. "Why? You like?"

"Oh yeah, I definitely like. I'm tempted to pay for a month with her myself afterward."

"What? His girlfriend? That's really kinky, Doug. Besides, my tits are much bigger than hers. You said you like that. You always say you like that."

"So? I'd have the two of you together. That way I get the best of everything."

"Together?"

"Yeah, I love a good dirty threesome. It'd be quite something, watching you and her going to work on each other."

"You know, I think I'll enjoy that. Roselyn always looks so sweet. It would be fun to fuck her. I bet she'd be really hot if you press all the right buttons."

Without the name Lawrence could have forced himself to believe they were talking about someone else. That this was some ludicrous, appalling coincidence. Two other people going on a skiing trip. A different girl his father fancied. Someone else. Not them. Not him. Not Roselyn.

Lawrence's trembling fingers pushed at the heavy wood door. His father was sitting behind the desk, with Miranda perched in front of him. The front of her dress was unbuttoned, allowing her breasts to spill out. Her right nipple was pierced by a diamond stud. Doug was slowly licking the bud of erect flesh. He looked up in dismay as the door swung back to reveal Lawrence standing there.

Miranda gasped and hurriedly pulled her dress together.

"Son?"

It was the first time Lawrence had ever seen his father flustered. The guilt and shock simply didn't belong on that ever-assured face.

"Oh, boy. Listen, what we were saying . . ."

"Yes?" Lawrence surprised himself by how calm he was. "What, Dad? It's not as bad as I think? Is that what you're going to tell me?"

Doug's political control came back with a rueful grin. "I don't suppose I can, really."

"You bought her."

"It's a little more complicated than that."

"How? How is it complicated? Did you pay for her?"

"Lawrence . . ."

Lawrence took three fast paces into the room, bringing him up to the desk. "DID YOU PAY FOR ROSELYN TO SCREW ME, YOU PIECE OF SHIT?"

Doug flinched back from the fury. "Look, you were losing it, all right? Your school grades were rock bottom, you didn't have any friends, the psychiatrist said you were borderline emotionally retarded, unable to connect with the real world. I was seriously worried. I am your father, however good or bad I am at it."

"So you bought me a whore."

"Son, you had to realize how much Amethi has to offer for someone like you. I couldn't have you throw all that away. And she connected you. Call her what you like. Blame me for the way you met, and I admit it was pretty low. But look at you now, look what she's done, how much she's straightened you out. You're top of the class, you play in all the A-teams, outside school you're the one everybody socializes with. She's shown you how much there is to life here. I promise you I never lied when I said I was proud of what you've achieved."

"Of course you're proud. I became exactly what you wanted. Why did you ever have me, Dad? Why didn't you just clone yourself?"

"Son, please, I know this isn't easy. I mean, hell, I never thought you would fall for her quite like this."

"Why not, she's hot, remember? What else was I going to do, a loser like me?"

"Lawrence, you'll get over this. Admittedly"—he shrugged reasonably—"you'll probably hate me forever, but I can live with that, because I know I did the right thing."

"No, Dad, you did not do the right thing." Lawrence turned round and walked out.

Lawrence didn't know how he got there. He didn't even know when he got there. But sometime later that day, or week, or year, he stood outside the door to the O'Keefs' apartment. Even when it finally came into focus and he recognized where he was, he took a long time before he brought his hand up and knocked.

It was a gentle rap with his knuckles. Lawrence barely heard it himself. He knocked harder. Then harder still. He pounded on the door, seeing it shake in the frame.

"Open up!" he screamed. "Let me in!"

The lock clicked back and he stopped hammering. His hand hurt. Drops of blood welled up on his grazed knuckles.

Lucy O'Keef opened the door. "Oh. Lawrence. It's you." Her shoulders sagged, presumably with remorse. "Your father called me earlier. He said you . . ."

"Where is she?" he growled.

"I don't think this—"

"WHERE IS SHE?"

Roselyn eased her mother to one side. She must have been crying a long time for her eyes to be so red.

At that moment, she'd never looked more vulnerable and adorable. He stared at her mutely. There was nothing he could bring himself to say. Because he knew now that it was all true. And the one thing he couldn't stand was for her to have to say it to him.

He walked off back down the corridor to the elevator.

"Lawrence." Roselyn came out of the apartment, following him. "Lawrence, please, don't go."

He walked faster. Then he was running. His hand slammed on the little silver button set in the wall. Mercifully, the elevator door slid open straightaway. He stepped inside and pressed for the lobby.

"Lawrence." She slapped her hand against the door edge, and it froze. "I'm so sorry, Lawrence. I'm so sorry. I love you."

"He paid you." His thoughts were in so much turmoil he had trouble getting the words out. "He made you do it."

"No." She was sobbing. "No, Lawrence."

"What then? He didn't pay you?"

"The money wasn't for me. You don't understand. It's not like that."

"Like what? What can I possibly not understand?"

"I said yes because of Mary and Jenny."

"Your sisters? What the hell have they got to do with this?"

"We had nothing left. Nothing. McArthur shares are just about worthless on Earth. Not that we ever had many. You can never know what that's like, to be poor. Not you. You're a golden child on a planet that's too young to know any form of decay. This was the only way we could escape Dublin, get off Earth. Me . . . doing this."

"You're part of it. You're the biggest part of his lie there was. I hate you for that!"

"I never lied to you, Lawrence."

He punched the lobby button again, wanting this torment to end. "Shut up! Shut up, you bitch. All of this has been false. All of it."

"Only the beginning." She leaned against the wall, utterly exhausted. "That's all, Lawrence. Just me saying hello. One little word. Not the rest of it. Everything since then was genuine. I can't fake loving you for a year and a half. You know it was real. You know that!"

The elevator doors slid shut. Roselyn's devastated wail stabbed clean into his heart.

Vinnie Carlton opened his apartment door to find Lawrence slumped against the wall outside. "What the hell happened to you, man?"

Lawrence showed no sign he'd even heard the question. He was staring ahead without seeing anything. Vinnie shrugged to himself and put a hand under his friend's shoulder, helping him up. "Let's get you inside before the cleaning robot shoves you into the rubbish chute," Vinnie said. "Come on, you look like you need a drink or ten." Lawrence didn't resist as he was steered into the apartment's lounge. A mug of tea was put into his hands. He drank it automatically, then sputtered. "That's disgusting, Vin. What's in it?"

"Rum. I like it."

"Oh." Lawrence drank some more, sipping it down carefully. Not too bad, actually.

"Going to tell me what happened?" Vinnie asked.

Lawrence glanced around uncertainly. He'd come here because Vinnie was the only person he could turn to without getting parents involved. Although Vinnie was a really good friend, Lawrence tended not to come to the apartment much. He'd never quite forgiven Vinnie for saying he and Roselyn couldn't use it to have sex.

Everything in his life was connected to Roselyn.

"You've no idea how lucky you are living by yourself," Lawrence said.

"How so?"

Lawrence told him.

Vinnie sat and listened to the entire story with his face running through a wide spectrum of emotions. "Shit, Lawrence," he said at the end. "This is going to sound stupid, but are you sure?"

"Oh, yeah. I'm sure."

"Christ. I don't believe it. I thought Roselyn was great. She was so . . . real."

"Right. Girls, huh?" Lawrence tried to make it sound as if he didn't care, as if this were just a standard-issue problem in any relationship. Happened every week. It didn't work. He was too close to breaking down again. Hated himself for that.

"Yeah, girls."

The feeling in Vinnie's voice caused Lawrence to look around the lounge, as if he'd just become aware that something was missing. "Where's Nadia?"

"Ha! We split after the party at Hillier's. She said she didn't want to know someone who was so embarrassing to be with in public. Bitch! What were we supposed to do? Let Alan get beaten to a pulp?"

Lawrence smiled briefly at the memory. "Well, he almost did, anyway."

"Yeah! I just don't have any respect for someone who acts like that."

The humor faded.

"What are you going to do now?" Vinnie asked.

"I don't know. I can't go home, not after this. And I can't ever face her again."

"Well, shit, Lawrence, you can stay here, you know that."

"Thanks. But I can't. I've got to move on. You know? Get clean away."

"You mean one of the other cities?"

"No. I mean right away. Listen, you came from Earth; was she telling me the truth about other companies still flying explorer starship missions?"

"Sure. There aren't many of them left, mind you. I didn't pay a lot of attention to that kind of thing. But she was probably right about Richards-Montanna, and she'd definitely be right about Zantiu-Braun. Hell, that company owns half the bloody planet these days."

"Then why isn't any of this in Amethi's datapool?"

"Oh, it'll be there. It's just that you haven't got the access codes."

"Okay. Then why restrict it? It's not that seditious."

"Who knows? Corporate paranoia, most likely. Don't forget this isn't a democracy."

"Yes, it is," Lawrence said automatically.

"Corporate stakeholding is a little different from the traditional model. Your vote is balanced according to your wealth."

"It has to be. You can't have the poor voting themselves more welfare money. That's economic suicide."

Vinnie pressed his hands to his temple. "Lawrence, I'm not arguing with you. I chose to come and live here, remember. Amethi is quiet and prosperous, a condition that it buys for itself with a heavy load of social hypocrisy. For all that, it has a lot going for it. All I'm saying is, if the Board want to guide our development steadily along the don't-rock-the-boat course they've mapped out, then there are some policy areas and activities best avoided. I'm taking a guess that they don't want anyone to consider the option of leaving. They would hardly be the first government to have that opinion. And the more new planets that are discovered and opened to colonists, the more options there are for people to leave and pressure to facilitate it. If there's nowhere to go, then you have to stay here and work for the Greater Good of the community."

"Bastards."

"It wasn't personal, Lawrence. They didn't notice your obsession with exploring new star systems and cut off all access to starflight information from the datapool."

"I have to leave," Lawrence moaned. "I just can't stay here. You understand that, don't you?"

"Are you talking about going offplanet?"

"Yeah. I want to go to Earth. If there's any chance, *any,* that I can get on an exploration program, I have to take it. I couldn't live with myself if I didn't, not now."

"Okay. I can see that."

Lawrence looked up, trying to maintain some dignity. He didn't want to beg; not to a friend. "Will you help?"

"How?" Vinnie was suddenly cautious.

"Nothing much. I'm rich: I've got a stake in McArthur, remember. Which came out of trust on my eighteenth birthday. I can do what I like with it now. And what I like is to buy a ticket to Earth."

"Your old man will never let that happen." Vinnie took a moment. "Is there enough? It cost my family a bloody fortune to send me here."

"There's enough. But I know what my father will do if I try cashing in my stake. That's why I want the name of the legal firm that runs your family's affairs. They're independent, aren't they? If anyone can help push this through, they can."

"Won't do you any good. Sure they're independent, but your daddy's on the Board. If he says you can't go, there isn't a lawyer or court on the planet that can have that overturned."

"Fuck it!" Lawrence could feel his muscles tensing up. So far he'd received every shock with amazing composure. But it wasn't going to last. Each time, the urge to lash out physically was stronger. "I have to go," he shouted at Vinnie. "I have to."

"I know." Vinnie gave him a dubious look, weighing up some invisible options. "Okay. I might be able to help. But if I do, and this doesn't work, you are going to be in seriously deep shit."

"You mean I'm not now?"

"Not compared to this, no."

Lawrence was suddenly very interested. He knew Vinnie well enough to know this wasn't the usual bullshit they fed each other. "What is it?"

"I have some software that I shouldn't have. And I really mean shouldn't. It's called Prime, and it's so powerful it actually has a weapons-grade classification on Earth. Taking a copy off the planet is probably a capital crime."

"No shit? What will this Prime do?"

"It's a quasi-sentient routine; you run it in any kind of neurotronic pearl and it'll be able to subvert every AS on Amethi. Not only can you block every askping your father launches to find out where you are, it'll also cover your tracks at the bank when you take your money out to buy a starship ticket. The first he'll know about you leaving Amethi is when you send him a video file of yourself on a Mediterranean beach sipping piña coladas."

"Damn, it's that good?"

"I'm not even going to risk giving you the top version, no way. But I'll let you have a version that can do the job. And, Lawrence, when you get to Earth, don't advertise the fact that you've got it. Prime is superior to anything on Amethi, but I've had my copy for a while. I expect Earth's datapool will be protected. Certainly the sensitive sections will have shields."

"Okay. I won't forget. And, thanks."

"That's okay. You've been a good friend to me here. I appreciate it. Just remember me when you're on your adventures." He grinned. "That is, right up to the point where you get caught with it."

CHAPTER EIGHT

IT WAS ANOTHER HOT, HUMID DAY IN MEMU BAY AS LAWRENCE led the platoon on their sixth morning patrol. They'd been on Thallspring for a week now, and this campaign was much worse than the last time he'd walked these pleasant, open streets. Ebrey Zhang hadn't used a collateral necklace yet, but Lawrence was sure it could only be a matter of time.

Not that this was as bad as Santa Chico, he kept telling himself. Be grateful for small mercies.

Platoon 435NK9's established patrol sector was the Dawe District. It was an inland area, mainly residential, where the sprawl of neat suburban homes encroached on one of the small hills at the foot of the fortress range behind the town. The streets were broad and clean, with tall Sitka spruces on either side, their branches twisting about wildly to produce a profusion of strange dapples on the pavement. Two tram routes linked Dawe's citizens to the center of town, the big clumsy vehicles trundling along their tracks with bells clanging brashly at the sight of any cyclist pedaling away ahead. Strangely, the only time the bell didn't sound was when a Skin suit appeared on the road in front.

Ostensibly the platoon were there to back up the regular police foot patrol. In reality their regular visibility was emphasizing Z-B's presence.

Platoon 435NK9 made their way up a street lined with small shops. Not many people were outside in the midmorning sun, and those who were stared resentfully as the Skins lumbered past. Taunts and obscenities dogged their every move. The constables they were supposed to be accompanying smiled at the shouts without any attempt to conceal their contempt.

"Oh, man, I hate this," Hal muttered. It was the hundredth time he'd complained that morning.

Lawrence checked the positional display that his suit AS was displaying. Hal was keeping pace on the right flank. "Just stay with it, Hal. They haven't done anything."

"Yeah, give the rest of us a break," Lewis said.

"But listen to them."

Lawrence hadn't been doing anything else. All morning he'd heard *KillBoy.* That one word was yelled over and over again, intended to provoke and intimidate in one hot blast of air. The alleged name of the sniper who'd shot Nic after landing.

KillBoy, already the Robin Hood of modern legend. A wounded, mutilated or persecuted victim of Z-B's last asset-realization mission to Thallspring—take your pick. He prowled the streets of Memu Bay looking for lone Skin suits. When he found one, superweaponry would cut through its carapace as if it were real human skin. Another vile invader would bite the dust, and all good Memu Bay citizens could walk taller knowing their oppressors were going to lose, and that there was justice in the universe.

Lawrence didn't like it at all. There was no KillBoy, not in the flesh. Just some shadowy resistance group, probably set up by the government, who'd been issued some nasty hardware. Rumor and tension fabricated the rest. But it gave the locals a solidly believable icon, a protector who would save them if they did step out of line. Not good, for that belief gave them a sense of invulnerability. Which they certainly didn't have against Skin. And Z-B's platoons were edgy after the disastrous landing. The situation could only get worse.

Music suddenly swirled out of an open bar, a dance track that quietened with equal speed. Three of the platoon had turned at the disturbance, only to be greeted with several young men lounging around the bar's door, giving them the finger.

"Guess we can cross that one off the list," Karl said. "It's not exactly welcoming."

"None of them are," Edmond said.

"Hell, it was never on my list to start with," Hal grumbled. "Man, what a dive. And there's no real action in this part of town. We've got to get us down to the marina for any serious pussy."

Lawrence grinned at them as he listened to their inane chatter. They were due some outleave tonight, finally getting away from their barracks. Z-B had commandeered a string of resort hotels just behind the marina to billet the platoons in. Physically, there was nothing to complain about. He'd got himself a double room in a four-star hotel. Big comfy bed, balcony facing out across the harbor; it had a decent restaurant downstairs, and a bar, games room and gym, swimming pool, even a sauna—which the bastard officers had monopolized. But they weren't permitted out. Not until things had quieted down, Ebrey Zhang declared.

By the end of the first week their commander had decided that time had come. There had been no more sniper incidents. The production levels at the biochemical plants had risen back close to their prelanding levels. They were becoming grudgingly accepted by the local population.

Last night some other platoons had tested the waters, and nothing too untoward had occurred. Tonight, 435NK9 would get its chance to paint the town red.

Lawrence thought it was too early. The junior officers must be feeding Zhang exaggerated reports of the patrol sweeps for him to think things were calm around the city. But nobody had asked his opinion. Still, he was glad the platoons were getting leave. He'd need two uninterrupted days at some time to go out into the hinterland and realize his own personal asset.

A TVL88 helicopter growled overhead, meandering around the edge of the foothills. Several Skins sat on the broad side door, feet dangling out above the skids as they watched the buildings below. Immobile, featureless gargoyles, ready to react to any trouble. The helicopters were Z-B's own KillBoy, visible support for the troops on the ground, providing invincible firepower backup. Several of 435NK9 waved as the machine passed by.

"For heaven's sake, you odious child," Odel was saying. "No Thallspring girl is going to look at you. When we go into a bar, we'll clear it faster than a swarm of hornets. I absolutely guarantee it."

"You tell him, cretin," Karl said.

"He's right, Hal," Lewis said. "Stick with a sim-suit running porno-i's. Those girls will do anything you tell them."

"I don't need none of that shit," Hal protested. "They ain't too fond of us back in Queensland, either, but I never had any trouble scoring down on the Cairns Strip."

"Didn't have much money left over afterward, did you, though?" Karl said. "And every morning after it's a trip to the surgery for an antidose."

The platoon's communication link filled with harsh laughter.

"This ain't funny!" Hal said. "My balls are going to explode unless I get some serious pussy tonight. And I'm telling you, it ain't going to be no trouble. Not for me. I'm younger than you guys. And I'm *built,* you know. I've got the look. The girls will go for that, no matter where we are in the galaxy. Being fit never goes out of style."

"Oh, give me a break," Lewis said. "If they go for anything, it's not going to be some punk delinquent working off a court rap."

"I fucking volunteered for strategic security!"

"What the chicks go for is a guy with some experience. Right, Dennis?"

"Bull's-eye. You've got tonight's tactics all wrong, kid. We have a certain novelty value: face it, technically we're aliens from another planet. The ladies will be intrigued by us. We can snag them with that. And the more planets we've been to, the more fascinated they'll be by us. Everyone apart from Hal will benefit."

"Hey!"

"Face it, kid, you just haven't got the staying power us mature guys have."

"That's a bunch of crap. You old farts can't even get it up, never mind keep it there. The girls know what they like, and tonight they're going to overdose on me."

"Let's keep this formation tighter," Amersy said before the bull got any worse. "Come on, Jones, you're falling behind. And, Dennis, close in; give Odel some support."

"You got it, Corp."

The platoon checked their relative positions and improved their formation.

Up ahead of Lawrence, the street opened out into a small square where a tiny central lawn was surrounded by neat flowerbeds. Clunky old gardening robots crawled along the edge of the white-and-scarlet salvias, rusty implements prodding at the soil. The constables slowed their pace, dropping behind. They did it every time there was a major junction, in case there was some kind of ambush around the corner.

Edmond and Lewis went wide, getting close to the shop fronts and covering the opposite sides of the square as they moved forward. There was no ambush. No *KillBoy*. The platoon crossed over the square with the constables ambling along behind.

"Do you reckon we should buy some clothes from around here first?" Hal asked. "I mean, to blend in with the fashions, and such. We don't want to come over as total dumbass aliens. You've got to look sharp in any bar."

"Hal," Lawrence said, "let's stay focused on current affairs, shall we?"

"Sure thing. Sorry, Sarge."

Lawrence walked off the grass and crossed the road. He didn't like to intervene with the normal platoon bull. But the kid was too boneheaded to take Amersy's hints. With a bit of luck, tonight he would actually find some silly tart who fancied screwing an alien invader. The kid needed some way of letting off tension. He was starting to irritate everyone.

Red icons flashed up over Lawrence's sensor grid. The suit AS spliced his communications into the link that Oakley's platoon was using. A 2D indigo city map expanded out of its grid, featuring deployment symbols blossoming with script orders as the headquarters tactical AS analyzed the incident.

The incident: one of Oakley's platoon was down, a squaddie named Foran. A stone wall had collapsed on top of him. Civilian datapool overlap showed some kind of traffic malfunction in the same location, a thirty-ton robot truck had gone offline. Foran's medical telemetry was intermittent from underneath the pile of rock, but the information so far showed that his Skin carapace had been breached in several places by the fall. Internal organ damage, broken bones and blood loss were showing.

Oakley's platoon was patrolling the sector adjoining Lawrence's.

"Dispersal pattern one," Lawrence told his platoon. It could be a classic diversionary tactic, in which case it was unlikely that the true assault would come quite so close by. But he wasn't taking chances, not in this environment.

The platoon exited the street with smart professionalism, going into the nearest buildings through doors and larger open windows. Lawrence himself darted into a small hairdresser's. The row of women sitting under tentacle-armed IR drying units went rigid with alarm. Both the constables were left alone outside, staring around in astonishment. Video telemetry grids showed Lawrence several outraged homeowners yelling at his troopers.

Lawrence switched to the command channel. "Oakley, do you need help?"

"Shit, dunno—! Get it, get it. That one! Come on, lift."

"Oakley, what's your status? Is this a prelim diversion?"

"No, it's fucking not! A goddamn wall has fallen on him. Shit, it's the size of a mountain. We're never going to shift it."

Lawrence saw the deployment icons representing Oakley's platoon all clustering in one spot. "You're getting dense. If that sniper's around, you're going to get punished. Suggest you pull some of your team back."

"Fuck you, neurotronic-brain Newton! That's one of mine under there."

"Newton," Captain Bryant said, "take some of your platoon and help the dig. We need to get Foran out of there."

"Sir, I don't think that's—"

"He's alive, Sergeant. I'm not allowing one of my men to die here. This was a traffic accident, not a setup for a sniper. Understand?"

"Yes, sir." Lawrence took a moment to compose himself, knowing full well what his own medical telemetry would be showing Bryant. Not that the captain would be looking. "Hal, Dennis, you're with me. Amersy, finish our sweep."

It was a narrow alley in an old commercial district. Vertical stone-and-concrete walls with white paint badly faded and peeling, scraggly weeds sprouting all along the base. The only windows were high up and covered with bars, glass too dirty to see through; doors were sturdy metal, welded up or sealed with thick riveted plates. Dust was still rolling out of the entrance when Lawrence arrived, thick gray clouds of dry carcinogenic particles that latched on tenaciously to his Skin carapace. Crowds of civilians were gathered around on the main street, several with handkerchiefs over their faces. They all peered into the gloomy alley. Two TVL88 helicopters were circling just above the rooftops, magnetic Gatling cannon extended from the noses like squat insect mandibles. Their rotors were exacerbating the dust problem.

Lawrence checked around quickly. There was no obvious high building providing a firepoint nest down the alley. His suit AS increased the infrared sensor percentage as he made his way into the dust; his visual picture lost all color apart from gray, black and pink—though the general outlines maintained their integrity. He saw rubbish piled up against the walls on either side of the alley: boxes, bags and drums all printed with the town's civic emblem, denoting it ready for collection. There couldn't have been a pickup truck down here for a month. In some places the piles were so big they actually sprawled right across the cracked tarmac. Lawrence had to clamber over them.

There was a kink in the alley, and he was abruptly facing the collapsed wall. He grunted in dismay. "Shit, this is a mess." A huge section had collapsed, leaving tattered shreds of tigercotton reinforcement mesh flapping along the jagged upright edges. The building behind had been some kind of warehouse, or disused factory, a big empty cube with aging metal beams and ducts running up the walls, now bent and twisted, whole strands torn free and dangling precariously. Its flat concrete panel ceiling had collapsed along with the wall, crashing down and shattering over the floor and a big crumpled truck. On the opposite wall at the front of the building, a roll-up door had been torn apart, showing a wide street outside that was clogged with stationary traffic.

Lawrence took only a second to work out that the truck had gone runaway, bursting through the door to ram into the wall. Exactly when Foran was standing in the alley on the other side.

That was quite extraordinary bad timing.

He didn't believe any of it. Instinct hardened and sharpened by the last twenty years was flashing up warning icons of a kind more potent than any AS symbology.

Skins swarmed over the massive pile of debris. They flung body-sized lumps of concrete and stone through the air as if they were made of feathers, digging out a wide crater above their fallen comrade. They possessed the desperate stop-go motions of hive insects synchronized for maximum productivity.

"Let's get to it," Lawrence told Hal and Dennis curtly. They joined the other Skins, prizing big chunks of masonry free. Grit and powdery fragments spewed off each piece like a dry liquid. The filthy deluge of dust made visibility difficult even with Skin sensors. Infrared helmet beams were turned up to full intensity, creating swirling crimson auras as if vanquished stars were expiring in the cloud.

It took nearly fifty minutes to excavate the rubble. At the end there was only enough room for two Skins to work in the bottom, carefully picking up lumps of stone and handing them to a chain of Skins to be carried clear. The crater walls were so unsteady it would take very little to trigger a further collapse. Foran's Skin was slowly exposed. Dust around him was clotted into mud with glistening scarlet blood. Bloodpak reserves and stored oxygen had kept him alive, though nearly half of his medical telemetry was in the amber, with several organ functions flatlined red. He was unconscious, too, when he was finally lifted clear.

All the paramedics did was hook his Skin umbilicals up to fresh bloodpaks. The Skin was providing the most stable physiological environment possible until they could get him into trauma surgery. They rushed him away to the medevac helicopter that had landed in the middle of the street at the end of the alley.

"I didn't think anything could get through our Skin," Hal said lamely as they milled around at the foot of the rubble.

The dust was settling now that the digging had stopped, cloaking the immediate vicinity in pallid gray.

"Believe it," Dennis said. "A hundred tons of sharp rock falling on top of you is going to puncture your Skin."

"Poor bastard. Is he going to be okay?"

"His brain's still alive, and oxygenated. So they'll be able to bring him up to full consciousness without any trouble. The rest of him . . . I don't know. He'll need a lot of replacement work."

"But we bring prosthetics with us, right?"

"Yeah, kid, we've got a whole bunch of biomech spares. I guess at least he'll be independently mobile at any rate. Whether he'll ever rejoin the platoon is another matter. You know how top-rate we have to be."

Even with Skin muscles augmenting every move, Lawrence felt distinctly non–top-rate right now. His own muscles ached from the effort of digging. For a moment, the mantle of cloying dust brought up an image of Amethi during the Wakening, when the slush stuck to everything, imprisoning the world in a decrepit winter. He looked around the narrow alley. The piles of rubbish were as wide here as they were at the end. Foran would have had to walk right next to the wall.

Lawrence slowly moved across the lower part of the rubble until he could see back into the ruined building. The traffic on the main road in front was moving again. Skins stood guard beside the wrecked door. A couple of techs were examining the truck, shifting the concrete slabs so they could get into the engine compartment. Captain Bryant was standing behind them.

"What happened to it, sir?" Lawrence asked over the secure command link.

"They don't know yet," Bryant replied. He sounded annoyed. "Damn, I really don't need accidents like this messing up my command."

"This wasn't an accident, sir."

"Of course it was, Sergeant. The truck went out of control and crashed."

"It crashed into one of us."

"Your concern for our personnel is commendable, but in this case it's misplaced. This is a traffic accident. A tragic one, I accept, but an accident."

"What did the traffic regulator AS log as the fault?"

"It didn't log anything, Sergeant. That's the problem. The truck's electronics crashed."

"The software or hardware?"

"Sergeant, you'll be able to read the report for yourself as soon as it's been made. We haven't even accessed the truck's memory block yet."

"But what about the fail-safes?"

"Newton, what the hell are you doing? What's the matter with you? He will recover, you know, he'll get the best possible treatment."

"Sir, I just don't see how this could be an accident."

"That's enough, Sergeant. It's unfortunate, but it happened."

"Not one fail-safe cuts in when the electronics crash. Sir, not even Thallspring technology is that shoddy. Then it veers off the road to hit a door square in the center."

"Sergeant!"

"And after that it demolishes a wall while one of our men is standing directly behind it. One of the few things that can damage a Skin suit. I don't buy it, sir. That's not one coincidence, that's about a thousand falling into line."

"*Enough,* Sergeant. It was an accident for exactly those reasons. Nobody could organize anything like this, nobody knew when Foran was about to walk down this alley. That is, nobody else knew. Of course, I was supervising this morning's deployment. Are you saying I was at fault in some way?"

"No, sir."

"I'm glad to hear that. The matter is closed."

The command link went dead. Lawrence shook his head. A fairly pointless gesture in Skin. The trouble was, he could understand why Bryant was reacting in this way. The captain was too weak to acknowledge an opponent who could organize such a beautifully elaborate trap. Accepting the fact that someone did have the knowledge and skill to bring it off was massively unnerving.

"If the Wilfrien were alive today, you'd think you were looking at an angel. They were the golden ones; to be in their presence was to adore them. At its height, the kingdom of the Wilfrien was among the most powerful members of the Ring Empire. Indeed it was one of the founders. Its people helped to explore the thick wreath of stars around the galactic core. They made contact with hundreds of different races, and brought them together. Their technology was among the best in existence. Wilfrien scientists developed fast stardrives that everyone else copied; they worked out how to create patternform sequencers that could reshape raw matter into machines or buildings or even living organisms. And they gave all this knowledge freely to the peoples they encountered, helping them to incorporate it into their societies, extinguishing poverty and the conflict that such disparity always brings with it. They were a wise and gentle race that were admired and respected by everyone else in the Ring Empire. They set a standard of civilization to which most aspired and that few ever really achieved. Every story of the Ring Empire includes them, for they were the shining example of what it's possible for sentient life to become. Whenever we say Ring Empire, more often than not we're thinking of the Wilfrien society." Denise smiled round at the children. They were out in the school's garden, relaxing on the lawn with glasses of cool orange juice and lemonade. Big white canvas parasols had been opened, throwing wide shadows across the grass. The children all sat in the shade, out of the burning morning sun. As always, they watched Denise with worshipful eyes as she invoked their sense of wonder.

"The Wilfrien inhabited over three hundred star systems. With their patternform sequencers they had constructed fabulous cities and orbital stations. They grew themselves castles in the depths of space; they had metropolises that soared among the storm bands of gas giants, more delicate and intricate in appearance than the twirls of the clouds through which they meandered; they even encased starburst towers inside lenticular force fields and sailed them across the furious surface of their suns as if they were nothing more than coracles on a woodland lake. Oh, they were impressive, the Wilfrien. They lived in such bizarre places almost for fun, to laugh and enjoy everything the universe had to offer, for they could be as wild and exuberant as they were thoughtful and dignified."

Her narrative never faltered as Prime monitored the progress of the Z-B platoons going about their morning patrols. Information gathered from the platoons' own communication links was insinuated into her mind by d-written neurons. She regarded their busy little icons and whirring scripts with mild contempt. So crude, when simply *knowing* the raw data was easy. Several Skins were approaching the alley. "Given their nature, not to mention their reputation, Mozark knew he would be visiting the Wilfrien even before he took off on his quest. Strangely, the closer he got to the kingdom of the Wilfrien, the less impressed the local people were by the magnificent race adjoining them. Eventually, when he arrived at their home planet, he found out why."

Simple time velocity equations provided a list of three

possible trucks. Prime programs installed themselves in their electronics, erasing their own datapool traces as they went.

"The Wilfrien were old as a species; even as individuals their lives extended for hundreds of millennia. They had traveled further and faster than anyone else in the Ring Empire. Their peerless technology had plateaued. Every race around them was content and wealthy thanks to their largesse. There was nowhere outward left for them to go, neither physically nor mentally. If they could be said to have a flaw, it was their impetuosity and interest in all that surrounded them. Yet now, there was no strangeness in their universe, no mystery. In olden times, men would write *Here be Dragons* around the edge of their maps, when what it really meant was: we don't know what's there. None of the Wilfrien starcharts had dragons; they were sharp and detailed all the way out to the end of the galaxy. The only journey left to them now was the journey back to where they came from. They turned inward.

"Mozark landed on the edge of a city whose towers put those of The City to shame. Some of them had tops that pierced the atmosphere; several were alive, like reefs of coral that had thrust up out of the ground; others were composed entirely from planes of energy fields. He even saw one that was made up of blobs of translucent sapphire, as if they were cells ten meters wide; they all slithered and slipped around each other at random, though they always maintained the same overall shape. But they were all empty, those dizzying spires and paradise palaces. The Wilfrien had abandoned them to live on the ground below, leaving them open for wild animals and creeping plants to claim them back."

One of the Skins was entering the alley. Mounds of rubbish that the cell members had carefully dumped over the last week forced him to walk close to the wall. Denise gave her final orders to the Prime that had taken complete control of the truck. It cut its link to the traffic regulator AS with a last emergency declaration call—broken as it hit the safety barrier. The empty warehouse doors were dead ahead. Inertia took over as the Prime erased itself, propelling thirty tons of truck through the door and onward toward the rear wall at fifty kilometers an hour. "Of course it would take thousands of centuries for any kind of decay to assault the fabulously strong materials that the Wilfrien buildings were made out of. For now they stood as tall and proud as always. But the signs of their inevitable future were already beginning to show. Leaves and twigs were accumulating around the base, mulching into a rich compost from which ever more vigorous plants grew; colors were losing their sheen below spores and grime. Hundreds of years of winds had blown soil and sand in through the lower floors, allowing the rot to begin around all the artifacts that were fabricated from simpler compounds.

"Hardly believing what he was witnessing, Mozark walked across fields of food crops that had been plowed into what had once been majestic parkland. The Wilfrien who were tending them left their labor beasts to greet him warmly. Bowing and stuttering in confusion, for they still inspired awe among those in their presence, he asked what had happened to their civilization, which had embraced over a thousand light-years. They smiled kindly at his lack of comprehension and told him they were done with it. Their battle for knowledge, they said, was won; they knew everything worth knowing. What they were, therefore, had no further purpose in the context of their achievement. They were now embarked on a completely different path of development, one last final application of their glorious heritage. Life itself would become pleasant and simple. Their bodies were modifying and adapting, melding to fit perfectly with a natural planetary environment. But unlike a primitive, pretechnology society, they would never starve or become ill, for this was a designed simplicity, taking advantage of everything their planet could provide. Their minds would quiet over the generations until the joy of a single sunset provided as much satisfaction as breaking down the barriers of space and time with the mental tools of mathematics and physics. They would raise their crops and their children and dance naked as raindrops fell from a wild sky. As the relics of their past crumbled and sank silently back into the earth, so they would become one with their world and be at peace with themselves.

"Mozark raged against such deliberate decay, forgetting both his manners and his earlier awe. He asked—begged—them to reconsider, to find new challenges. To become once more the golden Wilfrien he had worshiped from afar. They laughed sadly at what they saw as his simplicity in believing that progress could only ever be found in one direction, onward and upward. Their nature, they said, had led them to this point. This was what they were. This was what they wanted. Life without complexity. In this new-dawning milieu they would be happy without even trying. Isn't that what all life should be? they wondered. Did he not want to reach such a destination himself? they asked. When he told them of the quest he was on—for himself, for his own kingdom, and for Endoliyn—they laughed once again, but with even greater sadness. Travel far enough, they told him, and like us you will arrive at the place you started from. The universe is not big enough to hide what you seek.

"Mozark went back to his ship and took off immediately. He pushed his starship's engines hard, racing away from the Wilfrien homeworld as if it were filled with monsters. As it shrank away in the viewscreens, he cursed them for betraying their ancestors' monumental struggle. Everything every Wilfrien in history had achieved, they had thrown away like spoiled decadent children. He thought it to be a calamity of the highest order, made worse by the fact that only someone from outside could really appreciate its true magnitude. The Wilfrien couldn't see what they were doing was so wrong. Their rush into decline went against every belief he treasured. He hurt just thinking what Endoliyn would say if he returned home to tell her that true happiness could only be found in ignorance. For that was what he considered the Wilfrien were doing, closing themselves away from reality like a flower at the end of the day. Perhaps, he thought, they had been beaten by the universe after all, that its wonder was just too great for them. He knew that for all their splendor, his nature was stronger: he would never admit such defeat for himself or his people. In that alone, he had risen above his old heroes, although he was sure he would regret their passing for the rest of his life. A little of the magic had disappeared from the galaxy; the golden were tarnished now, never to regain their luster. But still he flew on, as determined as ever."

A bulky black helicopter thundered low overhead, the sound washing out Denise's voice. The children leaped to their feet and charged out from under the parasols to watch the alien warcraft pollute their sky. It streaked away toward the Dawe District, heavy, menacing guns sliding out of its nose cavities with smooth urgency that was almost a sexual motion.

Denise followed them out into the sunlight, watching hot fumes spilling from the invader's gill-like turbine baffles as it filled the air with its battle cry. She took hold of Wallace's and Melanie's hands as the children looked uncertainly from the racing machine back to her. "They won't sell many ice creams at that speed, will they?" She chuckled. The children broke into ebullient giggles, laughing and pulling faces at the retreating horror. "Come along then." She swung the hands high, allowing Melanie to twirl below her arm. "I've a tale to finish. We're almost done for today, and the nasties aren't going to spoil our fun, are they?"

"No," they all yelled. Getting back under the parasols became a race, with lots of jostling to be at the front. Denise let go of Melanie and Wallace, allowing them to sit at her feet with exaggerated self-importance.

"Miss, did the Ring Empire have people like Zantiu-Braun?" Jedzella asked.

Denise glanced around at the worried faces. "No," she assured them. "There were people who were bad, sometimes evil. But the laws of the Ring Empire were strong, and the police clever and vigilant. Nothing like Zantiu-Braun and this invasion could ever happen there."

Edmund turned round to his classmates and went *Phew,* wiping his hand across his brow. The children smiled again, content that the Ring Empire remained sacrosanct.

Denise hopped off the tram at its third stop along Corgan Street, several hundred meters behind the Skin platoon. She knew where they were without having to apply her d-written systems to the datapool. The noise of ragged voices keyed her in.

KillBoy's in the driving seat
Crash hit! Crash hit!
KillBoy's seen the meat
Crash hit! Crash hit!
Skins are in the bodybag
Crash hit! Crash hit!

She smiled behind her sunglasses. KillBoy wasn't something she could take credit for: some nameless poolpoet had invented him on landing day after the sniper shot. But he was rapidly becoming one of the cause's biggest assets.

It was youths who were doing most of the chanting. Respectable, responsible adults who would normally call for the police the moment two teenagers started drinking beer on the street were nodding silent encouragement as they walked along the pavement.

This was why she was here, to gauge the mood of the average Memu Bay inhabitant. It wasn't something she could determine from editorials and reports out of the datapool. Judging by this response, her fellow citizens had a vicious streak she wouldn't have necessarily assigned to the descendants of right-on liberals. Mocking people whose friend and colleague had just suffered a horrifying accident was a taboo she hadn't expected to be broken. It left her feeling just a little uncomfortable.

She caught up with the platoon, hanging around on the edge of the crowd that was following them, curious about their reaction. Her d-written neural cells intercepted their communication link, giving her full sound and vision intimacy. They were largely ignoring the chants and abuse hurled at them, busy making private unheard jokes about members of the crowd. Boyishly obscene observations about the girls (including her) were followed up by zooming in on the appropriate section of anatomy with their helmet sensors; sexual derision about the males and their imagined deformities concealed by strange folds in their trouser fabric. Quite the little counterpoint and morale booster.

The platoon crossed into a wide concreted area around the base of a big apartment block, which the local kids used for their games. A dozen or so skinny boys just into their teens were kicking a soccer ball about. Their game trickled to a desultory halt as they turned to stare at the invaders.

Most of the crowd began to turn back, heading for the shops and bars and haunts along the street, probably intimidated by the open space. Denise slouched on the corner by a shop, watching the platoon march away. Following them here would make her too visible; besides, she'd learned what she needed.

Suddenly the soccer ball was powering through the air. It almost hit one of the Skins, the sergeant himself no less, but he dodged back. Denise blinked as his foot shot out, stopping the ball in midflight. His toe nudged it about; then it was arching up. His knee came up underneath and bounced it twice; then he kicked it gently to another Skin. They started passing it to each other.

The boys who'd been playing were now all standing sneering, striking a variety of stubborn hands-on-hips poses designed to show how tough and unintimidated they were.

"Give us the ball back!" one shouted. He was the tallest, all gangly limbs and a thick beret of curly black hair.

"Sure," the sergeant said.

The kid took a half pace back in surprise at hearing the modestly amplified voice. Then the Skin was walking toward him, nudging the ball along in front. He got right up to the kid, who made the mistake of going for the ball. The sergeant neatly flicked it round him, and kept on going to the next youth. Another attempted tackle, another failure. The sergeant was picking up speed, and the other kids flocked toward him for their own moment of victory. He got around another three, then kicked the ball over their heads. It was a perfect arc that placed the ball at the feet of another Skin. He kicked it firmly, and it smacked against the wall between the two fading white lines that marked out the goal.

The sergeant held his arms high. "Easy."

"Yeah?" the tallest kid scoffed. "You're in Skin, asshole. Come out and try that against us."

There was a moment's pause, and the sergeant's Skin split open down his neck. The tall kid took a startled half pace backward as the head wriggled free of the split. His face and hair were shiny with a pale-blue gel, but he was still smiling.

Denise's hand flew to her mouth, smothering her gasp of surprise. The shock had overridden all her cause-dedicated calm. It was *him. Him!*

"Skin suits give us strength," Lawrence said cheerfully, "not skill. Still, not to worry. Some of you have a smattering of talent. Twenty years' time, you might come up to our level."

"Fuck off!" the kid cried. "You bastards would just shoot us if we didn't let you win."

"You think so? Over a soccer game?"

"Yeah!"

"Then I feel sorry for you. You're the ones shooting us, remember?"

The kid shrugged awkwardly.

Lawrence gave him a friendly nod. "If you ever fancy your chances on a level field, come and give us a game. Ask for me, Lawrence Newton. We'll take you on. Buy you a beer if you win, too."

"You're shitting me."

"So call my bluff." Lawrence winked and began pulling himself back into the Skin. "Be seeing you."

Clever, Denise thought as the platoon marched away, leaving the kids standing limply behind in a communal bewilderment. The platoon's communication link was roaring with a dozen variations on *what the fuck were you doing?*

But then, she told herself, you shouldn't have expected anything different from *him.* He was clever, and a bleeding-heart humanist. Someone like that would always try to build bridges with the enemy.

Thank goodness, a tiny traitor part of her mind whispered.

Denise's jaw hardened with determination. It didn't matter. *He* could not be treated any different from the others. The cause could not allow that.

She walked back down Corgan Street, planning how to turn the soccer match to her advantage. In war, which this was, *his* kindness was a weakness she could exploit.

Myles Hazeldine hated the wait in the anteroom. No matter how urgent the summons, and how irate Ebrey Zhang was, he always had to endure this ritual. He refused to show his temper, conceding the bitter irony. This was his study's anteroom, and he had always made his visitors wait, be they allies or opponents.

How obvious and petty it was, establishing the true authority figure. *Did they once laugh at me for such crudity?* he wondered.

The doors opened, and Ebrey Zhang's aide beckoned him in. As usual, the Z-B governor was sitting behind the big desk. And as usual, it galled Myles. The sharpest reminder of Thallspring's miserable capitulation.

"Ah, Mr. Mayor, thank you for coming." Ebrey's cheerful smile was as insincere as it was malicious. "Do sit down."

Keeping his face blank, Myles took the chair in front of the desk. An aide stood on either side of him. "Yes?"

"There was a nasty traffic accident today."

"I heard."

Ebrey cocked his head expectantly. "And?"

"One of your people was hurt."

"And in a civilized society, someone would say something along the lines of: Sorry to hear that. Or: I hope he's all right. Standard conversational procedure, even here, I believe."

"The hospital says he'll live."

"Try not to sound so disappointed. Yes, he'll live. However, he won't be returning to frontline duty. Not ever."

Myles smiled thinly. "Sorry to hear that."

"Don't push it," Ebrey snapped. "I'm going to have that accident thoroughly investigated. My people will oversee your transport forensic team. If they find anything suspicious, I'm going to use up some of my collateral. Still smirking, Mr. Mayor?"

"You can't be serious. A truck hit a wall."

"That's what it looks like. But maybe that's how it was meant to look. How often do your automated vehicles have traffic accidents, Mr. Deputy?"

Myles couldn't help frowning; he'd never actually heard of one before. "I'm not sure."

"The last one involving any sort of injury was fifteen years ago. For a fatality you have to go a lot further back. Even your antiquated electronics can manage to keep vehicles running smoothly. I find the timing highly suspicious."

"The odds pile up. Don't tell me your systems can do much better."

"We'll see." Ebrey activated a desktop pearl and waited for its pane to unfurl. He glanced at the script that began scrolling down. "Now then, I see the Orton and Vaxme plants still haven't got up to their proper capacity. Why is that, Mr. Mayor?"

"The Orton plant was undergoing refurbishment when you landed. You ordered it back into production status before the new components were properly integrated. It'll probably get worse before it gets better."

"I see." A finger tapped on the card's screen, changing the script pattern. "And Vaxme?"

"I don't know."

"But no doubt you'll find some engineering-based reason. After all, it could never be a human fault."

"Why should it be?" Myles asked pleasantly. He knew he was goading Ebrey too hard and didn't really care.

"Get its production back up," Ebrey said levelly. "You've got ten hours. Make it plain to them. I am not going to be dicked around on this."

"I'll see what I can do."

"Fine." He waved at the door. "That's all."

"Actually it isn't." Myles enjoyed the annoyance that washed over Ebrey's face. "I've made this request to your aides twice already today, but never even got a reply. It isn't as if I shout wolf every time we have a medical problem."

"What request?"

"I need some resources reallocated from the university biomedical department. You took our most qualified people away to help with those new vaccines you wanted formulated over at the Madison facility."

"I can't spare anyone to lecture some bunch of backward students with falling grades."

"It's nothing to do with that. There have been a couple of new pulmonary ward admissions at the hospital."

"So?"

"The doctors aren't sure, but it seems to be some kind of tuberculosis variant. It's not something we've seen before."

"Tuberculosis?" Ebrey asked; he made it sound as if Myles had told a sick joke at a funeral. "That's history. It doesn't suddenly resurrect on a planet light-years from Earth."

"We don't know what it is, exactly. That's why we need an expert diagnosis."

"Oh, for Christ's sake." He flicked the desktop pearl off. "You can have them for a day. But I'll hold you responsible if Madison falls behind."

"Thank you."

The Junk Buoy was modeled on a thousand waterfront resort bars that Lawrence had enjoyed in his twenties, and those had all been centuries out of date long before he even reached Earth. It catered for all sorts, although the sudden influx of Z-B platoons these last two nights had managed to repel most of the locals. When the first platoon came in and slapped on the bar demanding beers, the manager tried to refuse. They were ready for that; the sergeant had a communication card with a link already open to City Hall. A few words were said about licenses and there was no more trouble, only resentment. But the platoons were used to that, it hardly spoiled their evening.

Lawrence and Amersy sat under a thatched parasol out on the patio as the last crescent of gold-red sun sank behind Vanga peak. Both of them were sipping Bluesaucer beer from chilled bottles while the rest of the platoon spread themselves around the bar.

"Did you hear about Tureg's platoon?" Lawrence asked quietly. None of his own men were close, four of them were round the pool table. Edmond was in a corner booth, talking to a well-dressed local man—which made Lawrence frown briefly. Hal, of course, was sitting up at the bar, wearing a white T-shirt that was tight enough to outline every muscle and smiling at all the girls who came in.

"I heard," Amersy said. "The hatch nearly cut old Duson in half when they tried to open the lander pod. They reckon the thing was pressurized to ten atmospheres. Goddamn company using cheap suppliers again."

"That's bullshit, and you know it. No way a drop pod could pressurize like that."

"One of the RCS nitrogen tanks vented. The valve jammed. It happens."

"A valve jammed! Those things are supposed to be fail-safe. And nitrogen doesn't vent inside the pod, you know that."

"It can, if enough things go wrong."

"Ha!"

"What then?"

"Foran got caught by a runaway truck, didn't he?"

"Come on!" The patch of white skin on Amersy's cheek flushed darker. He leaned in closer. "You can't be serious," he hissed. "How could they sabotage a lander pod?"

"It was out beyond the boundary."

"So what: you're saying this *KillBoy* resistance group managed to change its descent trajectory?"

"No, of course not. It drifted off track, enough of them do. This one was sitting out there in the middle of the jungle for a week before we got around to dispatching a recovery sortie. Plenty of time for them to find it and rig the nitrogen."

"You've got to be wrong, man. The only way they could do that was if they could get around our software security."

"Yes."

"No way. We're talking e-alpha here. Nothing can break that encryption."

Lawrence tried not to dwell on the Prime program he still carried in his bracelet pearl. He'd never actually tested it against e-alpha, although it could certainly break Z-B's second-level software. "I hope not."

"It can't, Lawrence." He was almost pleading. "If they could break e-alpha we'd be wide open to them. Hell, we'd never even have made it down from orbit."

"Yeah." Lawrence took another sip from his bottle: it was his fourth, or fifth. Not a bad brew, based on some Nordic ideal of three hundred years ago with an alcohol percentage higher than he was used to. "I guess you're right." The sun had vanished now, pulling a veil of deep tropical darkness over Memu Bay. Streetlights and neon signs threw a rosy haze into the air above the marina. Farther down the beach, someone had started a bonfire. He took a slow glance around the bar, watching his men fooling about. "Will you look at that? We're commanding the biggest bunch of losers in the galaxy."

"They're damn good, and you know it. We just got all shook-up by Nic, is all."

"Maybe. But this whole outfit isn't what it used to be. There used to be enough of us to damn well make sure there were no screwups like truck crashes and pressurized lander pods. And nobody would ever have taken a shot at us like they did poor old Nic."

"Lawrence . . ."

"I mean it. I used to go along with it when I was younger. Now I'm old enough to know better. A lot better."

"Jesus, Lawrence, are you having a midlife crisis on me? Is that what this is?"

"No, that's very definitely not what this is."

"You got doubts about the job, Lawrence? If you have, then I'm telling you, you've got to sideline yourself. It ain't right someone with doubts leading us. You might—"

"Hesitate to shoot? I won't hesitate to shoot. I came to terms with that a long time ago. Our Skin is the one thing that stops our conscience being put on the line every day. We don't kill anyone; technology takes care of that. We knock them out and give them the mother of all headaches, but no scruples get trashed on the way."

"Then what the hell is this about?"

"My life. I shouldn't be here, you know. I made the wrong choice a long time ago."

"Ho fuck." Amersy took a big swig of beer. "Is this about that girl again?"

Lawrence's hand moved automatically to the small pendant under his T-shirt. "Fate, I was stupid. I should never have left. Never."

"I knew it! God damn! Who the hell keeps killing themselves over a girl for twenty years? Lawrence, man, I can't even remember the first time I got laid, never mind what her name was."

Lawrence grinned over the top of his bottle. "Yes, you can."

"Yeah, okay. Maybe. But Jesus . . . twenty years. I mean, your chick, she's got to be grossing out at a hundred kilos now, a housemom out in the burbs dosing up on antis to get through the day, with at least a couple of ex-husbands, not to mention some grandchildren knocking around."

"Not Roselyn. She would have made something of her life; she was never as dumb as me. And in any case, she was only a part of Amethi."

"You always go on about that planet like it's some kind of paradise. Why did you leave?"

"I told you, I'm a dumb fuck. The dumbest there is. I made

a mistake. I had it all, you know, I just didn't realize it at the time."

"Everyone's like that when they're teenagers. I mean, Christ, you've met my kids."

"Don't complain, they're good kids. You're lucky to have a family like that."

"Yeah, man. Guess so."

Lawrence couldn't help smiling. Hell, two guys getting loaded in a bar, talking about their families and how they'd screwed up their lives. How deeper in could you get? "Would you leave?" he asked slowly, trying to make it come over casual.

"Leave what?"

"The platoon. Strategic security. Z-B. Everything. Would you quit if you could?"

"Come on, man, you know I've got a family. My stake's not big enough to take care of them if I stop work. I can't quit."

"But if you could? If you didn't have to worry about your stake."

Amersy grinned wide. "Sure. If I could dump this shit, I would. Who wouldn't?"

"Good," Lawrence said in satisfaction. If he ever hoped to pull off his private mission into the hinterlands, he would have to have Amersy on his side. "Let's go get some more beers."

Edmond Orlov lurched into them as they made their way back to the bar. He clutched at Amersy, barely stopping himself from falling. His smile was beatific. "Hey, Corp, Sarge, how you doing? Ain't this the coolest place? Apart from the heat, that is."

He started giggling wildly. Lawrence hadn't really been paying attention, but he thought Orlov had just come out of the toilets.

"You know, it's still pretty early," Amersy said. "You've got to learn to pace yourself, man."

"Sure thing." Edmond threw a salute, almost missing his head. "You got it, Corp. But don't you worry. I'm on it." He tottered over to the jukebox, and after squinting, managed to slide his credit coin in the slot. A spiral video grid twisted up inside the juke's cylindrical pane. Edmond started muttering: "Oh yeah" and "you, baby, you" to the AS as his finger waved at various grids. "Gimme some of that. Oh brother, I want me a piece of that, too." Ska calypso music started to pound out of the overhead speakers. Edmond backed away from the juke, eyes closed, arms waving in a rhythm that didn't quite correspond to anything being played.

All of the locals were nudging each other and smirking at the solitary, swaying figure. His own platoon mates and several of the other platoons laughed and clapped as he began to speed up.

"I gotta have that beer," Amersy said, and broke for the bar.

Lawrence took a last backward glance at Edmond. Something was going to have to be done about him. But not tonight. "Pain level's too high," he whispered as he went after Amersy.

Hal was still on his prominent stool at the middle of the counter. His smile flicked on at every girl who walked in. It never lasted long. The girls who arrived in groups checked him out immediately, then giggled among themselves as they found an empty section of the bar away from him. He earned himself some hard warning stares from boyfriends. Single girls had seemingly all perfected the same dismissive sneer.

"I've been ripped off," Hal whined to Amersy as the corporal leaned on the counter and tried to attract one of the barmen. "Can we employ lawyers to sue people here?"

"What the hell are you talking about?" Amersy asked.

"This," Hal grunted. He flicked his glance downward.

Amersy peered at the trooper's feet. "Your shoes don't fit?"

"No! Not that!"

"What's happening?" Lawrence asked. "Hal, you still here? I thought you'd have scored by now."

"I've been sold a dud," Hal told them through clenched teeth. He held his left arm up. There was a slim black band round his wrist. "I haven't got a bleep out of it all evening. Eighty goddamn credits that son of a bitch took off me."

Lawrence had to forcibly hold back his laughter. "Is that what I think it is, Hal?"

"It's not illegal, Sarge," Hal protested. "The guy in the shop swore everyone here uses PSAs."

"Okay. Maybe there's just no one here with your . . . preference."

"There has to be." Hal lowered his voice to a desperate plaint. "I keyed in an open acceptance. That's like *anything* these girls are into, I'll go with it. The fucking thing still doesn't work."

Amersy finally managed to get in an order for some more bottles of Bluesaucer.

"Give it time," Lawrence advised.

"I've been here over an hour already. And Edmond told me about this place."

"What about it?"

"They like—" Hal swiveled his head from side to side, making sure no one was listening, then lowered his voice. "They're into threesomes here."

Lawrence groaned. He might have guessed his men would grab the wrong end of that local legend. "That's trimarriage, Hal. It's different."

"Yeah, but they've got to get used to it first, try it out."

Lawrence put a friendly arm round Hal's shoulder. "Listen, take my advice, kid, forget the bracelet and the threesomes for tonight, okay? Just be yourself. There must be a dozen girls in here. Go over and ask one of them if she wants a dance." He gestured at the dance floor, which probably wasn't the best illustration. Two squaddies were prancing around an oblivious Edmond, imitating his crazed movements with grotesque exaggeration. They were both holding on to their beer bottles, with the foaming liquid sloshing out. Their audience was cheering them on. "Or a quiet drink," Lawrence added quickly. "It doesn't matter what you say to them, as long as you say something. Trust me on this one."

"I suppose," Hal grunted sullenly. He glared at the PSA bracelet, willing its electronics to flicker into Technicolor life. The little display panel remained stubbornly dark.

"Good man." Lawrence and Amersy collected their beer and fled back out onto the patio.

After an hour, Jones Johnson had just about got the pool table figured out. One of the middle pockets had a worn cushion that you had to watch when you were shooting from the top, and there was a definite slope away from the bottom left corner. Now that he knew all that, he could maybe start hustling a little credit. Certainly from their fellow platoons, and if he got lucky from a local who thought he was king of the skewed table.

Most of his own platoon hung around as the evening wore on, cheering him, or groaning in sympathy as the balls refused to drop. The Junk Buoy began to fill up after sunset. Platoons who'd been here last night reported that the locals had stayed away. Not tonight.

The pool games went on. Three wins. Two losses (one strategic). Karl and Odel and Dennis ordered them all some surf 'n turf. They dug into the big platters, chugging down the too-sweet horse piss that passed for beer in Memu Bay, keeping their cue on the table.

After a couple of hours, Edmond's fix was depressurizing. He packed up the dance floor and slumped in a chair, arms hugging his chest and shivering as if the night had brought a front of arctic air in off the water. Jones was kind of pleased about that. Edmond's dancing was always embarrassing, but stoked up, someone had to watch him. And they'd all seen Lawrence give him the eye—before the sarge and Amersy settled down to get seriously hammered together. Not that it mattered; they all looked out for each other in here as much as out on patrol. That's what platoon membership was about.

Even the kid, who was now drunk enough to venture around the girls. Nobody could quite hear what his lines were, but he kept pointing at a black bracelet on his wrist as he staggered from one to the next. All the girls he talked to waved him on or turned their backs to him. The dance floor was heaving with people. And now that his cue aim was wavering from the drink, Jones quite fancied his chances out there among the sweaty strutting bodies. The Junk Buoy's DJ had taken over from the jukebox, and the mood of the crowd was already up and going higher. There were some seriously good-looking pieces of skirt out there, too. And the can-time had stretched on for way too long since they'd left Cairns.

Jones moved out onto the dance floor along with Lewis and Odel. Even with the beer buzzing him, he could move with a decent groove. And there was one girl in a scarlet T-shirt dress with a high hem. She kept returning his grins. She was way too young, still a teen. Which just made it hotter.

He danced with her for a couple of minutes, then put his arms around her and started making out. She was just as eager, letting his hands squeeze her buttocks while his tongue delved down her throat. Her own hand came round, closing on his balls. They'd still not said a word.

Shouting. Angry yells out on the edge of the dance floor. Bodies moving sharply, the way they always did when they were being pushed. Jones lifted his head to look round. "Oh fuck."

It was the kid. He'd made a play for a girl who was in a

group. Hadn't checked, or was now too drunk to notice, the boyfriend, who was being backed up by half a dozen youths.

Drunk or not, Hal was still trained enough to respond automatically to the shove. Going with the momentum of the impact, then spinning round, arm coming out, hand flat to chop. Screaming at the fuckers to back off. Them screaming their own fury about alien motherfuckers. Two of them closing fast. Hal dropped into a self-defense pose, arms and legs locking *just so.* Looking pretty silly as the oblivious dancers behind him kept jostling him around.

The first barrage of fists flew. A girl screamed at the top of her voice. Hal's knuckles crunched into a rib cage with a satisfying jolt running back up his arm. A fist slammed into his own cheek. Red flash. And he was staggering back into more people. Blood foamed out of his mouth.

Everyone in the Junk Buoy was suddenly aware of what was happening: locals seeing an invader—the perv who'd been pestering girls all evening—brutally assaulting one of our lads, platoon squaddies seeing one of ours being surrounded and smacked around.

An implosion of bodies rushed in toward the fight.

Jones levered his way through the barbarous crush. Elbows thudded into him. He kicked out. A broken bottle was stabbed toward his face. He ducked, spinning around, kickboxing the attacker.

Screams. Bloodlust. The DJ kept the volume cranked up big. Wild fists and feet. Random targets. Many people started chanting: "KillBoy."

A girl jumped on Jones and bit his ear. He bellowed in fury and slammed her into a pillar. She puked up as she fell away. He saw Lawrence staggering back into the room off the patio. A knife flashed.

"Sarge!" A chair registered as a blur of motion above and behind. Jones's arm came up to block, way too late. The solid wood backrest crashed into his forehead. Stars exploded. Very briefly.

Lawrence just managed to sway away from the knife blade as the man slashed at him. Somewhere in his brain there was a perfect countermove; a sort of physical chess maneuver that would enable him to disarm and subdue his attacker with a bent forefinger. Or something. He laughed joyously as he tried to work out how to slide into a fluid kung-fu–style stance. Unfortunately someone hit the floor behind him and bounced into his legs, sending him toppling backward. He thudded into the wall. "Ouch. Hey, that damn well hurt." He laughed again, then stopped urgently as he threw up. A girl on all fours beside him shrieked in disgust as he spewed over her short red dress. She slapped him hard and scrambled to her feet. Lawrence waved and tried to say sorry. That was important, he felt. He couldn't quite see where she was anymore, so instead he threw up again. It'd been ages since he'd been in a decent bar fight. Mind, he was pretty sure it had been more fun last time around.

Police, reinforced by two Skin platoons, arrived at the Junk Buoy within four minutes of the owner raising the alarm. By then the fight had spilled out into the street. Several people were in the water, thrashing about frantically according to how drunk they were.

"Stop this right now," the senior sergeant said. Even with Skin amplifying his voice, no one took any notice. Several bottles were thrown at the Skins.

The two platoons formed a loose semicircle around the brawl, with the police standing behind them. The senior sergeant took a bulky cylindrical canister off his belt and held it high, angled slightly toward the Junk Buoy. There was a dull thud from one end. Its web flew out, a mesh of fine fiber that seethed like a gray-silver nebula in the air as it expanded, then settled over the fighters. Strands stuck to clothes and flesh alike, stretching with every motion. Nobody noticed.

Several thousand volts were pumped through it. People screamed, muscles suddenly locking. Purple-white static flared around extremities, fingers and hair squirting out sparks. Then the fiber's conductive molecules disassociated and the current vanished.

In its wake it left a stunned silence and convulsed postures. After a second, those it had struck and immobilized juddered down gulps of fresh air. Limbs trembled uncontrollably. Nobody was fighting anymore. Locals regarded the picket of dark Skins with considerable trepidation. Squaddies who'd been caught by the web grinned nervously, holding their hands up.

"Thank you," the senior sergeant said briskly. "You are all under arrest. Please wait here." He marched toward the bar's main door. The spent web canister was dropped, clattering away on the stone-paved road. He pulled another one from his belt and stood in the doorway. "Pack it in!" he yelled. The new web canister was fired into the Junk Buoy.

Lawrence woke up knowing he must have only seconds to live. His head was obviously split open, allowing someone to pour boiling oil over his exposed brain. He groaned feebly, moving about. Which was a big mistake. He dry-retched. His hands waved about slowly, coming into contact with thin strings of vomit beading out of his mouth.

"Oh fucking hell."

The light was agonizingly bright and penetrating deep inside his broken head. He didn't so much blink as weep the world into focus. Not a very good focus, he had to admit.

Someone had dumped him in a very weird hell. He was lying on the thin gray carpet tiles of what looked like a brightly lit airport lounge. There were long rows of red plastic chairs screwed into the floor. People were slumped listlessly in them. Some of the men were injured, holding pressure dressings to cuts and bruised eyes, blood staining the white fabric. Girls in small tight dresses leaned against each other, either asleep or staring blankly. There were other people sleeping on the floor—at least he assumed they were sleeping; none of them showed any signs of movement. Several Skins stood guard around the perimeter of the room, imposingly silent and still.

Lawrence got it then, and memory oozed back. The fight. This was a hospital waiting room, then. Not hell after all.

Slowly, *very slowly,* he turned on his side, then levered himself up to a sitting position. Pain pounded away on the side of his head, making him nauseated again. He winced, dabbing at the spot with fingers. There was a huge tender lump just behind his left ear.

Amersy was sitting in one of the red chairs beside him. The corporal's white cheek had turned gray; both eyes were badly bloodshot. He was holding a chilpak across his forehead. His shoulders were trembling.

Lewis, Odel, Karl and Dennis were in the seats beside him; Odel with his right hand swallowed by a blue field-aid sheath, Karl with a busted nose and blood on his lips and chin. Edmond was lying on the floor, curled up at Karl's feet.

"Ho shit," Lawrence croaked. "What—"

"We got webbed," Lewis muttered. "The owner called the cops."

"Oh great." He paused, pulling down some more air. "Everyone okay?"

"Sure. We were kicking some serious butt in there till our own cavalry came over the hill and shot us. Fuck. I mean, whose side are they on?"

Lawrence wasn't going to give any sort of answer to that. "What's our status?"

"The kid's in with the doc right now." Amersy jerked his thumb toward the curtained-off cubicles at the back of the room. "Nothing bad, at least not broken. And we're on notified restraint until the medics clear us."

"Great." He looked round to see if there was some sort of pillow he could rest his head on. "Where's Jones?"

"Christ knows."

"That's good. He'll make his own way back." The effort of talking and thinking was incredibly tiring. "Let me know when it's my turn." He lowered his head back onto the carpet tiles again.

The nurse was surprisingly sympathetic. Lawrence had no idea what time it was when he was finally called into a cubicle to be assessed and cleaned up. Very early morning, he guessed.

She scanned the side of his head where the bump was, and the medical AS decided he wasn't concussed. "But I'll get a human doctor to examine the image when we've one free," she told him. "Just to be on the safe side."

"Thanks."

"It'll be a while. They're a bit busy right now." She laid him on his side and pulled the grubby T-shirt over his head.

"Sorry."

"Don't be. You didn't start it. Did you?"

"No. But I should have realized it was inevitable."

She started squirting some kind of cool cleaning liquid on his lump. Lawrence grunted at the sharp sting.

"Any fool could have told you that."

"I'm not just any fool, I'm supposed to be in charge."

"In charge, huh?" A gauze napkin was dabbed on his skin, soaking up the excess liquid.

"Yeah, I know. Listen, I don't suppose you've got anything for my headache, have you?"

"Headache or hangover?"

"Both. And they really don't like sharing space."

"Not surprised. Hold that." She took his hand and pressed it against the napkin. He could just see her shoes as she walked over to a wall cabinet.

"Anyone badly hurt?" he asked.

"Us or you?"

"Just anyone."

"Three deep stab wounds. One emergency regenerative procedure, a girl's face was cut up—"

"Aw shit."

"—several broken bones. And that electrocution weapon of yours has left a lot of people very shaky. Nobody dead, though. I suppose we should be grateful for small mercies." She handed him a couple of purple capsules and a glass of water. "Take these."

He swallowed them automatically. Only afterward did he realize how trusting he'd been. Strategic security policy was quite strict on receiving external medical assistance, especially in nonlethal situations.

The curtain was shoved back, and Captain Bryant stormed in. He was in full uniform, the light mauve fabric showing up his anger-heated skin. "There you are, Newton."

"Excuse me," the nurse said. "I'm treating this man."

"He's cured." Bryant held the curtain open for her. "That will be all."

She gave him an indignant look and walked out.

"Would you care to explain, Sergeant?"

"Sir?"

"What the hell happened tonight? I let you out for a quiet drink and the next thing I know you're restaging Santa Chico."

"There was some kind of argument. About a girl, I think. It sprang from that."

"Then it damn well shouldn't have *sprang*. For God's sake, you're supposed to stop this kind of thing."

"I wasn't actually there, sir. Otherwise I would have."

"You should have been there. You're their sergeant. I depend on you to keep order."

"We were off duty."

"Don't even start pulling that one on me. There's a damn sight more to your job than official duties, and you know it. And if you don't, you shouldn't have those stripes."

"Sir," Lawrence grunted with extreme petulance. If he hadn't been so unstable he would have said fuck it and simply smacked Bryant one.

"Now where is Jones?"

"Sir?"

"Jones Johnson. Remember him?"

"I thought he'd gone back to barracks."

"He hasn't reported in, and the police didn't take him into custody with the rest of you. Where is he?"

"I don't know, sir. Have you checked the hospital?"

"Of course I have."

Lawrence rubbed at his eyes. The capsules seemed to be having some effect. At least the nausea was fading. But he felt desperately tired. "Officially he doesn't have to report back until oh-six-hundred hours, sir."

"Don't play it smart with me, Sergeant, you don't have the IQ to pull it off. Jones is the only person unaccounted for, and he's under my command. Have you any idea how badly all this reflects on me? After this total debacle, I don't want further loose ends. Do you understand that?"

"What I'm saying, sir, is that if he got out from the fight before the police arrived, then he's probably with a girl."

"He'd better be. I want you to take that shambles you call a platoon back to barracks right away. You're on double house duties, and any breakages from the Junk Buoy will be met out of your pay. I shall also be loading an official reprimand onto your record. Now get your act together, Newton."

The curtain was tugged back forcefully as the captain strode out.

Lawrence gave his invisible back the finger, then groaned in misery as he sank back down onto the examination table.

* * *

Jones Johnson woke to a hot ache in his wrists and back. Despite that, he was alarmingly cold.

Not surprising. He was naked, spread-eagled with his wrists fastened in some kind of manacles that hung from an oval frame. Ankles, too, were held fast against the base of the frame. The rest of the room was empty. As far as he could see, it didn't even have a window, just a plain wooden door on his left. The walls were whitewashed concrete, the floor some kind of spongy black matting.

Instinctively he tugged at the manacles. Whoever had built this frame knew what he was doing. His freedom of movement was very limited.

The worst thing about it was, he simply *could not* remember how he'd got here. There had been some kind of fight in the Junk Buoy. He'd seen a knife flash. Combined with a chair?

What the fuck happened after that?

His brief struggle with the manacles left him panting. There was the dull throbbing on his forehead that indicated a big bruise.

"Hey!" he shouted. "Hey, can you guys hear me? Anyone there? Hey."

He watched the door for a while, expecting someone to come see what the commotion was about. Nothing.

It's a brothel, he told himself, *an S and M joint, that's all. I took a hit in the fight, and those turds Karl and Lewis paid for this. Some dominatrix will arrive any minute and start hitting my ass with a cane. The bastards.* "Hey, come on, guys, this isn't funny anymore."

Still nothing happened. He couldn't hear any traffic sounds, any voices.

Bastards.

He needed to pee, too. God damn!

And who would have thought that Memu Bay had a

cathouse that specialized in this kind of stuff. He stopped that train of thought straightaway.

Some time later the door opened.

"About fucking time," Jones yelled. "Come on, get me out of here."

A man came in, dressed in a dark blue boilersuit. He paid no attention to Jones at all. He was carrying a large, and clearly heavy, glass container, which he placed on the floor by Jones's restrained feet.

"Hey! Hey, you," Jones said. "What the fuck is this? Hey, say something. Talk to me."

The man turned round and walked out.

Jones shook himself about as much as he could. It was all pointless, the manacles never budged. But the door hadn't been closed.

"Look, whatever they paid, I'll match it."

The man came in again, lugging another, identical, glass container.

Jones found he was sweating now. His heart had begun to flutter in that way that acknowledged his subconscious knew something was deeply wrong. He just couldn't admit it to himself, because that would be when the panic and dread would kick in.

"Please," he asked. "What is this?"

But the man had left again.

He didn't want to think it. Not that. Not *KillBoy.* That this wasn't something Karl and Lewis had thought up for a laugh when they were drunk. That he'd been the dumbest fuck in the universe and let some fanatical resistance group snatch him.

"But I don't know anything," he whispered. "I don't."

Torture was centuries out of date. It really really was. There were drugs, all sorts of techniques. Available to all modern, well-equipped, properly financed police and security forces. Didn't Thallspring have them? Backward primitive Thallspring?

It didn't matter, he persuaded himself, because Z-B would be turning the town upside down in their search for him. The sarge would never let them stop. He looked after his men. Good old sarge. Any second now and the door would fly off its hinges, and the platoon would charge in to rescue him.

The mute man was back again, with a third container. This time he'd brought a load of clear plastic tubing as well, which he left looped round the container's short neck. Jones stared at it, bitterness and furious resentment contaminating his anger. The apparatus was for an enema. He was going to be raped. Gang-raped most likely. Part of the softening up. Part of breaking him.

He clenched his fists, pulling desperately. "God no. No. No." His contorted face so nearly let tears escape down his cheeks. "Why me? Why did you pick on me? It's not fair. Not fair."

The door closed again behind the man. Jones let out a sob, and the tension went out of his body, leaving him drooping painfully from the frame.

"Please," he told the empty room. "I'm nobody. I'm not important. You don't have to do this. Please."

He was sniveling now. Wretched and pathetic. Back on Earth, anti-interrogation training had gone through the routines for strengthening resolve. How to withstand tiredness and strain, how not to be caught out in lies. That was training. That wasn't real. Not when some bunch of psychotic terrorists have got you stripped naked and strung out like they're about to crucify you. Not when you are so utterly helpless that you would genuinely sell your soul to the devil you now want to believe in very badly indeed. Because there's no other way out.

Where were they? God damn it, where were the platoon?

"Everyone is important in their own way, Mr. Johnson."

Jones's head snapped up. There was a beautiful young woman in the room: her long flattish face was one that any man would find enchanting. Thick dark hair swung around her head as she stared at him. Her movement was birdlike, examining him from minutely different angles. She was twisting a gold ring on her index finger.

"Please," he entreated. "Just let me go."

"No." She said it with a finality that was horrifying.

"Why! What are you?"

"At this particular stage of our mission, I suppose you could call me a revolutionary anarchist. It is my task to bring chaos and disorder to Memu Bay."

"What?" he blurted.

She smiled gently and took a step closer. Her proximity was one he found alarmingly sexual. Then she picked up the tubing. One end was carefully plugged into the top of a container. She began to uncoil the rest.

"Don't," he begged. "Jesus, please."

"There will be very little pain," she said. "I am not a sadist, Mr. Johnson."

Jones clenched his buttocks as if he were going for Olympic gold. "I'll tell you anything. Just . . . don't."

"I'm sorry. You're not here for questioning. I already know more about the universe than you ever dreamed existed."

He stared at her, coldly shocked by the realization that she was no revolutionary, she was simply insane. Bug-eyed, dancing-in-the-moonlight crazy. It was one of the universe's most heinous crimes that a creature so beautiful should possess such a demented soul.

"People will die," he cried. "Your people, the ones you're supposed to be fighting for. Is that what you want?"

"Nobody will die. Zantiu-Braun will never know for certain if you are alive or not. It is a dilemma that will eat at their souls. That is what I want."

She brought the end of the tube up to his neck. With absolute horror, he saw the end was shaped exactly like a Skin circulatory nozzle. It clicked neatly into his carotid valve.

"It won't work," he said hoarsely. "If you want me dead, you'll have to do it the hard way. It's not that *easy,* bitch!"

"Good-bye, Mr. Johnson." She glanced at her ring.

Jones laughed in her face. Stupid bitch didn't know the valves were e-alpha protected. His laugh burbled away to a terminal scream as he saw his precious scarlet blood race down the tube and splatter into the container.

He actually saw her flinch. There were tears in her eyes, revealing shame. "Know this," she said. "Your essence will go forward to flourish in a world free of sorrow. I promise you." Then she turned away.

He cursed her to hell and beyond. He screamed. Pleaded. Wept.

All the while his blood flowed along the tube.

Fight it, he told himself. *The boys will find me. Don't lose consciousness. They'll rescue me. They will. My friends. There's time. There's always time.*

One of the containers was completely full. And still the tube was red as his heart pumped away faithfully.

Blood and world began their final fade into gray.

CHAPTER NINE

LAWRENCE'S FLIGHT TO EARTH LASTED SEVERAL WEEKS. HE didn't have any of the claustrophobic cabin restrictions and mind-rot routines that were the norm throughout his every subsequent flight. Passengers traveling from Amethi were a rarity; there were only eight on board the *Eilean* when it activated its compression drive. It meant only one life support wheel was active. But even then he had a whole family cabin to himself, and the rest of the place to roam through. The crew tended to ignore him, assuming he was some rich brat whose overindulgent Board family had paid for the flight and a tour around Earth. He never even registered with the other passengers, McArthur ultraexecutives who spent the whole time interfaced with their personal AS. He got to spend as much time as he wanted in the gym, while the rest of his waking hours were taken up accessing the ship's extensive multimedia library.

It should have been, he decided later, golden-age space travel, slow and leisured. The only possible equivalent was a voyage in a 1930s-vintage airship, although those would probably have better food. And a decent view.

Perfect, if he could just have forgotten about *her.* But the loneliness and relative isolation contrived to tweak every tiny memory into a full-on reverie. The color of a graphic display reminded him of a certain dress she wore, exactly that shade of turquoise. Food was a meal they had shared once. Menus on the multimedia library brought back the hours they spent accessing together, curled up in each other's arms on the couch in his den.

Starflight, the desire of his life, was made wretched by the love of his life. Ironies didn't come much worse.

Earth, however, did not disappoint. During the orbital transit flight from Glencoe Star, McArthur's Lagrange point base, he spent almost the whole time pressed up against one of the ship's four viewports watching the planet grow larger—blissfully unaware of the radiation threat. He'd thought his departing view of Amethi to be the most wondrous sight of his life, with its surface features of ocher, ash and white, and Nizana's dusky radiance reflected from Barclay's Glacier. But Earth with its vibrant montage of *living* colors made his heart ache as it grew closer and brighter. He landed in a Xianti, bitter at the spaceplane's lack of windows.

McArthur's principal spaceport was Gibraltar, the Rock's inhabitants still stubbornly clinging to their independence from Spain, if not the European Federation. Their government council had negotiated a deal with McArthur, involving liberal taxes in exchange for infrastructure investment and a mutual noninvolvement/liability clause.

The politics went clean over Lawrence's head. Once he was out of the spaceplane and through the arrivals hall (where he generated a brief flurry of interest from security—uneasy at allowing a Board family member to wander around unaccompanied), all he wanted to do was get outside and experience the romance of the Old World. He walked away from the terminal building that now occupied the site of the ancient RAF base and went down to the huge spaceplane runway that the company had built, a five-kilometer strip of concrete stabbing out into the Med. For hours, he simply stood in wonder on the boulders that skirted the concrete, looking out across the water. Spain was on one side, an unbroken swath of urbanization that ran along the shoreline right over the horizon, with the mottled brown slopes of low mountains rising up behind the tumbling sprawl of whitewashed buildings. Africa haunted the other side of the sea, a dark, featureless stripe highlighting the boundary between water and sky.

For some strange reason, given how he'd actually flown across half of Amethi, this planet seemed much larger. He simply couldn't get used to the scale of the elements. So much water, lapping and frothing around his feet. It gave off the most potent smell, hundreds of subtle scents blending together in a harsh salt zest. And the air . . . nothing had ever been so warm in his life, he was sure of it, not even the tropical domes. The heat and humidity made it hard to breathe.

It wasn't until the sun fell and the lights from the Costa twinkled over the swelling water that he turned around and made his way into Gibraltar town. Money wasn't too much of a problem; his Amethi credits were easily changed at the bank for EZ Dollars. With the resulting balance he'd be able to stay in any average hotel for several months before even thinking about getting a job. That wasn't what he wanted.

He remained in Gibraltar for several days, spending nearly all of the time accessing the global datapool, filling in significant gaps in his political education. The one thing Roselyn had been truthful about, he was relieved to discover, was Zantiu-Braun. Most companies were reducing their starflight operations to an essential minimum: keeping in contact with existing colonies and asset-realization missions on planets newly acquired from their struggling founder companies. While almost alone, Z-B maintained a small exploration fleet, and was still establishing colonies through their portals. Although even they weren't building any new starships. All the Lagrange shipyards had either been mothballed or turned into service and maintenance bays.

Starflight really was an era that was drawing to a close. But it wasn't over yet. Even at its current wind-down rate, there would be ships flying for decades.

A week after arriving on Earth, he took a train to Paris and walked into Zantiu-Braun's headquarters. The personnel division, like McArthur security before them, were slightly flustered by his origin. However, the AS and human supervisor managed to convince him his best way into the exploration fleet was through their general Astronautics Division. He didn't, they pointed out, have enough money to buy himself an initial stake in Z-B large enough to select his career path. What he should do, like hundreds of thousands before him, was get in on the bottom rung and earn the stake necessary to make the transition. As an added advantage, people who applied from in-house were always given preferential selection over those coming from outside. His lack of a university education was dismissed as unimportant at this stage: Z-B always offered educational sabbaticals to any staff member eager to progress up through the company structure. And as it happened, there were openings in an Astronautics Division that would serve as an excellent primer to their starship officer college. Had he ever thought of a career in strategic security?

Two days later he was on the train to Toulouse.

Eight months after that, he was in space again, heading for the Kinabica system. He and Colin Schmidt, the two newest members of Platoon 435NK9, were held in pretty high contempt by their fellow squaddies.

Kinabica was one of the earliest star systems to be settled, and in two and a half centuries it'd achieved a respectable socioeconomic status, with a high-level technology base. Quite how and why its founding company, Kaba, had divested itself of such a primary asset was never detailed in the briefings the platoon were given. Kinabica with its population of seventy million was now effectively self-supporting. All the principal investment had been made. There was no more heavy-duty industrial plant to be shipped out, no more biochemical factories or food refineries required, no mining equipment that couldn't be built locally. Everything was there, in place, wired up and chugging away merrily.

"It's because there's no dividend," Corporal Ntoko told Lawrence one day during the flight. Like every newbie, Lawrence was filling his day with questions, though he asked a lot more than Colin. Ntoko had taken some pity on him and supplied him with a few answers. It did at least stop the questions for a while. "Kaba has poured money into Kinabica ever since its discovery, and it's getting virtually nothing in return. The whole place is a rotten stake for investors on Earth."

"But it's a whole planet," Lawrence insisted. "It must be profitable."

"It is, but only within its own star system. Suppose they produce a memory chip with a density equal to anything on Earth: they still have to ship it across fifteen light-years to sell it. While any Earth factory producing the same kind of chip has only got a couple of thousand kilometers to reach its consumer population, and that's by train or bulk cargo ship. Which transport method is always going to be more expensive?"

"Okay, so Kinabica should produce something unique. That's how real trade works, an exchange of goods between supplier and consumer on both sides."

"That's the theory, sure. But what can Kinabica produce that Earth can't? Even if they got lucky and designed a neurotronic pearl way ahead of anything on Earth, it would only take a couple of months for any of our companies to retro-engineer it. At our current level of manufacturing technology, the only production that makes sense is local production. Starflight is just so goddamn expensive."

"Then why are we doing this?"

"Because asset realization is the one thing that can justify interstellar flight. On Earth, the concept is plain digital accounting, swapping figures around in spreadsheets. There's very little actual money involved. Z-B accepted Kaba's negative equity loading to help with its own starship operation funding problem; the two complement each other perfectly, provided you have the balls to see it through. That's why we're out here in a tin can flying faster than a speeding photon, to turn all that nice corporate financial theory into dirty physical practice. Z-B was in almost the same boat as Kaba was when it came to financing our starflight division; they'd laid out a trillion-dollar expenditure over the last couple of centuries and have precious little on the balance sheet to show for it, except for fifty multibillion-dollar starships with nowhere to go. Except now we have Kinabica's debt on our books, we can legitimately employ our own starships to collect some equity. As we've essentially written off the planet's founding investment debt, all we need is the products from their factories to sell on Earth. That way, the production costs are simply cut out of the equation, so now all the money realized by the sales of Kinabica's high-tech goodies goes directly into maintaining Z-B's starships, the strategic security division, and servicing the equity debt. If the accountants do their sums right, we also come out with a profit."

"Sounds like piracy to me," Lawrence said.

Ntoko laughed at the youth's surprise. "You got it, my man."

Platoon 435NK9 was scheduled to land on Floyd, a large moon orbiting Kinabica. While the rest of the Third Fleet platoons would be trying to keep a lid on Kinabica's resentful and resourceful population, they would be intimidating the three thousand inhabitants of Manhattan City.

Floyd was just large enough to hold on to an atmosphere, a thin argon-methane envelope that occasionally snowed ammonia crystals during midnight on the darkside when the temperature became seriously chilly. There were no seas or even lakes; its surface was covered with a spongy dull rouge vegetation, like a lichen with dendrite fronds. The claggy stuff covered every square centimeter of the moon, from the top of its few sagging mountain ranges to the bottom of crater basins. Not even boulders or cliffs remained free: its grip was pervasive and total. The locals called it Wellsweed, after the avaricious Martian weed in *The War of the Worlds.*

From the platoon's landing vehicle it looked as if they were gliding over an ocean of thick liquid, with strange crumpled wave patterns suspended in time, casting long, low shadows. They were having to use heavily modified Terran lunar cargo landers to get down to the ground. The vehicles were normally a simple cylindrical pressurized cabin, with rocket engines, tanks, sensor wands, thermal panels and cargo pods clustered around it in an almost random pattern, while three metal spider legs were flung wide underneath to absorb the impact of touchdown. Now the whole clumsy edifice had been encased in a lenticular composite fuselage designed to protect the vulnerable bulky core from the meager atmosphere during descent and deceleration. It was the closest the human race had ever come to building a flying saucer, though it certainly lacked the smooth elegance normally associated with the concept.

The sun had just risen above the low hills behind Manhattan City, beginning its seventy-five-hour traverse of the sky, when they wobbled in over the spaceport. Various strobe lights and guidance instruments ringed the patch of blasted rock that served the city (all currently dark and inactive). Noxious yellow flame belched out of dark holes in the vehicle's fuselage. Legs unfolded with labored jerky motions, allowing them to settle to the ground with alarming creaking sounds and the muted roar of the rocket jets drumming against the badly stained fuselage.

A second, then a third vehicle from the Third Fleet swooped in gracelessly and touched down beside them. Nothing marked them out as interlopers more than the local ground-to-orbit shuttle craft that were parked along the far side of the spaceport, silver-white spire rocketships standing vertically on curving scimitar fins, their pedigree taken direct from the dreams of the 1950s.

The platoon disembarked, edging clumsily down an aluminum ladder welded to one of the landing legs. On the ground, Lawrence's muscle skeleton AS struggled to compensate for the low gravity, restraining every movement. They jostled, bounced and slithered their way toward the main airlock of Manhattan City. Bulky impact armor worn over the muscle skeleton made it look as if they'd sealed themselves in puffball spacesuits to cover the short distance.

The smaller one-person airlocks only just allowed them to pass through.

It was the combination of Floyd's minerals and the odd biochemistry that Wellsweed employed that had justified the construction of Manhattan City. Essentially it was nothing more than a dormitory town for the refineries and processing plants that produced complex organic molecules that were used by Kinabica's medical and chemical industries—high-value, low-mass products, perfect for Z-B to reclaim and transport back to Earth.

Once the platoons were inside Manhattan, the mission proceeded along more standard lines. The commanding officer delivered his polite ultimatum to the city's administrator, who immediately agreed to all the demands. Technical support teams came in and started going through the inventory and refinery specifications.

There were plenty of suitable products that could be shipped back up to the starship orbiting Floyd. Unfortunately, there wasn't much of anything stored at Manhattan; batches were usually delivered straight down to Kinabica. For some inexplicable reason all Manhattan City's industrial facilities had been shut down five hours after the Third Fleet had emerged from compression.

Platoons were dispatched with Z-B technicians to "assist" city personnel in restarting the production lines with minimum delay.

On day two, Lawrence found himself with Colin, Ntoko and a couple of other 435NK9 squaddies, bouncing and tottering over the ubiquitous Wellsweed into a small crater a kilometer to the north of Manhattan City, where a chemical plant had been dug into the protective insulation of the regolith. They were escorting a pair of Z-B technicians and five of the chemical plant maintenance crew who had been assigned to restart the systems.

He scanned his image-intensifiers around, eager to absorb

as much as he could. His first Alien Planet. Admittedly it was different from both Earth and Amethi. He was just slightly disappointed that it wasn't more interesting. Wellsweed made it look as if the whole place had been meticulously foam wrapped ready for storage. He kept looking up at the huge, brilliant crescent of Kinabica hanging above the horizon, wishing he'd drawn that assignment. A genuine new world. The i-i's made it glow enticingly.

Apart from the spaceport, the crater was the first area they'd encountered where the Wellsweed was patchy. Dozens of crude tracks and wheel furrows crisscrossed the floor, cutting right through the vegetation. The center of the crater was home to a series of regular humps, each one a couple of hundred meters long. Ribbed cylindrical heat exchange towers stood on top, resembling the brick chimneys of the Industrial Revolution four centuries and seventeen light-years distant. The dirty soil that had been bulldozed on top of each bunker was speckled with dull rouge blooms of new-sprouting Wellsweed, stains that were gradually spreading and merging. In comparison to the torn carpet on the crater floor, none of the new growths seemed particularly healthy.

The airlock was large enough to hold the whole group. After it cycled, the inner hatch opened into a warren of concrete corridors. Long rectangular windows set in the walls provided views across chambers full of tangled machinery and piping. Blank steel doors led away into offices, workshops and vaults lined with deep storage tanks.

It was bewildering to the squaddies, even with their muscle skeleton HUD visors providing them with a full map of the installation. The technicians and maintenance crew were unfazed, making their way directly to the quiet control center. Within minutes, the management AS had begun speaking to them at half a dozen stations simultaneously, while the big status board began to light up with schematics as the plant came back to life.

"You'd better check out the rest of this place," the senior technician told Ntoko—a reasonably polite way of saying "get lost."

"We're with that," the corporal assured him.

"Check it out for what?" Colin asked as they left the control center.

"Revolutionaries and terrorists, I guess," Ntoko said. "Relax, my man, we pulled an easy duty with this one. Walk around the scenery for six hours, and we're back in the barracks with no harm inflicted."

"I thought there'd be more to it than this," Lawrence admitted.

"There never is, son," Ntoko said cheerfully. "The platoons are only ever here for that one anarchist hothead who doesn't give a shit about collateral and the gamma soak. Everyone else knuckles under and gets on with the job. They might not like us, but they don't cause any trouble."

"Do we ever use the gamma soak?"

"Never have. I doubt we ever will."

"Thank Fate for that."

"It's logic, not fate. If we ever got to a situation where it needs using, we've lost anyway. If things are so out of control you need to kill half a million people to frighten the rest into obeying you, there's not really a hell of a lot of point in using it. That kind of madness will never achieve anything except to twist the level of hostility beyond reason, and with it the probability that we'll ever make it home. Use gamma soak against a planet, and they'll throw everything they've got against the starships by way of retaliation and vengeance." His thick, armored hands tapped against his thigh with a sharp *clacking* noise. "In any case, I could never give that order. Could you?"

"No, sir," Lawrence said firmly.

"Course not. But you'll still have to shoot your scatter pistol when I tell you."

"Ready for that one, Corp."

"Good man. Now, you and Colin make a sweep through the two eastern bunkers. Make sure there's nobody lurking around avoiding collateral status. It's not that unlikely. Some people just don't trust their fellow citizens to behave. Sad but true."

They made their way along badly lit corridors, taking junctions at random. Infrared, motion detectors, i-i's and sound filters couldn't detect anyone else in the bunker.

"This is a total waste of time," Colin grumbled on the local frequency. "It's not like the planet, where people can hide out away from the cities. We know exactly how many people there are in Manhattan; it's listed in the AS memory."

"Quit complaining. Like the corp says, it's an easy duty."

"Yes, but how's it going to look on our records? I wanted to see some action, perhaps get the chance to earn a commendation."

"Will you relax? Keeping the whole of Manhattan City under control without ever having to fire a shot is like the universe's most perfect operation. And we're part of that. Now that's what'll get you a commendation. The company likes things that go smoothly."

"Possibly."

Overhead pipes began to gurgle and shake as fluids rushed down them. It had been happening all morning as the plant slowly came back to life. The ambient temperature had risen fractionally as the machines all returned to work. Even through the protective layers of armor and muscle, they could feel vibrations building in the walls and floor.

"Newton, Schmidt, get over here," Ntoko ordered. "Bunker three, section four."

"What's up, Corp?" Lawrence asked.

"Just get here." Ntoko's voice was flat.

"On our way."

They couldn't run. Any real strength applied through their legs would smack them straight up into the ceiling. Instead they moved with long, loping strides, arms raised ready to slap themselves down if the arcs became too high.

As they approached the door to bunker three Colin drew his carbine, taking the safety off.

"Are you crazy?" Lawrence hissed. "Those things are loaded with explosive shells. You could blow a hole clean through the wall."

"We're underground, Lawrence. All I'm going to kill is hostiles and rock."

"And chew up a billion dollars' worth of machinery." Lawrence pulled his own scatter pistol out. The magazine was loaded with toxin darts. "You know policy; assets have priority."

"Fine fucking policy that is," Colin grumbled. A further few words were muttered, which the helmet mike had trouble picking up. Lawrence guessed they were German anyway. Colin always reverted to his native language when he felt stressed. He paused and shoved his carbine back into his holster, removing a maser wand.

Lawrence didn't comment. He walked forward, and the bunker door slid open. The main corridor stretched on ahead of them, its tube lights flickering at an almost subliminal frequency.

"We're in the bunker, Corp," Lawrence said.

"Good, now get down here to us."

Lawrence's HUD flicked up the plans for bunker three. Section four was at the end of a side corridor eighty meters away. They started walking toward it.

"You reckon this is some kind of hazing?" Lawrence asked. He'd switched off his radio, using the armor's external speaker on low volume.

"Not sure," Colin murmured back. "You reckon the corp would pull that kind of stunt?"

"Dunno. He might want to see how we react."

"If he'd just tell us why he wants us."

"Maybe he's been captured."

"Oh, come on!"

"Well, it's possible. Why else is he being . . ."

Lawrence's armor microphone picked up a scuffling sound. His motion detector registered a fast airwave wash down the main corridor directly behind him. Both of them spun around, assuming a low crouch position, weapons searching for a target. The i-i scoured the walls and floor on high resolution, revealing nothing.

"What the fuck . . ."

Lawrence switched to the secure suit band. "Corp, is there anyone else in this bunker with us?"

"Nobody's been authorized by the AS, why?"

"Somebody moving around out here."

"Just a minute."

Lawrence and Colin straightened up, keeping their weapons ready.

"Could have been the machinery switching on," Colin said. "No telling what effect it'll have on the sensors."

"The AS should filter it."

"I've checked with our people in the control center," Ntoko said. "Everyone's accounted for. The local AS is relaying camera images to my suit. I can see you two, but there's no one else in here."

"We thought it might be the machinery glitching our sensors," Lawrence said.

"Okay. Keep a watch. And put your i-i's to medium resolution; high-rez produces some weird effects."

"Roger. With you in a minute."

They made the side corridor without further incident and started down. The door at the end was open. Lawrence couldn't see anyone inside, just another big chamber full of black-and-silver machinery, the kind of towering mechanical exhibition that could have come straight out of a steamship's engine room. Thin gases were leaking from pipes; the general noise level rose with every step closer.

An armor-suited figure appeared in the door. "Hi, lads," Meaney called. He raised his arm to wave. Something moved behind him, eclipsing one of the ceiling lights.

"Down!" Lawrence screamed. He and Colin thrust their weapons forward. A target circle flashed across his HUD.

Meaney froze, framing his suit in the doorway. His gauntleted hand suddenly made a move for the carbine holstered on his waist. The dark swirl bobbed about behind him, sliding away from the light. Then it was gone, slithering into the intestinal tangle of pipes and valves.

"Behind you!" Colin yelled.

"What—" Meaney was turning, his carbine half out of the holster. The other two were racing toward him, the AS angling their muscle skeletons so they were leaning forward at a sharp angle, providing a degree of balance in the low gravity.

"Where'd it go?"

"In there, in that gap."

Lawrence jumped up cautiously, gun held out in front, pointing into the metallic crevice ready to fire as his helmet sensors rose up level. The i-i's green tinge revealed a gap that was nothing but a dusky jumble of twisting pipes and looping cables. Infrared showed some of the pipes glowing pink.

He relaxed his trigger finger as he landed on his heels. "Shit! Missed it." His HUD display was registering a high heart rate. Adrenaline hummed eagerly in his ears. This was all way too elaborate for a hazing. He tried to concentrate on his training for unknown territory. *Be suspicious. Always.*

"What the fuck are you two doing?" Meaney demanded.

"Didn't you see it, for Christ's sake?" Colin said. "It was right behind you. Are your sensors screwed or something?"

Meaney's carbine waved around at the cliff of chemical-processing equipment. "What was behind me?"

"I don't know. Something up there."

"Where?"

"Jesus, what's wrong with your sensors?" Colin asked.

"Nothing's fucking wrong with them."

"Then you must have seen it."

"Seen fucking *what?*"

Ntoko emerged down an aisle formed by the hulking stacks of machinery. He was holding his scatter pistol ready in his right hand. "Okay, what do you two keensters think you keep seeing?"

"I'm not sure, Corp," Lawrence admitted. "We saw something moving about behind Meaney."

"My sensors didn't track anything," Meaney said.

"Something?" Ntoko said. "A person or a robot?"

"Well, it was up there, and smallish," Lawrence said, trying to recall the shady image.

"It didn't move like a robot," Colin said. "It was fast."

"Could have been a rat," Ntoko said.

"A rat?" Lawrence asked. "Why would Kaba import rats, especially to Floyd? They don't contribute anything valid to the ecology."

"They're not imported, son. They just tag along for the ride. Anywhere in this universe where humans are, you'll find them as well. Sneaky little sons of bitches, as well as vicious."

"There aren't any on Amethi."

"No? Well, then, you were lucky. Now get your AS to run a constant track for small object motion. If anyone sees anything, tell me straightaway. Got that?"

"Yes, Corp."

"Good, now come with me." He marched off back down the aisle he'd come from.

"What did you find here, Corp?" Colin asked, hurrying after him.

"Dust."

"Dust?"

"Yeah, dust. But wrong."

Not for the first time, Lawrence dearly wished he could shrug inside a muscle skeleton. The strange sightings had left him hyped; now the corp was telling them they were here for something different. He couldn't relax. Something else was in here with them, he knew it.

Ntoko led them into an open space at the end of the machinery, where Kibbo was waiting. On the other side from the refinery equipment were two huge cylindrical tanks embedded in the concrete wall. The domed end of the one facing Kibbo was five meters high, weld seams between the metal petal segments clearly visible. Bolts the size of a fist secured its rim to the end of the tank.

Ntoko squatted down, and beckoned Lawrence and Colin over. He pointed at the floor. "There, see?"

Lawrence upped his light amplifier sensitivity, knowing there must be some abnormality. The original gray concrete floor had been darkened from age and chemical stains. Dust lay amid the small ridges and pocks. He pulled the focus back. There was a broad track leading to the tank. Wheels and feet had been moving to and fro in what must have been a regular procession. Interesting but hardly alarming. He switched to the second tank, but the floor there had an even distribution of dirt.

"So?" he asked cautiously. "They serviced this one recently."

"Try infrared," Ntoko said softly.

The tank with the track leading to it was five degrees warmer than the other.

"That was what clued us in first," Ntoko said. "The signature is completely different. Yet according to the plant's AS inventory they both have exactly the same fluid inside."

"So what's—"

This time everybody's sensors picked up the movement. They swiveled toward the source as one, weapons ready. Against all training and instinct, nobody fired.

An alien was creeping out of the machinery three meters above the ground, hanging on to the conduits and support struts so it was ninety degrees to the vertical. Lawrence's first thought was of disappointment; it was unimpressive. A body the same size as a German shepherd, with six (or eight—he couldn't quite see) spiderlike legs, bent almost double round the knee-hinge joint, which ended in small horned pincers. Its fuzzy scale hide had the shading of a dirty oil-stain rainbow. The only gross abnormality, a true alienness, were the eyes, or what he assumed were the eyes: chrome-black buds along the flanks that were flexing about constantly. There was a head of sorts; one end of the body was bulbous, with a blank slit for a mouth.

It wore some kind of plastic bracelet on each of its limbs, right up by the body joints. They seemed to be fused with the flesh.

"General alert," Ntoko was announcing calmly. "Come in, Ops, we have a contact situation here. Ops?"

Lawrence's HUD was flashing red communication icons at him; the local net relays had crashed. He paid little attention. Another alien was crawling out of the machinery.

"Up there," Meaney croaked.

A third alien was walking along the ceiling above them, its limb pincers gripping the pipes with little effort as it picked its way along.

There were eight limbs, Lawrence saw at last. Definitely nonterrestrial, then. He watched the creatures with a mixture of elation and astonishment. The bracelets were high-technology artifacts. They were sentient! He was making first contact with sentient aliens.

This moment was everything he'd ever wanted from life. He let out a soft, nervous laugh; incredulous that this should be happening here and now. His hand was trembling. He hurriedly engaged the pistol's safety, then asked the skeleton AS to find the rules governing first contact. They ought to be in the memory somewhere.

"Corp, what do we do?" Colin asked, his voice high and excited. He was shuffling back toward the tank, keeping his weapon trained on the first alien.

"Just stay—"

The alien in front of him extended a limb. A maser wand of very human design was gripped in the pincer. Lawrence stared at it numbly.

"That's . . ."

The alien fired. Lawrence's HUD instantly displayed a schematic of his armor suit. Red icons clustered round like enraged wasps, indicating the energy impact pattern. Superconductor shunts were racing toward burnout as they tried to dissipate the beam.

"Move!"

Lawrence dived to one side, trying to break the maser's lock. The effort sent him flying close to the roof, limbs waving in panic. An automatic weapon opened fire below him, projectiles hammering into concrete and metal. The lights went out. Lawrence hit the floor and bounced almost a meter. His HUD reported that the maser was no longer on him. He waved his scatter pistol about ineffectually.

The space around him was illuminated by the muzzle flashes from two guns. Their topaz strobing revealed huge plumes of thick vapor screeching out of the processing equipment. More aliens were scuttling out of mechanical crevices. He saw two of them lugging a mini-Gatling between them.

"Ho fuck!" He rolled fast, abandoning the pistol and reaching for his carbine.

"How many?" Kibbo yelled.

"Where?"

"What happened to the lights?"

Lawrence's motion detector was rendered useless from the billows of vapor. Infrared struggled to acquire targets through the cloying haze. The i-i simply expanded the swirls in the fog, shading them a sparkly green. Another warning icon flashed, accompanying an audio tone, alerting him to the rapidly increasing toxicity level of the chamber's atmosphere.

A carbine opened fire, sending explosive rounds thudding into the dark. Detonations were swamped by the cloying gas, their flashes turning into a whiteout haze. Visibility was down to fifty centimeters and closing fast.

"Hold your fire."

Lawrence aimed the carbine in the direction he'd seen the aliens with the mini-Gatling and shot off twenty rounds. "They've got heavy artillery, Corp," he shouted as the weapon juddered in his arms. Explosions pulsed around the bunker. Someone else was shooting as well, small-arms fire. A round slammed into Lawrence's armor below his right pectoral. He was flung into the air, spinning. Red and amber icons traced elaborate circles around him like a protective swarm of holographic birds. Pain thumped straight through the hard outer shell and muscle skeleton to punish his ribs.

The radio channel was a caterwaul of yelling and bawled orders. Nothing made sense. Lawrence landed on his back, the impact pushing the last gasp of air out of his lungs. He dropped the carbine. Something moved under his legs, writhing about urgently, pushing his knees apart. Shock mobilized him instantly, allowing him to overcome his pained daze. He twisted quickly, bending down to grab at whatever was lifting itself up between his legs.

Excruciating pain bit into his leg, just above the knee. Icons reported both his armor and muscle skeleton had been sliced open. His hands fumbled into contact with a broad object, his brain telling him it was the same size as an alien's body. Through the swirling vapor he could just make out the ruddy infrared blur. It was an alien, and one of its limbs was holding a power-blade. Lawrence lunged for the knife, jerking it free from the pincers. He saw the alien's limb snap from the force of his grab, took a breath and punched the body with his free hand, delivering the full power of the muscle skeleton behind the blow. His armored fist ripped straight through the creature's hide, squishing into internal organs.

He almost vomited, pulling his hand free again, bringing with it dripping webs of membrane and gore. Another bullet struck him, sending him sprawling.

"Ceiling," Ntoko bellowed. "Shoot the ceiling out. Explosive shells. Do it now."

Explosions pummeled the concrete above Lawrence. The blastwaves made his battered armor creak from intense pressure stress. He fumbled his hand across the floor, searching for his own carbine. A maser beam washed over him again, prompting a flurry of scarlet symbols, and he jerked himself clear. Small-arms fire was raking the air above him. Some kind of gelatinous sludge was creeping across the concrete floor, slopping against his armor shell.

He found both the alien body and his carbine and rolled onto his back. Half a magazine of explosive shells smashed into the ceiling. Heavy lumps of concrete fell out of the fog in what seemed like slow motion, splashing into the sludge.

"What are we doing?" Meaney asked. "Why aren't we killing them?"

Lawrence slapped another magazine into his carbine. He'd been about to ask the same question.

"Breaching the bunker," Ntoko said. "I'm going to blow this gas and those alien mothers clean into space."

Lawrence opened fire again. Above the explosions he could hear a shrill whistling sound gather strength. Abruptly, the chemical fog was thinning, and the whistle increased to a tormented howl. A slice of sunlight prized its way down through the vapor, quickly expanding. Lawrence's i-i hurried to compensate, throwing up filters. He shifted his aim, letting the carbine shells chew the edge of the widening fissure in the ceiling. A huge, jagged section of concrete blew upward, pummeled by the bunker's venting atmosphere. The last of the gas surged upward, tugging Lawrence off the floor. Then he was tumbling down in absolute silence. Blazing sunlight shone into the bunker, revealing confusion and mayhem. The dense conglomeration of machinery had been torn to ruin, pipes ripped open, stolid processor units shredded. Sprays of fluid and gas were still pumping out, their ragged plumes curving upward before dispersing. Several alien corpses were hanging limply from metal fangs. They'd all been hit by weapons fire, pulping the tawny flesh.

"Come in, Ops, we have an emergency situation," Ntoko said. "Request backup immediately. Receiving hostile fire."

Lawrence's AS confirmed that they'd reestablished communications now that the roof had been split open. He clambered painfully to his feet as Ops began to interrogate Ntoko. Blood was leaking out of his knee where the power-blade had cut through, most of it, he was confident, coming from the skeleton muscle. Lines of pain flared along his torso with every move he made. He could see cracked dints in the armor; scorch marks had blistered the outer layer. "Oh hell," he groaned.

"We beat them." Kibbo's voice had a hysterical edge. "We beat the fuckers."

"What were they?" Colin asked. "Where the hell did they come from?"

"Holy shit, lads," Meaney said. "We've just fought our first interstellar war."

"And won the bastard! We kicked some ass, huh?"

"We did, man. They ain't gonna mess with this platoon again, that's for sure."

"I don't get it," Lawrence said. "What did we do? Why did they shoot?"

"Who cares?" Meaney said. "We are the masters now!" He let out a whoop, raising his arms in a victory salute. He froze. "Holy shit!"

Lawrence looked up. Aliens were crawling along the top of the broken roof, front limbs gingerly probing the blackened concrete edges. Several were easing themselves through the gap, gripping the twisted reinforcement struts. Maser beams stabbed down, playing over the squaddies. They returned fire, using carbines to chew away at the concrete.

"Get to cover," Ntoko ordered. He led them over to the wheezing bulk of machinery, firing as he went.

"They're natives," Lawrence said, shocked at the realization. "They don't need suits to survive, look. They have to be native."

"Big fucking deal," Meaney cried. "What did we ever do to piss them off?" He was shooting as he dodged behind a solid hunk of equipment.

"Stole their land and their women, I guess," Lawrence said.

"That's a real big fucking help, Lawrence," Kibbo yelled. "What is wrong with these alien freaks?"

Colin sent a whole magazine from his carbine roaring into the fractured ceiling, mauling concrete and aliens alike. "We didn't blow the bastards into space, we just let them in, for fuck's sake!" Concrete and flesh rained down over the squaddies.

"No more saturation fire," Ntoko ordered. "Let's conserve what we've got. Pick them off."

Lawrence ducked down into an alcove, then raised his carbine. A crossed targeting circle drew sharp violet bars across the ruined ceiling. He switched to single fire and located an alien. One shot blew its body apart. For aggressors, they were terribly vulnerable. That didn't make a lot of sense.

"How long before the cavalry arrives, Corp?" Colin asked.

"Any minute now. Just hang on in there."

For the first time in his life, Lawrence found he was praying. He wormed his way deeper into the too-small alcove, wondering if the God he knew didn't exist could be of any possible use. Asking couldn't possibly make things worse.

Simon Roderick hadn't expected to visit Floyd during the mission. As far as Z-B was concerned, the moon was simply a minor manufacturing location, easy enough to control and strip of its wealth. That was during the planning stage. Now those assumptions had changed drastically. And as a result, Simon was having to cope with low gravity and the uncomfortable indignity of a spacesuit.

The wretched devices hadn't improved much since the last time he was in one, eight years ago—an inner layer that exerted a fierce grip on his flesh, and a globe helmet that blew dry, dead air into his face, making his eyes water. The backpack weighed too much, which on Floyd translated into awkward inertia.

It was almost tempting to wear a muscle skeleton, as his three-man escort was doing. But he could never quite decide which was the lesser of the two evils.

His escort remained outside as he stepped into the chemical plant's airlock. After it cycled, he emerged into a drab concrete corridor. A reception committee had assembled for him, six squaddies in full muscle skeletons, carrying improbably sleek and dangerous-looking weapons hardware. Waiting with them were Major Mohammed Bibi, the commander of the Floyd operation, and Iain Tobay, from Third Fleet intelligence, along with Dr. McKean and Dr. Hendra from Z-B's biomedical science staff.

Simon's spacesuit AS confirmed the chemical plant's atmosphere was breathable, and he unsealed his helmet. "Are we expecting further trouble?" he asked lightly, his gaze on the stiff-at-attention squaddies.

"Not expecting, no, sir," Bibi said. "But then we weren't expecting this particular incident to start with."

Simon nodded approvingly. The major was probably overcompensating for the unexpected firefight, but it was prudent. He couldn't fault the response.

They clumped along more identical corridors to bunker three, section four. There was a noticeable difference in the air as soon as the steel door slid open. A mild chemical stew permeated the standard oxygen-nitrogen mix, with ammonia percolating to the top. He wrinkled his nose up.

Dr. McKean noticed the motion. "You get used to it after a while. We've brought in extra atmosphere scrubbers, but the processing machinery is still spilling some volatiles."

"I see." Not that Simon cared. Technical types always overexplained their world.

As he walked down an aisle formed by the machinery, the evidence of the fight grew more pronounced. Pools of dark tacky fluid were oozing out from underneath, while the smells strengthened. Metal became buckled and twisted; torn fangs blackened from explosive heat. When he came out into the open space at the end, the elaborate machinery was simply mangled scrap.

Temporary plastic shielding had been fixed across the broken ceiling, its epoxy adding another acidic fragrance to the mélange. Bright sunlight shone through the translucent covering, tinged pink.

The tank that had caused all the trouble was now open, its large cap hinged back against the wall on thick hydraulic pistons, like the entrance to a giant bank vault. A ramp had extended from inside. Several Z-B personnel were moving round in front, helping to clear up the mess and shifting trolleys of equipment up and down the ramp.

Simon saw a couple of them were moving slowly, every movement careful, as if they were in pain. He called up files via his DNI: Meaney and Newton. Both in the firefight, both injured and assigned light (noncombat) duties. He was mildly interested by Newton's background.

"How's it going?" he asked them.

Newton straightened up from a mobile air purifier and saluted. His eyes flicked toward Major Bibi. "Fine, thank you, sir."

"Yes, sir," Meaney said.

"That was a good job you did," Simon said. "Muscle suits aren't exactly configured for head-on military action."

"They're good systems, sir," Newton said. He was relaxing slightly now he knew they weren't being bawled out.

"Now that you've used it in combat, any suggestions?"

"Better sensor integration would have been a help, sir. In fact, better sensors altogether. We were operating blind once the AS opened the gas valves; that muck screwed up our i-i and the motion detector."

"That must have been difficult."

"Corporal Ntoko knew what to do, sir, he held us together. But like you say, if we'd been up against serious opposition we'd have been in trouble."

"I see. Well, thank you for your opinion. I'll see what I can do—not that the designers will listen to an executive, I expect. They don't hold us in terribly high regard."

"But you pay them, sir; they hold that in high regard."

Simon grinned. "They certainly do." He indicated the body of an alien, now covered with a sheet of blue polyethylene. "First encounter with an alien life, Newton?"

"Yes, sir. Shame it was under these circumstances. For a moment I thought they were real aliens."

"Real? I don't think you can get much more real than these."

"I meant sentient, sir. It's a crime what was done to these poor things, rigging them up like waldo robots."

Simon pondered the young man's idealistic dismay. Only the truly young could afford that kind of morals. No wonder Newton had rebelled against his background. "I suppose it is. Have you been in the tank, yet?"

Newton pulled a disapproving face. "Yes, sir."

"Ah, well, my turn now."

"Sir?"

"Yes?"

"Will they be punished, sir?"

"Who?"

"The . . . people who abused the aliens, sir."

"Ah, I see. Well, you must understand, Newton, while we're here, as well as enforcing the local law, we're also subject to it. That's what gives us the legitimate right to procure the assets that we do, because we work within their own legal framework even if they don't like or admit that. What we don't do is impose and enforce foreign laws on the indigenous population. If their constitution says it's okay to sleep with your sister, then that's what we let them get on with. So unfortunately, while enslaving and conducting experiments on animals or aliens is illegal in most countries on Earth, it isn't here."

"You mean they've done nothing wrong!"

"Not at all. They launched a serious assault on legitimate law-enforcement officers in the pursuit of their duty."

"So what's going to happen to them?"

"That's what I'm about to decide."

Simon paused as he was going up the ramp, looking down

on yet another discreetly covered alien corpse. "Have you learned anything about them?" he asked McKean.

"Not much," the doctor admitted. "They're native to Floyd. Mammalian. Socially, they're halfway between a pack and a hive. Their whole physiology slows down considerably during the night-time cold. They eat Wellsweed; in fact, they spend ninety percent of their time grazing. And that's about it."

"So they're not sentient?"

"No, sir. We're trying to mine some references to them from Manhattan's memory, but so far we've drawn a blank. It's obviously been deep encrypted. Certainly nobody on Earth knew about them. Which is surprising. From a xenobiological viewpoint I cannot overstate how important they are. Kaba should have been shouting about them from the moment of discovery."

"Kaba's Earth Board probably weren't informed," Simon said. "You never reveal a good poker hand, Doctor."

He continued up the ramp, with Major Bibi leading the way. The interior of the tank had been split into two levels, each subdivided into several compartments. The arrangement reminded him of a bomb shelter. Not a bad analogy, Simon thought. They were certainly zealous about their security.

"I take it you have debugged this?" he asked Major Bibi.

"Yes sir. I've had technicians sweep it for physical defenses, and the plant's AS has been dumped into a sealed storage facility. It was complicitous with the attack on the squaddies, we know that from the gas release. Our own forensic AS is taking it apart code line by code line to determine what kind of routines were hardwritten in. We suspect it was puppeting the aliens as well. I've also had wipehounds running through the plant's datapool to make sure there are no subroutine remnants lurking. But there's a lot of circuitry here, especially in the processing machinery; we should have an all-clear in another ten hours."

"And the Manhattan City AS?"

"Definitely part of the business. Wiping that is more difficult; it does supervise a lot of hardware functions in the city, including life support. So far I've settled for installing limiter and monitor programs in the datapool."

"Very well." They stopped in front of a heavy security door. There was an elaborate DNA lock panel on one side, but the slab of reinforced metal itself was retracted.

Inside the room, medical support equipment had been stacked into elaborate columns. Eight of them were spaced along the middle of the floor. Each one was topped with an opaque plastic sphere fifty centimeters in diameter. Wires and slim tubes wormed out of the equator to disappear into the stacks at various levels. Five of them were inert, while the remaining three hummed and whirred quietly, with small indicator lights winking above various components. A couple of Z-B technicians were busy taking apart one of the inert columns. Dr. Hendra silently signaled them out, and they left without a word.

Simon stood in front of the first active pillar, staring at the globe. "Your opinion on the procedure's viability, Doctor?"

"Oh, it's viable, all right. In fact, it's much more proficient than the kind of rejuvenation treatments that are employed on Earth."

"Really? I thought Earth led that particular field."

"Technically we do. But v-writing a whole human body is enormously complicated. You have to vector new genes into the individual cells of every organ and bone and blood vessel, not to mention skin. Those genes all have to be specific to their destination. The best we ever manage to revitalize in each organ is twenty to thirty-five percent of the gross. Enough to make a difference, but there are just too many cells for all of them to be revitalized. That's why there's no point in extending rejuvenation past the third treatment. You run smack bang into the law of diminishing returns."

"Depends how young you are when you have your first treatment," Simon murmured.

Dr. Hendra gave a complicit shrug. "As you say. But it's unusual for anyone to undergo the treatment before they reach sixty. These days it's far more effective to provide germline v-writing to inhibit the aging process. When you're only ten cells tall, all those shiny new improved genes can be vectored in without any room for error."

Simon smiled knowingly. "Of course." Dr. Hendra's file showed he was born of such a process, which, given the genetic engineering of the time, would give him a life expectancy of around 120 years. His parents had both been stakeholders in Z-B, middle-management level. In those days the company provided it for only the upper echelons. They'd been lucky to qualify. Now, of course, it was available to every stakeholder, regardless of the size of their stake. Another huge incentive to invest your life with Z-B, and one of the reasons they were one of the largest companies on Earth and beyond. "And yet you regard this particular procedure as effective."

"Indeed." Dr. Hendra gestured at the plastic sphere on top of the medical stack. "Isolate the brain, and you can repair at least eighty-five percent of the decayed neuron structure. As you don't have to worry about repairing anything else, it allows you to concentrate your resources most efficiently. After all, you are only rejuvenating one kind of cell, although admittedly there are many variants."

Simon used his DNI to activate the column's communication system. "Board Member Zawolijski, good morning."

"A good morning to you, Representative Roderick," the brain replied.

"That was most impolite of you to shoot at our squaddies."

"I apologize. My colleagues and I are somewhat set in our

ways. Your platoon's incursion alarmed us. The corporal had discovered this tank. Ours is not an aspect of Board family life we wish to share with the rest of the civilized galaxy."

"Indeed, and does that include the Board of your new parent company?"

"Certainly not. I speak only of the fact that it can be done. The . . . cost, in social terms, could be regarded as unacceptably high by certain human factions."

"That's very heartening. The Board that I represent would certainly appreciate a full and complete technical briefing."

"I'm sure that can be arranged."

Simon's personal AS had been scrutinizing Zawolijski's root links into Manhattan City's datapool caches. The brain was reluctantly acknowledging the retrieval probes, allowing access to sealed memory blocks. A file expanded in Simon's vision, indigo script flowering around a single full-color image. It was the Kinabica police and court records of Duane Alden, beginning with his juvenile arrest and cautions for shoplifting, vehicle theft, and aggravated assault. As he matured he'd swiftly progressed to narcotic violation, burglary, armed robbery, extortion and finally murder. The last crime was a holdup that had been bungled thanks to Duane's drug-ridden state. The whole sorry episode had been captured on a security camera. His court case had lasted a mere three days. An appeal had been dismissed a month later. He was due to be executed in another two weeks, a month after his twenty-first birthday. The intervening three months had been spent in a prison's hospital wing, where tough medics had thoroughly detoxed him, at the same time pushing him through an intensive health regimen. Duane had resisted at first, but warders always have methods of guaranteeing compliance among even the most recalcitrant inmates. His lawyer was currently lodging an "abusive treatment" complaint, but that was just going through the motions.

Observing the naked, full-length holographic image of

Duane Alden that appeared to hover in the air between him and the encased brain, the one phrase that came to Simon's mind was *Golden Youth.* Duane was physically flawless and distinctly handsome.

"Your new body, I take it," Simon inquired.

"Yes," Zawolijski said. "He's quite splendid, isn't he? Several centimeters taller than my last. And that face . . . so bold. I'm sure the ladies will be appreciative."

"I'm curious. Exactly how old are you?"

"Two hundred and eight years, Earth standard."

"And this body would be number . . . ?"

"My fifth replacement. I remained in my original until I was sixty."

"A new body every thirty years. That seems slightly extravagant."

"Not really. Twenty to fifty: the best years of a man's life."

"In the classical model, yes, but now that human bodies can be v-written for enhanced life expectancy, the period of primacy is considerably longer."

"Quite so. But such germline treatments are only just becoming commonplace on Kinabica, and as the parents invariably request additional modifications such as increased intelligence, such specimens are less likely to stray."

Simon canceled Duane's file and frowned at the brain. "You believe that enhanced intelligence ensures a noncriminal life?"

The brain chuckled. "Less likely to get caught, actually. Or if they do, then it's after a long and arduous investigation. By which time they're past their usefulness to the Board."

"You should use equally intelligent police officers to catch them."

"At the salary we pay?"

"I see your point. Which leads to my next question. Why not simply clone yourself a replacement body?"

"Ah, one of our race's favorite myths. Have you any idea

how difficult and expensive that is? Growing a human in vitro until—realistically—they're sixteen. How would you suppress the arrival of consciousness over that time?"

"Would that problem arise? I'd have thought the lack of external stimuli would eliminate any chance of thoughts germinating."

"Coherent thought, certainly. But even infants have a basic awareness, and more than that by parturition. Sensory deprivation for sixteen years produces a monstrously retarded consciousness. It doesn't quite qualify as a personality. But believe me, it's a problem sustaining a body in an amniotic tank for any time after its first year. It wants to be birthed and struggles against its confinement."

"Then clone a body without a brain. V-write it out of the genome."

"Oh, please, how would you replace the autonomic function control? Technologically? There are far too many subtleties involved for some kind of wetwired chip to regulate."

"What about growing parts separately? Accelerating a replacement organ's growth to its maturity is a proven procedure. After that you simply assemble them into a full body."

"That merely increases the original problem by two orders of magnitude. The number of separate parts in a body is incredible, and that's just the principal glands and organs. Don't forget the entire circulatory system, skin, a skeleton even. What order would you start stitching them together in, in order to make sure they stay functional during the procedure? How much surgery does it actually take to assemble an adult human being? No. The idea is pure science fiction. I assure you, we have explored all these avenues. The most efficient way to produce a human body is the old-fashioned method of unskilled labor. Until we can develop some kind of active nanonics capable of integrating cellular structures or resetting individual DNA strands, transplanting a brain into a criminal's body is the most reliable procedure to regain a healthy young body."

"Very well. But what about the neuron regeneration process you employ? There must be some memory loss."

"Not from the regeneration. My memory loss comes from standard brain decay. New neurons don't contain old memories. That's perfectly acceptable to all of us; in fact, it's essential. The brain is finite, no matter how many improvements we have v-written in each time we undergo rejuvenation. I have to have the capacity available to store my new life's experiences when I re-enter society."

"If you are forever discarding the past, then you have forgotten who you were."

"Never, that's the beauty of this procedure. I have complete continuity with the baby born those two hundred and eight years ago, which is the overriding psychological factor. The strongest memories anyone has are connected with identity. The events that define what you are, shape your personality and who you have become, are so powerful they are part of your essence. They have become instinct, retained no matter how much regeneration is required. I might not be able to remember the intimate details of a day one hundred and thirty years ago, but that is no longer relevant; I *know* that I am the individual who lived through that day. Continuity of consciousness rather than unbroken memory, that is the human soul, Representative Roderick."

"Then what of the biological imperative? Your body is not genetically yours. You cannot reproduce for yourself; any offspring you sire will be those of Duane Alden. What is the point of your existence other than sheer vanity?"

"And you accused us of relying on classical models? With so much v-writing these days, whose child is truly theirs anymore? But to answer your question, that particular aspect of rejuvenation has the easiest remedy. My balls are cloned and transplanted along with my brain into every new body. For females, we simply implant cloned ova. All of us take part in life to the fullest degree when we return. We are complete to a degree unachievable by ordinary living, twenty years old with the intellect of a centenarian."

"What do you return as, a distant cousin?"

"Whatever identity is most convenient. Family stakeholding is not scrutinized and analyzed, Board family trusts operate privately, executive Board members are not celebrities."

"The perfect system."

"To sustain us and our chosen way of life, yes. That's why we wrote the constitution the way it is."

"And now your Earth Board has sold you out."

"Please, Representative Roderick, you have no need to sustain your legal fallacy with us. Zantiu-Braun is here because it has the ships and the firepower to raid our world, filling its own coffers with complete impunity. We acknowledge the reality of your strength."

"I'm pleased to hear that."

"So what deal do you require?" the brain asked.

"Deal?"

"For us to continue our existence without interruption. We would be happy to accept your Board members into our fraternity. It is a good life here: Kinabica is a wealthy, advanced world with a stable society. They would lack for nothing."

"The Board I represent would not be able to accept that offer."

"I'm offering you virtual immortality lived as a plutocrat, and you're turning that down?"

"We have different goals and objectives."

"And you don't think these objectives can run in parallel to immortality? I find that hard to believe."

"That really isn't your concern."

"Then what do you want?"

Simon pursed his lips, regarding the isolated brain with a

weary disappointment. The techniques and ingenuity of the Kinabica Board were impressive, but their goals were so *old.* They'd be more suited to life in the Renaissance era, or maybe the British Imperium. They could have achieved so much more with what they had; instead they looked to the past for their template, building themselves an impregnable stone castle amid a stagnant society. All they'd done was secure what they already had. With a brand-new planet offering infinite horizons, no fresh possibilities had been explored, no impossible dreams attempted. It was truly pitiable.

"We want nothing from you," Simon said. "As you said, your planet is a wealthy one. It's in your Board's interest that you continue to keep it wealthy, and that coincides with our wishes."

"You have no objection to our rejuvenation method?"

"None. Keep your lives. We don't covet your banality."

CHAPTER TEN

TEN MINUTES IN, AND ALREADY THE DAY WAS NOT GOING WELL for Simon Roderick. He had eschewed taking over President Strauss's ceremonial office for the Third Fleet's tenure on Thallspring. That would be too clichéd, he felt. In any case, it was General Kolbe who was the official Z-B liaison to the planetary executive; he should be the one visible to the public. So while the hapless general tried to placate a bitter and resentful press and populace, Simon had found himself a comfortable office in the East Wing of the Eagle Manor, ousting the flock of presidential aides who had clustered around their chief, offering advice, analysis and general chicanery.

The Eagle Manor itself was situated on a slight rise at the center of Durrell, which provided Simon with a broad view out across the city. Normally, the mornings brought a brilliant sunshine beating down on the impressive buildings and lush squares of the capital. Today, thick, dark cloud was clotting the azure sky. A weak drizzle smeared the wide panes behind his desk, blurring the crisp lines of the distant skyscrapers. Vehicles on the circular highway ringing the Eagle Manor's expansive grounds were all using their headlights, nova-blue beams shimmering on the wet tarmac.

As soon as he arrived, his personal AS produced the summaries he used to monitor the state of life in the capital. Overnight, production at the factories designated for asset acquisition had fallen several points. That corresponded to a high number of staff failing to show up for their shifts and reduced supplies of raw material. Even traffic within the capital was light that morning, though when he glanced out of the window at the radial of wide avenues leading away from Eagle Manor's circular highway he couldn't notice any decrease in the volume. There were still lines at every junction. Then the indigo script of the medical alert file scrolled up.

He sat perfectly still in his high-backed leather chair as he read the reports. "Tuberculosis?" he asked incredulously.

"That is the diagnosis," his personal AS replied. "And there is little margin for error. Seventy-five cases have been identified in Durrell already; the projection is for double that by the end of the day, and rising after that. Reports of possible contagion are now arriving from outlying districts and other provinces across the planet. The strain appears to be a particularly vigorous one."

"Do they have a history of it here?"

"No. There has been no recorded case of tuberculosis since first landing."

"Then what the hell is the cause?"

"The preliminary conclusion by local doctors and public health officials is that we are the source of the infection."

"Us?"

"Yes. After conferring with our medical AS, I agree the conclusion is logical."

"Explain."

"This particular strain is the product of several hundred years of combating the disease with increasingly sophisticated medical treatments. Every time human scientists developed a new and stronger antibiotic to treat the tubercle bacillus, the bacillus evolved a resistant strain. By the early twenty-first century tuberculosis had evolved into one of the so-called superbugs; it was effectively resistant to all antibiotics."

"Which if I remember correctly was countered by the new metabiotics."

"That is correct. Metabiotics held the superbugs at bay for nearly a century. Eventually, of course, they developed resistance even to them. By that time, genetically engineered vaccines were readily available. They have provided an effective treatment ever since. For every new strain the bacillus evolves, we can simply read its genetic structure and provide a specific vaccine. This has produced a stalemate in terms of widespread contagion."

Simon stared out at the wet city with the somber realization of where this was leading. "But we still haven't eradicated the bacteria."

"No. That is not possible. Earth's cities remain a fertile breeding ground. Local health authorities are constantly alert for the emergence of new strains. When such cases are discovered it is possible to manufacture a vaccine within thirty hours. In this way, epidemics have been averted for two hundred years."

"And prevented on the colonies as well?"

"Colonists were rigorously screened for a broad spectrum of diseases before departure. If any of them were infected, they would be vaccinated. In all likelihood, the tubercle bacillus was never transported across interstellar space, at least not in an active state."

"So they don't have the same kind of health program in operation here?"

"No."

"In other words, we did bring it here."

"It is the obvious conclusion. The most probable scenario is that one of our personnel was exposed to an advanced bacillus, and was himself immune through vaccination, or he could have received germline v-writing, in which case his immune system would be enhanced and highly resistant. But he would still be carrying it. If that is what happened, then it was spread around the entire starship he was traveling in. Everybody on board will now be carrying and spreading the infection."

"Don't we screen everybody before a mission?"

"Not for specific diseases. Such a screening schedule was deemed too expensive given that the fleets were no longer used to found colonies. The platoons undergo constant medical monitoring from their Skin suits. So far, that has been considered adequate."

"Shit." Simon let his head sink back onto the seat's rest. "So it's not just tuberculosis, it's a superbreed of tuberculosis, and nobody on this planet is going to be immune."

"The medical AS believes the section of the population that received germline v-writing will prove resistant."

"Percentage?" he snapped.

"Approximately eleven percent have received germline v-writing, of which half are under fifteen years old."

"Okay. What does our medical AS recommend?"

"Immediate production and distribution of a vaccine. Isolate all confirmed cases and begin enforced medication treatment."

"Is it curable?"

"There are precedents. The medical AS has templates of metabiotics that have proved successful in the recent past. We can also combine that with lung tissue regeneration virals. Such a procedure will be neither cheap nor quick."

"Estimated time?"

"For full recovery: two years."

"Damn it. What about the time it will take to implement the vaccine production?"

"Production can begin within twenty-four hours once you issue a priority authorization. To produce it in sufficient quantities to inoculate the entire planetary population will take three weeks."

"What the hell will that do to our asset-production schedule?"

"An appraisal is impractical. There are currently too many variables."

His desk intercom bleeped. "President Strauss is here, sir," his assistant said. "He's demanding to see you immediately."

I bet he is, Simon thought. "Show him in."

"Sir."

"And ask Mr. Raines to come in as well, please."

When it came to someone who would soothe the way for asset realization and make sure Z-B's staff integrated well with the planetary legislature and civil service, President Edgar Strauss was not your man. The usual threats and coercion seemed to have almost no effect. He was rude, stubborn, uncooperative and in some cases actively obstructionist. Simon had even refrained from using any of his family for collateral: if they took after him they would probably welcome martyrdom.

Strauss stormed into the office with the same inertia as a rogue elephant. "You motherfucking fascist bastard! You're killing us. You want this planet cleaned out so you can stuff it with your own families."

"Mr. President, that's simply not—"

"Don't give me that, you little shit. It's all over the data-pool. You've released tuberculosis; some v-written type to boost its effectiveness."

"It is not v-written. It is a perfectly natural organism."

"Crap!" Edgar Strauss's gray eyes glared out of his hard, reddened face. "We're absolutely defenseless against it. You committed genocide, and condemned us to a long painful death. You should have done it with the gamma soak, you bastard, because this gives us the opportunity to slice your throats one by one. What use is your collateral now, huh?"

"If you'll just calm down."

The door opened again, and Braddock Raines slipped in. He was with Third Fleet intelligence; in his mid-thirties, the kind of man who could normally blend into the background of any scene, allowing him to assess what was happening with a minimum of interference from local officials. It was the simple knack of invoking trust in people. Everyone who talked to him would always say how pleasant he'd been, the kind of guy you'd enjoy talking to over a beer. Simon knew he could always be relied on for an accurate report of the most difficult situations.

"Who's this? Your executioner?" Edgar Strauss asked. "I know you'll never let me live now that I know the truth. Too scared of me. How are you going to do it, sonny, knife or a nice messy bullet to the brain?"

Braddock's jaw dropped. For once he was too shocked to respond.

"Shut up or I *will* have you shot," Simon snapped.

President Edgar Strauss sneered contentedly.

Simon took a long breath and sat down, waiting for his blood to cool. He couldn't remember the last time he'd lost his temper. But the man was quite intolerable. How typical of a primitive, backward planet like Thallspring to elect a blunt man of the people like Strauss. "Mr. President. I have only just been informed of this terrible outbreak myself. I am of course shocked and dismayed that such a thing could happen on this beautiful planet. And I would immediately like to go on record to assure you and the entire population that Zantiu-Braun will be doing everything we can to assist the local health authority to combat the disease. Templates for a vaccine and relevant metabiotics will be made available immediately. If all the necessary planetary resources are given over to dealing with the situation, then we're confident of a swift and effective end to the problem."

"It'll take a month before we can make enough vaccine to go round, sonny. How many people will die in the meantime?"

"We estimate three weeks maximum for a sufficient quantity of the vaccine to be produced. And with the correct procedures, nobody who has contracted the disease will die. However, that will require complete cooperation from your authorities. Are you going to assist with that? Or do you want your people to suffer needlessly?"

"Is that why you introduced this, to help subdue us?"

"It was not introduced by us," Simon ground out. "The tuberculosis bacilli have a long history of evolving new and unpleasant variants. Nobody knows where this particular one has evolved. Only a fool or a politician would seek to blame us for this." His personal AS informed him the president was receiving a stream of files from the datapool, all encrypted. Updating him on tuberculosis, no doubt.

"Oh yeah," Edgar Strauss said. "You and it arriving together is just a complete coincidence. What kind of screening procedures does Zantiu-Braun use on its strategic security personnel before departure? Huh? Tell me that, sonny. The people who come from Earth's big cities, where TB has been breeding away for centuries. You check them all out, do you?"

From the corner of his eye, Simon saw Braddock Raines wince. He kept his own face impassive. "We employ the same procedures that every starship leaving Earth has always used, as mandated by UN quarantine law. We wouldn't be allowed to leave Earth orbit without them. Didn't the Navarro house starships use them?"

"Of course they damn well used them. We've remained uncontaminated until you bastards started invading us."

"Then why didn't it happen last time we were here?"

Edgar Strauss's glare deepened. "So this vaccine is another improvement you want us to adopt. Another product that is more sophisticated than anything we have."

"And your problem with that is . . . ?"

"You're fattening us up for next time. That's what all this fallacious generosity is about. You even turn our misfortune to your advantage. These vaccines and metabiotics will be available for you to harvest on your next violence-crazed invasion, along with all the other advances. I've seen how many new designs you've released to our companies and universities. Neurotronics, software, biochemistry, genetics, even metallurgy and fusion plant design. You've made it all available out of the kindness of your heart."

"We want our investment in Thallspring to be successful. Naturally we help you in upgrading your technology and science base."

"But only for your profit. If we were still producing old-fashioned systems next time you come, you would reap no dividend."

"You think that?"

"I know that, and so do you."

"Then all you have to do is not use them. Go right

ahead." Simon gestured expansively at the city beyond the window. "Tell them that, Mr. President. Persuade them they don't need the latest version memory management software, tell them they don't need next-generation brakes on their cars. Best of all, tell them they don't need better medicines."

"You'll lose in the end. You know that, don't you? There are fewer starships this time. Where did they go? Why didn't you build replacements? One day you'll come here and we'll be strong enough to resist. We grow while you wither away like every other decaying society in history. This is our time that's dawning. An end to starflight will bring an end to tyranny."

"Did your speechwriters dream that slogan up, or did you actually manage to think of it for yourself?"

"My grandchildren will dance all over your grave, you little shit." Edgar Strauss turned on his heel and marched out. He whistled the first few bars of Thallspring's anthem as he went.

Simon watched the door swing shut behind him. "My grave doesn't exist," he whispered to the president's back.

"That was fun," Braddock said stoically. "Would you like him to have an accident?"

Simon permitted himself a dry laugh. "Don't tempt me."

"So why am I here?"

"We're going to have to start this vaccination program that the medical AS recommends. I want you to supervise inoculating strategically important personnel: everyone who is critical to continued asset production. Start with the factory staff, but don't overlook people who work at the power stations and other ancillaries. I want to keep any disruption to our schedule to an absolute minimum."

"You've got it."

The pump station was unimpressive—a flat-roof box of concrete measuring twenty meters on each side, tucked away behind a chain-link fence, itself surrounded by a hedge of tall evergreen thorn bushes. It was in the corner of a small industrial estate on Durrell's outskirts, invisible from the trunk road outside, ignored by the estate.

At night, it was illuminated by tall halogen lights around the perimeter. One of them was off, while another flickered erratically. Maybe it was the angle of their beams, but they seemed to show up more cracks in the concrete walls than were visible during the day.

From his sheltered position in the hedge, Raymond studied the gate in the fence. A simple chain and padlock was all that held it. Although they'd studied long-range images, they'd never been quite sure if that was all. Now he could confirm it. One padlock.

Security wasn't a large part of the water utility's agenda. Enough to discourage local youths from breaking in and causing petty damage. To that end, there were a couple of alarms and sensors rigged outside—at least, they were the only ones listed in the station's inventory.

Prime was probing every aspect of the little station's internal data network, examining each pearl and circuit for hidden traps and alarms. And not just the station: the local datapool architecture was being scrutinized for inert links leading to the station, secondary trip alarms that would link into the datapool only when an intruder activated them. If they were there, the Prime couldn't find them.

Caution could only be taken so far before it became paranoia.

Raymond told the Prime to go to stage two. Images from the visual and infrared sensors around the station's door froze as the software infiltrated their processors, although their digital timers kept flipping through the seconds, making the feed appear live. Another routine inserted itself into the lock. Raymond heard it *click* from where he was hiding.

He slipped out of the shadows and scrambled up the fence. A quick gymnastic twist at the top, and he landed on the unmown grass inside. It took another three seconds to reach the door and open it. Total elapsed exposure time, seven seconds. Not bad.

His d-written eyes immediately adjusted to the darkness inside, a tiny scattering of light gleaned from LEDs glowing on the equipment boards. There was only the one room. He could see the pumps, five bulky steel cylinders sitting on broad cradles. Thick pipes rose out of the concrete beside each one. Their heavy throbbing filled the air with a steady vibration.

He took the pack from his back and removed the explosives. Working quickly, he moved along the pumps, securing the small shaped charges directly above the bearings.

His retreat was as quiet and efficient as his entry. The lock *clicked* shut behind him. As soon as he was back over the fence, the door sensors resumed their genuine feed. The Prime withdrew from the Durrell datapool, erasing all log traces of its existence as it went.

The red-and-blue strobes were visible long before the pump station itself. Simon could see them through the car's windshield as they turned off the main road and into the industrial estate, throwing out planes of light that flickered off the walls of buildings. Over a dozen police vehicles were drawn up around the pump station. Electric-blue plastic *Police Crime Boundary* barriers had been erected, forming a wide cordon outside the shaggy evergreen hedge. Uniformed officers were standing around it, while forensic personnel and robots carried out a slow centimeter-by-centimeter search of the ground.

Skin suits moved around inside the barriers like guards

overseeing a chain gang, never physically mixing with the forensic team. A crowd of reporters was jostling the blue plastic, shoving sensors forward. There must have been twenty direct feeds diving into the datapool, delivering the operation direct to the public in every visual and audio spectrum acceptable to human senses. Even laser radars were being used to map out the scene in 3D. Questions were shouted at police and Skins, regardless of rank. A constant harassment, deliberately pitched to provoke a response of any kind.

Simon's DNI was providing him with technical results from the forensic team as soon as their sensors acquired it. The grid of indigo tables and graphs was depressingly devoid of valid data.

"Can you believe this?" Braddock Raines said. He and Adul Quan were sharing the car with Simon. They were both staring out at the rest of the spectators. Staff from the factories and offices on the estate had gathered outside their respective doorways to observe the police operation firsthand. They shivered in the early morning chill, stamping their feet and swapping gossip and rumor, most of it invented by themselves.

Braddock took over manual control of the car and slowed it, steering around the clumps of people standing in the road. Most of them seemed oblivious to traffic.

"You want to go in, Chief?" Adul asked. "It won't be very private."

Simon hesitated for a moment. True, i-holograms could provide him with the scene of the crime to peruse at his leisure. And he had an inbuilt reluctance to be identified as any sort of important figure—especially here. Yet there was something about this whole act of sabotage that unsettled him. He just couldn't work out why. Whatever he was looking for, it wouldn't be in a hologram, no matter how high the resolution.

"I think we'll take a look."

"Okay." Adul started to inform the platoon sergeant they were arriving, while Braddock parked the car as close as possible.

Reporters saw them pull up. Half a dozen made their way over as the doors opened. Three police officers and a couple of Skins moved to intercept them and clear a passage for Simon.

"Are you guys Zantiu-Braun's secret police?"

"Will you use collateral necklaces in retaliation?"

Simon kept a neutral expression in place until they passed through the cordon. When they made it inside the pump station his nose crinkled at the sight. Then he realized he was standing in a couple of centimeters of water.

Each of the pumps had been torn apart, their impellers bursting out of the casing. Chunks of metal were embedded in the concrete walls and the ceiling. No piece of machinery was left intact; even the control boards were buckled and shattered.

Simon's gaze swept from side to side. "Competent," he murmured. "Very competent." He saw the senior police officers, five of them huddled together. The sight amused him. He'd visited a great many crime scenes over the years, and anyone above the rank of lieutenant always sought out and stuck with his or her contemporaries. It was as if they were afraid they'd get mugged by the junior ranks if they were alone.

His personal AS interrogated the police AS and discovered the officer in charge. Detective Captain Oisin Benson. He was easy enough to identify: no other senior officer had hair that unkempt.

Oisin Benson caught sight of him at the same moment. He gave his colleagues a knowing look and came over.

"Can I help you?"

"We're just here to take a preliminary look, Captain," Simon said. "We won't get in your way."

"Let me phrase that better," Oisin Benson said. "Who are you, and why do you think you have the right to be here?"

"Ah. I see. Well, we're from the president's office, and we're here by the authority of General Kolbe. And the reason we're here is to determine if this was an anti–Zantiu-Braun act."

"It wasn't."

"You seem to have come to that conclusion remarkably quickly, Captain. What evidence have you got for that?"

"No slogans painted here. No statement released by freedom fighters. None of your people or operations were targeted. This is purely a civil matter."

"Are there a lot of terrorist explosions on Thallspring?"

Detective Captain Oisin Benson leaned a fraction closer and smiled coldly. "They're about as rare as tuberculosis, Mr. Roderick."

So much for being unobtrusive, Simon thought. "Actually, Detective, our operations were targeted by this. The pump station provides several factories with water. All of them will have to curtail their operations until supplies can be restored."

"Out of the seventeen factories supplied by this station, only five are being forced to provide your tribute. The utility company that owns this station, on the other hand, is the subject of several lawsuits concerning toxic spillage brought by the families of those afflicted. It's a court battle that is taking a long time to resolve, and the company so far has not made any interim payments to the victims."

"Has the company been threatened?"

"Their executives have received a great many threats, both verbally and in e-packages; they're normally directed against them personally or their families, but there have been a considerable number made against the company itself."

"How convenient."

"You don't like the truth, do you, Mr. Roderick? Especially when it doesn't coincide with your own agenda."

Simon sighed, resentful that he had to get involved in a public squabble with this petty official. "We're going to look around now, Detective. We won't take up any more of your time."

"How considerate." Oisin Benson stepped to one side and made a sweeping welcome gesture with his arm.

Simon splashed over to examine the first of the ruined pumps. He could feel the water seeping through his shoes to soak his socks. Two other people were studying the mangled machinery: an engineer who wore the utility company's jacket and a technician from Z-B. The technician gave the three security men a slightly forced nod of acknowledgment. The engineer appeared completely indifferent to them as he ran a small palm-sized sensor over the wreckage.

"Anything of interest?" Simon asked.

"Standard commercial explosive," the technician said. "There are no batch code molecules incorporated at manufacture, so I doubt the police will ever be able to trace it. Apart from that, I'm guessing they were all detonated simultaneously. That implies a radio signal. Could have come from outside, but more likely a timer placed with them. Again, very simple components. Universally available."

The engineer straightened up, pushing a hand into his back. "I can tell you one thing. Whoever did it knew what they were about."

"Really?" Simon said. "Why is that?"

"Size and positioning. They used the minimum amount of explosive on each pump. This station building is like all our others, the cheapest covering you can build, basically it just keeps the rain and wind off the pumps. Concrete panels reinforced by tigercloth, that's all this is. And the whole thing is still standing. Six explosions in here last night, and the only damage is to the pumps. I'd call that a remarkably controlled explosion."

"So we're looking for an expert, then?"

"Yes. They knew plenty about the pumps, an' all. Look." He tapped a section of casing that resembled a tattered flower, fangs of metal peeled back. "They went for the bearings each time. Once they were broken, the impellers tore the whole thing apart from the inside. They spin at several thousand RPM, you know. Hell of a lot of inertia bottled up there."

"Yes, I'm sure there is." Simon consulted a file his personal AS was scrolling. "How long will it take to get the station back online?"

The engineer sucked his cheeks in, making a whistling sound. "Well, you're not looking at repairs, see. This is going to have to be completely rebuilt. I know for a fact there's only two spare pumps in our inventory. We'll have to contract the engineering firm to build us the rest. You're looking at at least six weeks to build and install. More likely eight or nine, what with things the way they are right now."

Back in his office, Simon waited until his assistant had served himself and the two intelligence operatives with tea before he asked: "Well?"

"Clever," Adul said. "And on more than one level."

"There's definitely no evidence to justify using collateral," Braddock said.

"I doubt we'll be able to use collateral for some weeks to come, not with this wretched TB outbreak," Simon said gloomily. "It's going to be tough enough keeping control with the locals blaming us for that. Put collateral executions on top of contagion, and we'd be in serious danger of losing overall control."

"We can hardly leave ourselves wide open to them," Adul

protested. "They could pick off our asset factories one by one."

"Humm." Simon settled back in the deep settee and sipped his tea. "This is what's bothered me since I realized how well executed this attack was. Just exactly who is 'they'?"

"Government," Adul said. "Strauss put some clandestine group together and provided them with all the equipment and training they needed. It can't be anyone else: look at the level of expertise involved. Just enough to mess us up, and always short of invoking justifiable retaliation."

"I'm not so sure," Simon said. "It seems . . . petty, especially if Edgar Strauss is involved. Which he would have to be to authorize the formation of some covert agency. He favors the more blunt approach."

"Good cover," Braddock said ruefully.

"No," Simon said. "He's not that good an actor."

"It's worse, he's a politician. One of the most slippery, conniving species of bastards the universe ever created."

"It still doesn't ring true," Simon said. "Whoever they are, they know exactly what they're doing. Yet they're not doing anything except letting us know that they exist. List all the anti–Z-B acts here in Durrell since we landed," he told the office AS. "Category two and above."

The three of them read the file headings as they scrolled down the holographic pane on the table. There were twenty-seven, starting with the destruction of the spaceport's hydrogen tank during the landing, moving on to include a couple of riots aimed at platoon patrols, squaddies targeted for fights when they visited bars and restaurants at night, a truck driven into the side of a Z-B jeep, industrial technicians beaten up while the accompanying squaddies were lured away, power cables to factories cut and reserve generators shorted out, production machinery wrecked by subversive software, raw material vanishing en route and finally the explosion at the pump station.

"Twenty-seven in three weeks," Adul said. "We've seen worse."

"Categorize them," Simon said to the AS. "Separate out the incidents that have affected production of assets." He examined the results. "Notice anything?"

"What are we looking for?" Braddock asked.

"Take out the two times our staff were hospitalized by thugs at the factories, and the road crash that wrecked the cargo of biochemicals on its way to the spaceport."

Braddock ran down the list again. "Ah, the rest is all sabotage, and nobody has been caught. There are never any leads."

"Last night's attack on the pump station has the same signature. Whoever it was went through the door alarms and sensors as if they weren't there. There is absolutely no record of anyone breaking in."

"Could have been an employee at the last inspection."

"Eight days ago," Simon said. "And there were three of them. That would mean they all had to be involved."

"How effective has this sabotage been?" Adul asked the office AS.

The holographic pane displayed several tables, which rearranged their figures.

"Jesus wept!" Braddock exclaimed at the total. "Twelve percent."

"That's very effective sabotage," Simon muttered. "Catalogue any slogans at the scene or radical groups claiming responsibility."

"None listed," the office AS replied.

"The other incidents," Simon said, "the riots and fights, catalogue slogans and claims of responsibility."

The list scrolled down the pane again. A complex fan of lines sprang out from each file, linking them to other files. Simon opened several of them at random. Some were visual, showing graffito symbols sprayed on the wall in the aftermath of riots and fights, most of them with daggers or hammers smashing Z-B's corporate logo, while the rest were crude messages telling them to go home in unimaginative obscenities signed by groups who were mostly initials, though someone calling himself KillBoy was quite common. Others were brief audio messages, digitally distorted to avoid identification, that had been loaded into the datapool for general distribution, declaring that various "acts" had been carried out in the name of the people against their interstellar oppressors.

Simon felt a brief glimmer of excitement at the results. The notion of the chase beginning. And most definitely a worthy adversary. "We have two groups at work here," he said, and indicated the list on the pane. "The usual ragbag rabble of amateurs keen to strike a blow for freedom and clobber a couple of squaddies into the bargain. And then someone else." The AS switched the display back to the sabotage incidents. "Someone who really knows what they're doing and doesn't seek to advertise it to the general public. They also know where we're the most vulnerable: financially. There's a small margin between viability and debt on these asset-realization missions. And if our losses and delays mount up, we might not break even."

"I have a problem with this," Adul said. "This sabotage group might keep their activities secret from the rest of Thallspring, but we were always going to know."

"We know, but we can't prove it," Braddock said. "Like the pump station, none of them are directly attributable as anti–Z-B acts. There are always other, more plausible, explanations. And they have covered their tracks well, especially electronically, which I find disturbing."

"We know," Simon said. "And we were always going to know at some stage. They must have realized that."

"That's why they keep their attacks nonattributable."

"There's something missing," Simon said. "If they are this good, then why aren't they more effective?"

"You call twelve percent in three weeks ineffective?"

"Look at the abilities they've demonstrated. They could have made it fifty percent if they'd wanted."

"At fifty percent we would have used collateral, no matter what plague is killing the population."

"My God," Adul said. "You don't think they cooked up the tuberculosis as well, do you? That's going to have a huge effect on asset production."

"I won't discount it altogether," Simon said. "But I have to say I think it's unlikely. Suppose we didn't have templates for metabiotics and vaccines? They'd be exterminating their own people. That doesn't seem to be their style."

"But we are going to take a serious reduction in viability from their activities so far. They've been tremendously effective."

Simon shook his head. "They're holding back."

"Chief, the only thing they haven't done is declare all-out war."

"I want to think this through: they always knew we would uncover what they're doing, yes? That much is obvious. Very well, by clever deduction we discover there is a well-organized covert group intent on sabotaging our asset-acquisition schedule. What is our response going to be?"

"Hunt them down," Adul said.

"Of course, and?"

"Step up our security."

"Yes, which is going to tie up a great deal of our capacity, both in AS and human time."

"You think that's going to leave us open to their real attack? That this is all just a diversion?"

"Possibly. Though I admit I could be overestimating them."

"If what we've seen so far is just a diversion," Braddock

said, "then I don't want to think what their main attack's going to be like."

"Their ability is worrying, yes," Simon said. "But I'm more concerned by their target. Our presence here is tripartite: personnel, starships and financial. They've already struck at our finances. If they wanted to render asset-realization inviable, they could have done it."

"They'd face collateral," Adul said.

"Santa Chico faced collateral. It never deters the die-hard fanatics. Consider it from their point of view: five hundred, even a thousand people dead, in exchange for ridding themselves of us for good. Wars of national liberation have rarely cost so few lives."

"So you think it's either us or the starships?"

"Yes. In which case, my money's on the starships."

"They'll never get them."

Simon smiled at the younger intelligence operative. "I know. That's where all our faith is placed, our most impregnable fortress, as secure as e-alpha. The starships are invulnerable. We can detect and destroy any missiles. Our AS's will prevent any subversive software from infiltrating onboard networks. And nothing gets past spaceport security. We deep-scan every gram of cargo. And no natives are ever allowed to dock.

"But just imagine they did get through, or that somehow they have acquired Santa Chico's exo-atmospheric armaments."

"How?" Adul demanded. "Santa Chico's thirty light-years away. Even if they sent a maser message with the schematics, it couldn't have reached here by now. Besides, we haven't seen any of the spin generators in orbit."

"We always assume that Earth is the only source of starships, or even portals. If anybody else can construct them, then it will be Santa Chico."

"Dear God, if the Chicos are organizing resistance to the asset-realization missions . . ."

"Precisely. But I'm not convinced of that myself. I was on Santa Chico. Interstellar revolution doesn't fit with their societal goals. And in any case, that planet is closed to spaceflight now. I'm simply using them as an example, a warning against complacency. We are totally reliant on our starships. If they are eliminated, then we are effectively dead. Our nonreturn would damage Zantiu-Braun's interstellar operations permanently, possibly even to the point of shutting them down. That would be a catastrophe we cannot permit to happen. For all their ability to sustain themselves, the new worlds are dependent on us bringing them technological advances. Earth remains our race's intellectual and scientific powerhouse. However unwelcome our links are, they cannot be severed."

"Sir, I think you're overreacting," Braddock said, grinning nervously. "It's one thing to blow up a couple of water pumps. And I acknowledge they did it flawlessly. But from that to shooting down or blowing up starships . . . It's not going to happen."

Simon considered the operative's insistence. He'd known he would have trouble convincing them how serious this intangible threat was. Everyone in Z-B placed their trust in the dogma of the invulnerable starship, even Quan and Raines, by nature and profession the most suspicious members of the Third Fleet. Safeguarding this mission was going to test his skill and authority in ways he hadn't envisaged when they embarked.

He held up a hand, a soft smile of understanding on his lips. "Humor me for the moment. If nothing else we need to disprove the notion."

"Sir."

They both nodded eager agreement, relieved by his mild reaction.

"So, let us consider our strategy. We definitely need to tighten up security in the industrial sector. Parallel to that, we need to keep a close watch on possible sabotage routes that can lead to the starships. I'm open to suggestions."

The population of Memu Bay was giving the platoon more space as they moved along their patrol route. Odel Cureton had been on enough patrols now to notice the difference. Before today, the locals had never really bothered much with them. The adolescents had shouted and spat, adults ignored them, nobody ever moved out of the way on crowded pavements. Pretty standard behavior. He'd seen it on every asset-realization mission (Santa Chico excepted). Today it was as if he had some invisible force field projecting out around his Skin, snowplow-shaped, moving people aside as he approached. One thing hadn't changed: the stares of hatred and contempt; if anything they'd grown more intense.

A day after the TB warning, and their demon status was now irrevocable. Not only were they here to steal Memu Bay's hard-earned wealth, their very presence endangered everyone. Demons with killer breath, every exhalation releasing a new swarm of lethal bacteria into the town's humid, salty air.

He turned down into Gorse Street. Hal was on the other side, keeping level. There were no police with them today. The assigned constables simply hadn't turned up. Odel didn't care; he knew he could rely on Hal out here on the streets. For all the stick he took, the kid was actually a good squaddie. As he watched, he saw the kid's head turn slightly as a couple of teenage girls walked past. He smiled to himself, imagining the kind of sensor imaging that the kid was requesting from his Skin. Not that he needed much enhancement. The girls weren't wearing a whole lot to begin with.

It was about the hottest day since they'd landed. Not a cloud in sight. Every whitewashed wall seemed to reflect the full force of the sun. Several sections of his display grid were indicating just how the heat was affecting his Skin. The weave of thermal fibers underneath the carapace was working at high capacity, radiating the heat generated by both his own and the Skin's muscles. His gill-vents were siphoning heat from the air before he inhaled. Even the carapace had adopted a light shading, partially reflecting the sun's rays.

Tactically, it put him in shitty shape. A glowing beacon to just about every sensor going. Odel had never got the memory of Nic out of his head.

They reached the end of Gorse Street. "Sector eight clear," he reported in. There was a lot of comfort to be had from routine these days. None of the platoon bitched about the sergeant's insistence they stuck to the protocols. If anyone could get them through this and out the other side, it would be Lawrence Newton. After the last few missions, Odel knew his faith wasn't misplaced.

"Roger that. Continue the sweep," Lawrence told him.

"Got that, Sergeant."

Odel and Hal crossed the road and started off down Muxloe Street. It was another row of small shops sitting under tall, austere apartment blocks, most of them claiming to be general stores and packed to their dirty ceilings with junk. But the road was wide, with a constant stream of traffic. The sergeant had quietly dropped side streets and narrow alleys from their itinerary over the last few days. Busy streets and plenty of people made ambushes and booby traps difficult.

Pedestrians melted away with sharp, rancorous glances. One woman pulled her two young children to one side, shielding them with a protective arm, their high voices chirping questions as he passed by.

He had a strong impulse to stop and remonstrate with her

and anyone else who was listening—to reason logically, to explain, to prove he was a good chap really. The sergeant had done it with a bunch of children playing soccer the other day. But Odel knew he could never pull off anything like that. He didn't have the words, and people laughed at his accent.

He kept on walking down the street. Tactile sensors flashed up numbers in their designated grid, telling him how hot the pavement slabs were under his Skin soles. He'd heard of people frying eggs on rocks heated by sunlight. These weren't far off.

Several of Muxloe's shops were shut, or closed—five of them together in a dilapidated block whose concrete panel walls were crumbling away in big broken blister patches. Gray-green fungus thrived in the cracks. Their windows were covered with bent, rusty roller blinds. Paint on the signs above the doors was fading, leaving little indication of what they had once sold. Polyethylene waste bags and weathered boxes had been dumped along the outside wall. Near the far end was a big glass bottle full of a bright scarlet fluid. A green T-shirt had been tied to the fat neck.

Odel was almost past the derelict shops when he stopped and spun around. Nearby civilians stared at him fearfully, wondering what they'd done. The Skin helmet's visual sensors zoomed in, the T-shirt filling his vision.

"Sergeant!" he called. "Sergeant, I've found something. Sergeant, come and see this. Sergeant!"

"What is it?"

"You've got to see this." Odel untied the T-shirt. The white lettering on the chest read *Silverqueen Reef Tours Cairns.*

Inside his hot Skin, Odel started shivering. He switched his sensors back onto the bottle. The liquid inside . . .

Lawrence waited in the anteroom as various aides scurried in and out of the mayor's study. Every time one of them slipped in he wanted to barge past, to demand Ebrey Zhang's attention. Forty-five frustrating minutes so far.

Captain Bryant had finally lost patience with him after a fruitless hour in the barracks, which they'd spent arguing. "You've had my answer, Sergeant," he snapped. "I cannot authorize any further action at this point."

"Then who can?" Lawrence asked. Given the way Z-B's strategic security force was structured, you simply couldn't be more insulting to your senior officer. Both of them knew it.

Captain Bryant took a moment to compose himself. "You have my authority to raise this with Commander Zhang. Dismissed, Sergeant."

No matter how the meeting with Zhang went, Lawrence had blown it with Bryant. He found himself smiling on the walk over to Memu Bay's town hall. He couldn't give a flying fuck about Bryant and the report he would now be getting from the captain at the end of the campaign. He'd just gone and committed himself. Up until that moment his own private little asset-realization mission had been theoretical. The pieces were in place, but still he had held back from initiating anything. Then that one heated question had relieved him of any conscious decision-making.

Typical, he told himself wryly. *Every major turning point in my life is decided by flashes of temper.*

Thirty minutes into his wait, the City Hall lights flickered and went out. They were getting used to cuts in the hotel that the platoon had adopted as its barracks. The power supply failed most evenings when someone burned the cable, or lobbed a Molotov at a substation. But the fusion plant itself was always left intact: after all, the town would need that after Z-B left. It wasn't just the barracks that suffered; power to the factories was interrupted. Internal rumor had it that they were over 20 percent behind on their asset-realization schedule.

Lawrence smiled to himself as shouts of alarm and annoyance echoed around the spacious cloisters. The overhead lights glowed like dim embers for a minute; then about a third of them slowly returned to full brightness as the emergency power supply came online, leaving the rest dark. Shadows swelled up out of the ornate arches and alcoves. If City Hall was anything like the hotel, the cells wouldn't have managed to fully recharge since the last time. Their swimming pool had been emptied a week ago because of the power drain that the filtration and heating element placed on the hotel's reserves.

One of Ebrey Zhang's aides called him in. Lawrence pulled down the bottom of his dress tunic and went through the open doors. He halted in front of the big desk and saluted. All of the study lights were on.

"Sergeant," Ebrey Zhang acknowledged with a wearied tone. A hand waved the aide out of the room, leaving the two of them alone. Ebrey moved back in his seat, picking up a desktop pearl to play with. He smiled. "You've been giving Captain Bryant a hard time, Newton."

Lawrence had been hoping for the easy routine. He remembered Ebrey Zhang from a couple of campaigns ago, when he'd been a captain. The man was a good enough officer, a realist, who understood the principles of command. Knowing when to be a ballbuster and when to listen.

"Sir. It's one of my men, sir."

"Yes, I know that. But leave off Bryant. He's new, and young, and still finding his feet. I'll have a word with him this time, but that's all."

"Thank you, sir. And Johnson?"

"I know." Ebrey sighed reluctantly. "But be realistic, Newton, what can I do that Bryant hasn't done already? If you can give me any hint where to search, I'll chopper ten platoons there immediately."

"He's dead, sir. There's no point in searching. We have to

show them they can't get away with that. None of us will be safe unless you do something."

"Ah. *They.* I take it you mean this KillBoy character?"

"Yes, sir, it seems likely. It's his group that is organizing all this. You have to turn the citizens against them. Make everyone understand that he's going to get them killed if he doesn't stop. Without their support he's nothing."

"KillBoy, the conveniently phantom enemy."

"Sir, we've been shot at, booby-trapped, maimed, injured, put in the hospital. We're almost clocking up as many casualties as we did on Santa Chico. Half the platoons are scared to set foot outside the barracks. He's no phantom, sir."

"You really think it's that bad?"

"Yes, sir, I do."

"I know it's tough on the street right now, Newton. But we've faced tougher. I have a lot of confidence in people like you to get the squaddies through this and lift intact at the end of it."

"Do my best, sir. But we need help to keep people in order."

Ebrey turned the rectangular desktop pearl over a couple of times, staring morosely at the furled pane. "I do understand what you're saying, Newton. However, I have a problem right now. It's going to be very difficult to use collateral when this TB threat is still ongoing. Thallspring's population sees us killing them anyway with the disease. I have to be totally convinced that they have murdered Jones Johnson before I can activate a necklace."

"Sir. It's his blood. Four liters of it. DNA checks out one hundred percent."

"And that's my problem. Where's the rest? You see, he can survive that loss easily enough. Infusing artificial blood isn't even a difficult medical procedure. Any teenager with a first-aid proficiency certificate could manage it. So what happens after I flood the datapool telling Memu Bay that we're retaliating for them murdering one of my squaddies, and then he turns up alive after the necklace is activated? Have you thought of that? Because that's the situation here. This Kill-Boy can organize snipers and mysterious accidents. He can certainly hold on to a captive for a couple of weeks until we screw up. I simply cannot allow that to happen."

"He won't turn up, sir. They killed him." There were other things Lawrence wanted to mention. Like how the killers knew where to leave the bottle of blood in the first place. No one outside of Z-B knew the patrol route that the platoon would take, not even the local police. It was planned out in the operations center ten hours before they went out. Even he didn't get briefed until an hour beforehand. To his mind, e-alpha was totally compromised. Yet for all Ebrey Zhang's apparent reasonableness, he could imagine the commander's reaction if he blurted that out. Right now, it would be one conspiracy too many.

"You're probably right," Ebrey Zhang said. "And I've had personal experience on what it's like to lose a platoon member. More than one, in fact. So I know how you all feel right now. But I simply cannot take the risk. I'm sorry, Newton, genuinely sorry, but my hands are tied."

"Yes, sir. Thank you for seeing me, anyway."

"Listen, your platoon's had two casualties now. That will be making the rest edgy. Am I right?"

"They're not happy, sir, no."

"I'll speak with Bryant, have him assign you some extra relief time."

"Sir. Appreciate that."

"And you can tell your men from me: one more incident like this, and I won't hesitate in using collateral. They'll be safe on the streets from now on."

If he felt any irony at the time that he'd chosen, Josep Raichura didn't show it. One o'clock in the morning, and Durrell's spaceport was illuminated by hundreds of electric lights, making it seem as though a small patch of the galaxy had drifted down to the ground. White-pink light shone out from deserted office windows. Stark white light, with a bleed-in of violet, drenched the giant arboretum at the center of the terminal building. Vivid sodium-orange cast wide pools along the loops of road that webbed the entire field. Blue-star halogen fans burned out from the headlights of the very few vehicles driving along those roads. Dazzling solar cones were embedded within the tall monotanium arches that curved above the parking aprons like the supports of some missing bridge, illuminating huge swaths of tarmac where the delta-shape spaceplanes waited silently.

An embroidery of dapples, overlapping in some areas, leaving others in somber darkness, and none of them revealing any activity. The universal indication of human installations dozing through the night shift. It was home only to the basic maintenance crews closeted in the big hangars, tending the myriad machines in readiness for the dawn and its surge of activity. Moving among the inert structures, and even fewer in number, were the Skins—the ones who'd drawn the bad duty—surly inside their private, invulnerable cocoons, resenting the tedium that came from walking the empty perimeter, the boredom of checking with the crews hunched over diagnostic instruments, the frustration of knowing that even when their duty did end they'd be too tired to enjoy the day (as much as any of them could in the hostile capital). Sticking with it nonetheless, because they knew this was the one place that had to remain secure if any of them were ever to get off this godforsaken planet and return home.

The spaceport at this time, then, was a little enclave of doleful and miserable people, serving their designated hours with an efficiency well below par. A time when human body-cycles were at their lowest; the classic time for nefarious raids and excursions. A time of vulnerability recognized as such by every guard commander since before the fall of Troy. And still they remained unable to install any sense of urgency and heightened alert among the men they led.

So Josep, armed though he was with his d-written body and Prime software, kept with tradition and history and used the small hours to make his exploratory foray. The perimeter was easy enough to breach. There was a fence, and lights, and electronic alarms that certainly never suffered from the human malaise during the night, and sentry Skins. Had he wanted to, he could have wriggled through it all like a special forces commando, with even the nocturnal animals unaware of his passing. But, frankly, when there's a huge front gate, why bother?

At noon, he rode his scooter up to one of the eight main road barriers, his little machine jammed between a juggernaut full of biochemicals and a convoy of cars belonging to afternoon-shift workers. He swiped his security card through the barricade's slot and took his crash helmet off for the AS to run a visual identity check. Every received byte tallied to the profile that his Prime had loaded into the spaceport network the previous day, and the red-and-white-striped barrier post whisked upward, allowing him through.

He drove carefully around the small roads linking hangars, warehouses and offices on the northern side of the sprawling glass and metal starfish that was the terminal building. Thallspring didn't possess a huge space program, but the respectable number of projects and commercial ventures it did have were all supported by Durrell's spaceport. Fifteen standard low-orbit (six hundred kilometers) stations circled above the equator. Twelve were industrial concerns, churning out valuable crystals, fibers and exotic chemicals for the planet's biggest commercial consortiums; the three others were resorts, catering to very rich tourists who endured the rigors of surface-to-orbit flight to marvel at the view and enjoy zero-gee swimming and freefall sex (occasionally combined) in heavily shielded stations. A small flotilla of interplanetary craft were maintained, principally to support the scientific research bases that the government had established on several planets. And orbiting a hundred thousand kilometers above the equator was the asteroid, Auley, which had been captured eighty years ago, to which clusters of refinery modules were now attached. Thousands of tons of superpure steel were produced there each month, then formed into giant aerodynamic bodies that were flown down through the atmosphere to feed Thallspring's metallurgical industry. In addition hundreds of other, more sophisticated, compounds that could be formed only in microgee conditions were extruded from the asteroid's raw ores and minerals and shipped down by more conventional means. In total, all this activity had developed to a stage where a fleet of over fifty spaceplanes were required to sustain it.

The Galaxycruisers were an indigenous design, so claimed the Thallspring National Astronautical Corporation, a consortium of local aerospace companies that built them—though anyone with full access to Earth's datapool would have noticed a striking similarity with the Boeing-Honda Stratostar 303 that had first flown in 2120, of which eight had been shipped to Thallspring. Whatever the origin, the scramjet-powered spaceplanes were a success, boosting forty-five tons to low orbit, and capable of bringing down sixty tons.

Zantiu-Braun had diverted several of them from normal operational duties to lift its plundered assets up to the waiting starships. Given that most of the spaceplanes were already used to support the industry that provided the most high-value manufactured products of all, in the orbital stations, the number that could be taken out of scheduled flights was sorely limited. In any case, there were not enough passenger spaceplanes to lift the entire invasion force back up to the starships at the end of the campaign. So Z-B had brought forty-two of their own Xianti 5005 spaceplanes to augment the indigenous capacity.

It was these newcomers that interested Josep. He ate lunch in the maintenance staff canteen, sitting beside one of the picture windows that overlooked a parking apron. Only two cargo-variant Xiantis were parked there, with Z-B's own crews and robots working on them. The rest were flying. He chewed his food slowly, taking the time to examine the area, note where crates had been stacked, the shortest distance from spaceplanes to a building, location of doors.

After lunch he kept moving, either walking through the terminal or riding between sections on his scooter. People who look like they have a purpose can go unnoticed in the most security-conscious environments. All the time, he was correlating the physical reality of the parking apron layout with the electronic architecture that his Prime had trawled out of the datapool. He even risked sending it into the Z-B AS that had been installed in the spaceport command center to run their groundside operations, including security. Details of the alarms and sensors installed themselves in Josep's vision, a ghost diagram of cables and detectors locking into his visual perception, threading their way in and out of buildings and underground conduits. Schedules, timetables and personnel lists followed. He began to work through them all, slowly reducing options, finding the best-placed spaceplane, best route to it, optimum time, multiple escape routes. Afternoon faded into evening, and the spaceport lights came on as the gold sun sank below the hills fencing Durrell. There were fewer takeoffs now and more landings as the big machines returned home for the night.

By one o'clock all flights had ceased. Josep walked along the rear of a vast maintenance hangar, whose arched roof covered five Xiantis and three Galaxycruisers, and still had four empty bays. Inside, there was less illumination than there was outside; the lightcones fixed to the metal rafters were bright, but their beams were well focused, splashing the concrete floor with intense white circles. Beyond them, shadow embraced over a quarter of the hangar's volume. His path kept him on the fringe of the lighted areas, and well clear of the bays where crews were working. A couple of Skins were inside the hangar, wandering about at random, so he had to be careful there was nothing suspicious about his movements. Keeping out of the lights altogether would have drawn their attention.

Josep reached one of the unoccupied bays and moved forward. Just to the side of the massive sliding doors at the front was a smaller door. He reached it and put his palm on the sensor plate. The lock buzzed, and he pushed it open.

Twenty meters away, the sculpted nose of a Xianti pointed at the maintenance hangar. Solar cones shone far overhead, glinting off the pearl-white carbon-lithium composite fuselage. There was a service truck parked on either side of the spaceplane, with hoses plugged into various umbilical sockets along the underbelly. An airstair led up to the forward airlock.

Josep walked over the tarmac, concentrating more on the icons being relayed from the spaceport network than his eyesight. Four cameras covered the spaceplane. His Prime had infiltrated each one, eliminating his image from the feed to Z-B's AS. Three rings of sensors were arranged concentrically around the sleek machine. None of them registered his presence as he walked across them. No Skins were within five hundred meters.

The airstairs were protected by both a voiceprint codeword and a biosensor that registered his blood vessel and bone patterns. It was an effective security device, but only ever as good as the patterns that were loaded into the system's e-alpha fortress. Josep's codeword and body map corresponded to one of those on file, and the airstair door slid open. He took the steps two at a time. The airlock at the top had a simple manual latch. Pull and turn.

Secondary lighting came on, illuminating the small cabin with an emerald glow. This Xianti was one of the cargo variants. Its cabin was cramped, with minimum facilities and room for up to five seats for the systems officer and payload managers. At the moment there were only two bolted to the floor, with the brackets for the others covered in plastic sleeves. Josep went forward and sat in the pilot's seat. The curved console in front of him was surprisingly compact, with three holographic panes angling up out of it. The two narrow windshields allowed him to see down the length of the nose, but showed very little else. He could understand that. Technically there was no need for any controls or windshields at all. The human pilot would always be fitted with a DNI. And that was only used for efficient communication with the AS pilot, which really controlled the spaceplane. The console and its displays were emergency fallbacks, although many people preferred pane graphics to the indigo icons of DNI. Windshields were there purely for the psychology.

Josep took a standard powered Allen key out of his belt pouch and hunched down in the seat to examine the base of the console. There were several inspection panels underneath. He opened two of them and found what he was looking for. The neurotronic pearls that housed the AS were sealed units buried deep in a service module, but they still had to be connected to the spaceplane systems. He wormed his dragon-extruded desktop pearl into the narrow gap toward the fiberoptic junction, and waited while the little unit morphed itself, extending needle probes into the unit. Prime flooded in.

They might have managed to infiltrate a spaceplane AS pilot through a satellite relay, but the risk of detection was too great. It was a single channel, easily monitored for abnormalities by secure AS's on the starships. Either they attempted to take over every Z-B AS, or they established a direct physical link. The first option wasn't even considered.

The dataflow reversed, dumping the entire AS pilot program into the desktop pearl. They would examine it later, learning the minutiae of ground-to-orbit flight in the strange vehicle. Its communications traffic. Docking procedures. When the time came, Zantiu-Braun would never know that someone and something else was on board until it was far too late.

The desktop pearl card informed Josep that it had copied the entire AS. Prime began to withdraw from the spaceplane's pearls, erasing all evidence of its invasion. Needle probes slid out of the fiberoptic junction and melted back into the casing. Josep replaced the panel and tightened it up.

Despite all his preparation, planning and caution, the one thing they all accepted was that there could be no protection against chance.

Josep had already opened the secure door at the bottom of the airstair when his relay from the cameras around the parking apron showed him a man emerging from the maintenance hangar. He was dressed in the loose navy-blue coveralls worn by all the spaceport's engineering maintenance staff. Prime immediately ran identification routines. Dudley Tivon, aged thirty-seven, married, one child, employed by the spaceport for eight years, promoted last year to assistant supervisor, fully qualified on Galaxycruiser hydraulics. He didn't have DNI, but his bracelet pearl was on standby, connected into the spaceport network. Prime moved into the communication circuit, blocking his contact with the datapool.

There was a moment when Josep could have ducked down behind the airstair, out of Dudley Tivon's sight. But that was an unknown risk. He didn't know what direction Dudley Tivon would walk, or how long he would be milling round outside. Every second spent crouched down was a second of exposure to anyone else who came along from a different direction. There were three Skins currently in the vicinity.

Instead he walked straight for Dudley Tivon. That reduced the outcome to two possibles. Either Dudley Tivon would assume he was just another night-shift worker going about his business, and do nothing. Being seen didn't concern Josep. So far his visitation had left no traces. Z-B didn't even know they had to look for evidence of anyone penetrating their security. Or Dudley Tivon would question what he was doing. In which case . . .

For a few seconds, as he drew close, Josep thought he'd got away with it. Then Dudley Tivon's pace slowed to a halt. He frowned, looking first at Josep, then back to the foreign spaceplane.

Prime in the surrounding cameras immediately began generating a false image, showing four different viewpoints of Dudley Tivon walking on uninterrupted across the parking apron.

"What are you doing?" Dudley Tivon asked as Josep drew level.

Josep smiled, nodding at the hangar. "Gotta get over to bay seven, Chief."

"You came out of that spaceplane."

"What?"

"How the hell did you get in it? You're not from Z-B. Those things are wired up eight ways from Sunday. What were you doing in there?" Dudley Tivon began to raise the arm on which he wore his bracelet pearl.

Information trawled from the datapool came into Josep's mind. Dudley Tivon's wife had been fitted with a collateral collar.

The assistant supervisor was making an issue of seeing where Josep had emerged, and he could never allow acts of sabotage or dissent against Z-B. It might well be his wife's collar that was activated in retaliation.

"I was just—" Josep's right arm shot out, stiffened fingers slamming into Dudley Tivon's Adam's apple. The man's neck snapped from the force of the blow. His body lurched back, but Josep was already following it. He caught the limp figure as it collapsed and lifted it effortlessly over his shoulder.

The Skins were still out of sight. Nobody else was outside the maintenance hangar. Josep jogged quickly to the door he'd used on his approach to the spaceplane and slipped through.

There was an office fifteen meters away from the door, shut for the night. He reached it in five seconds, bundled the corpse inside, then checked to see if anyone had noticed. Neither the maintenance crews nor Skins had reacted, and no alarms were screaming into the datapool.

They even had a contingency for an incident like this. Priority had to be given to getting the body out of the spaceport for disposal. No suspicion must be attached to the area.

Josep called up a menu for cargo robots currently in the maintenance hangar.

Camera feeds outside continued to show Dudley Tivon walking across the parking apron. He opened a door into the neighboring hangar and disappeared inside.

"After eight years of flight, Mozark had traveled halfway around the Ring Empire, stopping at over a hundred star systems to explore and learn what he could in the hope of inspiration. He could no longer see his own kingdom; that little cluster of stars was lost from sight behind the massive blaze of gold, scarlet and dawn-purple light that was the core. Few of his kind had ever ventured into this part of the Ring Empire, yet he felt comfortable amid the races and cultures inhabiting this section of the galaxy.

"Mozark might not have seen any of these species before, but everywhere he traveled he was able to communicate with his new hosts and eventually able to learn their separate philosophies and interests and goals and dreams. In many ways this heartened him, that he had so many ideas at his disposal, all of which he was eventually able to understand. Some he regarded as magnificent, and he looked forward to introducing them to the kingdom when he returned home. Some were simply so alien that they could never be adopted or used by his own kind, although they remained interesting on a purely intellectual level. While some were too hideous or frightening even to speak of."

Edmund immediately stuck his hand up, as Denise knew he would.

"Yes, Edmund?" she asked.

"Please, miss, what were they?"

"The hideous and frightening ideas?"

"Yes!"

"I don't know, Edmund. Why do you want to know?"

"Coz he's horrible!" Melanie shouted. The other children laughed, giggling and pointing at the beleaguered boy. Edmund stuck his tongue out at Melanie.

"Enough," Denise told them, waiting until they'd quietened down again. "Today's story is all about the time when Mozark meets the Outbounds. Now this wasn't a single race: like the Last Church, the Outbounds attracted a great many people to their cause. In many ways they were the opposite of the Last Church. The Outbounds were building starships. Not just the ordinary ones that the Ring Empire used for trade and travel and exploration. These were intergalactic starships." She gave the children a knowing look as they *ooohed* with wonder. "The greatest machines the technology of the Ring Empire could devise. They were the largest, fastest, most powerful and sophisticated ships that this galaxy has ever known. The effort to build them was immense; the Outbounds had taken over an entire solar system to serve as a construction center. Only a star with all of its circling planets could provide them with the resources necessary. Mozark spent a month there, flying his own small ship around all the facilities, playing tourist amid these tremendous cathedrals of engineering. The Outbounds proudly told him of the ocean-sized converter disks that they'd dropped into the star, where they'd sunk down to the inner layers to settle amid the most intense fusion process to be found within the interior. That was the only place to generate sufficient energy to power the tens of thousands of industrial bases operating through the system. Behemoths in their own right, these bases were partially mobile, allowing them to swallow medium-sized asteroids in their entirety. The rocks were digested and separated into their constituent minerals, which were then fed into refinery towers. Biomechanical freighters that only operated in-system would collect the finished products and ferry them to manufacturing facilities where they would be fabricated into components for the starships.

"The shipyards they were built in were the size of a small moon. Each individual intergalactic ship was miles long, with silver-and-blue hypermorphic hulls that would gather up every speck of starlight falling on their spinshifted molecules and radiate it away again in a uniform coronal shimmer. When they were parked in orbit, they were smooth and egg-shaped. Then, when their engines came alive, flinging them into the nullvoid at hundreds of times the speed of light, they would instantly convert themselves into sleek rapiers sprouting long, aggressive forward-swept tail fins. It was as though the nullvoid where they now traveled possessed an atmosphere of elementary photons through which only their metasonic profile would fly.

"Mozark, of course, was enthused by the whole project. The Outbounds were the Ring Empire's final and greatest pioneers. The intergalactic ships were taking colonists to other galaxies. New empires would be born out there on the other side of the deep night. That would be a wondrous future flowering out there amid the unknown, replete with challenges and struggle. Life would not be smooth and complacent as it was amid the Ring Empire.

"He watched the ordinary passenger starships dock, bringing the tens of thousands of colonists who were searching for a new life for themselves and their descendants. They had come from kingdoms right across the neighboring section of the Ring Empire, hundreds of different species united by wanderlust. The first time he saw an intergalactic ship launch itself into nullvoid he felt nothing but envy. They were his soulmates, and he was being left behind. But such was his duty; he had to return home to his own kingdom. There and then, with his own ship still floundering for stability in the energy backwash of the intergalactic ship's drive, he wanted to bring word of this enormous venture back to his people. He envisaged the kingdom's resources being turned over to a similar project, carrying them all on a magnificent voyage to the future. It was only after the massive ship had vanished from his sensors that doubts and disillusionment began to creep into his thoughts. He had undertaken this quest voyage to find something that would benefit and inspire all of his people. Yet how many of them, he wondered, would really want to discard everything they had and gamble on a wild trip into uncharted reaches of the universe? Many would: millions, perhaps hundreds of millions. But his kingdom was home to billions of people, all of them leading a relatively happy existence. Why should he make them abandon that? What right could he possibly have to tear them away from the worlds and society they had built, and which served them so well?

"That was when he finally began to understand himself and his own dissatisfaction. Looking out of his own ship at its proud, giant cousins orbiting a nameless barren Outbounds planet he now saw only a difference in scale. Both he and the colonists were prepared to fly away into the unknown in order to find what they hoped would be a worthwhile life. They were probably braver than he, taking a bigger chance with what they would find and where they would end up. But for them it would be the flight itself that was the accomplishment. When they reached that far shore, they would have every ability and material advantage at their disposal that they had in the Ring Empire itself. There were no new ideas waiting for them out there, only space that was—one hoped—a little less crowded. They were taking the primary Ring Empire culture with them in the form of the technology and data that were their heritage. Just as the similarity that pervaded the Ring Empire was due to its monoknowledge base, so these fledgling seeds would sprout identical shoots. If anything, he decided, the colonists weren't as brave as he was: they were just running away. At least he was trying to help his people back in the kingdom."

Denise stopped, conscious of the way the children were regarding her with faintly troubled expressions. One or two of them were even resentful and impatient, picking at the blades of grass and throwing the occasional wistful glance out at the white town beyond the wall. This was no longer the story they thought it was going to be, a quest with terrible hardships to overcome and monsters to battle. All they were hearing was how Mozark kept turning his nose up at wonders and sights beyond anything they would ever know. A fine hero he made.

She rebuked herself for losing sight of whom she was telling this to and gathered up her memories of the story. There was much that she could discard: shorn of its abstracts and philosophizing, it could still be made to work for them.

"So when he was standing there in his starship, thinking all these thoughts about the Outbounds and the Last Church, and The City, and even the Mordiff, Mozark suddenly knew what he had to do."

"What?" one of the girls asked avidly.

"He had to go home," Denise said. "Because he knew then what he was going to say to Endoliyn, the thing he was going to devote the rest of his life to."

"What!" the chorus was yelled at her.

"It's a beautiful day," Denise said with a mischievous laugh. "You should be out there playing and enjoying it. I'll tell you what happened when Mozark returned to his kingdom soon."

"Now!"

"No. I said soon."

"Tomorrow, then."

"Possibly. If you're good."

They promised her they were and would always be.

She let them scatter and fling themselves about on the school's small, protected lawn. There was no need for her to check her big old watch; she knew what the time was. The goodwill soccer game was about to start.

Clusters of d-written neural cells connected Denise with Memu Bay's datapool. Several reporters were covering the game—not that there was much interest. Public access figures for the game were minimal. They were already lining their cameras up on the pitch, bringing the two teams into focus as they went through their prematch kickabout routines.

Lawrence stopped the ball firmly and tapped it with the inside of his right foot. It bobbled along the ground, rolling to a halt a couple of meters away from Hal, who gave him a disgusted look. The maneuver was supposed to be a deft pass, landing *just so* for Hal to kick into the defenders' goal area.

Instead, as Hal made a frantic dash for the ball, two of the lads they were playing against tackled him. For a moment Lawrence thought they were playing rugby by mistake. Hal hadn't quite reached the ball, and they were high, legs lashing out.

Hal yelped as he fell, his shoulder taking the full impact. "Fuck me," he grunted under his breath.

The ref blew his whistle.

Hal looked up at him expectantly.

"Free kick," the ref grunted reluctantly.

"What card are you showing them?" Hal asked indignantly. The ref walked away.

Lawrence and Wagner got their hands under the kid's shoulders and lifted him up. "He's got to be kidding," Hal cried. "That was a yellow card at least."

"Slightly different rules here," Lawrence said, hoping it would calm the kid down. Hal looked as if he was about to start a fight.

The two lads who'd tackled him were grinning happily. One of them showed a finger. "KillBoy says spin on it."

Hal lurched forward, snarling. Lawrence and Wagner just managed to hold on to him. There were a few desultory cheers from the touchline where the locals were gathered.

It wasn't different rules here at all. For the tenth time since the goodwill game started, Lawrence's Loafers versus the Avenging Angels, Lawrence wondered if this had been such a good idea after all. The locals saw this purely as a way to legitimately hack Z-B squaddies to pieces with the strangely long studs on their boots and tackles that would make a kung-fu master wince.

Just before kickoff, Ebrey Zhang had come over for a quiet pep talk with the team. After he'd finished spouting on about opportunities and enhanced community relations, he'd said to Lawrence: "We don't want to cause any sort of commotion here, Sergeant. Let's just take it easy out there, shall we?"

"Are you ordering us to lose, sir?" Lawrence had asked. He supposed in a way it was flattering, their commander assuming they would automatically win. But he'd seen some of the youths they were up against. Big and fit-looking. It should be quite a tight game.

"No, no," Ebrey said softly. "But we wouldn't want a walkover, would we? Bad feeling and all that."

"Got you, sir."

"Good man." Ebrey slapped him heartily on the shoulder and joined the rest of the Z-B supporters.

Goodwill had run out in the first five minutes. Not that the Avenging Angels had ever brought any to the pitch in the first place.

Hal took the free kick, sending the ball in a long arc over to Amersy. The corporal began his run down the wing. Lawrence ran level with him on the other side, two Avenging Angels marking him close all the way. Close enough to mistakenly knock into him when the ref happened to be looking the other way.

Lawrence skidded along the mud, almost losing his balance. Amersy had raced on ahead now, leaving Lawrence hopelessly misplaced to receive a pass. "Damn it," he growled. His markers were surprised when he elbowed them aside. Fortunately the ref was still watching Amersy as the corporal was tackled.

"Support!" Lawrence screamed at his team. "Support him, for fuck's sake, you pitiful assholes."

"Now, Sergeant," Captain Bryant's voice carried in faintly from the touchline. "No need for that sort of language."

Lawrence glared, managing to force out a few words under his breath.

Amersy was trying to lift himself off the ground as the victorious Avenging Angels made off with the ball. The hulking hooligans actually had good ball control, Lawrence admitted grudgingly. They nudged it between them, easily beating their way around the one midfielder who tried to intercept them.

Where the fuck was the rest of the team?

"Defense," Lawrence shouted desperately. His arms semaphored wildly.

At least his backs had some understanding of tactics. Two were coming forward to take on the Avenging Angels with the ball. Three were guarding the goal area. A midfield duo were heading to the other wing, marking the Avenging Angels striker who was dodging forward into position. Lawrence saw one of their midfielders heading for an open space in the center circle and ran to cut him off.

Not such a bad game after all, and his men could play tough too.

The land mine went off under the Lawrence's Loafer defender on the right of the goal area. It blew him three meters straight up into the air, taking off his legs and shredding his lower torso. Lawrence dived to the ground at the dull thudding boom of the explosion. An eerie moment of silence followed. Then the defender's upper torso thumped down, lifeless arms flopping about grotesquely from the jarring impact. His head twisted around to stare blankly at the goalmouth. Lawrence recognized Graham Chapell, a squaddie from Ciaran's platoon. Blood and gore splattered across half of the pitch. There was still no sound; everyone was too shocked even to scream.

Lawrence looked around wildly, seeing the steaming crater that had ripped out of the ground, understanding immediately what had happened. Everybody else had flung themselves down. He watched in horror as the ball rolled on, bouncing and juddering across the rucked grass field.

Stop, he implored it silently. *Oh fuck, stop. Stop!*

The damn thing was easily big enough to trigger another

mine if it passed over one. It was rolling toward Dennis Eason, who was watching it coming, his face drawn into a rictus of terror and fatalistic expectation.

The ball stopped half a meter from him. He let out a sob of relief as his head dropped back to the mud.

People were yelling and screaming now, spectators as well as players; they were all flat on the ground. Z-B personnel were all shouting at everyone not to move, to stay exactly as they were. Help was on its way.

Lawrence clenched his fists, pushing them into the mud, furious at how helpless he was. Waiting with every muscle locked tight in fright and suspense. Supremely vulnerable without his Skin. Open for death from any passing student revolutionary with a whim to be a hero that day. He *hated* KillBoy right then. Hated this whole fucking world. That had never happened before. Not ever. The best he'd ever come up with before today was animosity and contempt.

All they were doing here was playing soccer, for God's sake. Soccer. Their own people as well, few of whom were out of their teens. He could hear the young Avenging Angels around him, whimpering in terror, several of them crying.

What the hell was wrong with these people? He wanted to shout it out at them. They'd hear. They'd be here watching, relishing the distress and dread they'd created. Gloating as the knife was twisted.

But all he could do was grit his teeth and lie still, the muddy water seeping into his shirt and shorts. Waiting for the glorious sound of the helicopters.

Seven platoons were rushed to the park where the soccer game was being played. Their helicopters landed on the roads around the outside. The Skins advanced cautiously, sensors probing the ground as they came.

They reached Ebrey Zhang first, leading the commander away down a safe path marked out by beacon tubes that flashed a bright amber. His helicopter thundered away overhead as the remaining Skins spread out over the park, sensors playing back and forth. People were slowly led away one at a time, shaking with relief as they leaned on the squaddies. They reached Lawrence forty minutes after the helicopters arrived. He stood unsteadily, staring around. A confusing grid of amber lights were flashing all across the pitch. Three red lights gleamed bright among them. One was four meters away from where Lawrence had lain.

A medic squad was picking up pieces of Graham Chapell from cleared sections of the pitch, putting him in thick black polyethylene bags.

"Bastards," Lawrence hissed as the Skin eased him toward the waiting jeeps. "You utter bastards."

Dean Blanche was ushered into the mayor's study by one of Ebrey Zhang's aides. The commander only needed one look at the carefully blank expression on the internal security captain's face to know it was going to be bad news.

"So?" he asked when the doors were closed.

"They were our land mines," Captain Blanche said.

"Shit! Are you sure? No, forget that, of course you are. Goddamnit, how could that happen?"

"We don't understand yet. According to the inventory they're still in storage. We did a physical check, of course. Eight are missing."

"Eight?" Ebrey asked in alarm. "How many were planted in the park?" He was never terribly at ease with land mines. Z-B policy required them to be available in case the situation on the ground became troublesome, and the squaddies had to protect strategic areas from outright aggression. Effectively that meant the spaceport during their retreat. He was thankful that he'd never had to order their deployment. The damn things were a lethal legacy that could last for decades, completely indiscriminate in choosing their victims.

"We found five. With one detonated . . ."

"Oh, Christ." Ebrey went to the small drinks cabinet on the rear wall and poured himself what the locals laughingly described as bourbon. He didn't normally drink in front of his junior officers, and certainly not those from internal security, but it had been a long, bad day, and this wasn't a happy ending. "Want one?"

"No, thank you, sir."

"Your choice." He stood at the French windows, looking up into the night sky. It was three o'clock in the morning, and the stars were twinkling warmly. After today, he was seriously beginning to wonder if he'd ever make it back up there among them. "So we've got three mines planted out there somewhere in town waiting for us to step on them."

"Two, sir."

"What? Oh, yes. Two unaccounted for. Any chance the platoons could have missed them in the park?"

"It's possible, sir. I'm going to order another sweep in the morning, when it's light."

"Good man. Now how in Christ's name did they get them out of the armory?"

"I'm not sure, sir." Blanche hesitated. "It would be difficult."

"You mean difficult for anyone outside Zantiu-Braun."

"Yes, sir."

"I can't believe one of our own people would do this. There's no grudge or vendetta worth it." He looked around sharply at the deeply uncomfortable captain. "Is there?"

"No, sir. Nothing that serious among the platoons."

"We're missing somebody. Jones Johnson, the one whose blood they found. Could he have . . . I don't know, defected?"

"Possible, sir."

"Is Johnson capable of getting into the armory?"

"I don't know, sir. A lot of the squaddies tend to know shortcuts through our software."

"Damnit. We have safeguards for a reason. Especially on weapons."

"Sir. I do have one possible lead."

"Yes?"

"The other mines were on standby, and the soccer teams were running all over that field for thirty minutes before the explosion. It must have been activated just before Chapell ran over it."

Ebrey brightened. "KillBoy transmitted a code."

"Yes, sir. If it went through the datapool we can try to trace it. Of course, it could have been an isolated transmitter. In which case, someone had to be close enough to send the code. I can review all the memories from every sensor in the district. The AS may be able to spot someone who fits the right behavior profile. But somewhere in today's data there should be some evidence."

"Whatever you need, as much AS time as it takes, you've got it. Your assignment has total priority. Just find this piece of shit for me. I don't care how long it takes, but I'm going to see Mr. KillBoy swinging from the top of this Town Hall before we leave."

CHAPTER ELEVEN

EARTH. ONCE MORE.

The brilliant white-and-blue world continued to fascinate Lawrence as much now as it had during his first arrival five years ago. As always during the transfer flight down from

Centralis to low orbit, he spent as much time as possible staring at the real-time images provided by the interorbit ship's visual sensors. As they curved in over the Americas he watched wide swirls of cloud twisting with soft grace out across the western Atlantic, congealing into a single storm spiral, pure white around the ragged edges, but darkening swiftly toward the dense high center as if night were erupting out of its heart. Within days the Caribbean islands would be cowering from winds and waves and stinging rain, unbound elements stripping the leaves from every tree and washing the land into new shapes. Once again their population would hunker down and wait for the howling winds to pass. And then afterward they'd carry on anew, treating the event like an unwelcome holiday. The palms would sprout new fronds, and people would sport and swim on the clean white sands. He smiled down at them from his angel's perch. Only on a world so teeming with life could such acclimatization occur, he thought. A world where life belonged, where symbiosis between nature and environment was the governing evolutionary factor. Unlike Amethi.

He still held a nest of feelings for his old homeworld. They weren't as strong now as they had been when he first arrived, and most of them remained antagonistic. But every now and then, he could recall times and places from that world when he'd actually been happy, or enjoyed himself. None of those times were with Roselyn. He still shielded himself from those recollections. There was too much pain involved, just as sharp and bright now as the day he left.

His hand went to the pendant under his shirt. He'd almost flung it away the day he left Amethi. Then he decided to keep it so he would never forget all of the treachery at loose in the universe. Nowadays it was a kind of talisman, proof he'd survived the very worst life could throw at him.

The Xianti 5005 carrying them down landed at Cairns spaceport in the middle of another of Queensland's baking-hot afternoons. There was no one waiting to meet Lawrence. He walked past his platoonmates as various families rushed forward in the arrivals hall. Wives and long-term girlfriends flung themselves at their menfolk, clinging tightly and trying not to cry. Until the starship arrived back from Quation two days ago, none of them had heard how the asset-realization campaign had gone; who was alive, who was injured, who wasn't coming back. Relief and fear echoed through the big air-conditioned hall. Children milled around the embracing couples, smiling and happy that Daddy was home again.

There had been a local girl called Sandy whom Lawrence could reasonably claim to be a regular girlfriend in the time between Floyd and Quation. Sandy had promised to wait for him, but that was just over nine months ago now. She was twenty-one; he never seriously expected her to hang around.

So he walked out of the terminal building into the clean sea air, taking a long minute to look around at the scrub-covered hills behind the spaceport, looming dark as the sun sank behind them. The humid breeze blowing in from the ocean. Gulls squawking. Another spaceplane splitting the air overhead like slow thunder. He smiled around at all of it, welcoming the scene as he might an old friend. He would always associate the sea and its smell with Earth.

The taxi rank was at the south end of the terminal. Lawrence walked down to it and slung the only luggage he had, his shoulder bag, into the backseat of the elongated white bubble. It had a human driver rather than an AS, an old Chinese man who wanted to talk about how Manchester United was playing this season. He thought Lawrence's accent was British.

"Never been there," Lawrence had to admit.

"But you know about Man-U?" the driver asked anxiously.

"I've heard of them."

"Of course you have. Most famous team on the planet. I access every game. I installed a horizontal hologram pane in our apartment so I can watch the whole pitch. My wife doesn't like it."

"No kidding?"

"Yeah, she wanted a new sofa. I access through membranes as well. The last three seasons I've paid the team's media agent for multi-player-viewpoint feed. It costs, but it's worth it. This way I can see what's happening on the ground as well as get an overview. I like to stay with Paul Ambrose as my viewpoint when the first eleven play, he's got good ball sense."

"Sounds great."

"First eleven only play once every four days. I have to make do with second eleven and third eleven in between."

"Uh-huh."

"Afternoons, I access the under-twenty-one side. Sometimes I have to record them when I'm working. My friends in the other cabs, they have fun trying to tell me who wins. I turn my datapool access off and they drive up next to me and shout the result. I always have to shut my ears those afternoons. One day, when I save up enough, I'm going to Europe to watch them play live. My wife, she doesn't know that."

"Really." They had cleared the spaceport to merge with the short highway into town. To his left, Lawrence could see the thin strip of protected mango swamps running along the coastline. On his right, suburban apartments had colonized the land almost up to the foothills.

"You just down from Quation?"

"Yep."

"Your wife not meet you?"

"Not married."

"Wise man. You enjoy yourself while you still can, my friend. When I go to Europe, I won't take my wife. So you got anywhere to stay tonight?"

Lawrence could have returned straight to barracks; it wouldn't have cost him anything. But the whole fleet was on four-week leave, and the bonus pay for the campaign was sitting in his bank as well as the whole nine months' back pay. He'd made no plans at all. Some of the other single guys on the starship were talking about sailing between Pacific islands and raising hell on every beach resort they landed on. Colin Schmidt had invited him on a tour of the casinos in Hong Kong and Singapore. Others promised Perth still rejoiced in its claim to be party capital of the Southern Hemisphere, an easy train ride away. "No," he said. "I don't have anywhere to stay." He pressed the window key and let the glass slide halfway down; wind and highway noise rushed in. Up ahead the glare of lights from the Strip was already flickering through the town's outlying buildings. Lawrence laughed at the sight of the gaudy neon and holograms beckoning him back greedily. He'd never been so perfectly content. No cares, no obligations, plenty of money and lots of time to spend it in. Life didn't get much better.

"I know some places," the taxi driver said, giving him a hopeful sideways glance.

"I'm sure you do. Okay, what I want is a decent hotel, maybe one with a pool. Not too expensive, but somewhere with a wideband datapool feed and twenty-four-hour room service. And where they don't mind me bringing a guest home for the night. Got that?"

"Ah!" The taxi driver nodded happily. "I know just the place."

The hotel was just on the seedy side of its two-star rating. But it did have a pool, and Lawrence's second-floor room had a tiny balcony that looked over the gray geometrical sprawl of southern Cairns. He checked in and wandered down the nearest shopping street, a broad glassed-over concourse whose bargain customers had successfully repelled the bigger chain stores from investing. He bought some clothes in the small shops. Nothing too sharp, just something that he could wear out on the Strip, and that didn't have a Z-B logo.

He scored with a girl early on that night. A great roundabout walker in her late teens, out on the road with her friends, backpacking their way around the coast of Australia. She was pretty, and slim, with olive skin and her dark hair arranged in tight braids that had colored phosgene beads dangling on the end. When she moved her head quickly, they twirled round like a rainbow halo. He sweet-talked her away from her friends before they all hurried back to the hostel and their prebooked cots. She was fascinated that he'd actually been born on another world, appreciative of the classy foreign bottled beer his money bought, and showed a keen interest in the fact he'd spent months away from Earth.

"Deprived, huh?"

"I guess you could say that," he admitted. "The natives weren't very friendly."

Back in the secluded shadows in his room she screwed like an energetic kangaroo, pounding away up and down on top of him. For the first hour he was sure the dilapidated old bed was going to give way under them. He poured more of that expensive beer over her chest and licked it off before she pushed his head down between her legs. They accessed a thrash-rock feed and tried to fuck in time to the thumping music, eventually collapsing in howls of laughter as the codpiece-endowed vocalist screeched out lyrics about giving his baby some hardassed lovin'. Room service delivered club sandwiches with more drink, and they sat cross-legged on the sagging mattress feeding each other. Then they watched a non-i comedy show before fucking again.

She left first thing in the morning to join up with her friends. They were heading farther north, hoping to get some casual work in Port Douglas to pay for the next leg of their great middle-class adventure. By midday Lawrence had to think hard to remember her name.

That next night, it was another girl. She liked highballs instead of beer, and electric jazz rather than rock, but she was just as randy.

The whole of his first week passed the same way. Sleep during the day. Have a decent meal in midafternoon. Take a walk before the evening started. Hit the Strip after the sun went down. Some days he ran into other squaddies from the fleet, and they'd have a few rounds together, maybe shoot some pool or spend an hour in one of the game arcades. He never got drunk; there was no percentage in that, given his endplay. Once or twice he went out on a club's dance floor. Each time it was because the girl was keen to dance first.

Seven days after he landed, his bracelet pearl received a message from fleet administration ordering him to report to the base. His application for starship officer college had been processed. He was going to be forwarded to Amsterdam for entry assessment.

He sat up in bed, holding his glasses up in front of his face, reading the message again with a slow-growing sense of delight. His life was finally coming together the way he wanted. His father, Roselyn, Amethi, that was paying his dues. He'd *earned* his place on Z-B's starships.

The girl lying in the bed beside him lifted her head and peered around the hotel room in classic morning-after confusion. She blinked at Lawrence. Her expression changed to one of recognition. "Hi," she grunted.

"Morning."

"Good news?" She nodded at the glasses he was holding.

Lawrence considered the question. The obvious thing to do would be blurt out the assignment, tell her about what it meant to him. It was the kind of thing that should be shared, leading on to a happy day spent together, perhaps a good meal with a bottle of champagne. But, truthfully, the only person he could tell who'd appreciate what it meant was Ntoko. And he was pretty sure the corp wouldn't want his own family vacation interrupted by a babbling Lawrence Newton bragging how he was leaving the platoon behind.

That was when he admitted to himself just how lonely he'd become. There really was nobody to call. Nobody on this whole planet who knew him, nobody who cared about him.

He dropped the interface glasses back on the bedside table, then pulled the sheet back off the two of them. A few blades of morning sunlight had crept round the curtains, falling on the bed to illuminate their bodies. The girl gave him an uncertain little smile as he gazed at her. For all of their intimacy during the night, he felt nothing, no connection, no urge to try and *make it work.* The only reason she was here was for sex. He didn't even feel guilty about that. She'd been eager enough.

To think, after one night with Roselyn he'd been ready to spend the rest of his life with her. God, how stupid had he been back then? Talk about being straight off the farm. He could teach her a thing or two now.

As always, that treacherous little thought sprang up: *I wonder what she's doing now.*

"Nothing important," he said brusquely, angry with himself for the weakness. Then he rolled closer and put his mouth over the girl's ear, and in a throaty demanding whisper told her what he wanted from her. With a slight show of reluctance she positioned herself over the edge of the bed the way he instructed, so he could celebrate with her the only way it was ever going to happen between the two of them.

Lawrence took one of Z-B's twice-daily flights from Cairns to Paris, a big subsonic passenger jet that refueled at Singapore. From Paris he transferred to a train that whisked him across the heavily forested European countryside to Amsterdam. He arrived at the old Central Station that backed on to the harbor in the middle of the city.

Cairns with its eternal heat had made him forget it was only spring in the Northern Hemisphere. He pulled his full-length coat on as he walked out of the station, but didn't bother to do it up. The sun was shining out of a clear sky, warming the air.

Outside, Prins Hendrik Kade seemed to be a twenty-lane road given over entirely to bicycles. He'd never seen so many of the machines in one place before. They were all the same silk-white color, with the city emblem embossed on the central spar. Bells rang all around him, making him twist his head about in alarm. Twice he had to jump sharply out of the way as cyclists sped toward him. They obviously weren't going to swerve.

Optronic membranes threw up a city map, and he set off down Dam Rak, the long broad road opposite the station. Trams trundled along rails embedded in the cobbles. He'd never seen machines that looked so ancient, although they were in perfect condition. It was good to be walking through a city and for once not have to be on guard. Quation had not given Z-B a joyful reception. But here, the citizens smiled warmly when they saw his mauve uniform.

He wasn't surprised. According to the briefing he'd downloaded from the company memory, Z-B was a heavy investor in Holland. And where their primary installations were based sprang up a host of smaller companies to provide support, both specialist and general. The country had a high prosperity index, even by European standards.

His first hint of disenchantment came right outside the officer college. Z-B's Amsterdam headquarters, containing the college, was a big five-story stone building that was eighty years old, though its exterior had been crafted in nineteenth-century bleak, with tall vertical slit windows. Squatting across a broad cobbled square from the fortresslike Royal Palace its architecture was more than appropriate.

A small group of demonstrators were clustered around

some kind of stall twenty meters from the main entrance. Potatoes were baking in what had to be the most primitive oven on the planet, a cylinder made out of solid iron. Charcoal glowed behind a grate on the front end, while a black chimney stack at the rear made the whole thing look like the boiler from some kind of steam engine. The sign above the stall was offering the potatoes with a dozen different fillings. Very cheaply, too, Lawrence noticed. There was an emerald-green circle at one end of the sign, with a stylized white bird emblem in the middle, its wings swept wide.

None of the pedestrians filling the square seemed interested in buying the potatoes. The demonstrators, mostly young people, were singing in none-too-tight harmony, which was presumably putting off the prospective customers. Lawrence didn't know the song; it seemed to be some kind of folk chant, with the ragged voices rising defiantly for the chorus:

Give us back to ourselves
Take back your money
Give us back to ourselves
Turn back your starships
Give us back to ourselves

Several of them were carrying hologram panes on long poles, blazing with anti–Z-B slogans. A couple of bored police officers were standing fifteen meters away, watching over them. They catcalled and jeered anyone walking up and down the broad stone stairs to the entrance of the big headquarters building. Z-B personnel scurrying in and out studiously ignored them.

When Lawrence started up the stairs they directed several insults at him. He smiled and waved cheerfully, knowing how much that always annoyed their type. His gaze found a girl in the middle of the group, more attractive than any of her cause sisters, with compact dainty features amplifying her intent expression. She was wrapped up in an old-fashioned navy-blue duffel coat with wooden toggles, its hood down to show off raven hair that had been frizzled into a thick mass of short curls. Their eyes met, and he broadened his grin to a male-ape invitation. He laughed heartily at the angry scowl she fired back at him.

Minority-cause fascists, no sense of humor.

Three receptionists sat behind a curving teak desk in the vast, empty lobby. One of them gave him directions to the officer college, in an annex of its own at the rear of the main building. "What are they here for?" he asked, pointing out through the tall glass doors at the protesters.

"Regressors," she said. "They want for us to go away and stop influencing 'their' lives with 'our' policies."

"Why?"

The receptionist gave him a pitying look. "We're not democratic."

"But anyone can buy a stake in Z-B."

"Tell them."

The officer college was a modern glass cube connected to the headquarters building by a couple of bridges on the third and sixth floors. Lawrence walked across the lower one, trying to damp down his trepidation. If all went well he'd be spending the next three years here learning everything from life support engineering to astrogration. Although quite why the flattest country in the world had been chosen as the training ground for starships was a question that his downloaded briefing had never covered. Someone somewhere in the company must have had a strong sense of irony.

He reported in to the corporal in the foyer, saluting sharply. The man gave a disinterested wave back and entered Lawrence into the administration AS.

"Turn up at oh-seven-fifteen hours tomorrow," the corporal said. "You will receive your introduction to the assessment week. This is your accommodation warrant." He handed over a small card. "You're staying at the Holiday Inn. This entitles you to a single bedroom, along with breakfast and dinner. Don't try ordering room service or beer with it. You have your lunch here in the mess. You're in group epsilon three. Don't be late." The corporal returned to the pane displays on his desk.

"Thanks. Uh, how many others in the group?"

"Thirty."

"And how many places are we competing for?"

The corporal gave him a tired look. "We process one group per week. And the annual intake is one hundred officer cadets. Work the odds out for yourself."

Lawrence made his way back through the main building. On average they'd take two from each group. A one-in-fifteen chance. *No,* he corrected himself. *Nothing here is down to chance. I'm going to make it.*

When he walked into the Holiday Inn half of the people in the lobby were from Z-B, and several of them were obviously in town for their officer assessment. He could spot them from a long way off. In their early twenties, fit, serious expressions, well-cut clothes, trying to hide fluttering nerves. He guessed they could spot him just as easily.

That afternoon he went down to the basement pool and swam a mile. As always, his fitness had suffered on the starship back from Quation, and the last week hadn't exactly been dedicated to healthy living. He climbed out, reasonably pleased with his time. The exercise gave him that extra degree of confidence for tomorrow: thanks to their own training, Z-B had kept him in top shape for the last five years.

Lawrence couldn't stand the idea of having his supper in the hotel restaurant. The place would be full of all the other candidates, forcing themselves to be polite to each other. So he set off on a short walk through the old city as dusk fell. Amsterdam's heart had been beautifully preserved, with marvelous old houses lining the canals, each with its own hoist on the top. The antique mechanisms still worked, hauling furniture up so it could be brought in through the windows. Houseboats were tied up on the still black water between the arched stone bridges, ranging from tiny cruisers to barges with double decks and roof gardens. Berths had become so valuable that the city hadn't issued a new houseboat license for over two centuries; his briefing had mentioned that some had stayed in the same families for over eight generations now.

The bar he eventually found on Rembrandtplein served a decent menu of hot food, and beer that claimed to replicate the recipe of an original Dutch lager. It wasn't the classiest place in town, but it had a lively atmosphere, and a hologram pane was showing a sport feed. He sat at a table near the back and ran through the menu. It took him a moment to work out that the last ten items on the sheet were narcotics, three of which were quite hard. There was an option to have some of the lighter ones as garnish on your food.

His waiter took the order and delivered some of the supposed original-tasting lager. Lawrence settled back and took a look around. The big pane on the far wall was showing Manchester United versus Monaco. He chuckled and took another sip of his lager.

The girl from the protest group was sitting up at the bar, giving him a cool stare. He did a double take, then smiled and raised his glass in salute. She looked away hurriedly.

Too bad, he thought. She was with a couple of other girls, no male companion in sight. Her duffel coat was slung over the back of her stool. She was wearing a thin scarlet rollneck sweater, with an impractically long scarf wound loosely round her neck, and baggy olive-green trousers held up with a broad rainbow bead belt. With those clothes, and an age he estimated at three or four years younger than himself, she had to be a student. Philosophy, no doubt, he decided, that or sociology. Something utterly useless for the real world.

His food arrived. Pasta with a three-cheese sauce and smoked ham. A side order of garlic dough balls. Sprinkling of ground pepper. Hold the hashish.

He wound the first strands around his fork.

"Killed anyone today?"

He glanced up. The girl was standing by his table.

Just like Roselyn, appearing out of nowhere to talk to me.

Somehow, he thought the motive would be different.

"Not today, nor any day," he answered, politely casual. Her nose was too broad to make her a classic beauty, but she had what people called a fierce intelligence lighting her eyes, analyzing and judging everything she saw. It made her very appealing: that and the raw hostility. Getting her into bed would be quite a challenge.

"You're one of the cybersoldiers," she said. "I can see the blood valves on your neck."

She had an accent he couldn't quite place. "And you're a welfare princess. I saw you standing in the Dam square while everyone else was working for a living."

Her cheeks darkened in anger. "I devote my time to achieving something worthwhile: your downfall."

"Had any success?" Lawrence had heard of opposites attracting, but this was ridiculous. He was sure she was about to throw her drink over him. Except her glass was back on the bar. She couldn't be carrying a weapon. Could she?

"We will," she said.

"So who do you plan to control our factories and revitalization projects once you've driven us out of your country? Yourself and your friends, perhaps?"

"We'll close down your factories. We don't want that kind of society."

"Ah, green anarchy. Interesting ideology. Good luck convincing everyone to adopt it."

"I'm wasting my time. You're not allowed to think: you just recite company dogma. Next you'll tell me to buy a stake if I want to change the way things are."

Lawrence closed his mouth before he said, *Well, yes, actually.*

"Are your career and your stake worth so much that you have to build them up on the destruction of others?"

She looked so damn earnest when she asked him. It was the worst kind of student politics: we can change the whole world if we can just open a dialogue. Try opening a dialogue to a mob flinging Molotovs at you. "I've never destroyed anyone," he said lightly.

"You've taken part in the campaigns to pillage other worlds. If that's not destruction, I don't know what is."

"Nothing is destroyed. And our campaigns help fund the greatest human endeavor there is."

"What's that?"

"Establishing colonies on new worlds."

"My God, you're worse than a cybersoldier, you're an ecocide advocate."

"It's even worse than that, actually. I'm here in Amsterdam to join the starship officer college. I'm going to find lots of new planets we can ecocide."

Her head was shaken in soft disbelief. "Why?" she asked, genuinely puzzled. "Why would anyone do such a thing? That's what I never understand about your kind. Why do you always think that you can only achieve anything by violating what's right and natural? If you have this urge, why can't you channel it into something creative?"

"Exploring the universe is the most creative endeavor there could possibly be. It's the culmination of a thousand years of civilization."

"Starflight is the most appalling waste of resources and money. Z-B is practicing interstellar imperialism with its expansion program. There's no worthwhile outcome. We have a planet here that desperately needs our help in just about every domain you can mention, and we can't provide that help because you're bleeding us to death."

"Z-B provides almost as much funding for ecological and urban revitalization projects as it does for starflight."

"But they're *your* revitalization projects. Revitalizing in your image, spreading the dead corporate uniculture into weaker societies."

"Look, I can see where you're coming from. You want money devoted to issues you think are important. That's perfectly natural politics, convincing governments or corporations to pay for your own pet projects, or convincing enough people to win you the popular vote. Fine. Keep on campaigning and raising people's awareness. But you will never, ever, get my vote, because I will always vote for more starships. And the only practical way I get those is through a stake in Z-B. Sorry, I'm not going to be converted. I'm already doing the one thing I believe in the most."

"It's wrong, and in your heart you know that."

"I do not know that. I'm afraid all your arguments fall down with me, for the simple reason that you can't look above your own horizon. You have no sense of wonder, no drive. You've limited yourself to seeing only the smallest pixel in the picture. You practice the worst sort of parochialism."

"I see this whole world and how it's hurting."

"Exactly. This world and no other. Without starflight I would never have been born. I'm not from this planet." He smiled at her frown of confusion. "I'm from Amethi. And we don't practice ecocide there. We're regenerating an entire living biosphere. Something I happen to think is worthwhile in the extreme."

"You weren't born on Earth?" she asked.

"That's right."

"Yet you came here to join Z-B so you could fly starships further into the galaxy?"

"Yep."

Her short laugh was of pure incredulity. "You're crazy."

"Guess so." Lawrence grinned back. "So are you going to wish me luck for my assessment tomorrow?"

"No. That I can never do." Her expression was sorrowful as she turned away.

"Hey," he called. "You didn't tell me your name."

For a moment he thought she was going to ignore him. Then she glanced back over her shoulder, hand running through her buoyant hair as she made the decision. "Joona," she said at last. "Joona Beaumont."

"Joona. That's good. I like that. I'm Lawrence Newton. And I wish you a happy life, Joona."

Finally, just before she reclaimed her barstool, she allowed him to see a slight smile tweak her lips.

Breakfast was as depressing as Lawrence expected it to be. The Holiday Inn restaurant was full of his fellow candidates, all being hearty and cheerful. He joined in, putting on that same mannerly facade the way he'd learned back home when his father had other Board members at the house and he had to be a proper little Newton. It was surprising how easily the deceit came.

The other hopefuls were mostly from upper-management families with big stakes in Z-B, fresh out of college, or with a few years spent in one of the company's various spaceflight divisions. Dressed in his strategic security uniform, and with his starflight experience, Lawrence soon became their focal point. They kept him busy answering questions throughout the meal. He was still telling them about Floyd and the aliens when they walked en masse over to the headquarters building. Lawrence looked around the square, but there was no sign of any protesters. Not that he'd expected them there quite so early in the morning.

Group epsilon three's morning started with the introduction, a half-hour talk from a captain about what Z-B looked for in its starship officers. The usual bull about devotion to duty, comradeship, professionalism. Lawrence got a different version from a strategic security officer every time the platoon was put through a new training course. The captain ended with: "We expect you to give us better than your best."

Day one was devoted to testing their reflexes. The college's i-environment was the most sophisticated Lawrence had ever experienced. They were given full stim-suits to wear, a tight-fitting one-piece made from a fabric of piezoelectric fibers; then led into a big anacoustic room with three rows of gyro-seats. Once they were strapped in, the AS started off with simple coordination tasks. It was easy to begin with, three-dimensional grid alignments, like being inside a hologram pane graph, lining up the glowing green-and-scarlet symbols. They soon progressed to steering fast cars through a maze, and different wheel limitations and engine fluctuations were gradually introduced. Crashes became progressively more violent. After lunch they were given full aircraft simulations, taking up single-seat jet trainers. That was when the AS began to put them under stress, giving them engine flame-outs, failed flaps, spins that were so fast they threatened to make Lawrence vomit. Equipment malfunctions at critical moments. Cockpit fires, with real smoke blowing in through the suit helmet vents and heat searing their hands and legs.

When it was finally over, Lawrence had to grip the gyro-seat's support pillar while his legs regained their strength and stopped shaking. There was a noticeable lack of jovial esprit de corps in the locker room afterward as they all showered and changed.

It was raining when they came out of the headquarters building, a thin, cold drizzle whipped up by the erratic gusts blowing out of the streets surrounding the square. Joona Beaumont was standing outside, her duffel coat hood up against the weather, stamping her feet on the cobbles. There were only three other protesters with her, and the potato stall was absent. They propped up their panes, but couldn't summon up the enthusiasm to shout anything.

Lawrence gave her a quick nod, but she didn't respond. He wasn't even sure she saw him.

An hour later it had stopped raining, and he made his way back to the bar on Rembrandtplein. He didn't bother with a table this time, just sat up at the bar and ordered a mixed mango and apple juice.

Joona arrived a few minutes later. She saw him immediately, and Lawrence offered the empty stool beside him. There was a moment's hesitation; then she came over, shaking the water from her coat.

"You look frozen," he said. "Can I get you something hot?"

She signaled to the barman. "Tea, please. Put a gram in."

"It's bad for you, you know," Lawrence said.

"What, it glitches your circuits? I don't suppose you'd like to lose control, would you?"

"Nothing to do with it. It's a poison, that's all."

"All medicines are to some degree. That's how they kill germs. It's perfectly natural."

"Right. So how did your day go?"

"We made our point."

"Did anybody listen?"

"Being there is our point."

"Then I guess you made it well."

Her tea was delivered. She gave the barman a smile of gratitude.

"You going to ask how my day went?" Lawrence inquired.

"No."

"Okay." Lawrence dropped a ten-EZ-dollar bill on the counter, stood up and walked out. And just how cool is that?

He sort of blew it at the door, when he looked back to see how she'd reacted. She hadn't. She was sitting with her elbows resting on the bar, holding the cup of tea to her mouth with both hands.

He shrugged and stomped off into the night.

Day two was all about puzzles. The AS controlling the i-environment put him on a small tropical island four hundred meters long and barely seventy wide. A few palm trees and spindly bushes grew along the central strip, but it was otherwise desolate. He was in charge of a five-strong party that had been diving along the offshore reef. One of them was badly injured to the extent he couldn't be moved and needed medical care urgently for decompression sickness and unspecified internal organ damage. There were three islands nearby, one with a resort complex, the second with an abandoned plankton harvest factory, and the third also deserted, but with another diving party visiting it. The resort was farthest away, the plankton plant was known to have an advanced first-aid store with a quasi-AS diagnostic. He only had one boat, which couldn't make it to the resort before the injured man died. There were no communications systems.

Lawrence took a quick look at the map, comparing island positions. He left two people to look after the injured man and set off in the boat to the third island and the other divers. He told them to go to the plankton factory and take the medical equipment to the injured man, then set out by himself to the resort. With just himself onboard, he jettisoned all the surplus equipment he could find, allowing the boat to go as fast as possible. In theory, the medical equipment collected by the other divers should allow the injured man to stay alive while he made the long trip to the resort to alert a helicopter rescue team.

The AS allowed the scenario, although the chopper paramedics rebuked him for making the boat trip to the resort by himself. There was an experienced sailor in charge of the other diving team who could have made a faster trip. But the injured man survived.

For his second expedition he was in a deep, rocky canyon in a jungle. His little team was moving a lot slower than anticipated because of the difficult terrain; they were starting to run out of food. The canyon walls were too high to be climbed.

Lawrence asked them for their skills and found one member who was proficient with canoes. The team set about chopping down trees and building a makeshift raft. The canoeist was dispatched downriver to contact their base camp.

After two kilometers the canoeist encountered rapids too severe for the raft. He had to wait until the rest of the team caught up on foot and helped rebuild the raft so that it could be taken apart and carried around difficult sections. Right idea, not enough thought for the method.

An Arctic wilderness came next, with Lawrence by himself at the center of a ring of various equipment caches. To get to the food, which was on top of a pressure ridge, he had to collect the climbing equipment to reach it, but the climbing gear was too bulky and heavy to carry in the backpack; he needed the sledge, which was on the other side of a bottomless gorge. The collapsible bridge to get over the gorge needed the sledge to move it.

He just couldn't work that one out. But he did his best, fetching a single coil of rope from the climbing cache and trying to swing across the gorge. He wound up tumbling down into the black abyss when his ice axe anchor broke free.

After that came a classic cell/maze. The AS put him in a room with five doors, each of which led to another room with five doors. The hazards were mostly visible, with hinged flagstones, spikes stabbing out of the walls, flames, a pendulum, lions, walls that closed in, cutting wire at neck level, electrified segments, stones that fell from their ceiling cavities, tripwire-triggered darts, moss with an acid sap, rat swarms—though there were others like gas and ultrasonics that he didn't find until he was already well into the room. The doors all carried clues to what was in the room on the other side, sometimes numerical; then there were symbols, star signs, even poetry.

He was allowed five goes. The farthest he ever got was eight rooms from his starting place.

He was put in a starship just after it had suffered a meteor collision. Environmental support systems were failing, air leaking, power dropping, network glitched, no spacesuit, few tools. He had to make his way from his own badly damaged section to the lifeboat capsule halfway around the life support wheel.

After that the AS dressed him in a spacesuit that was low on oxygen and power reserves and left him clinging to a small asteroid with his ship on the other side. There were different types of survey sensors dotted across the surface, which he could cannibalize for components and gas as he tried to crawl his way back. The rock's microgravity field was just enough to stop him from achieving orbit by muscle power alone, and weak enough to leave him with all the maneuvering problems of freefall. He actually expired within sight of the little silver craft.

The locker room that evening was even more subdued than the previous night. The candidates all looked dazed and shell-shocked. Conversation was all: "But what do you do after that bit where . . ."

He couldn't see any protesters in the square. And the weather was a lot better that evening, high clouds and a dry wind blowing from off the land. It was still cold. He quite fancied a hot potato.

Joona was in the bar when he arrived, sitting at her usual

place, with empty stools on either side. None too sure of his status, he left a vacant stool between them, and ordered his mango and apple.

"Shouldn't you have something stronger?" she asked. "I'd say you've had a hard day."

"Alcohol isn't going to help. I've got an even harder day tomorrow. Have to keep a clear head."

"Is it worth it?"

He took a long drink from his tumbler. "Oh, yes."

"Doesn't seem it to me. Look at the state of you. What did they do to you in there today?"

"Put it this way. If you ever crash-land on a frozen desert populated by flesh-eating zombies, then stick with me, I'll get you out. Piece of cake compared to what I went through."

Joona cocked her head to one side, giving him an interested look. "And how does that help them select their officers, exactly?"

"It's testing our ability to think under pressure. They put us in all kinds of impossible situations today." He rolled the glass between his palms, regarding it with a miserable expression. "I didn't do very well. I lost count of how many times I got killed. Then again, the others were just the same, judging by what they said."

"How good are you?"

"What do you mean?"

She slid her hands across the bar, pushing the tea cup ahead of her, moving with feline grace as she leaned in toward him. "I mean, you're a . . . you're a soldier who's seen action. You've been in bad situations for real on those other worlds you plunder, right?"

"Yes. But we're trained in how to deal with hostile crowd or ambush situations. I know what I'm doing."

"Right, but what you're basically taught is how to keep cool under fire. And today they simply turned up the heat.

Were those situations genuinely impossible, or did you just flunk them?"

"You don't take many prisoners, do you? I suppose I could have done better in some of them, if I knew more about engineering and stuff."

"Has it occurred to you that these tests were actually dual purpose? It sounds to me like they were testing your character as well as your ability to think."

He slumped down on the stool. "Probably. I'm really up shit creek, then."

"Why is that?" with lazy amusement.

Lawrence realized just how stoned she was. "I have no character. You said so yourself."

"I didn't say you had no character. I said you had the wrong character, which for the purposes of today's experiment will serve you well. You're what they want."

"Let's hope so. Are you okay to get home from here?"

She straightened up again. "Oh, I don't need any help from you. I have a citybike card. I'll just take one off the rack, and zoooom, I'm home." She caught the barman's attention and wagged a finger at her cup. "Same again."

Lawrence drained his juice and stood up. "Take care." He walked to the far end of the bar where the barman was preparing her tea. "Do something for me," he said quietly to the barman. "When she leaves, call a cab for her. This should cover it." He put an EZ twenty on the counter.

The barman nodded and pocketed the bill. "Sure thing."

Day three was linked teamwork. The AS split them into groups of five and dropped them into a shared i-environment. There were to be eight tests. For the first five, they would rotate the leadership, while the last three were to be a group effort.

Lawrence's group was given a river to cross for its first task. It was running through a hot, unpleasant jungle, complete with insects that bit exposed limbs and reeking marsh-sulfur air bubbling out of the mud along the foot of the banks. Crocodiles peered at them from midriver, occasionally snapping their jaws in anticipation. Ropes, oil drums and wooden planks were stacked up on the bank. Even laying all the planks end to end, they weren't long enough reach across the water.

Their designated team leader started snapping out orders. He wanted to build a platform that would go halfway across the river; they would take up the section from behind them and rebuild it out in front to the other bank. Lawrence helped willingly enough, even though he knew they were wasting their time. The scheme was overelaborate. They should be building a raft.

He briefly toyed with the idea of slacking off, or maybe not tying off his rope as tight as it needed to be. Not active sabotage exactly, but as the idea was doomed anyway . . . There were only two places, after all. But he guessed the AS would be watching for anything less than 100 percent commitment.

Sure enough, when they turned the bridge into a platform in the water and started trying to build the last section, two of them wound up falling in the river along with several planks. The crocodiles moved in eagerly, huge jaws hinging open.

For his own command, Lawrence was given the last stone in a henge to erect. He took a quick inventory of the equipment they'd been given, which was mainly poles, shovels and ropes, and issued his instructions. They measured the length of the stone, and the height of the others. That told them how deep to dig the pit at the base of the stone. With that done, they set about tipping it in, rigging up levers and crude pulleys. This was the part that required a high level of coordinated teamwork, and everyone played his part perfectly, following each of the orders that Lawrence shouted out. Eventually the massive block tilted upright. Lawrence had a nasty moment when it rocked about, but it stayed upright.

It was the final three tests that made him irritable and disappointed. There was just too much competition between the group members for them to have their own idea adopted. Lawrence reckoned the AS had deliberately structured the tasks so that there were multiple solutions to each problem. His fellow candidates began to question him and each other, whining and bitching, especially when their own proposals were turned down. When Lawrence was convinced he had the most efficient solution to the second task he had to shout to make them listen, which they resented. They were competing, not cooperating. The simulations were deviating from the way people behaved in real life. Drawing from his own time in action with the platoon, Lawrence knew there would be a better level of rapport.

Hardly anyone spoke when they left that evening. Lawrence heard that there had been near fights in other teams during the last three tests. At least his own group had managed to keep reasonably civil. That must count in their favor.

Joona was in the square. The potato stall was back, along with a larger number of protesters. She caught sight of him, and intercepted him. Lawrence tried to smile off the startled looks of the other candidates, though he knew exactly what they must be thinking.

"Yours," Joona said curtly, pressing an EZ twenty into his hand. "I don't need your charity."

"It wasn't charity. I was concerned about you, that's all."

"Did I ask you to be?"

"How could you? You didn't know what planet you were on."

She turned quickly and started walking back to her friends. "I've survived in this city long before you got here, space boy."

"Sorry I cared," Lawrence shouted after her.

He had dinner in the Holiday Inn that evening.

Day four was interviews and evaluation. First time up, Lawrence was quizzed by two college officers about his background and motivation and likes and dislikes. He knew he had to be courteous and slightly self-deprecating and honest and relaxed and show he had a sense of humor as well as being overwhelmingly interesting. Tall order cramming those traits into ninety minutes while you're telling them your life history and slanting it so that your inquisitors believe they cannot possibly afford to let you slip out of the college.

The second interview was with an assistant to the deputy principal, a cheery old woman who dressed in clothes a century out of date, presumably to give her an authoritative schoolmarmish air. They sat on opposite sides of a steel-blue desk in her office, a fourth-story room with a good view out over the canal.

Data was scrolling down her desktop pane, which was just at the wrong angle for him to read it.

"You did very well on the simulations," she said. "Good reflexes. Good spatial instinct—whatever that is. High proficiency on logical analysis. Integrated well with the group command dynamics. Fast thinker. Care to comment on any of that, Mr. Newton?"

"We were a mess in the last three simulations yesterday. Too much competition."

"That's right. That's why we include them. Think of it as a measure of how unselfish you can be."

"And was I?"

"You certainly showed awareness of the situation. It was a mature reaction. You have the potential to be an officer."

"Excellent." Lawrence couldn't help his hungry grin.

"Which gives me something of a problem. You see, it's more than proficiency we're looking for this week. Your stake also has to be taken into account. And, frankly, there are candidates with an aptitude equal to yours who have a much larger stake in Z-B than you."

Lawrence managed to hold on to his expression of polite respect. "I suspect they all had inherited stakes. It's not actually possible for someone of my rank in strategic security to earn a higher stake than the one I have. A lot of the fleet platoon members opt for a much lower percentage. That should tell you all you need to know about my level of commitment to Z-B."

"It does, Lawrence, and it's very impressive, as was your commanding officer's report. But the figures speak for themselves. And we have to stick to our chosen method of selection. You understand that, don't you?"

He nodded sharply. *This is a hatchet job,* he realized. *She's turning me down. I've failed. Failed!* His fingers closed tightly around the end of the chair's armrests.

"Good," she said. "What I'd suggest is that you reapply in another couple of years. With the scores you've accumulated over the last three days we'd welcome you back again for another assessment. And by then your stake should have risen to a suitable percentage."

"Thank you." That was what it boiled down to. *Thank you.* His life's dream denied. *Thank you.* Five years devoted to the company, putting his life on the line. *Thank you.* He'd left his world behind, his life, his family, his one love. *Thankyou. Thankyou. Thankfucking YOU.*

It was sunny and cold when he marched down the stone steps to the square, a cloudless deep azure sky overhead. He blinked at the sharp light, which was what must be making his eyes watery. It was normally dark when he came out of the headquarters building. People got in his way as he walked. He pushed past them, heedless of their protests. Trams, too, they could fucking wait. Bastard cyclists always in the way.

Fortunately the bar was almost empty. But then it was only

three o'clock in the afternoon. When the evening crowd arrived, Lawrence planned on moving back to the hotel where he could call up room service for the rest of the night. He opened the front of his coat and claimed a barstool. "Margarita, one glass, one jug." He slapped a couple of EZ twenties on the bar. "And that's a proper glass, with salt."

"Yes, sir." The barman wasn't going to argue, not yet.

Lawrence dropped his head into his hands and let out a painful sigh, surprising himself by not shrieking in anguish. "Shit! Shit, shit, fuck it!"

Someone pulled out the stool next to his and sat down. Like they didn't have the whole fucking place to choose from. He jerked round angrily to tell them to— "Oh."

"I thought I'd better check on you," Joona said in mild embarrassment. "You nearly got run over by a couple of trams."

He turned away. "Enjoy your moment of triumph."

"Suffering in others is not a cause for rejoicing."

"In that case, give the hippie philosophy a break. It pisses me off."

"They turned you down."

"Yeah. All right? They turned me down. Bastards."

"Did they say why?"

"I'm not rich enough. That's what it was in the end. My stake in the company isn't enough. For fuck's sake, I've got a thirty percent investment in Z-B shares. A third of everything I earn goes straight back into the company. What the fuck else do they expect from me?"

"I don't know. What did you expect from them?"

"A fair chance. No, not really. I should have known. Me of all people. I know how companies really work, what really counts."

The barman put his margarita jug down in front of him, pushing a coaster forward for the glass. It was a proper margarita glass, with a thin rime of salt around the rim.

"What does count?" Joona asked.

"Internal politics. You want one of these, or have you got to run back to shout at my fellow corporate cyborgs?"

"We're not exactly on timesheets and shift work."

Lawrence nodded to the barman. "Another glass, please."

Waking up was accompanied by its timeless twin: where am I? Lawrence opened his eyes to see a long room with a desk and a couple of worn comfy chairs at one end. The floor was bare wooden boards, with a couple of rugs thrown down, one of which he was lying on. Opposite him was a broad arched window, with thick old curtains drawn. Scraps of streetlight shone around the edges, casting a dreary sodium-yellow illumination against the walls. Several large prints had been hung above the small fireplace, posters for various exhibitions and poetry recitals decades out of date. Definitely student digs. Brighter slivers of light silhouetted the door. When he lifted his head he could see a bed at the other end of the room. Joona was sitting on it, her back against the tarnished brass railings. She had a quilt wrapped around her shoulders. A reefer dangled from one hand, its end glowing morosely in the gloom.

"Oh, hell," he muttered. At least he was still wearing his uniform. "How did I . . . ?"

"I brought you here," she said. There was a current of humor in her voice. "My turn to rescue you from the bar."

"Thanks." He sat up gingerly. "Do I owe you a twenty?"

"No, a friend helped get you into the tram. There's a stop close to the end of this street."

"Uh, right." He didn't remember much after the third jug of margaritas. Just bitching on about Z-B and how he would have loved to be the first person to land on a new world. He ran his dry rubber tongue around the inside of his dry mouth. The taste was awful. Apart from that he wasn't too bad, just stiff from the floor. "How come I don't have a hangover?"

"I made you take aspirin and vitamin C, and a couple of liters of water."

"Right. Thanks again." The mention of water made him want to pee. Badly. Joona told him where to find the toilet, just outside and down the corridor.

"Try to be quiet," she said as he hurried out. "Everyone else is asleep."

His watch said it was quarter past two.

When he got back she was still sitting at the end of the bed, the reefer down to its last half-centimeter. "Want some?" she asked.

"No, thanks. Us cyborgs don't, remember?"

"Of course."

"Look, thanks again for taking care of me. I'd, er, better be going."

"Really?" She took a deep drag. "What's waiting for you?"

"Nothing much, I guess. I've still got three weeks' leave due. I just don't want to impose on you any more tonight."

"If I'd thought you were imposing I wouldn't have brought you here."

A sharp tingle moved down Lawrence's spine. He walked over to the bed and knelt down. She didn't say anything, just kept gazing at him with wide eyes. He took the last of the joint from her fingers and inhaled the way he'd seen it done on the i's. The smoke was bitter enough to make him cough.

Joona started to laugh. "I win."

"Win what?"

"I got to you."

"Yeah." He grinned and took another drag before handing it back. "You got to me. But then you were never going to run off and join the officer college with me, were you?"

She shook her head as if she'd been admonished and pouted. "No."

"Can I stay here the rest of the night?"

Joona nodded.

"With you?" he asked softly.

She opened the quilt. She was naked underneath.

When Lawrence woke up in the morning his earlier confusion was replaced by something close to embarrassment. Classic case of *now what?*

He was lying along the edge of the bed, the quilt covering him, with his back pressed up against the wall. The mattress really wasn't wide enough for two. Joona was curled up beside him, looking a whole lot more fragile than she had last night. She was thin, skinny enough for her shoulder blades and collarbones to be prominent, and a lot shorter than he recalled. She must have been wearing heels before. Funny he'd never noticed that.

When he tried to pull the quilt up gently around her shoulders she stirred and woke. Pale blue eyes, he saw, a contrast to her darkish skin.

"Well," she said.

"Morning."

"Yes, it is."

She snuggled up closer, closing her eyes.

Again: now what?

"So, er, what time do you have to get up?"

Joona's eyes stayed shut. "You're always in a rush to go nowhere, aren't you?"

"That's me."

"I was going to take a break from college. It's getting heavy there for me right now. I hadn't got a plan for getting up."

"You're at college?"

She sighed and sat up. "Yes, the Prodi. It's a complete shithole. They don't even have enough funds to stop the building from falling apart, and the lecturers are all fifth-raters who couldn't get an appointment cleaning the toilets at a decent university." She got up out of bed with a sudden energetic motion and padded over to the window, pulling the curtains back with a quick tug.

Lawrence didn't point out she was nude; he would have sounded like his mother. But the window was smeared with dribbles of condensation, only a few vague gray shapes of buildings were visible. Joona shivered and rubbed her arms. The air in the room was cold enough to make her breath show as thin vapor.

"Are you leaving me?" she asked.

"Like you, I don't have any plans."

"Actually, I was thinking I'd go to Scotland."

He couldn't figure out if that was an invitation. She certainly wasn't his usual type, not with all this twitchy energy and commitment to her stupid cause. He couldn't imagine her ever walking down the Strip at Cairns, hunting a good time as the sun went down. Come to that, he couldn't even imagine her laughing heartily. He'd never seen her do more than smile wryly every now and then. But then again, she definitely knew her own mind. Just like Roselyn. Unlike Roselyn, she wasn't happy with life. There was a lot of anger bottled up inside that small frame—a stupid form of anger, though he would never tell her that to her face. She was far too wrapped up in her issues to welcome contrary observations. He guessed that might make her kind of lonely.

The room had a singular imprint that was all her own. It wasn't just the air that was cold. Most people, he thought, would instinctively keep their distance.

So why didn't I?

Two lonely people. Maybe that was why they'd kept dancing around each other in the bar. They weren't opposites attracting after all.

"I've never been to Scotland," he said.

Joona was bending over the heatstore block that sat in the ancient fireplace, turning up its output. The black surface began to glow a deep orange, as if there were still embers in the grate. She gave him a fast, nervous smile. "You want to come with me?" There was surprise and hope in her voice.

"Sure. If you want me to come with you."

"I don't mind. It would be nice."

For a moment he thought she was going to jump back into bed with him. Instead she grabbed a big red-and-green-check nightshirt from the back of a chair and struggled into it.

"I'll put some coffee in the microwave," she said. "Then I have to do my yoga: it helps me center myself. We can go after that."

"Okay," he said, trying to keep pace with events. "I can pick up my stuff from the hotel on the way to the station."

"Will you book the train tickets? I hate using the datapool. I can pay you."

"Sure." He hunted around for his clothes, wondering what he'd gone and said yes to.

Lawrence and Joona took an express train straight out of Amsterdam direct to Edinburgh, traveling in a big U, down south to Paris, across to London, then up north again to the end of the l-pulse line at Waverley. To start with, Lawrence was impressed by Holland. The old canals were still draining the land. Windmills stood guard along the straight-edged waterways, although little wind now reached their sails, thanks to the extensive forests that had grown up in the last two centuries across the old farmland. There was a huge variety of trees, but with the canals slicing through them they formed such a regular grid it made them look like nothing more than fields. In a sense they were, not that they were cultivated, but the land management teams maintained them carefully. Even now, the drainage system couldn't be allowed to fall into disrepair, and the roots were a big potential hazard. It gave him an impression of an artificial environment barely one step ahead of Amethi. He thought that in a way Holland must be the first example of large-scale terraforming; human engineering and ingenuity wresting a livable nation out of an alien environment.

Lawrence soon tired of the fenlands, especially as their speed blurred details. "So why Scotland?" he asked.

Joona put her feet up on the table, ignoring the disapproving looks of the other passengers in the carriage. "My grandmother is Scottish. We're going to stay with her."

"Where, exactly?"

"Fort William."

He put his interface glasses on and accessed the datapool to find where that was.

"You spend a lot of time trawling, don't you?" Joona said.

"My education had a lot of holes. You must do a fair bit of accessing yourself."

"As little as possible. I prefer books."

"There's a time and a place for hard copy. My dad had a thing for books, too. I guess that's why I never use them." He grinned at the face she pulled. "What's your subject at Prodi?"

"I'm taking ecological management."

"Right." It wasn't what he expected. "Doesn't that mean you'll wind up working for a company?"

"There are companies, and there are companies. And then there are government agencies, at least by name. In practice they're another branch of corporate reclamation and revitalization divisions. But I won't take a job with any of them. There are still some private landowners who use the land in the traditional fashion. They farm, or log timber or run stables. That's what I want to help keep alive."

"Farming?" he said skeptically. "I thought that's what damaged the land in the first place?"

"Industrial farming did, yes. Pesticides and nitrates were poured over the soil in the quest for higher yields and to hell with the consequences. Agricultural machinery actually got so big and so heavy that it compacted the subsoil. By the end, in the developed nations, topsoil was little more than a matrix that suspended chemicals and water so the crop roots could absorb them. Then the companies developed protein cell technology and killed farming altogether."

"And stopped us raising and slaughtering animals for food. I mean, can you imagine how barbaric that was? Eating living things. It's disgusting."

"It's perfectly natural. Not that people think that way today. And I didn't say protein cells are a bad thing. After all, it means no one on Earth starves. But, as always, they went to extremes and eliminated every valid alternative. All I'm asking for is to keep a few pockets of independence alive."

"You mean like working museums?"

"No! These are havens for people who reject your corporate uniculture existence. There are more of them than governments and corporations like to admit. More of us."

"Ah, right, communes of back-to-the-earthers. So will you also be refusing the kind of medical technology that comes out of our wicked corporations?"

She gave him an exasperated stare. "That's so typical, denigrate something you know nothing about. I never said I was rejecting technology. It's the current global society that I refuse to obey. Technology doesn't have to come only from corporate labs, to be exploited for profit and policy implementation. It could come from universities where it would be made freely available to benefit everyone. Even small independent communities could support researchers. If we all had free access to data we could build a culture of distributed specialization."

"The old global village idea. Nice, but you still need factories and urban centers. You should know that culture always flourishes at the heart of society."

"The datapool is the heart of our society. You're still thinking in physical terms when you talk about cohesion. You can live in a cottage in the middle of a forest with every need taken care of, and still be totally in tune with the rest of the world."

"But why live there, when you can also live in a city, and interact with people, and go down to a bar in the evening and have a laugh and a drink? We don't all want to be hermits."

"I know. But your companies don't want anyone to be a hermit or anything else. According to them, we all have to fit into this uniculture they're trying to establish like neat little blocks on a circuit board. I don't want to be a part of that. I want my freedom."

"I think you're exaggerating."

She pointed to a badge on her coat lapel. It had a single eye at the center. "Open your eyes."

He managed to steer the conversation off politics and got her talking about music, which was always a relatively safe topic. You could disagree about bands, performers and composers without storming out or throwing things. She enjoyed orchestral symphonies from several classical and modern composers; from postelectronic music she listened to what he thought of as ballads and street poetry. Although she had thousands of hours of tracks loaded into her multimedia player card, she became animated about live concerts, telling him about all the venues she'd visited, the bands and orchestras she'd heard. As far as entertainment went, she was scornful of the i's, although she admitted to watching several current soaps. The i's, she claimed, were something she grew out of. And she really hated AS-generated dramas, preferring to visit theaters. Amsterdam had a host of small nonmainstream theaters where her student status got her reduced rates, she said, and the city had hundreds of performance groups eager to put on their works.

Lawrence almost pointed out that having so many groups evolve in a city proved his argument about culture. But he still wasn't sure how she'd respond to that kind of teasing.

Even after lunch in the buffet car, when she drank over half a bottle of wine, she was still tense.

That afternoon she asked what he enjoyed, and he was foolish enough to admit accessing *Flight: Horizon*. It was the first time he'd ever seen her truly laugh.

"I can't believe we export that kind of crap to other worlds," she chortled. "No wonder you have such a screwed-up vision of starflight. My God, and that ending."

"Ending?"

"The last episode. Unbelievable! Pretty hot, though."

"You saw that?"

"Yeah. Told you I was into i's when I was a kid. Why?" Her eyes narrowed, giving him a curious gaze. "Didn't you see it?"

"No," he said lamely, unwilling to admit the associations the show had for him. Even though he knew he was totally and completely over Roselyn, he'd somehow never quite got round to accessing those last few episodes. "We only ever got a couple of series on Amethi."

"Oh wow. You've got to access it now you're here. You missed a treat."

"That part of my life is over. I can do without revisiting it, thank you."

Her eyebrows rose at the finality in his voice. "Okay."

Fortunately she didn't pursue it, or even try to tease him. Their conversation rambled on. The one thing that they never mentioned was sex. He found that strange. It was as if last night simply hadn't happened. At least for her. They talked around just about everything else. As he was taking his cues from her, he didn't try to bring it up.

He wanted to. Joona was good company. Not necessarily pleasant company. If their opinions clashed she would argue until he gave up. That made her interesting, as much as her diametrically opposed worldview. When he thought about some of his barracks conversations he couldn't believe how dumb they were in comparison. It was that quality he'd first noticed in her, the fierce intelligence, that's what attracted him. So he wanted to know where they stood, which basically meant was she coming to bed with him tonight, and every other night of this jaunt? At one point he decided she was saying nothing in order to tantalize him, an intellectual's idea of foreplay. Though there were always doubts about that theory. She was too highly strung to avoid talking about anything important in her life. Which made her silence on the subject slightly puzzling.

He'd booked a double sleeper cabin for them that night. When she'd paid him for the ticket there had been no mistake or misunderstanding, she saw exactly what he'd got for them. The thought stayed with him all that afternoon. They'd slung their luggage in there as soon as they got onboard, his shoulder bag, her rucksack. The cabin was tiny, its fittings as compact as modern design allowed.

All the time they spent talking in the main coach he knew that she knew they'd be going back there after dinner. They'd strip off in the confined space, then climb into the low bunk together. The prospect was highly arousing. It would be almost like their first time. Last night, from what he could recall, had involved little passion and hadn't lasted long anyway, some perfunctory fumbling brought to a swift climax. First-time sex was always hot. And here on the train it was inevitable, which added that extra twinge of excitement as he spent the afternoon looking at her.

They went to the restaurant as the train slid out of Paris. Joona ordered a bottle of red wine. Lawrence had two glasses; she finished the rest and ordered another. Her conversation, which had arrived at the global uniculture's contamination of Africa, allowed less and less opportunity for him to say anything. Eventually it became a bitter rant. Lawrence didn't have any of the second bottle. Joona ordered a brandy for herself, which she finished before they left for their sleeper cabin.

When they got to it, they found the conditioning was faulty, leaving the little room chilly. Joona swayed about, looking at him with a lack of certainty in complete contrast to her usual attitude. She gave him a brief who-cares grin and started to pull her clothes off. It was Lawrence's turn to hesitate.

"Look," he said reluctantly. "You've had a lot to drink."

"I can handle it. This is nothing." She got the sweatshirt off over her head, then put an arm out to steady herself as she undid her jeans.

"I'm sure. I'm just saying, we don't have to do anything tonight."

"Yes, we do." Her grin widened into something close to defiance as she slipped her briefs down her legs. "Don't you get it? We have to. We must." She began kissing him. The smell and residual taste of the wine was off-putting. He put his arms around her in a mechanical fashion, trying to respond with the same intensity.

"We're building a bridge," she mumbled. "The two of us, two worlds joining. That means we're human after all."

He wanted to ask what she thought she meant. But he was busy freeing his own shirt, and she'd sat down heavily on the edge of the bunk. The cold air didn't help his mood, it actually raised goose bumps on his skin. He climbed into the bunk beside her, quickly pulling the thin quilt over the pair of them.

She started kissing him again, ranging over his face and neck. A hand closed round his cock. One elbow rested uncomfortably on his sternum. What she must have intended as a suggestive caress felt more like an irritable tickle down the side of his ribs. The whole event was completely unerotic. He couldn't believe it; not after he'd spent most of the day anticipating this moment.

Finally he managed to roll the pair of them around so she lay underneath him. He could barely keep his erection going; to help he had to keep thinking about a couple of the girls from the Strip last week, how lively they were. Joona smiled up drunkenly at him and groaned as he slid farther inside her.

Fortunately, the whole miserable entanglement was concluded quickly. "God, I love you," she said. "This is what I want."

"What is?" He managed to find a space on the bunk that didn't squash the two of them together, even though he was in danger of falling off. When he looked back at her she was already asleep. She started snoring.

He found a thick T-shirt and put it on, then spent ages lying beside her, staring up at the cabin's invisible ceiling, unable to sleep. Nobody's fault, he kept telling himself, the circumstances were wrong, that's all. The cabin, the air-conditioning, the wine: an unfortunate combination. Tomorrow will be better.

The express terminus at Edinburgh Waverley had been dug underneath the original station, leaving the surface structure untouched. They hauled their luggage up the escalators to the big old sprawl of platforms underneath their arching glass-and-iron roofs, and found the local train over to Glasgow. Old-style induction tracks still threaded through the center of the city, passing below the ancient castle perched on top of its rocky pinnacle. Lawrence watched it slide past, fascinated by the massive stone blocks and wondering how the hell the builders had moved them into position without robots.

Once it was outside the suburbs, the train accelerated smoothly up to two hundred kilometers an hour. That was the fastest it could manage: the track in this part of Scotland was still using the same route that it had for centuries, laid down in the first decades of steam engines. It followed the contours of the rugged Highlands, curving too sharply to allow the train to reach its usual top speed. Even though there were no more farms, the district parliament had never obtained enough funds to straighten the route through the wild glens and restored woodlands. The cost of drilling new tunnels through hard Scottish rock and constructing viaducts over broad valleys was simply uneconomical given the volume of traffic. So anyone traveling the Highlands had almost the same journey as the Victorians who'd originally pioneered the route. There was even an old iron rail laid alongside the induction track, where enthusiasts kept a couple of old steam engines chugging up and down the coast, pulling early-twentieth-century first-class passenger coaches along behind them. A huge tourist attraction in the summer.

As they had arrived at Edinburgh station in the early morning, Lawrence was able to see the countryside in clear daylight. Queensland and some sections of Europe he'd seen were just as rugged, but nothing on any planet he'd been on was as green. With spring coming to the Northern Hemisphere, the trees were fresh with new leaves. Heavy rains had soaked the ground, giving the grass a healthy, vigorous start to the season. He took the window seat and pressed himself against it, smiling contentedly.

This section of the journey was the one that he enjoyed the most.

They reached Glasgow in the middle of the morning and changed trains for Fort William. If anything this journey was even slower. But the scenery made up for it. He couldn't believe the long, rugged glens, and the lochs with their dark mirror water that went on forever. Their splendor made him aware of how much humans belonged in this environment.

Joona sat beside him with her arm through his, pointing out various landmarks. Ever since she woke up she'd acted differently. Attentive and eager, as if their night together had allowed them to reach some new level of understanding and commitment. He didn't know what to make of it at all, though the affection was enjoyable. It made it seem as if they were more of a couple. Certainly anyone walking through the coach would assume so.

Fort William was the end of the line, its station just above the shore of Loch Linnhe. They stepped out onto the platform and Lawrence tipped his head back to look up at the mountain looming over the small town. The entire slope was covered with pine trees, their dark shapes packed tightly together.

"Is that Ben Nevis?"

"No," Joona said brightly. "That's Cow Hill; the Ben's away behind it. You'll be able to see it from Grandma's over in Benavie, if the weather stays fine." She glanced out at the choppy gray water of the loch. Dark clouds were streaming in from the southwest. "Rain's on its way."

Her grandmother was waiting in the station parking lot. Joona let go of his arm and waved frantically as she ran forward. Preconceived notions Lawrence had built up about Granny Beaumont being some quiet little old lady with her gray hair wrapped in a bun and wearing a long tartan skirt vanished there and then. The woman was only as tall as Joona, but she was a picture of health, with dark red hair only slightly tidier than her granddaughter's. She was wearing tough cord trousers and a long olive-green coat splashed in mud. He didn't see how she could be a grandmother to anyone in their twenties; she couldn't possibly be past fifty.

"So you'll be Lawrence, then." Her accent was thick, but easy enough for him to understand. They shook hands.

"Yes, ma'am."

"And we'll have none of that nonsense. I'm Jackie. Now come along with the pair of you, into the van. I've to pick up a few things first; then we'll go straight home." She ushered them forward. The van was a three-wheel pickup, with an egg-shaped driver's cab ahead of an open cart section. It must have been twenty years old, composite bodywork fraying along the edges with fiber strands bristling out of the cracks and odd-colored patches epoxied over the larger splits. The cab's curving windshield had yellowed with age and ultraviolet, reducing the visibility considerably. There was no steering wheel, just a broad handlebar.

"Still works, then?" Joona said.

"Of course. Sugarhol is the easiest fuel you can brew, and no duty on it, either. That's why they stopped making cells that could burn it."

Lawrence kept a perfectly straight face as he climbed into the rear of the pickup. The chassis rocked about as he tried to find a reasonably clean piece of floor to sit on. Joona passed their luggage over the tailboard to him, then climbed up. She pulled a thick woolen hat out of her coat pocket, then produced some gloves. "Don't worry, it's not far."

"Great." He zipped the front of his coat right up to his chin, then jammed his hands into the pockets.

Jackie climbed into the cab and fired up the converter cell. A cloying smell of burned sugar burped out of the exhaust, swirling around the vehicle. Lawrence wrinkled his nose up; his eyes started to water.

"There's a still up at the cottage," Joona said. "She ferments her own fuel. See what I mean about people hanging on to their independence?"

"Absolutely."

She laughed and gave him a hug before the pickup lurched off. Jackie Beaumont drove them north along the A82, which ran through the middle of Fort William. This section of town had large civic buildings on both sides of the road. He saw the hospital first, a modern two-story complex with an arched silver roof, its geothermal power turbine housed in a small igloo at the side. Two search and rescue helicopters were parked on their pads behind the accident and emergency annex. The information and heritage center was opposite. Next to that were several sports pitches, each covered by big translucent domes, similar enough to Amethi's nullthene to give Lawrence an unexpected twinge of nostalgia. The town administration office resembled a Georgian mansion with its vivid ginger brick and broad white stone windows. Only arches leading to the underground vehicle bays gave away its true century of origin. A line of buses had drawn up outside the secondary school. Kids in smart gray-and-turquoise uniforms were running around them, chasing balls and snatching each other's bags.

Just past the theater, Jackie turned into a parking lot serving a single-story wooden building that resembled a long barn. It had wide windows under the overhanging eaves, with displays of just about every kind of camping and walking gear ever manufactured. A carved sign over the door said *Grimmers.* Jackie hopped out of the cab and headed inside. Lawrence and Joona climbed down and followed her in. A cleaning robot was rolling through the nearly deserted parking lot, sweeping up leaves and mud.

"Looks like a good little community," Lawrence said as they went through the door.

She nestled up beside him. "A rich one, you mean. They can afford the facilities. A lot of companies combined to build the reclamation plants. There's a lot of tourism as well. The other end of town is virtually all hotels. Between them, they bring a lot of money into the area."

"That's good, surely?"

"Only if you've got a stake in it."

Jackie was picking up several boxes from a counter while the assistant chatted to her. Lawrence hurried over and took a couple of the boxes from her. She smiled her thanks and gave him a further two. They were heavier than he expected. The labels said they contained some kind of dyes.

"For the wool," Jackie said as they went back out to the van.

"Wool?"

"I'm part owner of a flock."

"Of sheep," Joona said, grinning.

"Right." It was a bizarre notion.

The rain had arrived, a thick, heavy downpour driven by strong winds. Turbulent clouds boiled overhead. Lawrence could see the last of the day's sunlight shining over the mountains on the other side of the loch. There was no sign of a rainbow. He put the boxes next to his own bag and clambered back into the van.

It was another ten minutes' drive back to Jackie's cottage, which was a few kilometers out of town. She drove them along the side of the Caledonian Canal before finally turning off onto a dirt lane that led through woodland of silver birch, oak and sycamore. Her cottage sat in a rambling garden, a long building with sturdy stone walls and lead-rimmed windows. A diamond brick chimney stack at one end was crowned by tall clay pots, with smoke curling away into the darkening sky. He expected to see a thatch roof, but the blue slate was just as acceptable.

Jackie drove the van into a wooden lean-to outbuilding on the gable end, which served as a workshop and garage. Water was overflowing from the guttering, sending a thick curtain pouring down across the doorway. The final deluge finished the soaking that the rain had been persevering with. Lawrence squelched down onto the concrete floor.

"Inside with the pair of you," Jackie said as she swung the big doors shut. "Go on now."

Joona led him through a side door into the cottage's kitchen. It was a wide room, taking up at least a third of the ground floor. The brick hearth was filled by a four-door Aga: its racing-green vitreous enamel had darkened down the decades, and there were plenty of little chips in it. But it was still functioning, throwing off a welcome heat. Joona shrugged out of her coat and went to lean against it, gripping the tarnished chrome bar along the front.

"Good to be home," she said, and beckoned him urgently.

He stood in front of the ancient iron monstrosity, not quite sure what it was. Jackie took his hands and pulled him closer to it. Heat crept back into his dripping fingers.

"That's better," he said. "That wind chill was killing me."

"I spent many an hour drying out in front of the Aga. We've even used it to save some lambs."

"Huh?"

"If their mother dies, the plate-warming oven is just right for keeping them warm. Poor wee things need all the help they can get for the first few days."

"This is a stove?"

"It is that," Jackie said. She stood in the doorway, taking her boots off. "Installed over three centuries ago, I'll have you know, and still going strong. The burner ring was modified to use methane, but other than that, as sound as the day it left the factory."

Lawrence gave the stolid behemoth a suspicious look. If she was telling the truth, it predated the human settlement of Amethi. Amazing.

"You two had better have a hot shower and change your clothes," Jackie said. "You've turned blue. There's plenty of hot water. I'll have some tea for you when you come down."

Joona nodded. "This way." She took Lawrence's hand again and started to lead him playfully out of the kitchen.

"He's a big healthy lad you've got yourself, there," Jackie called after them. "You'll be needing the double bed tonight."

"Gran!" Joona yelled back. But she was smiling up at Lawrence, hunting his approval. He managed to smile back.

A kettle was whistling away on top of the Aga when he came back downstairs. He'd managed to find a clean T-shirt in his bag, and Joona had given him a thick apricot-colored sweater to wear on top. The arms were only a few centimeters too short.

He sat at the big oak refectory table in the center of the kitchen, watching Jackie make the tea. She used a china pot, spooning in dark flakes before pouring the water in. He'd never seen tea textured that way before.

"It takes longer, but it tastes better than your microwaved cubes," she said when she caught him staring. "Life's not so fast up here, we've time to let tea brew as it should."

"Fine by me, I could do with some slow living."

Jackie sat in a chair in front of a new-model desktop pearl. Its pane was showing a sweater with an elaborate pattern of bright colors. She told it to switch off, and the pane folded itself back into the casing. "I expect our Joona has been filling your head with stories of glorious revolution."

"Not really. She just has a real bug about the companies."

"Aye, well, she blames the companies for splitting up her parents. Her mother used to work for Govett; they handle a lot of the transport for the reclamation plants in town. Trouble was, Govett has an enlightened social policy; they move their personnel around every five years so they don't get stale or deadended. Her father, my Ken, there was no way he was going to leave the Highlands. How that woman didn't realize his commitment to the area I'll never understand." She sighed. "Then he went and got himself killed over in Glen Coe, a skiing accident. Joona was twelve at the time."

"And after that you brought her up here by yourself?"

"Aye. She refused to have anything to do with her mother. Stubborn, she is. Her mother helped us out with money and got her into the Prodi, but that's the only contact they've ever really had."

"I can see why she's so attached to this place."

Jackie poured some milk into a big mug, then used a

strainer to add the tea. "Not just to Fort William, it's our whole way of life she's devoted to."

He waved a hand round the kitchen with its age-darkened wooden furniture and scrubbed flagstone floor. Plates, cups and glasses stood along the shelves of a big Welsh dresser, probably all antiques. Copper pots and pans hung above the Aga, along with bundles of dried rosemary that gave off a mild scent. Despite the room's old-fashioned appearance, he could see a modern dishwasher and fridge built into the fitted cupboards. There had been a small cleaning robot in the lean-to outside. The only thing the kitchen really lacked was a texturalizer unit to make up basic food from raw protein cells. He suspected Jackie simply bought precomposed packets from town. A lot of people couldn't be bothered with home preparation these days. "You seem to be doing all right. I was worried I was going to be spending my holiday in a mud hut."

"I've a few interests, but the flock brings in enough to get by."

"How does that work?"

"There's a lot of land up here that they can't plant their damn forests on, you know. So we still have mountain sheep, and shepherds, and even sheep dogs. That part of our lives is the same as it has been for centuries."

He frowned. "You don't like the forests?"

"Oh, I don't mind them. But there's a difference between restoration and *uberretrogression.* These days if the ecological agency finds a clear piece of land bigger than a patio they want to plant a tree on it. It's a direct continuation of the Greenwave policy that came in after protein cells were developed. The old radical Greens saw that as their chance to finally repair the damage that farming had done. It's all a load of bull. Farmers were good for the countryside; they took care of their land. They had to, they depended on it. And I swear there was never so much woodland in Europe before, no matter how far back you go into prehistory to try to justify today's acreage. What we've got now is no more natural than the intensive arable farming that went on in the second half of the twentieth century and first half of the twenty-first."

"What I've seen of it looks magnificent."

"It certainly does, aye. Though you've no idea how many walkers get lost around here every year. And that includes the ones that have full navigation and communications functions in their bracelet pearls. Idiots, every last one of them. Our rescue teams are busy practically the whole year round. And we reckon to lose at least fifty sheep in the trees each season. There's supposed to be fencing, but even the robots can't keep it all maintained."

"Don't forget the wolves," Joona said. She came into the kitchen wearing a baggy blue robe, a big green towel wrapped around her hair. She sat next to Lawrence and gave him a quick kiss on the cheek. "They take dozens of sheep each year."

"Aye," Jackie said ruefully as she poured tea for her granddaughter. "Another species reintroduced courtesy of the environmental agency. As if we didn't have enough to contend with up here. But we get by. There's a fair weight of fleeces collected each year, and they keep the likes of me busy."

"You turn the fleeces into wool here?"

"Not directly. There's a couple of small local mills, cooperatives that we keep going; they wash the wool and weave it into yarn. Then it gets sent out to me and all the other crofters that are left up here. I knit it into sweaters like that one you're wearing. There's some that do blankets, and ponchos, and hats, and gloves. All sorts, really."

Lawrence looked down at his sweater, feeling it. "You knit these?"

Jackie laughed. "We're not Luddites, you know. I design

the patterns. I've an old barn down at the end of the garden where I've got three cybernetic knitting machines. They do the hard work. I know how to maintain them, mind. I'm quite handy with an Allen key and diagnostic program, I'll have you know."

"All the tourists buy them," Joona said. "Natural wool sweaters sell for a premium. Those damn factory texturalizers can't get the artificial fibers quite right, they just feel wrong. And Gran's designs are best of all."

"So where's the problem?" Lawrence asked. "You're doing what you want, and the rest of society appreciates that."

"The corporations and district parliament tolerate us because there are so few crofters," Joona said. Her earlier good humor had faded. "They wouldn't welcome too many of their kind taking up our lifestyle."

"Now don't go getting her started, young Lawrence," Jackie said. "And you, my lady, can forget about politics for an evening. I get quite enough of that at the association meetings. Boring old farts, they are. So, Lawrence, were you really born on another planet?"

The evening that followed was one of the most pleasant he'd had for a long time. There was no pressure on him at all, no worry. He didn't have to go out hunting girls or drink. It was, he thought, what a real family evening should be like. Nothing like the ones he'd endured in his own home back on Amethi, not forced and littered with expectation. More like the ones he would have wanted his own family to enjoy, the ones they would have enjoyed if things had worked out between himself and Roselyn.

He gave Joona a quick, guilty look. But she just smiled at him. She was helping Jackie make the pasta for supper.

"Traditional Scottish spaghetti," Jackie had announced. They both laughed when he nodded eagerly and said: "Great."

They turned down his offer of help, for which he was quietly thankful. He was left to stroke a huge black cat called Samson, while they busied themselves up at the long counter. A bewildering variety of ingredients were produced from big earthenware pots that were fitted with wide cork lids. The bolognese was mixed, cooked, tasted, remixed.

He did make himself useful lighting the log-burning stove in the parlor. It was soon roaring away, throwing out such a heat he had to take his borrowed sweater off. Jackie produced a malt for supper, which he had to water down to drink.

The spare bedroom—with the double bed—had an uneven floor. When he walked across it cautiously he realized the oak boards were so old they'd hardened into something approaching steel. They creaked occasionally, but they were totally solid. There was no quilt on the bed, only sheets and blankets, which he was dubious about. But the blankets were obviously produced by Jackie and her fellow crofters, brightly colored with a thick weave, so he expected they'd be warm enough. A single light fitting hung from the low ceiling, its cone a lambent yellow, casting mellow shadows. Wind soughed stealthily round the cottage's gable end; he could hear the trees rustling around the garden.

He grinned expectantly at Joona after she shut the door, and hurriedly started pulling his clothes off. Her own movements as she undid the buttons of her blouse were hesitant, which he took as modesty. Which was arousing. By the time she'd finished undressing he was already waiting for her on the bed, determined that, finally, tonight should be good fun.

"Are we having the lights on or off?" he asked.

A troubled expression fluttered briefly over her face. "Off." There was an unspoken *of course.* She flicked the switch by the door. The faintest moonlight seeping through the curtains allowed him to see her as a dark, flowing shape as she moved toward him. The bedsprings bent and shifted as she climbed on.

Lawrence reached for her immediately, sliding his hands over her body. He cupped her small breasts and began teasing the nipples with his fingers. He licked at her neck, her shoulders, her face. Her breathing quickened and they kissed, his mouth smothering hers.

It wasn't that she didn't respond, exactly. She just wasn't as active as the girls he was used to romping with. He took that as his cue to start whispering suggestions and compliments, telling her of the acts and positions he wanted from her, promising how marvelous she would be performing them. Silently, Joona followed his directions.

Lawrence woke to the sound of some deranged bird being throttled very noisily just outside the bedroom window. Even the old peacocks back home never made so much racket.

At least the night's wind and rain had stopped. Daylight fluoresced the curtains a radiant jade.

Joona was sitting up with her back resting on a mound of pillows. A microsol tube was dangling loosely from her fingers, just like a reefer. She wasn't looking at anything within the room.

He wondered if he should say something about it. Sure, he liked a drink himself. But only when he was out for a good time. Her habit seemed to be on the wrong side of casual.

He settled for stretching elaborately and giving her a broad smile. Truly, there was nothing better than waking up in bed with a naked girl after a night of hot sex. He could feel his erection stirring already at the sight of her little breasts. "Morning," he said, and there was a lot of happy lechery in his voice.

Her focus came back inside the bedroom walls. "Now do that to yourself." Her voice was as calm and dense as the loch outside. "That's what you said."

"I, er . . ."

"The only time I've ever heard someone say that before was in a porno."

"Ah. Well, it just seemed right. Then." His face was hot as he tried to remember exactly what he had asked her.

"Some of those things you had me do; I don't even know the names for them."

Lawrence wanted to wake up. Now, please. This was not the way it was supposed to be the morning after. A few bashful grins exchanged when you're off-guard and reminiscing, silent acknowledgment how you both got carried away in the heat of it all, but as we're civilized folk we won't actually mention it. Certainly we don't talk details out loud.

"It's never been that way for me before," she continued. "You were so demanding."

"You . . . Why didn't you say if you didn't like it?"

"I didn't dislike it. You're my man. We have to meet on that level as well. I wasn't ready for so much at once."

You're my man. What kind of thing was that to say? Hell, this was excruciating. He hadn't a clue what to say. Any normal girl would tell him outright if he'd gone too far. A simple *no* would have sufficed. He wasn't an animal, he respected other people. "Sorry," he mumbled. And that just came out like he was sulking.

"I felt left out," she said. "That's what hurt me the most. You were having this fantastic time with me, with my body. And I played no part in it."

It was an effort not to put his hands over his ears. He just wanted her to shut up, which was the absolute last thing he could ask right now. Guilt verged toward being a physical pain. He'd been so proud of himself during their lovemaking. And he thought he'd roused her as well. "You should have said. You didn't say anything." Even to his own ears that sounded desperate and defensive.

She put a hand on his arm. "Of course not."

What? He didn't get it, he really didn't. He eyed the microsol again, suspicions bubbling through the turmoil of thoughts. "We won't do anything like that again. Okay?"

"That will be denial. Which is wrong and stupid, and would mess us up. The whole time, I'd just be thinking of what you really want to do to me." Her voice was the kind of sharp monotone used by prosecution lawyers.

Actually, what he really wanted to do right then was get out. Out of bed, put his clothes on, and walk back to Fort William where there'd be a train back to the real world. But he didn't want to leave her. Not just from the extra guilt he'd suffer from running away after last night. There had been good times in the last few days, times when they'd connected, times when they'd cared about each other. That was something that hadn't happened to him since Roselyn.

And didn't all couples have problems? Admittedly not quite as raw as this . . . "It won't be denial," he said slowly. "It'll be inclusion. Sex should be for both of us." *Hey, fast thinker, Lawrence.* It was a good block. She'd obviously accessed way too many self-help pop psychology manuals.

"Yes," she said seriously. "Yes, it would, wouldn't it? We must discuss what we are going to do first. That way we'll know each other better."

He managed not to shudder at the prospect. Sex should be spontaneous and fun, not analyzed clinically before. But if it meant ending this conversation . . . "That's that, then." He leaned forward and gave her a quick, awkward kiss.

"Do you want to start now? We could do one of last night's positions again, if you tell me which one."

"No. I think, er, breakfast is good for me right now." *It's not cowardice,* he told himself, *it's just polite and practical.*

Lawrence had a distinct sensation of déjà vu when they walked into the kitchen. Joona had become clingy again, laughing and smiling, giving him a quick kiss every minute. Touching him for reassurance that he was still there.

He suddenly wondered if the family were Catholics. Roselyn had always said nobody could beat orthodox Catholics when it came to guilt from the enjoyment of sex.

Forget about Roselyn, he told himself firmly. He kissed Joona back and received a bright adoring smile.

"Oh, you two," Jackie chided with a smile. "Cover your eyes," she told Samson.

It was a sunny morning, and when Lawrence accessed the forecast he was assured of clear skies for the rest of the day. They cycled into town, though as soon as they emerged from the woodland around the cottage Lawrence jammed the brakes on so hard he nearly skidded his wheels out from under. Ben Nevis was directly ahead, presiding over a quarter of the skyline. Its peak was still covered in snow, which broke up into jagged ribbons over the massive north-facing ridges of gray-brown rock. Long ribbons of glistening water slicked the near-vertical face. At the base of the rock, scree had spread outward like an invasive tide across the grassy slope.

"Now that is impressive," Lawrence said, and meant it. The sun was shining off the snow, making him squint against the glare. He was intimidated and challenged by the scale of the damn thing, wanting to know what it would be like to stand up there and look down. "You must be able to see half of Scotland from up there."

"We'll take a walk up it if you'd like."

"You're kidding. I'd never get up there without a muscle skeleton. Those cliffs look lethal even for technical climbers, and that scree is damn steep as well."

"You don't go up from this side, silly. There's a walkers' path that leads up from the glen. It only takes a few hours."

"Yeah, right." He gave the mountain a hard look before getting back on his bike.

Jackie had given them a list of things she needed from the

town. He suspected it was makework, allowing them to wander around together. He didn't mind.

"Nice town," he said as they walked along the pedestrianized main street. The buildings with their little shops on the ground floor either dated back four centuries, or were good replicas.

"It is now," she said. "The council has cleaned up and refurbished a lot of our old important buildings. There's enough money for that kind of urban regeneration now."

"Hey, does that mean you finally agree that the big companies are good for the economy? They're the ones who generate that money in the first place."

"I knew you'd approve. Fort William's very ordered now it's surrendered to the uniculture. Just how you like things to be."

"All this is a bad thing? I've seen towns in a much worse state than this and I've only been on Earth for five years."

They reached the southern end of the main street, where the main road had been diverted along the side of the loch. The rest of the town was composed almost entirely of houses, spreading back up the shallow slope from the water for over a quarter of a mile. Each one sat in its own lush garden, large enough for several trees. From where they were standing the intense verdant green of new silver birch leaves vied with the cotton candy swarms of cherry blossom to produce the most luminous array. Daffodils and tulips had colonized most of the lawns, speckling the grass with masses of yellow and red flowers.

"Oh, no," Joona said quietly. "This is a lovely place to live, even in winter. All these fine houses are well built and well insulated, and if you're ever invited inside one, tastefully furnished, too. Something like ninety-five percent of the town's housing was built in the last two centuries. They leveled the old housing estates that were put up before the building industry started using robotics; those kind of high-density houses were never made to last—not like Gran's cottage. So now we've got one house where there used to be two or three."

"Money, again."

"Yes. But that's not the only factor. The town's population is down almost twenty-five percent since the twentieth century."

"I thought the rural population has been declining ever since the start of the Industrial Revolution."

"It has. But I don't mean that. The total population is down, and still falling. That's why you can have bigger houses and gardens these days without putting pressure on the environment."

"Not having farmland helps, too, I'd imagine."

"Yes. It all fits together neatly, don't you think?"

The way she said it betrayed how scornful she was. He didn't reply.

Joona led him into a quiet café on the main street. The young waitress behind the counter greeted her warmly, and the two of them had a few quiet words. Lawrence found a free table near the window. Their hot chocolate arrived a minute later, along with some fresh-baked muffins. A small paper bag was passed to Joona, who vanished it into her coat pocket. She put three EZ tens on the table. There was no change.

Lawrence blew across the top of his mug. "Does Jackie know how much of that stuff you use?"

"You mean, does she care? Half of this is for her, Lawrence. Our kind of lifestyle has always included narcs of one kind or another."

"I still think you should ease off a bit."

Her blank face clicked on, as if she'd already inhaled a microsol tube. "Thank you for the interest. It's not necessary."

* * *

That night they did talk about what they would do in bed. It wasn't as bad as he was anticipating. Actually, it was quite arousing, almost as if he was her tutor, a reasonable enough male fantasy. At least it put their relationship back on what he considered a more even footing.

The next few days were spent in and around Fort William. They visited the theater: twice to watch live plays, once to see a cinema screening of Cameron's *Titanic.* Lawrence helped Jackie out around the garden, which had suffered the usual winter's worth of neglect and damage. A few broken branches needed sawing off. Fenceposts had snapped. He spent an entire morning stripping down and cleaning her ancient gardening robot, trying to get the rusty mechanical components to run smoothly again. The blades on the mower attachment's cylinder had to be taken to one of the shops in town for sharpening. Another morning was spent helping out with the knitting machines. They were housed in a barn at the end of the garden, a stone building as old as the cottage, with an open truss roof that was elegant in its simplicity, sturdy beams of thick untreated oak holding up the thin lathing that the slates were nailed to. But it was dry inside, if not terribly warm. The three machines clattered away enthusiastically, slinging out their finished sweaters every few minutes. They changed over the bales and refilled the dye chambers, then packed the finished sweaters into boxes ready for collection.

At the start of his second week, they climbed the Ben as Joona had promised. It was a short bike ride from the cottage to the visitor center perched on the banks of the River Nevis, which meant they were among the first to arrive that morning. They locked the bikes into the rack, then pulled on their walking boots.

The trek was a lot easier than he was expecting, just as she said it would be. Once they crossed the small bridge by the visitor's center, they picked up a simple track running along the side of the hill, heading steadily upward. It was paved with rough stone, with neat steps cut in on the steepest parts, which seemed slightly incongruous for a supposed wilderness walk. Joona told him that the Scottish Environment Agency had to maintain it at this standard to prevent erosion. It had to cope with thousands of walkers during the course of the year.

As they climbed he could see more and more of the glen with its astonishingly green vegetation stretching away below him. The path had already started to lead through the huge swath of bright, fresh bracken that had sprung up along this section of the hill. Small wooden bridges took them over narrow fissures.

It wasn't long before the path curved around into a deep, grassy cleft with a stream at its head, white water coursing noisily through the rocky gully it'd cut into the slope. They walked toward the water, then suddenly switched back to climb the steepening slope at a reasonable angle. Another turn brought them out to a marshy saddle with its own lochan of dead, peaty water. Lawrence took a look at the vast scree-smothered slopes looming above them and sighed in mild dismay. He still couldn't see the actual top of the mountain yet. They stopped for a while above the lochan to drink some tea from their flasks and put on another layer of clothing. It was getting colder with every meter they ascended. The air below the cleft had been perfectly clear, giving them grand views across the beautiful Highland peaks. Here the mountain was accosted by thin strands of mist pushed along by the constant wind, reducing visibility.

For the next stage the path zigzagged up a steepening scree-covered slope. The tufts of grass and heather became less and less frequent until it was just stone and raw soil under their feet. Each sharp turn in the path was marked out by a cairn. Slush began to build up on Lawrence's boots as he trudged onward. Patches of snow appeared more frequently on either side of the path. The mist was thickening. He couldn't see the bottom of the glen anymore.

"It's so clean up here," he said as they stopped for another rest. "I love it."

Joona eased herself onto a boulder and pulled the flask out of her backpack. "I thought your whole planet was clean."

"It is. But that's a different sort of clean. I was expecting Scotland to be different. You had so much heavy industry around here, I thought there'd be more . . . I don't know, remnants. Streams that are half rust from all the old machines dumped in the lochs, mounds of slurry out of abandoned coal mines, that sort of thing."

"Scotland's heavy industry was mostly down south. Besides, you saw the reclamation plants outside town; they're busy little bees."

"Yeah." He'd noticed them on the first morning when they cycled into town, gently disturbing the landscape on the other side of the River Lochy from Benavie: underground factories strangely reminiscent of the chemical plant on Floyd, long flat-topped mounds covered in lush grass. This time there were no heat exchange pillars on top, only rows of black vents that could have easily been overlooked. The real giveaway to how much industry was hidden below the earth were the pipes running down the rugged side of Creag Chail above them: twenty wide concrete tubes that emerged from the mountainside a couple of hundred meters up only to vanish into the ground behind the mounds. They carried enough water down from the Highlands to power the whole reclamation site.

Joona told him the site had grown up from a single aluminum plant that had been built there in the twentieth century to take advantage of the hydro power. As the Brussels parliament of that time slowly started to introduce stricter legislation governing recycling, the plant had expanded, with subsidiaries springing up to reclaim other types of materials.

Now almost all of Earth's consumer products were designed so that at the end of their lives they could be broken down into their constituent elements, which were then fed back into the start of the manufacturing cycle.

Fort William handled just about every breakdown procedure, from the original aluminum cans to electronic components, glass to concrete, and the whole spectrum of polymers. One of the most modern facilities of its kind in the world, it employed everything from smelters, catalytic crackers and v-written enzyme digestion right up to ionic fission for toxics. Junk from all over Europe arrived by train, ship and canal barge to be sorted and extracted.

"I guess there's not much pollution these days," he said.

"Not in the industrialized nations, no, not after the Greenwave. Even the nonindustrial regions like Africa and Southern Eurasia are relatively clean as well. It's not in the corporate interest to foul up their future territory."

"Joona, you've got to stop looking at everything so cynically. Just because people have different goals from yours doesn't automatically make them evil."

"Really?" She gestured down the glen. "One day if they have their way the whole world will be just like this. Everyone living in their big cozy house in their tidy suburban estate."

"Yeah, terrible. Imagine that, everybody having to put up with low crime and good medical benefits."

"But no freedom. No difference. Just the corporations and their uniculture."

"That's bull," he said. "People have been complaining about multinational companies and creeping globalism since the middle of the twentieth century. The world still looks pretty varied to me."

"Superficially it is. But the underlying trend is unification. National economies are becoming identical, and it's all due to the corporations."

"Fine by me. I have no objection to them investing money in poor countries and spreading their manufacturing base. It gives everyone a chance to buy a stake."

"There is no chance. If you want to get any kind of decent job, then you have to join up. And once you're in, so's your family."

"Your family benefits from the stake, yes. You get a say in what school your kids go to, everybody receives medical benefits, there's a good pension at the end. Stakeholding is a great social development. It involves, motivates and rewards."

"It destroys individuality."

"Taking a stake is the choice of the individual."

"A forced choice."

"Life choices usually are. Look at me, I took my stake in Z-B because it's the only one with a decent policy on interstellar flight. Other companies have different priorities, the choice is endless."

Joona shook her head wearily. "I will never sell myself out for a fancy house and full medical coverage."

She was rejecting everything her mother was a part of, he realized. "Then I'm happy for you. Your principles make you what you are. And that I like."

She gave him a brief grin, and sat up. "Come on, not much farther now."

After the last zigzag in the path, they were walking over a vast field of loose stone. The route ahead was easy enough to see through the thickening mist; a thousand footsteps had worn the thick covering of snow down to a compacted slushy brown trail. As they moved forward, the mist became patchy, with the wind propelling it along. Nothing else seemed to change. The path was the same ahead as it was behind. Occasionally, large boulders poked up through the snow. Other people on the path would appear as dark shadows in the brightly lit vapor before resolving into focus.

Abruptly, the ground fell away. They were standing at the top of a cliff. The base was invisible in the mist below.

"Almost there," Joona said cheerfully.

A few hundred meters brought them to the top of the Ben. Lawrence held back on his disappointment. It was just a flat uninspiring patch of snow-covered ground close to another section of the cliff. The mist meant they couldn't see more than fifty meters. Over the centuries there had been several structures built around the concrete survey marker that was the absolute pinnacle. Broken walls of stone protruded from the snow, outlining these ambitions of the past. Not one of them had a roof. The only intact building was a rescue center, a modern composite igloo that had a red cross on the side, and a small aerial protruding from the top. It was almost buried by snow. Lawrence spotted several small flat stones that had been laid carefully against it. When he bent to examine one he saw an inscription had been scratched on the surface. A couple of lines of poetry that he didn't recognize, then a name, and two dates, ninety-seven years apart.

"Not a bad place to be remembered in," he muttered.

They made their way over to the survey marker and climbed up it, just so they could say they had actually reached the top. The mist was starting to thin out when they made their way over to one of the collapsed walls where other walkers were huddled. Once they hunched down out of the wind they opened their lunchboxes. Jackie had packed them some thick beef sandwiches. Lawrence wasn't particularly hungry, the cold had taken his appetite away, but he munched away at one of them anyway.

Then the mist cleared completely and he stood up to look at the view. "Oh wow." You really could see half of Scotland. Mountains and glens and forests stretched away into a hazy horizon. Long tracts of water sparkled dazzlingly in the brilliant sunshine. He stared at it in a mixture of wonder and hopelessness. How could Amethi ever hope to achieve vistas such as this? All that effort . . .

Joona cozied up beside him. "When it's really clear you can see Ireland."

"Yeah? Have you? Or is that just a local myth for gullible tourists?"

She slapped at him playfully. "I have seen it. Once. A few years back. I don't come up every day, you know."

The sun was bright enough to make him squint. And the wind was bringing tears to his eyes.

"Stay here."

She said it so quietly he thought he was mistaken at first. Then he saw her expression. "Joona . . . you know I can't."

"Yes, you can. We're that new society you're looking for, Lawrence. This is where you can have your fresh start. Down there in the glens are free people building their own lives and doing what they want with them."

"No." He said it as gently as he could. "This is not for me. I've loved being here, especially with you, but I have to go back eventually. I'm too different."

"You're not," she insisted. "Your precious officer college rejected you, and you found us, me. It's inevitable. You must see that."

It was that earnestness of hers again. Sometimes it made her the strongest character he'd ever known. But there were occasions when it betrayed a worrying degree of vulnerability. She really didn't understand what went on around her, insisting on her own interpretations of events.

"Don't do this," he said. "We've had a great time together, and there's still another week to go."

"You have to stay, Lawrence. I love you."

"Stop it. We've only been together a few days."

"But don't you see how well you've fitted in here?"

"I'm a guest," he said in exasperation. "What the hell could I do here? Carve statues of Nessie for tourists?"

"You're a part of our lives. You lived with us. You made love to me. You even ate real food. All of this you welcomed."

"Joona, I stayed a few days. We're having a holiday romance, that's—" His subconscious sent out a disconcerted warning, almost like a physical jolt. "What do you mean I ate real food?"

"Real food." Her entreating smile never wavered. "Vegetables grown from the soil."

"Oh shit!" His hand came up to cup his mouth, and he stared aghast at the half-eaten sandwich. "Is this—is this?" He couldn't even bring himself to ask it. Not *that.* In his schooldays he'd always been revolted by the notion his ancestors had been forced to farm so they could eat—all the history class had.

"Aberdeen Angus beef," she said. "The best there is."

"Is it real?" he yelled.

"Well, yes," she said, oblivious to his horror. "Old Billy Stirling keeps a herd of them down past Onich. He slaughters a couple every month. There's quite a demand for it from the crofters. Gran always gets her meat from him."

Lawrence's legs gave way, pitching him forward. He vomited onto the snow, his whole stomach heaving violently. The spasms lasted for ages. Even when there was nothing left to bring up, his muscles were trying to squeeze out the last drops of acidic juices.

Finally, when he was through, he was on all fours with his limbs shaking unsteadily. He scooped up some snow and wiped it across his forehead, then tried to chew it to take the taste from his mouth.

"What's the matter?" Joona asked.

"What?" He looked up to see her frowning in concern. Several other walkers had come over to see if they needed help. "Did you say what's the matter?"

"Yes." She looked confused.

"You gave me a piece of a fucking animal to eat, and you ask me what the fucking matter is. An animal! A living creature. You're fucking crazy, that's my problem. You fucking . . . oh hell. How long have I been eating this shit?"

Her expression became pained. "You've lived our life with us, Lawrence. What did you think we ate?"

"Fuck it." He thought he was going to vomit again. The muscle reflex was certainly there, the inside of his mouth sopping wet, but by now there really was nothing left to bring up. He smeared some more snow against his head and slowly rose to his feet.

"Lawrence." Her voice was urgent, becoming shrill. She held out a hand to steady him.

He twisted from her reach. "Stay away from me. You hear? Stay away, for fuck's sake." He stumbled away from her, then managed to get his legs under control and picked up speed. Joona took a few paces toward him. "Lawrence!" she cried. "Lawrence, I love you. You can't go."

He started jogging down the track of compacted snow. "Don't call. Don't come after me. It's over." He stopped and turned to face her. "Over! Do you understand that? It's over. And I am leaving." He glared at their small bemused audience. "Thank you, and good-bye."

By now he'd regained almost full coordination. He ran. Ran down to the zigzag section of the path. Slowed slightly as he pounded over the slippery loose rocks and scree. Kept on jogging until he was long past the stream running down the cleft. Even then, when he was exhausted and dizzy from effort and shock, he kept moving fast along the final descent.

He took his bicycle from the rack at the visitor center, and pedaled to the train station in town. From there he caught the late-afternoon train to Glasgow. Changed for Edinburgh Waverley, where he could get an express to Paris. He had to wait two days in the French capital until there was a seat on a Z-B flight back to Cairns. He spent most of it drunk, moving from café to café in the old artists' quarter, trying to blot out the memory of the madwoman and everything he'd eaten at the cottage.

He never tried to contact Joona again. There was never any message from her, either.

CHAPTER TWELVE

EBREY ZHANG HAD FINALLY IMPOSED A BAN ON Z-B PERSONNEL leaving their barracks after eight o'clock in the evening. It had been yet another fight in a marina nightclub, resulting in another squaddie with serious stab wounds, that had eventually forced his hand. He knew it was going to be unpopular and bad for morale. But he didn't have any choice. No matter how well supervised the platoons were (and his first diktat had been that they had to be accompanied by their NCOs when they went out), there was always a disturbance of some kind, invariably resulting in injuries, and property damage, and worsening public relations—not, he was the first to admit, that they could get much worse.

So he'd called a staff meeting and announced his decision. Predictably enough, the officers had voiced their concerns. He'd said he understood, and that as compensation they could increase the amount of drink available in the bars of the hotels they'd taken over as barracks. Platoons on night patrol, though, were now under orders to arrest any Z-B personnel they found outside.

That one order had completely wrecked Hal Grabowski's

life. Memu Bay was bad enough when he was allowed to get out and blow off steam every few days. But this was like the end of the world. Bringing more beer into the hotel bar was no use at all. Hal had never been one for getting wildly drunk every night, and certainly it was no substitute for getting out. He hated being in the same building the whole time, with the same people, bitching about the same things, eating the same menu day after fucking day. The barracks hotel was worse than prison.

But he might just have managed to tolerate that if it hadn't been for the one thing completely absent from his life. What he wanted most, as he told everyone who would listen, was pussy. And lots of it. Their current existence was like being fucking tortured. Every day when he was out on patrol, the streets would be full of girls wearing next to nothing in the bright hot sunlight. Laughing, smiling, having a good time right in front of him. He wasn't supposed to say anything to them: the Skin meant he couldn't even smile on the off chance he earned a smile back. And now his single opportunity to get to meet a girl had been snatched away. The sarge had been sympathetic, but he said he couldn't bend the rules for anyone. Sorry. Hal thought his head was going to explode; right after his dick. He didn't even care about the order, that was nothing. The fact that it had to be broken was obvious. His only problem was how.

He had to wait until eleven o'clock when the hotel's main kitchen had finished for the night and the staff had all gone home. A squaddie from Wagner's platoon, a guy his own age and with a similar problem, had told him about the route out. The kitchen had a door that opened into a small backyard. There was only one security sensor covering the area, a motion tracker wired straight into the AS. Armed with the codes, which the squaddie had also provided, Hal spent half an hour that afternoon infiltrating the sensor's management program. He hadn't shut the little unit down; that would have put everyone's life on the line. Instead, he'd altered the diagnostic routine, making it repeat two hundred times instead of its usual once; the check that normally took three seconds now took over three minutes, with the sensor itself inactive while its support circuitry was analyzed. The diagnostic automatically ran at twelve minutes past the hour, every hour. His alteration would operate for that night only, then wipe itself after 3:00 A.M., allowing the program to revert to its default setting.

There was nobody left in the kitchen. He made his way past the stainless-steel benches and waited by the back door until the clock function on his bracelet pearl read twelve past eleven. He opened the door and stepped outside. There was no alarm. The yard measured three meters by fifteen: it was used as a store, with empty boxes and beer barrels stacked up against the walls ready for collection. Hal hurried down to the far end and scrambled up the boxes, to peer over the top of the wall. Nothing moved in the dark alley on the other side. He swung over and dropped down.

His luck was in. A taxi was parked on the side of the road twenty meters from the alley. The driver was reading something on his media card; but the yellow *vacant* light was on. Hal opened the back door and sat down.

The driver looked up, examining Hal in the mirror. "Where to, sir?"

"Marina district." Hal pulled his collar up, hoping it hid the valves on his neck.

"Sure thing." The driver spoke to the car's AS and they pulled away from the curb. His hands rested lightly on the wheel, allowing the AS to steer.

"Hey, er, you know this town pretty well?" Hal asked.

"Sure. I was born here. Took a trip to Durrell once. Didn't like it."

"I've not been out much recently, not since I got here. That is, not by myself, you know. I haven't gotten to meet too many people. You see where I'm coming from here, man?"

"Guess so. You want to meet people. Marina's the best place for that."

"Right. But I want to meet a girl. I want to be certain I meet a girl. You know anywhere a guy like me with some money in his pocket could be real certain about that?"

The driver grinned into the mirror. "Hey, relax, buddy. We're all human here. I know a decent cathouse that'll take care of you." He disengaged the AS and began steering manually.

The house was in one of Memu Bay's better residential streets, a big three-story building set back from the pavement by a narrow front garden. Hal opened the gate in the cast-iron fence and gave the taxi driver a glance. The driver gave him a thumbs-up and drove off. There was no one else on the street. "Fuck it," Hal grunted. He went up the three steps to the glossy black front door and rang the brass bell.

It was opened by a middle-aged woman in a glittery red cocktail dress. She had just too much makeup on for her to be a respectable house owner. At least the driver hadn't been jerking him around. Hal grinned. "Evening, ma'am."

She pursed her lips, looking him slowly up and down. Her gaze lingered on the valves now jutting over the top of his collar. "Can I help you?"

"I hope so. I'm looking for some company tonight."

She took a half pace forward and looked both ways along the street. "Are you recording, Officer?"

"I'm off duty, and getting my ass busted is the last thing I need right now."

"Very well." She gestured him in. "We have to be careful, you understand."

"Yes, ma'am. Same in my hometown." The hall had a marble tile floor and a high ceiling. A big crystal chandelier hung from a long brass chain, shedding a bright light. It could have been an ordinary house, except that someone had draped gauzy white fabric from just about every surface, giving it an odd chintzy appearance. A broad staircase led up from the rear of the hall, curving into a landing that ran around the second floor. Two girls wearing simple white cotton dresses with lace-up fronts leaned on the rail, peering down at him. One of them winked. It was all he could do not to wolf whistle at them. This was the right place all right. Classy.

"Humm." The tip of the madam's tongue licked at her lavender-colored lips. "We've never had an alien here before."

There was a nasty moment when he thought she was going to be the one. Not that he would have minded too much, but he was kind of looking for someone younger. Hal grinned roguishly. "I might be an alien, but I'm still compatible, ma'am."

"Before we go any further, I'm afraid there is the question of money. Which also has to be compatible." She told him a figure that made him hesitate. Goddamn locals, they knew he was desperate—but then everyone who came here was. Under her level gaze he pulled out a thick wad of bills and handed over most of them.

"Are there to be any unusual requests placed upon the lady in question?" the madam asked. "Understand, we can provide almost anything you ask for. But I have to be informed in advance. It will avoid any alarm and subsequent unpleasantness."

"No, I just like it kinda straight, you know. Nothing too weird."

"I see. And you are a young man. A virile man."

"Hey. You know. I keep in shape."

She arched an eyebrow suggestively. "I can see that. There are several of my ladies who probably have the stamina to keep up with you. Though certainly not all."

Hal knew he was grinning like a baboon. Didn't care. He was getting hard already.

"Micha, perhaps," the madam pondered. "Although she is very experienced. Perhaps that is off-putting to you?"

"Anyone who knows what they're doing in the sack is okay by me."

"Perhaps." The madam tapped a manicured finger to her lips, as if Hal were an exceptionally vexing problem. "Yes. I think Avril for you. She is very young, which is exciting, no?"

"Oh yeah." It took a lot of control not to yell.

"Very well. This way." The madam beckoned and started up the stairs. Hal followed hot on her heels. Both of the girls at the top of the stairs pouted at him as he went past.

The madam opened one of the doors around the landing. When he saw what was waiting for him, Hal almost shoved her out of the way. He couldn't believe the girl standing by the big bed inside. Back home she probably wouldn't even have been legal. Avril was lean and tanned, with fine shoulder-length chestnut hair framing a coy smile. She was wearing sports gear, very short running shorts and a Lycra halter top that was tight enough to hold up her pert little breasts and outline their nipples.

"Jesus H. Christ," Hal growled between his teeth.

The madam bowed slightly. "Until later." She closed the door.

Hal spent a long moment staring at Avril as his breath grew hotter, then moved purposefully across the room.

To start with, it was a standard missing-person report. Gemma Tivon waited for three hours past the time her husband usually arrived home from the night shift before trying to open a link to his bracelet pearl to ask where he'd gotten to. There was no reply; the datapool communications management AS reported there wasn't even a standby link to his bracelet. It was switched off. He never did that.

Gemma called the spaceport and asked if Dudley was working some unexpected overtime. The department supervisor told her no, then got the security people to check the parking lot and the gate log. Dudley Tivon's car wasn't in the lot. The gate log showed he'd signed out at seven minutes to six that morning, slightly earlier than usual.

Because they were conscientious employees, the spaceport immediately called the police and sent someone around to Gemma Tivon. The police accessed the local traffic regulation AS and used its log to track Dudley's car after it left the spaceport. As usual, he'd driven along the main highway back to the city; then his routine changed. He'd turned onto Durrell's outer beltway and continued on eastbound for another three junctions. After that he'd taken a minor road, then turned off that for an unmonitored track leading through a forest. There was no record of the car coming out of the forest on any of the approach roads.

As the spaceport was pressing hard, the police sent a couple of patrol cars into the forest and dispatched a spotter helicopter. It took them two hours, but they eventually found Dudley's car under a big pine tree. The interior had been soaked with some inflammable liquid and set alight. A forensic team was immediately sent to the scene, along with three more cars.

The Zantiu-Braun AS that monitored all capital zone police activity tagged the case for attention anyway because of the strange circumstances. The Third Fleet intelligence agency AS also tagged it, but for a slightly different reason: Dudley Tivon was connected with spaceflight, and Gemma was collateral.

Five minutes after the patrol car officer informed her dispatcher they'd found the car and it had been deliberately set alight, the relevant case datapackage was delivered to Simon Roderick's DNI from his personal AS.

"A cold trail, unfortunately," he said as Quan and Raines arrived in his office. "Over four hours now since the car was abandoned and torched."

"We can divert some of our helicopters to the forest to help with the search," Quan suggested.

"No," Simon told him firmly. "We don't initiate anything. I don't want us to draw attention to our interest. If the police want our assistance they can apply for it through the appropriate channels. In any case, simply finding poor old Tivon's body will prove little."

"Forensics might give us something."

"I doubt it. In fact I doubt we'll ever find the body. If our opponents have any sense, and as far as I'm concerned, they have plenty, the body won't even be in that forest. Besides, I'm not interested in how he died, only why."

"He'd delivered something to them and wasn't necessary anymore," Raines said. "Or he'd delivered something to the spaceport for them—a bomb, perhaps—and thought he was collecting his payoff."

"That's very crude for these people," Simon said. "In any case, you're overlooking his wife, Gemma. He isn't going to get involved in any venture that could jeopardize her. No, I think he was in the wrong place at the wrong time." He glanced at the two intelligence operatives. "Run up a week's timeline profile for me. Access every sensor log in the datapool, starting with last night and working backward from there. When you come back to me, I want to be able to watch his whole day, every single minute of it. And once you've done that, set up a secure link to the *Koribu* and correlate as much as you can from last night with skyscan."

It took them another five hours, but they were smiling when they returned to Simon's office. "We found it," Quan announced. He had the confident tone of every underling bringing good news to his boss. "Dudley was murdered, and that's only the start of it." His DNI routed the first set of files to the big wall-mounted pane opposite Simon's desk. A time-synchronized split image appeared. On the left was the recording from a camera overlooking the spaceplane parking apron. On the right was a skyscan picture of the same location.

Simon sat back in his seat watching as a man emerged from the bottom of the Xianti 5005's airstairs at the same time Dudley Tivon walked out of the maintenance hangar. On the left Dudley continued across the apron into another hangar. On the right, the two men confronted each other, and a second later Dudley was dead.

"I don't know how they did it," Braddock Raines said in admiration. "But there isn't a trace of software subversion anywhere in the spaceport's network. I even had one of our people go out there and physically pull the memory circuits for the cameras around the parking apron so we could go over them here. Nothing. So we know they can play that network like a maestro. Whatever they've got, it's damn impressive."

"Is that camera memory e-alpha protected?" Simon asked sharply.

"No, although it is protected by some excellent encryption. However, the backup memory in the AS is inside an e-alpha fortress," Raines said. "But we think that the subversion occurred at the camera itself, or at least its connection to the network. They had to have their own AS online to generate the false image in real-time. In itself, that's interesting. They subverted four cameras that we know of, and that takes up a lot of bandwidth. Our AS should have picked up that quantity of subversive dataflow within the spaceport network. The fact that it didn't is highly suggestive."

"You mean e-alpha is compromised?"

Raines screwed his face up, unwilling to make a commitment. "It is possible to do what they did without breaking through e-alpha forts. But it's difficult. Of course, so is subverting e-alpha. If they can't actually do that, their capability is close enough to make very little difference to us. The fact that their man got into the spaceplane is proof of that." He ran an earlier skyscan file, showing the intruder walking across the apron and straight into the airstair. "No record of any entry during the night," Raines commented as the figure approached the big delta-shaped spaceplane. "And there, see, when he arrives at the airstair he doesn't need to physically tamper with the security lock. The software's already been configured to admit him."

"You drew up a timeline for the intruder, of course," Simon said.

The two intelligence operatives swapped a mildly worried glance. "We tried. We couldn't even establish when he entered the maintenance hangar, let alone when he arrived at the spaceport. The only sensor data we can trust is from skyscan, and that's too limited to build up any kind of detailed profile."

Simon cursed quietly. There had been many times down the years that he'd suggested a greater satellite surveillance capacity during asset-realization campaigns. It had never moved past the proposal stage. If he was honest with himself, even he couldn't justify the expense of such coverage. He was just used to having that resource available. But Earth with its swarms of low-orbit satellites was unique. Out here, the best Z-B could offer its strategic security forces was enough satellites to provide constant coverage of the most strategic sites. Inevitably that meant the spaceport and the headquarters in the capital. The ground footprint did allow some overlap around each zone as the satellites orbited overhead, but not much. Looking at the fuzzy image of the intruder's head, he was thankful that the small skyscan flotilla had gone unnoticed by the resistance group. So far, it was turning into Simon's sole advantage.

"You must have tracked Tivon's car leaving the spaceport."

"Yes," Raines said, happy to appear positive again. He directed the requisite files onto the pane. Skyscan showed a small cargo robot trundle along the deserted parking lot at five o'clock. The intruder was walking toward Tivon's car from a completely different direction. They arrived at it simultaneously. The intruder opened the trunk and the robot deposited a sealed crate into it before rolling off along its route. Elapsed time was five seconds. The robot had barely halted.

Simon watched the intruder close the trunk and get into the car.

"He sat in there for forty minutes before he left," Raines said with respect. "Driving out at five might have drawn attention to the car. So he waited and left a few minutes ahead of Tivon's usual time. How's that for keeping your head?"

Simon kept staring at the pane. "Car profile?"

"It left the skyscan footprint twelve kilometers beyond the spaceport. He just kept driving along the main highway without stopping."

"Do you have an image of his face?"

"Not really. He tended not to look up; I'd guess he's quite surveillance-smart." A picture appeared on the pane, looking down on the intruder in the parking lot. He had tipped his head back slightly to study something a little higher than he was. It expanded into a collage of blurred pixels the size of golf balls. "And that's with AS enhancement. It drew us several possibles from that." Five high-resolution faces appeared on the pane, each time a man under thirty, and all with the same general bland features.

"They won't be any use." There were disappointingly few

distinguishing features in the extrapolations, Simon thought, even the characters in AS-generated i-soaps were more real than this. "And he's not even wearing a hat," Simon said thoughtfully. He gave Adul Quan a pointed look. "Remember the last time we had an incident like this?"

"The bar in Kuranda," Quan said. "Just before we left Earth. Do you think they're related?"

"Difficult to see how. Anyone who wants to keep below our horizon must invariably use the same tradecraft." He grimaced at the row of five impassive faces, impressed at the audacity and resourcefulness of the intruder. In all the campaigns he'd been on, Simon had never encountered a threat quite like this one. He couldn't help wondering why Thallspring of all places should produce this style of quietly lethal resistance movement. "No, I'm not quite ready to believe in interstellar conspiracies. We need to focus on the immediate threat. How long was he in the Xianti for?"

"Seventeen minutes," Quan said.

"Long enough for anything. Has it flown today?"

"Yes, sir. It took a cargo up to the *Norvelle* this morning. Landed at thirteen-thirty-five. No problems filed with flight control. It's undergoing standard preflight checks and refueling ready for another cargo run, scheduled for eighteen-twenty." Quan looked directly at Simon. "Do you want us to stop it?"

"No. Which starship is it scheduled to dock with this time?"

"The *Chion*, sir."

"Change it to the *Norvelle* again. If it has taken anything hostile up there, I want any possible contamination to be as restricted as possible."

"Yes, sir."

"After it's unloaded its cargo I want a mechanical fault declared, something that entails its docking in the *Norvelle*'s maintenance bay. Braddock, I want you to get up to the

Norvelle on the following flight. You are to carry instructions from me directly to the captain. Once you're up there, I want you heading a small army of the best technicians we've got. You're to rip that damn spaceplane apart, take it down to its individual molecular strings if you have to. But I want to know exactly what our friend was doing in there, and what he's left behind. Understand?"

"Yes, sir."

"Very well. And from now on we're to operate under the assumption that e-alpha has been compromised—that includes our communications. The one thing we cannot afford now is to tip our hand to them."

The police on Thallspring, as with police forces on all the human worlds, had very precise regulations on how to deal with every situation and crime their individual officers encountered. This body of knowledge had been painstakingly assembled over decades and was in a constant state of revision thanks to several factors such as legislation, failed court cases, successful court cases, devious lawyers, advances in forensics, pressure groups, previously botched procedures, human rights and human failings. Each officer had been trained to follow these procedures to the letter, especially for serious crimes. Cutting corners invariably jeopardized court proceedings.

So when the young girl came staggering into Memu Bay's marina police station at twenty-five minutes past two in the morning, weeping and screeching hysterically that she'd been raped, the desk sergeant knew exactly what had to be done. Detectives with specialist training were summoned, along with a female doctor. The victim was gently led to an interview suite by a female constable, and the whole event recorded.

Procedure insisted that a preliminary statement be taken as soon as possible. Ordinarily this was to ensure that if the alleged perpetrator(s) could be identified a patrol car could be dispatched immediately to the crime scene. A forensic team would also be dispatched to gather evidence.

This time, something unexpected occurred. The girl kept shouting: "He was an alien. I saw the things on his neck."

The detectives who had arrived to take the statement immediately called the precinct commissioner, who promptly called the mayor's office. That was where the second aberration slipped into the smooth running of the system, creating a great deal of anger and shock among the people dealing with it.

A lot of very senior staff from both Zantiu-Braun and the civil administration were woken up and advised what was happening down at the marina precinct station. From there another set of calls went out. The two lawyers regarded as Memu Bay's best were quickly retained by the victim's family (although they offered to waive their fee) and hurried to the precinct station. Inevitably, given the number of people involved at this stage, the media were alerted. All of the news companies respectfully withheld even the slightest rumor concerning the victim's identity from the datapool, though they did give her age as fifteen. What they made extremely clear was that an alien was the chief, and only, suspect.

Once the principal officials and the girl's distraught father had arrived, she was taken to a small examination room. In the presence of a lawyer, the detective in charge of the case and the Z-B legal representative, the doctor took samples of what the media referred to as "genetic evidence" of intercourse. Cameras also recorded her superficial bruising, grazes, torn clothing and swollen cheek. With that ordeal over, the nurse was finally allowed to treat her physical injuries.

The girl was sent home and assigned a social worker

trained in victim counseling. The precinct detectives would interview her in more detail once she'd had some time to recover.

Meanwhile the genetic samples were sent to Memu Bay's Medical Forensic Laboratory for immediate analysis, accompanied by the senior detective, the victim's lawyer, the police magistrate, the Z-B legal officer and a Z-B medical technician. The head of department herself had been called in to handle the analysis to make absolutely sure it went correctly. Even she was nervous as she placed a sample of the genetic evidence onto the scan array. It took the AS eight minutes to acquire the full DNA signature.

The detective first ran it through Memu Bay's central criminal register. No matches were found. The police magistrate then authorized a suspension of privacy warrant, which allowed the opening of the town's medical records for a comprehensive search. Again, no matches were found. The detective then formally requested the Z-B legal officer to run a check against their personnel files. Having no grounds to refuse, and bound by Thallspring's laws, he agreed.

Although Z-B's AS could have run the search in seconds and relayed the results back to the group while they were still in the MFL, the detective and his partner, accompanied by the Z-B legal officer, took a car over to the block of hotels that Z-B was using as its barracks. The detective received a full procedural briefing from the magistrate through his bracelet pearl on the drive over. The police commissioner was absolutely determined that justice should not be blocked by some technicality thrown at them by a Z-B legal smartass.

It was 5:32 A.M. by the time all the relevant parties assembled in front of the barracks duty officer. He listened to the detective's request and came out with a formulaic "full cooperation" statement. The file of the suspect's DNA was handed over to an assistant and loaded into the barracks AS.

Seventeen seconds later, a perfect match was confirmed.

Ebrey Zhang had been sitting in his office since half-past-three, drinking bitter coffee and munching nervously on stale croissants. He'd been given briefings from a legal officer and the civil administration AS on where they stood on jurisdiction. He'd had an unpleasant interview with General Kolbe, bringing him up to speed. The only bright spot of the morning was that he hadn't yet received a call from Simon Roderick personally.

But then, as he kept telling himself, nothing was proven yet.

Two cameras covered the scene in the barracks for him. His optronic membrane scrolled the search results as they happened. When the positive result emerged his whole body tensed up as if he'd been struck. He threw his desktop pearl across the wide study as hard as he could. The casing broke when it hit the far wall. "FUCK!"

His aide tried to remain impassive. It wasn't easy. News was pouring into the datapool about the incident. Three reporters were already outside the barracks. Fortunately, it was still early, but it wouldn't be long before a crowd gathered. This was shaping up to be one long, evil day.

On the big sheet screen facing Zhang's desk, the detective was requesting custody of the suspect from the barracks duty officer.

"Sir?" the aide queried.

"Okay," Ebrey said in defeat. "Hand him over."

The aide instructed his personal AS, which relayed the message to the duty officer.

"Get me five platoons in Skin and on duty immediately," Ebrey Zhang said. "I want the police station where they're going to take him to be completely secure. Make that very clear to our dear commissioner, too; I don't care how many of his precious constables he has to take off other duties. There's to be no lynching."

"Yes, sir." He snapped out a quick list of instructions to his AS.

Ebrey watched the scene in the barracks. Everyone was remaining so unnaturally civil it was almost comical. But not with this crime, he told himself. *Dear God, this oaf couldn't have hurt us harder if he was in collusion with KillBoy himself.* Only then did he think about the girl, and shudder. Ebrey Zhang had a daughter of his own.

"Send someone round to her house," he told the aide. "Get that fucking collateral necklace off her."

"Yes, sir."

Hal stirred in discomfort when the ceiling light came on. There were a lot of excitable voices nearby. A hand shook his shoulder.

"Piss off," he mumbled. He was still half dreaming about Avril.

"On your feet, Private!"

He lifted his head. Sergeant Wagner was standing above the bed, his face hard and contemptuous. Captain Bryant stood just behind him, looking furious and possibly just a little bit scared. There were other people crowding into the hotel room, two of them in local police uniforms.

"Whaa— Sir." Hal pushed the quilt off and clambered to his feet. He didn't salute. He only had his shorts on; it would have looked ridiculous. His heart started hammering. *Oh, shit, they found out that I broke curfew.*

"Detective," Bryant said with a sharp nod at one of the policemen.

The detective came forward. "You are Halford Grabowski?"

"Er, yes, sir." He glanced at Wagner, hoping for some kind of support. The sergeant's stare was fierce.

"I am arresting you on suspicion of rape."

"Ung." Hal's jaw dropped in astonishment.

"In accordance with the Perlman declaration I am advising you to say nothing at this time. I am entitled to take you from this place to an officially sanctioned holding area, where you are to be questioned with your legal representative present. Please put some clothes on."

"You've got to be fucking joking. Sir?" He turned to the captain.

"Get dressed," Bryant ordered.

"I didn't do nothing. Not that!"

The detective produced a pair of handcuffs. "Come on, son, don't make it any worse."

"You can't do this!"

"Oh, yes, I can."

Hal turned to the captain, pleading. "Tell him."

"While you're here, you are operating under local civil law, Grabowski. We made that quite clear during your briefing. Now put some damn clothes on, or you'll be taken to the station as you are."

Sergeant Wagner held up a pair of trousers. Hal managed a dazed laugh and took them from him.

"I want this room sealed," the detective was telling Captain Bryant. "Our forensic team will need to examine it later."

"I understand. Nobody will come in here."

It wasn't real, Hal was telling himself, none of this was real.

Lawrence Newton came into the room, dressed in a short gray toweling robe. He raked at his disheveled hair as he yawned widely.

"Sarge," Hal yelled. "For Christ's sake, help me."

Bryant held up a warning finger. "You are not active in this case, Newton. When an arrest is being made, nobody from the suspect's platoon is to be involved. Standard procedure. Now get out."

Lawrence gave the captain a reasonable nod, as if the facts

were obvious. He turned to the detective. "I'm the kid's NCO. What's the charge?"

"Newton!" Bryant stormed.

"We're taking him for questioning concerning an alleged rape."

"Really. When was it supposed to have taken place?" Lawrence asked.

"Early this morning."

"Okay." Lawrence looked directly at the frantic kid. "Did you do it?"

"Be quiet," Bryant demanded. "Grabowski, I'm ordering you not to reply."

"I didn't, Sarge. Not that. For fuck's sake, you gotta believe me."

Lawrence studied his face for a moment. "I do."

"Oh, thank you, Jesus."

"Hal, finish dressing," Lawrence told him. "Then you'll have to go with the police."

"Sarge!"

"Do it. We'll get a lawyer sorted out for you at this end. Clear this crap up quickly. Meantime, you do as you're told. Understand?"

"Yes, Sarge."

Hal finished dressing and reluctantly let the detective cuff his hands. As he was led away down the corridor all his platoonmates were waiting outside their doors. They shouted encouragement, slapped him reassuringly on the shoulder, told him they'd be on his case right away, no worries. He even managed a few sheepish grins. The last thing he heard as the elevator doors shut on him and his escort was Captain Bryant hissing furiously: "My office, Newton. Five minutes."

There was an angry crowd outside the police station. Hal could hear them from his cell. The chanting. The shouting.

Everyone had been polite to him since they arrived. It was an act, though, he could tell that much. A Z-B lieutenant had ridden with him in the police car, introducing himself as Lannon Bralow.

"I've been assigned as your legal representative," he told Hal.

"You mean you're my lawyer?"

"Yes."

Hal relaxed slightly.

After they got to the station, Hal was shown into a medical examination room and told to take his clothes off. They were put in a polyethylene bag and taken away. Then a doctor arrived and wanted to take samples. Lannon Bralow told him it was okay, and to cooperate. So Hal lay down on the couch and let the doc prod and poke. He only kicked up when the guy started to examine his dick. His dick, for Christ's sake! But Bralow was there and kept saying how it was okay, and everyone needed it done. Hal let it happen, but made the lawyer promise he wouldn't tell anyone else from 435NK9. Jesus, he'd never live that down.

Once it was over, the police gave him a one-piece overall to wear and took him down to the cells. What seemed like hours later, Lieutenant Bralow came in to see him.

"So, like, where are we?" Hal asked. He was a little pissed the sarge hadn't come.

"They're about ready to interview you."

"For what? I didn't do anything."

Bralow forced a smile. "Hal, the girl that's making the allegations . . . They found traces of you inside her. I was there when they took the samples. Our own AS identified your DNA."

"It's wrong. I never raped no one. I ain't no fucking animal."

"Hal, we've been running our own inquiry at the barracks.

We know you broke the curfew last night. Morkson told us all about the backyard and the motion sensor."

"Shit!" Hal groaned. Goddamned Morkson. What an asshole.

"Hal, now listen, you have to be level with me on this one. Half of Memu Bay is outside howling for your blood. The asset factories are on strike. There's a barricade outside the airport gate so our cargo trucks can't get through. The platoons are being attacked in the streets; we've had to use darts nine times already today, and it's not even noon yet. What the hell happened last night?"

"I went to a goddamn whorehouse. Okay? I mean, Jesus, it's been months since I got me some pussy. I was, like, on fire. And this curfew . . ."

"Right." Bralow sounded relieved. He opened up his desktop pearl. "Start at the beginning."

The room they interviewed him in was a large office with a big wooden desk and leather swivel chairs. Hal knew for sure this wasn't the usual place for interrogating prisoners. But then there were more people than he was expecting sitting in chairs waiting for him.

The detective, Gordon Galliani, was sitting beside a lawyer he introduced as Heather Fernandes, who he said was representing the victim's family. Two other men were sitting at the back of the room, one in a smart police uniform. Hal had been around long enough to recognize a senior officer when he saw one. The other wore an expensive, conservative suit. His eyes were puffy and red, as if he'd been crying. He was looking everywhere around the room except at Hal.

Lieutenant Bralow sat beside Hal. Captain Bryant was there as well. Which Hal could have done without. He wanted the sarge, or even some of the guys from the platoon. At least Bryant seemed to have calmed down since the morning. He even said a brief hello.

Hal sat down opposite the detective. There were a couple of desktop panes in front of him, each with a holographic pane unfolded and running a test pattern.

"Mr. Grabowski, we're here to try to establish exactly what happened last night," Galliani said with a friendly smile. "This interview is being recorded and can be submitted as evidence in any possible trial. Now, as you know, a very serious allegation has been made against you."

Hal leaned forward on the desk, his hands opening to the detective. "I never raped anyone, okay? I'm telling you the truth, here. And I can prove it."

"Really?" Galliani was momentarily thrown. "How do you intend to do that? We have gathered a lot of evidence that incriminates you."

"Look, I jumped the barracks curfew, okay? I admit that. But, shit, I didn't rape no girl. I went to a whorehouse to get laid. I paid for it, fair and square. Cost me a goddamn packet, too."

"You're saying you visited a brothel?"

"Yeah."

"What brothel? Where is it?"

Hal flinched. "I'm not sure. This taxi took me there. The driver knew it. It's only a few minutes' drive from the barracks."

Galliani waited in silence for a moment. "That's it?" he asked eventually. "That's your proof?"

"Yeah."

"I'm sure if you were to pursue this alibi you would soon establish its validity," Bralow said smoothly. "My client is trying to cooperate."

Galliani sat back and smiled at Hal. "Son, you've had three hours and full access to a smartass lawyer. This bullshit is the best you can come up with?"

"It's not bullshit," Hal said hotly. "I went to a brothel. It was a big smart house, they all were down that street; there was a little garden along the front with iron railings. I don't know the number, but I'll know it again when I see it."

"What time did you leave your barracks?" Galliani asked.

"Twelve minutes past eleven."

"And when did you return?"

"Twelve past two. That's when the sensor was inactive, see? Twelve minutes past every hour."

"If you don't know where this brothel is, how did you get back to your barracks?"

"The madam called a taxi for me. I got back to the barracks about quarter to two. I had to hang around and wait before I could actually get in."

"Did anyone see you?"

"No, man, I wasn't supposed to be there. I hung around in the alley; I guess there weren't even many people on the street that time of morning. But the taxi driver can vouch for me."

"Was it the same driver who took you to the brothel?"

"Yeah."

"I don't suppose you know his name, or even which taxi company this was?"

Hal shrugged awkwardly. "No. But I think he was using AS control when we left. You'll be able to trace him through the traffic regulator logs."

"We'll certainly check."

"And so will we," Bralow murmured. He met the detective's gaze levelly.

"So," Galliani said. "We've established you were out on the streets at the same time this alleged rape took place, and that no one can actually confirm exactly where you were."

"The taxi driver can, the madam can, Avril sure as shit can—" Hal was checking them off on his fingers.

"Avril?"

"The whore I spent half the night screwing. There were a

couple of other whores I saw there, as well. Don't know their names, though."

"But you'll recognize them when you see them?"

"Yeah, no problem."

"So all we have to do is find this taxi, and the brothel, and you're in the clear?"

"Yeah." Hal smiled happily. "Yeah, you got it, man."

"So how do you explain your semen being found inside the victim's vagina?"

Hal's smile dried up. "I don't know. It's a sting. A frame-up. It can't be anything else."

"And the girl's story? That you attacked her in Sheridan Park? That you threatened to set off her collateral necklace if she didn't do what you wanted?"

"Hey, that's your bullshit, man. None of that crap happened. None of it. I wasn't in Sheridan Park. She's lying. She's a part of all this."

"All this? So it's a conspiracy, then?"

Hal glanced at Bralow.

"Zantiu-Braun personnel would be the obvious victims of any rogue criminal elements in Memu Bay," the lieutenant said. "And we both know there are some."

"You've been having a tough time from our hooligan element," Galliani said. "But there's no organized resistance group, is there?"

Captain Bryant cleared his throat. "No. There is no organized resistance group in Memu Bay."

Hal twisted around in his seat to stare at the captain. "You've gotta be jerking me off. You were at the fucking soccer match, for Christ's sake. You saw Graham Chapell get blown to shit by KillBoy's bastards. You saw that!"

"We're still investigating the soccer game incident," Bryant said to Galliani. "We're not yet sure what happened."

"Jesus fucking wept."

"So there may or may not be someone, or some group of people, capable of setting you up for rape," Galliani said.

"Damn betcha there is," Hal told him. "It's that bastard KillBoy you should be looking for. Not me."

"Which means the rape victim must be part of the conspiracy?"

"You bet. You call her in here and give her the third degree. She'll crack."

"Strange how this comes back to the oldest conflict the human race has."

"What do you mean?"

"One of you is lying."

"It's her, man, I swear it. She's jerking you around. She's saying everything KillBoy told her to."

Galliani paused, as if considering something. Then he called up a file from one of the desktop pearls. Its pane displayed a girl's face. Hal was very aware that the detective was watching him closely.

"For the record, Mr. Grabowski, have you ever seen her before?"

Hal frowned, not quite understanding what was going on. "That's Avril. How did you get her picture?"

"Avril?"

"Yeah. The whore at the brothel. You do know where it is. Why did you say you didn't?"

"Let us be quite clear about this. You're saying that girl is Avril, whom you met in a brothel last night?"

"Yeah. Have you known this all along?"

"Mr. Grabowski, did you at any time last night have sex with the person you call Avril?"

"What, that's not her real name?"

"Did you have sex with that girl?" Galliani's finger tapped impatiently on the pane.

"Sure. I got my money's worth. I keep telling you. She's the one. I was there in the whorehouse with her last night."

There was another moment of silence. The detective appeared almost embarrassed.

"Mr. Grabowski, did you notice anything out of the ordinary about Avril?"

"Like what?" Hal wasn't committing himself. There was something badly wrong about this, he knew it. Damn but he wanted the sarge to be here.

"Did she, for instance, have a collateral necklace fitted?"

The question surprised him. "No. No way."

"You're sure about that?"

"Hey, I got to see a damn sight more than her neck, man. She wasn't wearing no necklace. What is this crap?"

"I think I've heard all I need to at this point, thank you," Galliani said. "We'll take a break. And I really think you need to have a long talk with your lawyer, Mr. Grabowski."

"Just what the hell is going on?" Hal demanded. "Okay, so I fucked some whore. That's not a crime. She wasn't even much good. I should have had a refund, man."

Someone in the office roared wildly. Hal searched around for the noise, just in time to see the man in the expensive suit charging at him. His face was red and contorted in feral rage, arms held out straight in front of him, hands ready to tear and throttle. He jumped at Hal, who didn't have time to move aside. The two of them crashed to the floor, thrashing about. Then Galliani and the senior policeman were pulling him off. Bralow hung on to Hal, who was game for getting back up and decking the old maniac.

"What the fuck . . ." he shouted.

The man was quickly hustled out of the door. He was sobbing now, a wretched gulping sound that was clearly audible even after the door was shut.

"This place is a fucking loony bin," Hal announced. "What the hell is going on here?"

Bralow sat down, sighed, then pulled the desktop pearl toward him. The girl's face was still on its pane. "She's the . . . the alleged victim," he said.

"Avril? No way, man. No goddamn way. I paid for her!"

"That's not her name."

Hal looked at the closed door, suddenly curious. "Who was that? The guy that went for me?"

"Her father. The mayor of Memu Bay. And she does have a collateral necklace. Ebrey Zhang put it on her himself."

"Oh, Jesus fuck," Hal whispered. He sat down heavily beside Bralow as real fright took hold. None of this was making any sense, goddamnit. "Lieutenant, you've got to get me out of this."

"That might be difficult now."

The *Norvelle* was in a thousand-kilometer orbit around Thallspring, its inclination of five degrees providing it with line of sight on Durrell each time it passed through the planet's prime meridian. At ten-fifteen in the morning it rose above the capital city's horizon. As the sensors acquired the sprawl of buildings, a low-power laser was fired from one of the huge vehicle's five communications bays, seeking out the East Wing of the Eagle Manor. It was detected by a small electronic receiver unit on the roof, which immediately sent an answering laser pulse back to the starship. With the beams locked on their respective sensors, their width reduced until it was less than two centimeters at the target point, providing a link that could not be intercepted. The rooftop receiver unit was connected to a module in Simon Roderick's office by an armored fiber optic cable. Again, splicing into the cable was impossible. The system provided him with the most secure link possible to the starship. Only five people knew of its existence.

Simon had been waiting for the call since he arrived at the

office that morning. His usual routine of administrative work had been delegated to his assistants and personal AS. Instead, his time had been spent reviewing information filed under the generic name "The Opposition." As he ran through it all he conjured up probable attack scenarios, which grew steadily more exuberant as the morning progressed. It didn't matter how fanciful he made them, he still couldn't determine what they were actually planning. Nothing quite fit into what was clearly an impressive capability. The more he went over it, the more he was convinced they were holding back, waiting to deliver the hammer blow.

The secure communication module chimed melodically, and a sheet screen on the wall lit up, showing one of the *Norvelle*'s cabins. A man was sitting in front of a freefall work bench, with straps holding him down in the light gravity field. He looked into the camera and gave it a thin smile. "Good morning. It looks very sunny and warm down there today."

Simon settled back behind his desk and looked at the face on the screen. It was his own, but fifteen years older. That particular batch of clones, the SF9s, were notorious for their phlegmatic temperament. Each generation tended to have its own quirk, which they put down to the individuality of the creche nursing staff and the inevitable influence they exerted during the clones' formative years. The SK2 batch, to which the Simon in the study belonged, were often regarded as the more peppery of the breed. Although they were positively mild compared to the short-tempered SC5s (whose proclivity had sparked a wholesale review of creche staff screening procedures). But whatever their behavioral nuances, they were all totally dedicated to the company that they controlled.

"Morning," the SK2 Simon replied. "So what's the result?"

"Well, the good news is it wasn't a bomb."

"I never expected it would be. Far too coarse for our friends."

"Young Braddock Raines was most thorough. The spaceplane cabin was scanned and analyzed down to a molecular level. He also had the accessible systems removed and reviewed in the starship's lab. There was no detectable foreign genetic residue. However, somebody had opened an access panel. There were metal traces in the Allen screws. The alloy doesn't correspond with the tools issued to our maintenance people."

"Thank heavens for that. I was beginning to think they were almost infallible."

"Quite. The panel gives access to several electronic components, including a major network junction. None of the components had any trace of tampering, except the junction. And that took some finding. The nuclear macroscan revealed some very peculiar stress patterns in the casing's molecular structure. Our so-called solid state physics experts are apparently baffled. They don't know what could have caused it."

"Interesting."

"The word is alarming. I don't like the idea of Thallspring having technologies that we don't understand. Especially when they're being used against us."

"Their development has been very well hidden. We've run all the usual financial audits through the Treasury network. They couldn't spot any kind of government funds being diverted for clandestine technology projects in the last ten years."

"Hardly surprising when you consider we're talking about people who can walk into our spaceplanes whenever they feel like it. Whatever they've got, it's real enough."

"Assuming whatever the intruder did to the junction gave him access to the spaceplane's network, what do our experts think he achieved?"

"The theory they're throwing around up here is total subversion. The IT boys have dumped the spaceplane's entire AS program into a storage core for analysis. So far they can't find a single extraneous code line. The best conjecture they can come up with is a hidden command compressed into the original code."

"In other words we don't know for certain what the hell the intruder did."

"Absolutely."

"Damn." The SK2 didn't waste time considering the puzzle. That was the advantage of having multiples working on the same problem: whatever solution his clone sibling came up with, it would be the same as the one he would eventually arrive at. And the SF9 had been thinking about this for over an hour already. "Recommendations?"

"This intrusion has to have been some kind of reconnaissance mission. The interest our friends have in the spaceplane demonstrates they want to get up here in one form or another, and as it's a Xianti they must be targeting the starships as a final destination. If they could fly up already, it would have been done. Therefore, they're still in the preparation stage. For myself, I believe he copied the AS to study our procedures."

"I see. So what else do they need?"

"For a hijacked flight to pass unnoticed, the only other consideration will be communications. We must hope they haven't already been there."

"I'll make the necessary preparations. I take it you consider e-alpha to be compromised?"

"Completely."

"That will have to be taken into account."

"Of course. I'll leave the details to you."

"Thank you. Send Raines down today, please. I'll need him for implementation."

"He'll be on the next flight." The SF9 Simon glanced at a

pane, reading the text. "So what policy have you decided to apply to Memu Bay?"

The SK2 used his DNI to consult his personal AS. The daily summaries poured into his brain, neatly tabulated in bright indigo. "Damn," he muttered as the Grabowski rape case scrolled down. He should have run the usual morning review. Problems like this should never arise in the first place. "What the hell is Zhang doing over there?"

The SF9 smiled, content with his little victory. The Simons always enjoyed scoring points off each other. It was healthy competition.

"I'll give it a full investigation," the SK2 said.

"No need. They've already got an appalling asset-acquisition record. Things have to be calmed down. Sacrifice Grabowski to the mob. Then get Zhang to crack down hard."

"Fine," he said dismissively. He was irritated at being caught short on an incident like this.

The SF9 cut the link, chucking contentedly.

When his car arrived outside the marina police station Ebrey Zhang seriously wondered if he should have put his Skin suit on. The protesters were eight or nine deep on the road and hyped up badly. He shivered as he read some of the slogans they'd sprayed on nearby buildings, saying what they wanted to do to Grabowski. Ten people in collateral necklaces were standing directly in front of the station entrance. They'd chained themselves together. Crude signs hung round their necks saying:

Death before Rape
So
Please kill me now

Stones, cans, bottles and what Ebrey hoped was only mud began to rain down on the car, making curiously dull thudding sounds as they dented the bodywork. Ten Skins and a batch of police in full riot armor pushed angry people aside, creating a route for the car.

"Holy shit," Ebrey grunted. A huge brown lump hit the windshield and spread wide. Definitely a turd. The driver had to use the wipers and a whole load of cleaning fluid to smear it away.

"This isn't getting any better," Lieutenant Bralow said. "There's at least as many here today as there were yesterday."

"Same with the asset factories," Ebrey admitted reluctantly. "They're still on strike."

"What does the general say?"

"Get it over with, and quickly."

"Easy for him."

"He's got a point. There's more than just Grabowski to consider." Ebrey indicated the mob outside. "This whole thing has got to be defused. We can't even implement the TB vaccination program right now. How crazy is that?"

"They should be able to move to trial in a couple of weeks."

"Weeks? That's no damn good. Haven't they finished their investigation yet?"

"Almost. Enough to thoroughly trash Grabowski's alibi. We've run our own checks in parallel, of course. Our AS can't find any taxi that could possibly have taken him anywhere, never mind the brothel."

"It exists?"

"No. We think the street he claims he went to was Minster Avenue. They're all private homes for the reasonably wealthy. There's no brothel."

"In other words it's all bullshit."

"Sir, he raped Francine Hazeldine. The best help I can be to him is by making a leniency plea."

"Ah. That was the other thing that the general told me."

"What?"

"We don't leave anyone behind, no matter what the circumstances."

Lieutenant Bralow gave his commander an agitated look, then nodded. "Yes, sir."

The car made it through the station's perimeter and swooped down into the underground garage. Detective Galliani was waiting for them. He said hello politely enough and told them that Margret Reece was waiting upstairs for them.

Ebrey Zhang kept his face composed even though he was seething. He was governor of Memu Bay; he was the one who summoned officials to him.

Not a chance, he told himself bitterly.

The only time Myles Hazeldine had slept at all in the last forty-eight hours was when the family doctor gave him enough sedatives to knock him out. Even those few hours had been twisted with nightmares and helpless fury. The same as his waking hours.

He knew he must keep calm for his beloved Francine's sake. But it was so terribly terribly hard. To make matters worse, she was the one who kept apologizing to him, saying she was sorry she'd stayed out late at the club with her friends. Sorry she hadn't called him, or a taxi when she left.

It was almost as though she was the one comforting him. Which was wrong. Another example of how bad a father he was.

And so the hours dragged by. Pathetic helplessness alternating with utter primitive fury. He never wanted to let Francine out of his sight ever again, wanted to keep her next to him where she could be protected. He also wanted to rip that piece-of-shit alien bastard's heart out of his rib cage and hold it up to the sun, crying out in victory as scarlet blood showered down.

Don and Jennifer had taken over the day-to-day running of the mayor's office. Actually, they'd insisted. The same as dear old Margret Reece had insisted he was never to be allowed near the suspect again. He'd managed to pull rank that first time, playing on sympathy to get into the interview. Damn, but that had been a moment of sweetness, actually getting his hands round the laughing, sneering kid's neck. However briefly.

But no way could they keep him out of this meeting. The first time Memu Bay would be able to stand up to the invader scum and insist on things being played by the book. They wouldn't like that, their own bogus legitimacy being used against them.

He was waiting in the marina station commissioner's office, not far from the one where they'd had the interview, as it happened. The commissioner was there, along with his boss, Margret Reece, and the police magistrate handling the case. Everybody seemed reluctant to look at him, much less talk. That didn't bother Myles. He had nothing to say. And their earnest sympathy only served to remind him what darling Francine had undergone. If he thought about that he would probably break down again.

The door opened and Galliani showed Ebrey Zhang into the room, along with the Z-B legal officer, Bralow.

Zhang nodded politely. "Mr. Mayor." He put his hand out.

Myles just wanted to smash his fist into the bastard's nose. Margret Reece had warned him, but he hadn't forgotten who'd put that necklace around Francine's neck. The chief of police was watching closely, as was the precinct commissioner.

Zhang stepped back, slightly subdued.

"Thank you for coming, Governor Zhang," Margret Reece

said. "I asked you here in your role as Halford Grabowski's senior officer."

"I understand."

"My officers have amassed enough evidence to formally charge him with the rape of a minor. The magistrate here has assigned a preliminary trial date. As his commanding officer, I'm asking you to sign him over into full civilian custody for the duration of the trial. I believe that is the requirement that Zantiu-Braun strategic security forces operate under."

"That is correct," Zhang said.

"Good." Margret Reece signaled the magistrate, who walked forward and offered a desktop pearl to the governor. Its pane scrolled a long legal script.

"Thank you," Zhang said. He glanced at the script. "This trial date is in three weeks' time."

"Yes," the magistrate said.

"What is the possible maximum penalty if Grabowski is found guilty?"

"I'm sure you already know," Margret Reece said. "But it happens to be lifetime imprisonment."

"Of course. There is an alternative."

"No, there isn't!" Myles snapped. "I knew it, I fucking knew they'd try and slime their way out of this."

"Myles, please," Margret said. "What alternative?"

"We court-martial him ourselves," Zhang said. "It will be quick, and the proceedings will be fair."

"Are you suggesting ours aren't?"

"Not at all. But neither you nor I want his lawyer to appeal on the grounds that he was given a biased jury. Which given the current state of affairs in this town will be a valid point."

"In other words, you want him judged by your officers?"

"Yes."

"No fucking way!" Myles shouted. "You authorize that custody order. Do it."

"Your police lawyer will be able to join our prosecution

team," Zhang said. "That way, you can be assured the case will be made correctly."

"I don't understand," Margret Reece said. "Why a different court? It will look like you're trying to load the judgment in his favor. Or . . ." She paused thoughtfully. "Are you considering a prison sentence that will be served on Earth? Is that it? If you find him guilty you take him home with you rather than hand him over to our prison service?"

"That isn't actually what will happen."

Myles sat up at that. Despite his turmoil, he was still politician enough to see a deal being offered. "What's the penalty under your court-martial laws if he's found guilty?"

Zhang looked right at him. "Capital."

Myles had never thought of that. The death penalty was expressly excluded from all court action by Thallspring's constitution. How strange that he, the custodian of the founding fathers' liberalism, should now be given the opportunity to go against their original creed. He should of course refuse: it violated everything his culture stood for. "In that case," Myles said, "we agree."

Nearly a third of the children were missing from school that morning, which saddened Denise. It was another beautiful day, with a hot sun already high in a clear, deep blue sky. A breeze from the sea was just enough to cool Memu Bay's baking streets for people to walk down without being too uncomfortable. So it wasn't the weather keeping them away.

Today was the day Halford Grabowski's court-martial began. The population of Memu Bay was holding its collective breath. After all the unrest and the huge emotional outpouring of anger that accompanied it, people had actually taken a step back. Perhaps they were shocked by the prospect of capital punishment—not that any protested against it. Whatever the reason, the trams were running normally, and most of the shops seemed to be open. There was no sign of any Skin patrol walking the streets. Quite a few people were down on the beach, enjoying the sand and the water. And Denise knew that the hurriedly formed workers committees at the major asset factories were meeting today to discuss going back to work.

Even so, some parents were obviously reluctant to let their little loved ones out of their sight again so soon after days when the whole town seemed ready to explode. Ironically enough, Melanie Hazeldine had been one of those who appeared. Francine had brought her, the two of them riding in the back of a big limousine with darkened windows.

Denise watched from the kitchen as the two sisters kissed good-bye, and Melanie raced on into the school to greet her friends with excited shrieks. She hadn't been in school for a week.

"How are you?" Denise asked quietly when the two of them were alone together.

"Okay." Francine managed a small, brave smile. "I'm really worried what this has done to Daddy. I didn't know he'd take it so hard."

"You can tell him afterward, if you'd like, after the bastards have left."

"Do you think I should?"

"I'm not sure," Denise said honestly. "You'll be shocking him again, that his sweet daughter got involved with a resistance group, and did what she did to help the cause."

"Has it helped? I don't know. I just wanted to hurt them for what they did to William. He was my brother, you know. I never really knew him—I was young the last time round."

Denise put her hands on the younger girl's shoulders, squeezing softly, wanting the contact to emphasize the gratitude she felt. "Oh yes, you helped. Take a look around at what you've accomplished. They can't walk the streets anymore. Do you know what that means to people, not having to bow down and get out of the way because a bunch of arrogant bullies are swaggering down the pavement? Their precious pillaging has taken a financial knock it'll never recover from. They won't make money out of us. You made all that possible."

"Yes." Francine straightened her back and smiled properly. "I did. Poor Daddy."

"Tell him if you think it will help. Let him blame me; he might find that easier. We shouldn't come out of this as the victims, because we're not. They are, now."

"Thank you, Denise." Francine leaned in close and kissed her. "You're so strong. We need you to defeat them. I don't want my children to fear the stars like I've always done since William."

Denise hugged her friend. "They never will. I promise."

After all of the children had arrived, Denise called them together and dished out the media pads. That was always a popular activity. She played them some sing-along tunes as they created a host of wondrous shapes in a rainbow of colors. Each one was held up for her inspection as she walked along. She offered a few suggestions and spoke thousands of words of praise and encouragement.

The children took a break for juice and cookies. Denise sat with them, drinking her own tea.

"Finish the story, miss," one of them asked. There was a chorus of "Please!"

With a small show of reluctance, Denise put down her tea. "Not everyone's here, I'll have to tell it again later." That earned her a round of cheers.

"Okay then," she said with a pantomime sigh. "I suppose I will." *After all,* she considered, *I don't know how much longer I'll be here.* The thought dampened her eagerness; even with a future of astounding possibilities opening to her, she would miss these smiling, mischievous faces. Trust and admiration were given so easily at their age. She felt like a fraud for receiving it, which troubled her.

"Mozark took seventeen years to complete his voyage around the Ring Empire. Weeks before he arrived, the kingdom had heard he was returning. There were celebrations on every planet, and when Endoliyn was told the news she burst into tears, she was so thankful that her prince was coming home. No word of the ship and its quest had been heard for over ten years. Which is a long time for even the stoutest mind to hang on to trust.

"But now the day had come, and millions of the kingdom's subjects gathered around the giant landing field. They waited for hours, staring up into the sky where the galactic core stretched from horizon to horizon, blazing with silver light. Finally they were rewarded with the sight of a tiny black speck high in the air. It grew and grew until its shape became clear, and the mightiest roar of greeting arose from the crowd. The starship settled on the very pad it had left from, which had stood empty these seventeen years waiting for its return.

"Mozark emerged first to be greeted by his father, the king, who wept openly at the return of his son. The rest of the crew were then welcomed with great honor, for although they were not as renowned as their prince, they had played no less a part in the voyage.

"And then, when all the public celebrations were done, Mozark traveled to the palace where Endoliyn lived, and asked for her forgiveness."

"Why?" Jedzella asked in wonder. "What had he done wrong?"

"Which was almost Endoliyn's reply," Denise said, grinning at the youngster. "She also wanted to know what he had done. To which Mozark said he had spent seventeen years away from her, which was unpardonable. They had been in love when he left, and such a separation was not something that should ever be inflicted on a loved one. She laughed, and told him he was silly, and that she loved him even more for embarking on his quest. What other person would sacrifice so much for the ideal he believed in? Then she asked the question that had been on her lips for seventeen years: what did you find?

"Mozark was so shamed by his beloved's question that he bowed his head. Nothing, he confessed, I have found nothing out there that we cannot build or think of for ourselves. In that I have failed you completely. All of this supposedly noble journey was for nothing. I almost wasted those seventeen years.

"Endoliyn was heartened by this small glimmer of hope. And asked him what he meant by 'almost.' He replied that there was indeed one thing that he found. A small, trivial, selfish thing, he said, which is mine and mine alone. And what is that, my love? she asked. He gazed upon her and said, I have come to realize that life is the most precious thing. It matters not where you are, or who you are. All that matters is how that life is lived; and lived it should be, to the full. And I know that my life can only be lived in such a fashion when I am with you. That is the only knowledge that I have returned with. I care not whether my kingdom rises into glory or falls into the abyss, all that I ask is that I share that prospect with you.

"Endoliyn laughed for joy, and said that of course she would live her life with him. And Mozark was overwhelmed with gladness. In due course they were married and Mozark became king, with Endoliyn at his side as his queen. They ruled the kingdom for many a long year. Few could remember a monarch so wise and kind. So, of course, the kingdom did not fall into the abyss, but prospered and continued to provide comfort and protection for all its peoples with serenity and grace."

The children waited for a moment, until it was obvious that Denise had finished. Not a few sullen glances flew among them. No outright resentment, but she knew they felt cheated.

"Is that it?" Melanie asked.

"I'm afraid so," Denise said kindly. "So what have you learned from Mozark's voyage?"

"Nothing!" one of the boys yelled. The others giggled.

"That's not quite true," Denise said. "I've learned a lot, and I think you have as well. The moral of the story is a simple one: This technology we have, and I'm the first to admit that it is a fabulous technology, should not be allowed to blind us to ourselves. Science is not an answer to our problems. By itself it cannot provide happiness, it can only ever help to light the way. We must find that happiness in our own way in the short time we've been given to walk this universe. When you grow up, you should concentrate on what's important to you as a person. In Mozark's case it was his love for Endoliyn, and it took him a voyage around the galaxy before he realized that. Only when he searched for a solution amid the science and minds of his peers did he see how empty such a quest was. The universe is centered on you, for that's the only perception of it you'll ever have. You are the most important thing in it, every one of you."

Mollified, but only just, they sprang to their feet and raced off to the games in the garden.

"Very good, my dear."

Denise turned to see Mrs. Potchansky in the doorway to the school's small kitchen. "Thank you."

"I've been interested to hear how it was all going to end," the old lady said.

"You approve, then?"

"Oh, yes. I still don't think it can quite be rated up there with the classics, I'm afraid. It would need some polishing and sharpening for that. But I'm glad you told it to them."

Denise looked out at the children as they caused their usual mayhem in the garden. "Perhaps I should have given them an ending with a little more oomph."

"If that's the true ending, then that's the way it must be told. Never sell yourself short, my dear."

Denise smiled and stood up briskly. "I never have."

"I know."

There was an edge to the old lady's voice that made Denise pause uncertainly.

"I'm proud of you, Denise," Mrs. Potchansky said. "You've done a wonderful job these last few weeks. Circumstances have not been easy. It gives me some hope for the future."

"We'll prevail, don't you worry."

"Of course you will. Thallspring will." Mrs. Potchansky went back into the kitchen.

Denise listened to mugs and plates being loaded into the dishwasher and wondered exactly what they'd just been talking about.

Michelle Rake had done well in getting the lab assigned to Josep and Raymond. It was in the botanical sciences block, one of the older university buildings away on the edge of the institution's parklike campus. An avenue of ehretia trees connected it to the central cluster of faculty buildings, their long dark green leaves casting a dense shade over the path underneath. The other students all said that when the panicles of white flowers bloomed they put out a delectable scent, but that wouldn't be for a few months yet. The tall trees and still air gave this section of the campus a secluded air, as if the important work had moved onward and outward, leaving just a few aging academics to tend their plants among the gently decaying facilities.

It was a perfect location for Josep and Raymond to run i-simulations—isolated, yet in the heart of Durrell. The botanical sciences block itself had a network several generations out of date, yet perfectly adequate for their needs. And even now, when life in the capital had settled down to more or less normal, over 30 percent of students were still taking an extended leave of absence. That left few people questioning their presence in the building; botany was never an oversubscribed discipline to begin with.

Cold climate chambers ran along one wall of the lab, shedding a pale violet light through misted-up glass. Their refrigeration cabinet rattled and buzzed in the corner. The two wooden benches that ran down the center of the lab were supporting a fair amount of glassware, resembling a sophisticated high-school chemistry set. Tables under the long windows were covered in clay pots filled with dozens of ugly, bearded cacti growing in bone-dry soil. The ubiquitous black cube of a datapool node was hidden under one of the tables, with three tiny orange LEDs on its front glowing amid the shadows. A Prime program had fortressed it from the rest of the network without alerting the management AS. Inside the cube, its array of neural pearls were generating an image of the *Koribu* as seen from an approaching Xianti. Josep and Raymond perceived the simulation through their d-written cell clusters, eliminating the need for stim-suits. Sensations slipped directly into their brains as they sat side by side in a pair of old leather armchairs, even managing to give them a modest impression of freefall. With their eyes closed they weren't aware of the botany lab. To anyone observing them, it would have looked as if they were in a deep REM sleep.

Inside the shared environment, Josep was in the Xianti's pilot seat while Raymond and a simulated Denise had begun suiting up behind him. The *Koribu* was visible through the windshield, a broad conglomeration of machinery, three hundred meters ahead of the spaceplane's nose. Two more Xiantis were also approaching the giant starship, their cargo bay doors folded open. Small one-man engineering shuttles were gliding out to meet them, ready to retrieve their valuable cargo.

"No physical contact with the big mother," Josep said wistfully. "So at three hundred meters' distance we produce the malfunction alert."

Amber graphics appeared on the console panes as well as the windshield, reporting a hydraulic failure in the Xianti's payload bay door actuators. The *Koribu*'s flight controller queried them. Josep kept his conversation within the guidelines they'd gleaned from the spaceport mission data logs.

After the starship's AS had received and confirmed the spaceplane systems data, they were cleared to dock in the starship's maintenance bay. For most minor equipment malfunctions, the spaceplane would be sent back down to the ground for inspection and repair. Maintenance in freefall was a difficult and expensive business: there were few situations that justified it. Having an inaccessible cargo was one of them. It was a relatively simple procedure to provide auxiliary hydraulic power to the spaceplane, allowing the payload bay doors to open and the cargo to be removed. After that the doors could be closed and locked against atmospheric entry. The spaceplane would be dispatched back to Durrell for a thorough maintenance overhaul. This scenario also had the added advantage that no one on the starship would actually see what was in the Xianti's cargo bay.

While Josep flew them around the *Koribu*'s cargo section, Raymond finished putting on his suit. It had been specifically tailored for him out of a silver-gray fabric no thicker than paper. In its inert condition the fabric was slightly elastic, allowing him to push his limbs into the sleeves with relative ease. The suit's hood was thicker, almost like a protective sports mask. He slipped it on, and small tubules melded with his nostrils, supplying him with air. His lips were engulfed by what felt like soft dry sponge, absorbing every exhalation. The suit sealed up along the front, then contracted. For a moment his skin felt as if it was being pinched everywhere. Then the shapemorph finished, and he couldn't even feel it anymore. Its surface changed color as its thermal conductivity altered, shedding excess heat from his skin, keeping his body at the temperature he was most comfortable with. His d-written neural cluster received sensory information from sensors on the outside of the hood, complementing his vision.

Finally, he picked up a black, sleeveless jacket and fastened it over his chest. It was primarily a harness for several weapons, their power packs and ammunition, although it also had several small gas jet nozzles around his shoulders and waist, giving him a degree of freefall maneuverability.

Denise had also finished suiting up. She was holding a handgrip near the airlock.

"Ready," Raymond told Josep. He and Denise airswam into the airlock. The inner door swung shut. Denise activated the cycle.

"Stand by," Josep told them over the secure band. "We're fifty meters away."

The outer hatch opened. Raymond could see the starship's engine section sliding past, its cryogenic insulation fiercely luminous in the bright sunlight. He gripped the hatch rim and pulled himself out.

The Xianti was heading for the starship's drum-shaped cargo section. A hundred empty silos opened into blackness. The sunlight was at the wrong angle to shine directly into them, though he could just make out a few that were filled with cargo pods. Up ahead, past the silos, the huge doors of the maintenance bay were hinging apart to reveal a rectangular metal crater. Mechanical cradles were extending their mandibles, preparing to grip the spaceplane.

"Go," Josep ordered.

Raymond tensed his leg muscles and pushed off. The jacket nozzles fired immediately, adding to his velocity. He flew fast and true into the end of an open silo. The nozzles fired again, killing his velocity. He closed his hand on the metal grid structure and hauled himself along to the top of the silo. Denise was in a silo twenty meters away. She raised an arm and pointed. He gave her a thumbs-up and began to move again.

Hand over hand they crawled along the cargo section until they reached the rim. Fifteen meters in front of them, the first life support wheel was turning silently, a wall of crinkled foam sliding along quickly enough to be mistaken for flowing water. Then they followed it around and the curves became apparent, shattering that illusion.

They both started to crawl up the side of the cargo section toward the axis. After twenty meters, they were level with the top of the wheel. Raymond's suit had to increase its visual sensitivity, the gap between the cargo section and the wheel was so gloomy. Hardly any light was reflected off the wheel's coating of foam.

Once he had a reasonable field of vision, Raymond fed his location into the suit's function control pearl and then entered the wheel's relative velocity. He let go of the cargo section's metal framework and drifted over to the edge of the life support wheel. When he was barely a meter above it, the jacket nozzles fired again, accelerating him in the same direction as the wheel's rotation. From his point of view, it seemed as if the wheel was slowing as he moved in toward it. There were innumerable protrusions amid the foam: conduits, pipes, even ladders. He grabbed at one, a thick metal loop, and the bogus gravity took hold, pulling him abruptly down onto the top of the wheel. A sixth of his weight had returned, holding him securely.

He saw Denise had landed a quarter of the way around the wheel behind him. She gave him a thumbs-up. Raymond stood, taking care not to make any fast movements. He had a very clear impression that beyond the edge of the wheel was now *down.* If he slipped, centrifugal force would fling him clear of the starship and into its sensor field. He took a few careful steps in toward the middle of the wheel and examined the structure below his feet with deep senses. Particle resonance located a clear patch, and he took a loop of energy focus ribbon from his jacket. When it was laid out on top of the foam, the ribbon formed a circle about two meters in diameter.

A hundred meters around the wheel, Denise was doing the same thing. Raymond stood away from the ribbon and activated it with a pulsed code. The foam underneath the ribbon flash-vaporized along with the carbon titanium alloy below. A two-meter circle of the wheel fuselage slammed upward, punched by a seething column of air. Bright white light shone up out of the hole. Paper, clothing, electronic modules and wildly oscillating sprays of liquid filled the column of air, hurtling into the darkness of the axis far above.

Raymond waited until the blast died down, then hurried forward and dropped down into the wheel. He was in some kind of lounge, with the detritus from the decompression swirling senselessly around him. Bright red strobes were flashing. The emergency airlock panel had sealed the hatchway. His suit pearl found the wheel's internal network frequency, and Prime poured into the nodes.

He walked over to the emergency airlock panel and used the Prime to override its safety locks. It slid open and he stepped into the chamber beyond. The panel slid shut behind him, and the hatch opened in front. Starship crew were running round in chaos. Prime shut down all other internal communications, then extinguished the lights. It didn't make any difference to Raymond: he could see equally well in infrared and laser radar. He brought up an EC pistol and started killing.

After the i-simulation Josep and Raymond opened their eyes, grimacing against the bright afternoon sun pouring through the botany lab's windows. Josep got up first and stretched elaborately.

"Not bad," he said. "I think we should start running more adversarial versions, though."

"Yeah. I guess so. It is a little easy at the moment."

"We can begin with you being spotted when you leave the spaceplane."

"Oh great."

Josep grinned and checked his watch. "We've got a couple of hours until Michelle gets back."

"How's that going?"

"Fine. Being an activist has sharpened her outlook. She likes courier duty: it makes her feel she's achieving something. How about Yamila?"

"I could never get her involved, not even at basement level," Raymond said. "She's too timid. Even suggesting it would frighten her off and I'd be left looking for new cover."

"Not at this stage, we can't afford it."

"I know. As it is, she thinks I might be seeing someone else. All those nocturnal absences."

"Speaking of which . . ."

"Yes." Raymond filled two cups with water and dropped a tea cube in each before sliding them into the microwave alcove. "We need the communication keys." It had come as a surprise to them when they analyzed the data from the Xianti. They'd known that the spaceplane communication traffic was encrypted, although they'd never bothered to examine it before. Had they done so they would have found that not even Prime could decrypt it. Theoretically, given enough processing power and time, any code could be broken, but Z-B used a particularly strong four-dimensional encryption technique for its spaceplanes and changed it every time. Even with the resources Raymond and Josep had available, they could never crack it inside the timeframe they needed for a successful operation.

"Shame the keys are physical. Z-B seems to take its spaceflight security very seriously."

"Prime keeps trawling up obscure references to Santa Chico," Raymond said. "I don't know what happened there exactly. But it's possible they may have lost a starship to some kind of weapon."

"No wonder they're protective. Onetime dimensional encryption indeed." Josep shook his head in admiration. "I'll collect them from the spaceport in a few days."

"Has the fuss over Dudley Tivon blown over?"

"Just about. The police have downgraded the case to a level-five resource funding. Prime picked up some activity in Z-B's security AS; it was flagged for senior staff attention. I presume they were interested because Tivon worked at the spaceport. But there was never any follow-up."

"We're in the clear, then?"

"Looks that way."

"Good. From what Denise has been saying, things are just about ready at her end."

CHAPTER THIRTEEN

THE FIRST TIME LAWRENCE NEWTON VISITED THALLSPRING HE already considered himself a campaign veteran. By then his attitude was relaxed enough to allow him to enjoy the planets that Zantiu-Braun sent him to. In this case, it helped that the population put up no serious resistance. He didn't even mind being assigned to Memu Bay rather than the capital. The coastal town was small enough to be easily controlled and large enough to boast extensive leisure facilities. Z-B's platoons had made full use of the clubs and bars along the marina since the first week after they landed. Even the locals had reluctantly started to welcome their spending power in the absence of the regular tourists.

The campaign had all gone reasonably well up until the fifth week when some lunatic rebel had firebombed two of the local food production refineries. Now the Z-B governor had been forced to impose rationing on everyone and activate three collateral necklaces in retaliation. The mood in town had soured, although the biochemical factories that were being asset-realized hadn't been affected.

So Lawrence hadn't grumbled too much that evening when Sergeant Ntoko announced 435NK9 had been assigned a hinterland patrol. They assembled early the next morning outside the hotel that was serving as their barracks. A convoy of eight jeeps to carry the three platoons, accompanied by five ten-ton trucks that would bring back any assets they found. They rolled out through the center of town and onto the eastbound start of the Great Loop Highway.

Although most settlers on colony worlds lived in towns and cities that were built on gamma soak patches, some had chosen to establish themselves out among the native vegetation and animals. These smaller townships and homesteads were almost always founded to harvest a valuable native crop or mine some mineral. Out in the mountainous hinterland behind Memu Bay there were several dozen such settlements, all of them linked by the Great Loop Highway that ran in a rough oval around the Mitchell Mountains, a series of high volcanic peaks dormant for thousands of years.

Thirty-five kilometers from Memu Bay the Great Loop Highway was still a wide, level tarmac road that had just cleared the modest barrier of mountains that encircled the coastal town. The Mitchells were rising out of the thick jungle ahead. Lawrence sat in the front passenger seat of the jeep while Kibbo drove on into the foothills country. He could see the range stretching away into the vanishing distance. Vulcanism had pushed an enormous plateau ridge up out of this side of the continent, running parallel to the coast for over two hundred kilometers. The table of the plateau was reasonably level, a kilometer and a half above sea level. Because of its size, it had a microclimate all its own. Amid the continent's pervasive tropical heat its domination of wind patterns pulled in a cooler, moist air that irrigated the whole area. Some of the most vibrant vegetation on the planet ran rampant around the plateau's lower slopes. Two major rivers flowed down from its heart, along with hundreds of smaller watercourses. But it was the peaks themselves that dominated the skyline, varying from small rounded mounds to giant jagged rock cones over seven kilometers high. Snow gleamed on over half of them, astonishingly bright in the clear air.

"Anyone ever climbed those mothers?" Kibbo asked.

"I think so," Lawrence said. "I saw some tour offices in town that ran trekking holidays up on the plateau."

"I hope the poor schmucks wear some kind of power suit. It looks tough up there."

"Mount Horombo is the tallest, eight kilometers. You wouldn't need a power suit for that, just really good thermal underwear. And an oxygen gill as well, I'd imagine."

"You fancy trying it?"

Lawrence laughed. "Not a chance."

"I wouldn't mind a go," Kibbo said. "It must be a fantastic sight from up there."

"I bet it's covered in cloud most of the time."

"Jeez, Lawrence, you're such a pessimist."

Lawrence had a private smile at that. It had been long

enough since that miserable, emotionally confusing time that had been born out of his assessment in Amsterdam. The memories no longer hurt when he brought them out to examine them. In fact, now he could look back in wonder at how he'd ever fallen for a girl as weird as Joona in the first place. Fate below, the signs he'd ignored!

There were even times when he thought about reapplying for starship officer college. Z-B might be run by a bunch of pricks, but it was still his only chance of realizing his old dream. Despite everything that had happened to him over the last few years, he'd never quite let go of the hope. And he'd notched up a damn good record with strategic security. Sergeant Ntoko said he was going to recommend him for a corporal's stripe once this Thallspring campaign was concluded. And he was damned certain his stake was large enough to satisfy the college's deputy principal now.

Life was good for him at the moment. Pessimism played no part in it.

The convoy started to wind its way up the plateau's slope. As the climb progressed, so the trees on either side of the road became progressively taller. Their branches were swamped with vines, enormous webs of them strung between boughs and trunks in a thick, shaggy lattice, sprouting cascades of gold-and-black flowers. Ripe gray fruit was dropping all around the vehicles, making the tarmac slippery with their pulp. Humidity closed in around the convoy, with layers of warm mist coiling between the tree trunks. Their Skin was almost white as it repelled the heat.

"Great Loop, my ass," Sergeant Ntoko grumbled from the lead jeep. The road was now down to a single band of tarmac, whose edges were being remorselessly chewed away by tufts of aquamarine grassmoss. He was often slowing for fallen branches, using the jeep's front grid bars to push them aside. Even the surface was cracking open, revealing dusty red earth underneath. Insects similar to terrestrial termites were busy building their soil castles up around the base of trees. The tiny creatures secreted a chemical cement, bonding the minute grains of dirt together so the odd-shaped tumuli glimmered with a metallic purple-and-blue sheen under the intense sunlight.

The air was noticeably cooler when they finally drove out onto the top of the plateau. Ahead of them, the trees were thinning out, although the individual specimens seemed to be even larger than those on the slopes, reaching thirty to forty meters high. In between them, the ground was carpeted in monster plumes of spiked crown reeds, their withered leathery seed pods swaying three or four meters into the air. The Great Loop Highway degenerated to a heavily compacted dirt track with deep wheel ruts that had been burned through the reeds. Sooty black clumps lined the sides where the highway maintenance robots had incinerated any living frond that crept back across the designated route. To prevent any possible misdirection, slender metal pillars were spaced every kilometer, wearing a high collar of solar cells to power their beacon lights and transponder.

Lawrence went back to mountain-gazing again. There were more Mitchells visible now, thrusting bluntly out of the lush subtropical plain, the awesome monuments of tectonic petulance. Long strips of dark cloud scudded round them, raining hard on the lower slopes.

The convoy reached Rhapsody Province first—an area marked out by dozens of slate-gray slag heaps thrown up across the plain where mining equipment bit deep into the plateau's stratums, hunting out bauxite deposits. The spoil was fresh and dark, slick with dank chemicals that prevented anything from growing on its treacherous, shifting sides.

Aluminum was not one of the metals that Auley supplied to the planet. And capturing another asteroid that had it in quantity wasn't cost-effective given the amount used by Thallspring's industrial concerns. Most cities had secured their own source, and Rhapsody Province even produced enough of a surplus for Memu Bay to export to other settlements.

Located almost dead center of Rhapsody was Dixon, a mining town, or at least an engineering center where the mining machinery was maintained and repaired. A full quarter of the town was given over to huge sheds of corrugated composite where cybernetic tools and humans worked alongside each other to service the mining and processing equipment. There was even a small fusion plant sited a kilometer away, a squat white concrete hexagon with a slightly convex roof. It was surrounded by pylons that carried bright red power cables out to the active mining sites.

The houses, shops and offices that made up the rest of Dixon all had a prefab appearance. Their layout and size were highly individual, but every wall was made from the same insulated composite paneling, and each roof was a matte-black solar collector. Even the air-conditioning cabinets that stood outside were the same make, all with their fans whirring away behind rusting chrome dissipater fins. The plain's dusty volcanic soil hazed the air above the grid of streets, frosting every surface with a dark ocher patina.

As soon as the convoy arrived at the end of the main street they had to stop and reverse. One of the massive excavator processors was being delivered back to its site after a spell in the maintenance sheds. The unit was twice as long as a locomotive, and three times as wide. It was sitting on an even bigger low loader whose caterpillar tracks must have been as wide as Lawrence was tall. He whistled with respect inside his Skin helmet as the massive rig crawled past, shaking the nearby buildings.

Captain Lyaute, who was commanding the convoy, ordered the vehicles to draw up in the town's central square. By the time they'd parked in a circle and the squaddies jumped out they'd gathered quite a crowd. It was the first time Z-B had visited the plateau; people were curious. They were also suspicious and sullen, standing well back from the Skins.

Lawrence hoped they weren't going to have to demonstrate the weapons capability of their Skin. It had taken several unpleasant days to convince the citizens of Memu Bay that they were invincible and everyone should just knuckle under and cooperate. But this bunch were tough engineers working hard for a living. They also had a quantity of hardware and tools that could damage Skin if correctly and creatively misapplied.

Lyaute snapped out a few quick orders, and three Skins snatched a civilian each. Before anyone could react, they'd been fitted with collateral necklaces. The captain started to speak to the crowd; he quickly had to crank up the volume on his Skin's speaker as the crowd shouted abuse and insults back at him. They were furious at what they were being told, that the squaddies were going to go through their town and help themselves to anything remotely valuable. Any resistance would result in the collateral necklaces being activated.

After walking just a couple of streets, Lawrence decided the convoy was a waste of time. There really wasn't much in Dixon worth taking. Not that the town saw it that way. As soon as the Skins went into the cavernous maintenance sheds they found the articulated trucks that brought the aluminum down to Memu Bay. Except they hadn't been used since the day the starships flew into orbit overhead. Every one of the big trailers was filled to capacity. But that was only a fraction of the hoard. Lawrence and Amersy walked into the first of the big sheds, only to stop in amazement. Aluminum ingots were piled up as high as the roof. Nobody was going to send the town's one product to the coast where it could well be stolen by the invaders and taken away on their pirate starships. Amersy laughed at the metal mountain. "What kind of idiot thinks we can afford to transport a shitload of aluminum on a starship?" he asked.

Lawrence didn't share his mockery. Thallspring had never heard of asset realization before this first campaign had arrived. Out here in the hinterland they certainly didn't know what was regarded as valuable. They were playing safe, trying to protect what they'd worked for. He could appreciate that.

When Dixon's AS was scrutinized, the logs showed that the excavator processors were operating at minimum capacity, and had been for weeks. The only reason the operators hadn't stopped them altogether was that it was more trouble to start them up again than keep them ticking over like this.

Captain Lyaute explained the financial reality of asset realization to the mine managers, trying to tell them they were wasting their time by the go-slow. They just glared at him.

A jeep was sent over to the hospital. Some of the more advanced medicines and vaccines were loaded into it. A truck was driven out to the fusion plant, where it could stock up with expensive spare components. Lawrence and Amersy helped shift excavator cutting heads from their storage racks in one of the big sheds, heaving them into the back of a truck. The bulky cones were studded with long compression-bonded diamond blades that Z-B would strip off back in Memu Bay before boosting them up to the waiting starships.

"That's our livelihood you're killing," one incensed technician yelled at them. "How can we buy food if we can't work, you bastards?"

Lawrence ignored him.

"The guy's got a point," Kibbo said. "This does seem kind of petty. The blades, okay, they're high-tech and expensive. But medicines from the hospital?"

"It's the same deal for everyone on the planet," Lawrence said. "They'll produce replacements as soon as we leave. We're not taking the factories with us."

"Still not quite what I thought we were about."

"Being seen up here is what we're about," Ntoko told

them. "We're flag-waving, that's all. The hinterlands have to know we're here, and we're real. It happens on every campaign. You send a convoy round all the backwoods settlements to prove they're not immune. If we didn't, places like this would be a haven for refugees and resistance movements. And the way to pay for these convoys is—"

"Is with valuable goods," Amersy finished. "Asset realization in miniature."

"You got it."

Lyaute decided that the convoy wouldn't be spending the night in Dixon. Anger was running on high voltage through the townspeople, and there were too many of those tempting heavy tools available.

When he got back into the jeep, Lawrence watched several Skins from the other platoons stuffing jewelry and household cards into their personal bags.

The convoy camped out on the plain that night, thirty kilometers past Dixon. They got to Stanlake Province the next day, where waterside villages were strung out all along the shores of the lake itself. They harvested strange aquatic weeds for their complex organic compounds, which were used down in Memu Bay's biomedical factories. Assets here were even scarcer than in Dixon. All the villages used solar panels and wind turbines to generate their electricity; there was no fusion plant. Only three of them had a doctor's office—serious patients were taken to Dixon, or air-ambulanced out to Memu Bay. Electronic systems were years out of date. In their raw form, the organic compounds were worthless. Lyaute did check that all of the harvest was being sent to Memu Bay. It was.

They drove on past the lake, deeper into the high plain. On the third day they reached Arnoon Province. Several of the Mitchell peaks were clustered together here, creating deep, meandering valleys between them. Dense forest had colonized the sheltered saddles between the high slopes. Slim curlicues of white cloud poured down from the craggy snow-covered peaks to writhe amid the treetops. The Great Loop Highway led straight through the thickest section of vegetation. Trees and vines blotted out the sun for long periods. Flat tree stumps lined the route where the highway robots had cut the path, with bulbous fans of bright coral-pink fungus growing out of the damp, rotting wood. But not even the robots could cope with the creepers that twined across the gap. Despite the jeep's all-terrain suspension, the journey began to get rough.

"What do this load of hillbillies do?" Kibbo asked as they rocked and shook their way through yet another tunnel of vegetation. Two hours in the cool forest, and they still hadn't seen any sign of human habitation.

"I think they grow tigercotton," Amersy said.

"That's Laeti Province," Lawrence said. "Arnoon collects willow webs. It's a vine that only grows in this forest, similar to wool apparently. They have a load of cybernetic looms that churn out clothes and rugs. The cities pay a premium for it." He grunted at the sudden echoes of memory. Isolated crofters living amid the mountains, selling their crafts to the rich city folk so they could buy the few items necessary to maintain their independence.

Either Great Loop Highway marker posts were being stolen, or the transport office had decided to space them farther apart in the forest. Up in the lead jeep, Ntoko was relying heavily on his inertial guidance and maps uploaded from the Memu Bay Town Hall, which were at least a decade out of date. They kept coming across junctions and forks in the muddy road that simply didn't exist in the old files.

Sometime around midafternoon, Lawrence caught sight of what he assumed to be a willow web. A small jade ball of what looked like tightly packed velvet gossamer was hanging from a high bough by a single sapphire-blue stem. The wood where the end of the stem was attached had swollen to three times the bough's ordinary width. In contrast to the rest of the forest vegetation, which was slick and damp, the furry surface of the ball was completely dry, as if it repelled the alternating deluges of rain and mist that inundated the trees.

They began to appear regularly after that. The first one had been a baby. Some he glimpsed deeper in the trees were as tall as himself, their surfaces mottled with dark lichen and plagued by ordinary creepers. Now that he knew what to look for, he could often see the shriveled stems protruding from distended branches, their ends cut cleanly.

Ntoko had actually driven into the village before he realized it. To start with he thought it was just a broad clearing. Then there were small children racing in front of the jeep, and he slammed the brakes on. Wheels slid on the carpet of wet mossgrass as the children squealed. They stopped, mercifully without hitting anyone. Smiling faces were suddenly surrounding the vehicle, hands waving as excitable young voices chattered.

Lawrence, who admittedly wasn't wearing his safety belt, had been thrown forward into the front seats. He straightened up and scanned his helmet sensors round the clearing. It was like a multiphase image that resolved only when you looked at it in the correct focus. Where before he'd seen trees with drooping branches tangled by vines, there were now wooden A-frame houses with roofs of elaborate reed mats that were obviously still alive. What he'd taken as a random scattering of shaggy bushes were actually compact gardens.

"Holy crap," Ntoko muttered as he climbed down. "This place is like the most perfect camouflage."

"You're thinking military, Sarge," Lawrence said. "This is just the way people like this live."

"And you'd know all about people like this, space boy."

Adults were emerging from the foliage, or doors, to gather around the convoy's parked vehicles. They hung on to their lively children, waiting expectantly. Captain Lyaute launched into his standard speech about Z-B acquiring ownership of the planet's founding corporation and requiring a dividend. Instead of the anger and resentment rife in Rhapsody and Stanlake, the Arnoon villagers grinned at him, amused by the whole concept. Even putting collateral necklaces on two of them didn't curtail their open derision.

The squaddies started to search the arboreal A-frames. Lawrence began to revise his opinion of the villagers as simple crofters. The inside of the houses were far from primitive. They had electric lighting, and air-conditioning, and running water; kitchens that were fully equipped with every modern convenience, fridges, washing cabinets, microwaves. Lounges that had full AS desktop pearls with large panes and sheet screens to play the thousands of hours of multimedia recordings stored in extensive libraries.

Domestic convenience items aside, it was the decor that impressed him the most. The villagers had put a lot of effort into fashioning their homes according to individual taste. Many of the frames were carved, then painted or bound with willow wool in vibrant primary colors to emphasize curves and textures. Carvings invariably followed the Hindu or Buddhist school, with many-armed gods and goddesses serenely contemplating the village as they sat astride powerful serpentine dragons. Inside the houses mood decoration was favored, with children's playrooms formed inside nests of brash, exciting patterns; lounges came in elegant abstracts or classical ornate, making them cozy and welcoming, bedrooms in cool, subtly blended pastels. He began to wonder how many of the villagers were artisans. Somebody had to take care of the practical stuff.

Spaced among the houses were carpentry shops where the furniture was fabricated along with the trusses and internal structure for the A-frames. There was a pottery as well. Another craft shop made jewelry. Lawrence found that one easily enough. Skins clustered around it like bees swarming over their queen. Few of the pieces were made from gold or silver or platinum, and none had precious gemstones, but the bracelets and necklace pendants and earrings were all beautiful, handcrafted with care and precision. Most of them had cavities to hold neurotronic pearls. They were all being stuffed into Skin pouches.

Lyaute held court with the village council in what passed for the meeting place, a pavilion created from ten big snowbark trees planted in a circle, their upper branches closely interlaced to form a broad dome of contiguous leaves that admitted slender ultramarine sunbeams while remaining closed against the rain. The captain was alarmed at the lack of anything valuable enough to qualify as a valid asset. The village did have a doctor, but a quick search of her surgery produced only five or six packets of medicines, all close to their expiration date. As in Stanlake, all serious accidents and illnesses were immediately shipped down to Dixon or on to Memu Bay. When Lyaute asked how many people lived in the village, the council said about six hundred. That was a lot more than the file from Memu Bay's Town Hall indicated. Even so, there weren't enough A-frames to house them all. Because many of us live out in the forest where we gather the webs, the council replied. Directions to such homesteads were not given by grid reference, they were in the form of *walk half a kilometer along the northwest path, take a right-hand turn and walk for another kilometer, then ford the stream and head for the second peak to the south . . .*

Lawrence was pretty sure the villagers were quietly making fun of the captain. But he had to admit this kind of community was unlikely to have anything Z-B could want. Like Dixon's aluminum ingots, sweaters, however colorful, were not a viable starship cargo.

The captain decided to send Platoon 435NK9 out to the wool center to check on the technology level employed. As it was a little way outside the village he called for guides. In spite of the now-blatant looting that was going on in the A-frame houses and craft workshops, the group of villagers accompanying them remained cheerful and polite. When their great-grandparents first settled this area, they told the Skins, they had brought bulldozers and concrete and dammed one of the larger streams fed by the snowfields on Mount Henkin high above. A hydro plant built into the base of the dam provided the village and the wool center with all the electricity they needed. Their own small tool shop could fabricate any replacements for the hydro system, making them almost self-sufficient.

The wool center was history with a vengeance for Lawrence: five airy wooden barns filled with old-fashioned machinery busily whirring away. Piles of big willow webs were being combed out and spun into yarn. Dyeing vats bubbled away. Bobbin winders clattered.

When it was compacted into strands and knitted together, willow wool made excellent water-resistant clothing. A fleet of small vans took the sweaters and ponchos and blankets down to Memu Bay, coming back with food and consumer goods. Unlike the aluminum trucks from Dixon, they'd continued making the run after the starships arrived. When he looked at a few of the sweaters coming off the knitting machines, Lawrence thought the patterns were conservative, nothing like as bold as the ones Jackie designed.

Three of the Skins draped sweaters around their shoulders. Lawrence didn't bother. He was scanning the area, still suspicious. The whole village idyll setup was just a little too perfect.

"Where does that path lead?" he asked one of the villagers. Where the road to the village headed off from the wool center, a small footpath disappeared into the forest.

"Just to the lake."

"It's well used." His sensors were showing him multiple

overlapping footprints in the drying mud, and the smaller branches of undergrowth on either side were cut back.

"What have you got, Lawrence?" Kibbo asked.

"Path to the lake, he says."

"What's at the lake?" Kibbo asked the villager.

Lawrence watched the slight smile form, then vanish on the man's face. He was going to say water, Lawrence knew.

"There is a temple at the lake, that's all."

"A temple?" Kibbo said. "What sort of temple?"

"It is a place of tranquillity, where one goes to meditate in solitude."

Kibbo conferred with Ntoko for a minute. "Okay, let's go check it out," the sergeant said.

"As you wish."

The villager's name was Duane Garcia. He was in his late forties, with thick curly black hair and a slightly rounded face to which a smile came easily. He was healthy and fit-looking in that way all people who led and relished outdoor lives appeared to be. Thinking back, Lawrence hadn't yet seen a villager who lacked vitality. Even the elderly ones seemed unrestricted by their age, while the little kids were like a gang of unruly miniature angels.

It started raining heavily as the four Skins and two villagers trudged down the path. Mud splattered Lawrence's legs up to his crotch. The droplets messed up half of his helmet sensors, producing blurred visual images.

Duane Garcia pulled his sweater's hood up over his head and whistled happily.

"Who's this temple dedicated to?" Kibbo asked.

"We don't worship gods," Duane said. "The universe is a natural phenomenon."

"Amen to that," Lawrence said.

"So why the temple?" Kibbo persisted.

"It's not a temple in the standard sense. We call it that because the architecture is a homage to some of Earth's historic buildings. The man who designed and built it was a good friend of my grandfather. Apparently he was quite upset when people started calling it the temple."

They topped a small rise, and the forest fell away along with the ground. Beneath them was an alarmingly steep slope that led down into a heavily forested little valley. It was vistas like this that gave rise to the notion of Shangri-la. And Lawrence could well understand why. Mount Kenzi, the second-largest of the Mitchell peaks, stood guard at the far end, a tremendous rugged wall of rock whose upper reaches were cloaked in a thick layer of snow. Below the frostline, waterfalls tumbled hundreds of meters down its sides to vanish into the upper strata of forest with a continuous explosion of white spray englobed by rainbows. The valley itself was the gulf between two of Kenzi's buttresslike foothills, with a river running its length. Tributaries slithered along the base of every crease in the land.

Directly below where they stood was a perfectly circular crater lake that had bitten slightly into the northern wall, resulting in a crescent cliff. A small island rose out of the center, like the back of some slumbering marine giant. At its highest it could only have been a few meters above the surface of the lake. A few trees grew around the edges, their sun-bleached roots struggling for purchase among the boulders.

There was a simple structure in the middle of the island: five columns of black-and-white fluted marble that supported a wide arched stone roof. Underneath it were two tiers of circular stone seating that could probably hold about twenty people. The whole thing did look distinctly Hellenistic.

A gravel path led away from it to a small wooden jetty. An identical jetty had been constructed on the shore of the lake opposite. A rowboat was tied up to the end.

"That's it?" Ntoko asked.

"Yes," Duane Garcia said.

The sergeant scanned his helmet sensors around. The path in front of them switchbacked down the risky slope. In several places, where the drop was sheer, the villagers had built handrails. Seventy meters below, the path wound into the dense forest again to emerge by the jetty.

"Okay, we've seen enough." Ntoko turned around and started walking back to the wool center. The other Skins went with him.

Lawrence remained on the top of the ridge. He still had that feeling that the villagers were having them on somehow. The rain was easing off now, boisterous clouds rolling away to the south, retreating from Mount Kenzi's imposing bulk. He requested a full-spectrum sensor sweep from his suit AS and targeted the temple. Nothing registered. There was no electromagnetic activity down there. No heat. It was just inert stone. Large gray-and-white birds flapped sedately through the air, their reflections keeping pace on the still black water.

"Hell."

As his sensors shifted their focus back, he was mildly surprised to see Duane Garcia was still waiting for him. "Checking on me?"

"Certainly not. It's a difficult path back, and there are several forks. We wouldn't want you to get lost."

Lawrence chuckled as they started walking. "Funny, I'd have said that was exactly what you wanted to happen."

Duane Garcia acknowledged the gibe with a slight grin. "I admit your arrival here is not the most welcome visit we've ever had. But I really don't want you having a genuine accident out here, if for no other reason than I doubt your commanding officer would believe it was an accident."

"True enough. Can I ask you something?"

"Certainly."

"Where's your jail?"

"A jail? I'm sorry, we don't have one."

"So in a settlement of at least six hundred people, there are no sinners. Sounds like paradise."

"I'm afraid not. We do have miscreants, of course, every community does. It's just that we don't believe in incarceration as a form of correction or punishment. Other penalties are applied. Restrictions, both physical and material."

"Humm. For the record, I don't believe all this Zen bullshit you people are selling the captain. This whole community is way too nice. Normally, by the third generation, any community founded around a single principle has developed a lot of dissenting voices."

"You have seen the way we live. There is little here to complain about. And if you do, you are free to leave."

"Nope. I still don't buy it."

"You're very adamant about that. Why?"

"I was born a third-generation colonist myself. I know all about the resentment directed toward obsolete restrictive ideals."

"That might just be you. Or perhaps our ideals are more appealing than those of your homeworld."

"Touché." *But I still know you're hiding something,* he thought.

Captain Lyaute decided that it was safe for the patrol to stay in the village for the night. The villagers clearly didn't represent the kind of threat evident in Dixon.

Families were temporarily evicted from various A-frame houses to make way for the squaddies. Lawrence was billeted with Ntoko, Amersy and 435NK9's latest recruit, Nic Fuccio. Their A-frame was one of those overlooking the central park where the convoy vehicles were drawn up. Five comfortable bedrooms, three bathrooms, a lounge, study, reception room, dining kitchen, a family room full of toys; all arranged in a T-shape. As he walked through it, Lawrence thought about some middle managers from Z-B he knew whose apartments were a lot more cramped. He claimed a bedroom with a big sliding glass door and stripped off his Skin. The bulky suitcase of field-support equipment extruded eight umbilical cords, and he plugged them into the suit. Blood and other fluids began to cycle through the flaccid synthetic muscle.

A warm shower washed off the blue dermalez gel, and he dressed in an olive-green sweatshirt and gray shorts to join his housemates on the balcony. Amersy had already found the drinks cabinet and mixed a jug of some lemon-based cocktail. Lawrence went for a can of Bluesaucer. The beer tasted better than it ever had down in Memu Bay.

He hadn't realized it before, but the village was situated on a gentle slope. Half of their A-frame was supported on thick wooden stilts to keep it level. From the balcony they could see out over a broad shallow valley where the forest formed an unbroken dark blue-green cloak.

"Do we ever get any runaways, Sarge?" Nic asked as he settled back in a cushioned sun lounger.

"No. We're too obvious. Why, you thinking of it?"

Nic gestured round the clearing. Eight of the convoy's squaddies remained in Skin, guarding the trucks and jeeps. It was an easy duty. The kids were hanging around the vehicles, with the Skins letting them sit in the driving seats. Several girls had appeared, in their teens or early twenties. Lawrence was sure they hadn't been around before. He would have remembered. Like the tourists at Memu Bay, they didn't wear much, T-shirts or halter tops, and shorts. From his angle, most of them looked cute to beautiful. They belonged perfectly to the idyll image. The duty Skins were very busy talking to them.

"Got to admit," Nic said. "It's tempting. I can see myself living like this once I've earned a big enough stake."

"I couldn't," Lawrence said.

"Why the hell not? A place like this, you've got everything you could possibly need. Hey, I wonder if they go in for that trimarriage lark? Country folk always stick with the original traditions longer than the townies."

Ntoko chuckled and pushed his cocktail glass toward the tall, healthy girls gathered round a jeep. "Two of them together would finish you off, man."

"There are worse ways to go."

"This whole living with nature in the forest crap is a dead end," Lawrence said.

"Whoa there, the man's got a bug jammed up his ass." Nic laughed. "What could be wrong with this, Lawrence? Do a couple of hours' work each day, then spend the rest of the time lying about drinking and screwing. Look at 'em. They're all smiling, none of them are stressed. They know they're on to a good thing."

"I've seen this kind of setup before. It appeals to us because we see it as a break from our job. But you can't live like this for eighty years. You'd die of boredom after six months."

"Oh hell," Amersy groaned. "Here we go, the starship captain speech again. We're all meant for higher things."

"It's true," Lawrence insisted. "This kind of existence contributes nothing to the human experience. It's a retreat for people who can't handle modern society. And the irony is, they're utterly dependent on that society. Villages like this rely entirely on the industrial products made down in the city."

"That's always been the way, Lawrence," Ntoko said. "Different communities live different lives and produce different things. Trading between them generates wealth. Centuries ago it was different nations; now we've evolved microcosms of that, with communities that are going down highly specialized routes. This kind of lifestyle wasn't possible before modern communications and transport. These villagers are as much a development of our society as Memu Bay is."

"They're dreamers who need a good dose of reality to wake up and take part in what the rest of us are building."

The sergeant raised his cut crystal glass to the sinking sun. "Well, this is the kind of dreaming I like. Now have yourself another beer and chill out, Lawrence."

"Yes, Sarge." Lawrence grinned and fished round in the icebox. A group of children walked past the end of the house's garden. They yelled something unintelligible, and Lawrence waved back. Places like this, he conceded, did have their uses. He'd never managed to relax quite this much before on Thallspring, not even clubbing down on the marina.

If he could just work out what was wrong with Arnoon . . . Which was when he saw one of the children, a boy, slip his hand into one of the bushes that marked the boundary of the garden. His fingers slithered casually through the chubby blue-green leaves and found one of the fruits hanging within. It was a smallish globe, with a satin orange sheen. He plucked it with an easy twist of his hand and bit into it. Juice dribbled down his chin.

"I knew it!" Lawrence hissed. "Did you see that?"

"See what?" Ntoko asked.

"He's eating fruit. Real fruit. Off a bush. They're all bloody Regressors."

Ntoko frowned at the boy over the rim of his glass. "You sure?"

"I saw him."

"Filthy habit."

"Fancy making your kids do that."

Nic pulled a face at the liquid slopping around the bottom of his own glass. "Hey, you don't think they've given us any, do you?"

"They'd better not have," Amersy growled.

Lawrence slumped back down in the sun lounger again.

He felt a lot happier now that he'd discovered the village's dirty little secret. *I knew nothing was this perfect.*

The fridge in the A-frame's kitchen had been filled with food ready for them to cook. He made a mental note to check the packaging that tonight's meal came out of. Thank Fate there weren't any animals grazing around the A-frames. At least the villagers weren't that twisted. They ate out on the balcony, microwaving pork barbecue ribs and baked potatoes. Nic even mixed up a couple of TexMex sauces from some sachets he found. Each of the packets had unbroken Memu Bay food refinery seals. Dessert was double-chocolate-chip ice cream.

They sat in the loungers, watching the sun going down behind the huge mountains. The village was dipped in shadow from late afternoon onward. Twilight lasted at least a couple of hours, silhouetting the peaks against a luminous amethyst-and-gold sky. Stars began to shine early on, twinkling brightly through the cold, thin air above the mountains. Eventually, the Milky Way blazed like a fat comet's tail across the night.

Lawrence wasn't really drunk when he went to bed, although he'd had just enough beer to keep his thoughts buzzing. He slept fitfully, waking every few minutes to twist and turn, thumping his pillow. About one o'clock in the morning, he heard the scream.

It was cut off almost immediately. For a moment he thought it might have been the confused end to some dream. Except he thought he'd been awake now for a quarter of an hour.

He lay there, wide-awake alert. It had been a female scream, he was sure of that. Now that he concentrated he could hear some kind of scuffling. Footsteps on wooden stairs. Another cry, muffled this time.

Lawrence came off the bed fast, snatching up a pair of interface glasses. He slipped them on and told his bracelet pearl to give him their light amplification function. The glasses didn't have a particularly advanced capability, certainly nothing like his Skin sensors. But they showed him the darkened bedroom, pulling it into focus with sparkling blue-and-gray tones. He slid the broad patio door open and went outside onto the veranda. His room was facing away from the village clearing, looking along the line of A-frames. Stars glared down on the village, banishing shadows.

A girl, maybe eight or ten years old, was running around between the A-frames. She was barefoot, wearing only a baggy white nightshirt. Her legs and knees were streaked with mud and grassmoss juice. He could see tears streaming down her cheeks.

"Jacintha," she called, then sobbed again. "Jacintha, please, where are you? Jacintha."

Lawrence jogged down the narrow steps from his veranda, asking Fate that Jacintha was her cat, or some other pet.

The girl saw him coming and cowered back. "Please, don't hurt me. Please."

Caught in the silver rain of starlight, she looked just like his sister Janice. *She must be twenty-one . . . Fate no, twenty-two, now. I wonder what she's doing?*

He held his hands out toward the little girl. "It's okay, nobody's going to hurt you. I just want to know what's going on. Can you tell me?"

She took a couple of paces away from him. "Nothing. Nothing's happening."

"Well, now, I'm not so sure, I heard a shout. Was that Jacintha?"

"I don't know."

"Listen, er . . . I'm called Lawrence. Can you tell me your name?"

She sniffled loudly. "Denise."

"Okay. Denise. That's a nice name. So are you going to tell me who Jacintha is?" He was looking round, trying to spot any motion in the village. Several A-frames still had their lights on: he could see the windows glowing around the edges of the curtains, as if they'd been bordered in neon. The convoy vehicles were dark outlines in the middle of the clearing. He could see a couple of Skins standing guard. The fact that they weren't showing any interest in him and the girl made him edgy.

"She's my sister," Denise said.

"Okay. How old is she?"

"Seventeen."

Lawrence swore under his breath. He had a pretty good idea what was happening now. Damn Captain Lyaute for his lack of discipline, and damn Z-B, too, for employing lowlifes as its squaddies. "Tell me, Denise, did somebody take her away?"

"Yes," Denise said meekly. "We were all sleeping together in Paula's home." She pointed at one of the A-frames. Lawrence could see several young faces pressed against one of its windows, staring out at him.

"Go on."

"Two of you came and said they wanted to ask her some questions. That it was about state security. They said she had to go with them."

"Where? Did you see where they all went?"

"Not really. It was this way, though."

She was pointing along the row of houses. And the scream he'd heard must have been fairly close. "Were they in Skin? You know, the big dark suits?"

"No."

"Good." Lawrence started running in the direction she was pointing. "Now you just wait here."

Denise hesitated, her lips quaking.

"You'll be fine." Indigo script scrolled down his glasses, giving him the convoy's current security status. It was level seven, no alerts or irregularities. He told his bracelet pearl to open a link to Ntoko and wake him. There was no light on anywhere inside the first A-frame as he ran past. The second A-frame had one window illuminated. Lawrence dashed up onto the balcony. Three squaddies were inside, sitting around a table playing cards.

The third A-frame had a light on. Its curtains were shut tight. Lawrence took the balcony stairs two at a time, heedless of the slippery dew under his bare feet. He could hear a murmur of voices from inside. The tight, guttural syllables that came from harsh, expectant men.

He pulled the wide patio door open and shoved the curtain aside. It was just as he was expecting. The girl, Jacintha, was lying on the floor, her long T-shirt pulled up round her neck, a pathetic, terrified expression on her face. Three squaddies stood around her: Morteth, Laforth and Kmyre—all from Platoon 482NK3. Laforth already had his trousers off, exposing his erection. Standing between the girl's ankles, he was using his feet to shove her legs farther apart.

All three of them turned to face Lawrence. Their shock and guilt twisted into relief when they realized it was one of their own.

"Jesus, Newton," Laforth spat. "What the fuck is the matter with you?"

"Close the goddamn door," Morteth said.

Lawrence pushed his glasses up so that Jacintha could see his face. "Have they raped you?" he asked.

She shook her head quickly. "No." Her voice was almost a squeak.

"Okay, come with me." He held a hand out and beckoned.

Kmyre stepped between Lawrence and Jacintha, put his hands on his hips and smiled challengingly. "This is our prisoner, Newton. Now either join in or fuck off."

Lawrence could smell the liquor on his breath. "Don't you get it, fuckhead? This is over. Finished. Understand?"

"How can this be over? We haven't started yet, buddy."

"You're not going to start. We're not here for this." He moved to one side. Jacintha was still lying on the floor, staring around uncertainly. Laforth was equally doubtful now; he glanced at Morteth, who was glaring at Lawrence. Jacintha managed to sit up and pull her T-shirt down over her breasts.

"Come on." Once again, Lawrence put his hand out for her.

Kmyre pushed it aside. "Get the fuck out of here, or I'll see to it that you're this terrorist's first victim."

Lawrence bent forward as if he were reaching for Jacintha. As he expected, Kmyre went for a kick to the back of his knee. He spun easily and caught Kmyre's foot as the kick went wide, pushing up hard. Kmyre yelled as his foot was propelled toward the ceiling, sending him toppling backward.

Morteth roared, lunging at Lawrence, arms outstretched. Lawrence stepped inside the bearhug and nutted him. The roar was cut off by the sound of bone snapping. Blood squirted out of Morteth's nose. Jacintha screamed.

Laforth's fist caught Lawrence just to the left of his sternum. He stumbled back from the impact of the blow, catching sight of Kmyre coming at him. This time, he went for the bearhug. A good move, but he didn't quite manage to pull it off. Kmyre predicted him, chopped at his right arm, finding the dead spot perfectly. Lawrence howled at the pain, but kept on pushing, using his momentum to take the two of them into the flapping curtain. It tore free from the rail in a storm of brass rings and they crashed onto the balcony with the thick fabric wrapping round them. Kmyre kicked out. Lawrence kicked back. Without shoes, he had little impact on the other man.

The two of them wrestled around for a moment. But with Lawrence's right arm still useless, Kmyre quickly managed to get on top. Lawrence's knee hit him in the back of his neck. He flopped away just as Laforth caught hold of

Lawrence's leg, twisting hard. Lawrence went with the turn, bringing his other leg round to thud into the man's ribs. Laforth fell over, his inertia taking both of them down the stairs.

It was a bad descent, and there was very little Lawrence could do to slow it, not while he was tangled with Laforth. Elbows and knees managed to hit just about every step on the way as they tumbled. His head caught a glancing blow as well, which more than doubled the number of visible stars. They crashed onto the muddy mossgrass and broke apart.

Lawrence was aware of several people approaching at a run. Half of them were children; the others were adults from the village. He couldn't see any of his platoon. Jacintha was still screaming and light from the patio doors was splashing across the A-frame's garden where they'd landed. The whole village must have been attracted by the commotion.

It didn't bother Laforth, who aimed a kick at him. Lawrence rolled aside easily and swung a punch. His accuracy wasn't too good with the pain distracting him. Laforth half ducked, receiving the blow on his shoulder, and tried to tackle Lawrence. As he closed in, Lawrence kneed him on the jaw, sending his head snapping back. Lawrence grinned down with savage satisfaction as Laforth fell heavily, barely conscious. Then Kmyre landed on his back, and they both collapsed onto the damp mossgrass beside Laforth.

"That's him," Denise yelled. "That's the man."

Great, Lawrence thought, as he blocked Kmyre's chop to his Adam's apple, *they'll think I'm the rapist.*

"They're like savages," a man's voice called out.

"Stop them!" Denise cried. "Stop them. They're hurting him."

"Jacintha? Jacintha, where are you?"

Kmyre kicked Lawrence in the ribs, flipping him over. He rolled twice and crouched. Charged at Kmyre, sending them both sprawling again.

"Stop it!" Denise yelled. "Please, someone."

"Jacintha?"

"Father. Father, I'm here."

"Jacintha."

"Call the dragon," Denise said. "It'll make them stop."

"*No,* child!"

"Are you all right? Jacintha, did they harm you?"

"Stop it, stop it!"

"I'm all right, Father."

With Kmyre pressed against him, hands clawing wildly, there was little Lawrence could do except claw back—the worst kind of wrestling, two drunks writhing round in a gutter. Legs scrabbled and jerked against each other as they rolled again and again.

"Oh please!" Denise wailed.

Brilliant white light stabbed down. Both Lawrence and Kmyre froze. Skin hands closed around them, pulling them apart forcefully.

"What the fuck is going on here?" Ntoko demanded.

Lawrence wheezed down some air, happy to let the Skin support him. He wasn't sure his own legs could do that right now, given how badly they were shaking.

A lot of people had gathered, illuminated by Skin helmet lights: villagers clutching their children; squaddies in shorts, blinking sleep from their eyes; still more villagers arriving.

"Well?" Ntoko asked.

"They—the girl," Lawrence gasped. "I heard a scream."

"Uh-huh." Ntoko glanced at Jacintha, who was in her father's arms, while her mother and Denise clung to her. "Shit," he murmured, and looked at Kmyre in the Skin suit's hold. He was smeared in mud and blood. Laforth was trying to climb to his feet, a painful process. Morteth stood on the balcony, one hand clamped over his nose, pinching the nostrils. His shirt was soaked scarlet by the astonishing amount of blood that had poured out of his mashed nose.

Ntoko beckoned the duty Skin sergeant over, and the two of them put their heads together. Lawrence could hear the occasional murmur.

The two sergeants faced the crowd. "Okay, people, that's it for tonight," Ntoko said. "You three"—and his finger stabbed out at Morteth, Laforth and Kmyre—"back into your billet, where you *will* remain until oh-seven-hundred hours. Travers, you have sentry duty. If they come out before the designated hour, you are authorized to use maximum force."

"Sir." One of the Skins saluted.

Ntoko went over to Jacintha and her family. "Ma'am, do you require medical assistance?"

"No," her father said. His arm had tightened around her. "Not from you." Jacintha gave a miserable nod of confirmation.

"Very well. Could you please return to your house for the rest of the night? And I can assure you, there won't be a repeat of this incident."

"Thank you."

Lawrence was impressed by the way such a simple phrase could be made to project so much derision.

The Skin holding Lawrence handed him over to Nic and Amersy. The two of them had to give him plenty of support as he limped back to their A-frame. Everyone else was heading back to bed, talking in low tones.

Denise suddenly appeared in front of Lawrence. She smiled up shyly. "Thanks." She raced away back to her family before he could say anything.

Nic laughed. "Got yourself a new girlfriend, Lawrence. A looker, huh?"

"Give me a fucking break."

Ntoko appeared where Denise had been a moment before. He wasn't smiling. "What the fuck do you think you're doing, man? You want to be a hero, do it on the company's time."

"Come on, Sarge, like you'd play it differently."

"I would have got me some backup. Don't you learn anything in training?"

"I called you."

"Jesus."

They arrived at their A-frame and went up the stairs to the balcony. Lawrence had to grip the handrail. Now that the adrenaline and endorphins were wearing off, he seemed to hurt everywhere. As soon as he was in the lounge, he fell into the sofa. "I need a drink."

Kibbo cracked a can of Bluesaucer and handed it to him. Lawrence took a sip and decided that was just too macho. Ntoko sat beside him, and opened up the first-aid kit. "Hold still, hero."

Despite the medicines, Lawrence was stiff everywhere when he woke the next morning. He took a hot shower, which eased things a bit. One ankle was badly swollen. Both legs were grazed. He had bruises everywhere. But Ntoko had insisted that the injuries were all superficial. "Nothing that gets you out of duty."

He had breakfast with Nic and Amersy, who had a good time joshing him over the fight. The sergeant didn't eat with them; he'd left the A-frame while Lawrence was still in the shower. He arrived back as they were finishing the meal. "You two get lost," he told Nic and Amersy.

"So what's happening about last night?" Lawrence asked.

Ntoko poured himself some coffee and sat down opposite Lawrence. "I've been talking to the captain about last night. He wants this shut down fast."

"What does that mean?"

"It means no righteous heroes kicking up a fuss when we get back to barracks."

"What are you saying, that those three get off free? They

were going to gang-rape a seventeen-year-old, for Fate's sake. I'm not eating that kind of shit sandwich."

"We all know what they were doing, and they're not getting away with it. But there are ways of dealing with situations like this where we don't all have to lose out."

"How?" Lawrence asked with deep suspicion.

"Okay, let's take it your way, clean and honest, all our dirty laundry out there in public. Morteth, Laforth and Kmyre stand trial. They're guilty, obviously, they get shipped home under guard and serve their fifteen years. Fair enough. But after the facts get read out in court, there's going to have to be an inquiry to find out why it happened."

"They're a bunch of drunken bastards. That's why it happened."

"Sure. But specifically, why didn't Lyaute have enough discipline over his men to prevent them even thinking about this? Why didn't Four-eight-two-NK-three's sergeant stop what was going on? It's the NCOs who take the immediate blame, you know that. How come the Skins on guard duty didn't see what was happening and step in?"

"They should have."

"I know, man. But things are damn slack around here. You've seen how everyone's been helping themselves to everything these poor old hillbillies have. Lyaute should have stepped in hard and fast right at the start. But he didn't, because he wanted a quiet life. So it just keeps getting worse until those three assholes pull a stunt like last night's and land all of us up shit creek. If this ever gets on an official report, half of the convoy is going to have a reprimand on our files."

Lawrence drank some tea, which was getting colder than he liked. "You mean if I do what's right and testify to the commander I'll screw everyone else?"

"Like I said, there are ways of dealing with this. Lyaute can operate through different channels, if you let him."

"What sort of channels?"

"Okay, I'll lay this on the line for you. Say nothing, and you'll come out of this campaign with a commendation on your report that's better than anything you could get for saving the general's ass in full combat and there'll be a stripe on your shoulder for sure. Morteth, Laforth and Kmyre will be quietly shitlisted once we get home. They'll be discharged or given latrine duty for the rest of their lives, and they certainly won't get any sort of campaign bonus, neither will they get any kind of reference from Z-B. Without that, no employer on the planet will touch them. It's the slammer without walls."

"And Lyaute gets himself out of this with a clean record."

"Yeah. Along with a whole bunch of other people who don't deserve a bad rap because he screwed up. And next time, he'll know how to run a command properly. That's got to be worth something, Lawrence. You and I know we've got damn few decent officers."

"Don't, Sarge. Don't try and sell this by telling me I'm going to be improving that useless asshole."

"Okay, man. You look at it any way you want to. But it's your call, and you've got to make it now. This can't be ducked. And if it helps any, I'd have done the same thing last night. It was the right thing."

"Something you haven't mentioned."

"Yeah, what?"

"The girl. Jacintha. What about Jacintha?"

"What about her?"

"Three of Z-B's finest tried to rape her."

"But they didn't rape her, did they? Thanks to the hero of the hour. She had a nasty shock, which is never going to be repeated, because we're never coming back. She can get on with her lotus-eating life again. Six months' time, we'll just be a bad memory fading away."

"That's it? She doesn't count?"

"This is politics, man. Her stake in this isn't as big as ours. So what's your decision?"

Lawrence grinned, even though the bruise on his lip hurt when he did it. At least the sarge was being diplomatic, pretending he even had a choice. He knew damn well if he carried this through and screwed Lyaute and the other sergeants he'd be the one who wound up shitlisted.

That was the way the companies worked. The way they'd always worked. The way they always would work.

He drank some more of his cold tea. "I guess I must have got these bruises falling downstairs last night."

CHAPTER FOURTEEN

THE COURT-MARTIAL WAS HELD IN THE BANQUETING SUITE OF the Barnsdale Hotel, which was barracks to eight of Z-B's platoons and half of the industrial technology corps. There was a dais at one end of the long room, normally used by a band. Today it had a single table with three chairs for the presiding officers, of which Ebrey Zhang was president of the court. Arranged below them, on the dance floor, were another two tables. One was occupied by the prosecution team, led by the Z-B attorney, who was being supported by the Memu Bay police magistrate, Heather Fernandes, and two more high-powered legal assistants. The defense table had two chairs, where Hal sat with Lieutenant Bralow.

Behind them, fifty plastic chairs had been arranged in rows to seat Z-B personnel, selected members of the public and a few media representatives. The first row was reserved for the mayor and whoever he chose to have with him—a couple of old friends, Margret Reece and Detective Galliani. Ten Skins were standing guard around the room, being pointedly ignored by the civilian audience. For once, the power supply was uninterrupted, allowing the lightcones to shine at full intensity.

When Lawrence arrived, escorting Hal, he was disgusted by the weighting. The kid had taken one look at the layout and virtually cringed.

"It's a fucking show trial," Lawrence growled at Bralow while Hal was distracted. The lieutenant answered with a slightly guilty shrug.

Lawrence took a chair from the audience section and brought it up to the defense table. He sat on it and gave Hal a solid, reassuring pat on the knee. The kid responded with a pathetically grateful smile.

Nobody remonstrated with Lawrence. He was wearing his full dress uniform, displaying more decorations than most of the officers in the room. If he wanted to stand by a squaddie under his command, none of the NCOs helping with the court arrangements were going to stop him. Bryant saw where he was and glared before sitting with the other officers.

The sergeant major called for silence. The presiding officers marched in and took their seats on the dais.

Lawrence couldn't fault the procedure. Prosecution made its case well. The details of the case were explained to the court. Selective sections of various police interviews with Hal were also played. Twenty minutes in, and already it looked bad.

Detective Galliani was called to the stand and told the court about Hal's alibi, which the kid had stuck to the whole time.

"Did you manage to trace the taxi that the defendant claims he took?" prosecution asked.

"No, sir," Galliani answered. "The traffic regulator AS has no log of any taxi being used on that street at that time of night. And Mr. Grabowski was most insistent on the time he left the barracks. In fact, we pulled the logs for every taxi in Memu Bay that day. None of them were unaccounted for at either of the times when Mr. Grabowski said he was traveling to and from the alleged brothel."

"Ah yes," prosecution said smugly. "The brothel the defendant says he visited. Does it exist, Detective?"

"No, sir. Mr. Grabowski himself identified Minster Avenue as the street where this alleged brothel was situated. We investigated every house. They are all private residences."

Lawrence had visited Minster Avenue himself two days ago. Not in Skin, he wore civilian clothes, a shirt with a high collar to cover his valves. Before he went, he trawled images of the street from the town hall planning department and showed them to Hal, who'd pointed unhesitatingly to number eighteen.

Standing outside the house Lawrence took his time looking around. There was the neat little front garden with its wrought-iron fence, just as Hal described. It guarded a squat white stone facade, with big windows, the paintwork clean and bright. Like all the others along the street, a home for the upper-middle classes. Lawrence activated his bracelet pearl and called up his Prime. A complex indigo image slid across his optronic membranes as the quasi-sentient program decompressed from its storage block. Perhaps it was his imagination, but it seemed brighter than the bracelet pearl's standard icons.

He opened a link into Memu Bay's datapool and told the Prime to trawl the household AS and the local traffic logs. Information began to scroll up almost at once. Whatever software the KillBoy resistance group used to cover their tracks, it was excellent, which strengthened his suspicions that they had compromised e-alpha.

Number eighteen's household AS told him nothing, because it had been inactive for over a week. The system was still waiting repair. Smaller independent sections of the house's network were functioning on autonomous backup mode, but they didn't have memory logs. Strangest of all, the security system was also offline; its sensors weren't even drawing power.

Minster Avenue's road traffic logs confirmed there had been very few vehicles driving along the street during the night Hal claimed he'd visited. Certainly no taxi had pulled up outside number eighteen. But the Prime dug deeper into the local transport network. Between 1:48 and 2:10 the network dataflow had increased by a small percentage.

After Hal had left.

There was nothing in the logs to account for the increase.

"Shutting up shop," Lawrence muttered to himself. The minute data abnormality wouldn't convince a court that had Hal's DNA sample taken off the girl. He wasn't even sure if it would be admissible in court. But it was good enough for him: an electronic graffiti roughly equivalent to spraying *KillBoy was here* across the front of the house.

Lawrence walked over the road and rang the brass bell. It took a minute before the black front door swung open. A woman in an apron stood in the hallway, giving him a suspicious stare. "Yes?"

"Elena Melchett?"

"Yes? Who are you?"

"Lawrence Newton. I'm covering the alien rape case."

Elena Melchett didn't look as if she wanted to cooperate with the media. "So?"

"Ah, the alien suspect claims he was somewhere in this street when the incident happened. It's his alibi. I was wondering if you had seen anything?"

"Mr. Newton, that obscene crime took place at one o'clock in the morning. I was in bed asleep. I certainly didn't see any alien thug hanging around outside."

"I didn't think so, thank you. Er . . ." He fished around in his pockets while Elena Melchett grew increasingly impatient. He found his media card and activated a visual file. "Sorry to be such a pain, but do you recognize this man?" The card's screen showed a picture of Hal.

Elena Melchett studied it. "No."

"Really? That's odd."

"What do you mean?"

Lawrence told the card to switch to another file. "This is a blueprint of your hall, isn't it?" He peered past the woman at the big staircase that curved up to the second-floor landing.

This time Elena Melchett barely glanced at the image. "It's similar."

"I'd say it's identical. Even down to the marble tiling."

"What do you want, Mr. Newton?"

"That alien suspect, he put this image together with an architect program. How would he know what your hallway looked like if he'd never been here? You did say you didn't recognize him, didn't you?"

"Get out!" Elena Melchett ordered him in a strident voice. "Out, and don't come back. If I see you around here again, I'll call the police." The glossy door slammed shut.

The prosecution had got Hal up on the witness stand. Lawrence could finally appreciate the saying about someone being his own worst enemy. It wasn't going well. In fact it was excruciating just being in the same room.

The prosecution wanted to know why he'd jumped curfew.

Hal—good old honest fresh-from-the-farm Hal—said he did it because he was desperate for sex.

The prosecution wanted to know where he'd gone that night to hunt for sex.

Hal told them the brothel on Minster Avenue, doggedly sticking to his version of events. Lawrence presumed it was because his mother had always told him to tell the truth.

The prosecution tore that version of the fateful night to shreds, and there wasn't any evidence that Lieutenant Bralow could produce to back Hal up. Then they went on to ask about the genetic samples. Hal claimed the girl was a whore, and that the rest of it—the rape allegation, the nonexistent brothel—was all a setup by KillBoy.

It didn't go down well. Francine Hazeldine's haunting statement had already been played back to the court. Lawrence had watched the presiding officers as her fragile voice had described what happened that night, detail by agonizing detail.

The more the farce carried on, the more Lawrence admired KillBoy's strategy and resourcefulness, and the more angry he became. Hal was just too easy. He wanted to stand up in the banqueting suite and face the locals, asking: "Why don't you try this one with me?" But then, the devastating effect that the trial would have on Z-B's morale was the final triumph of that elegant strategy.

He was also haunted by the terrible specter of responsibility. There should have been a trial very similar to this last time he was on Thallspring. The fact that it had never happened was in no small part due to him. Justice then had been circumvented rather than served. Now justice was coming back to strike them with a vengeance.

Lawrence spent most of the time wondering if the two could possibly be connected.

Only by a God with a very twisted sense of humor, he decided.

After five hours of testimony and witness examination, the presiding officers recessed the court so they might consider their verdict. They took ninety minutes, which Lawrence thought was a diplomatic enough length of time given that they'd already decided that verdict before the court-martial even began.

Hal stood in front of the dais facing the presiding officers, his shoulders squared, as Ebrey Zhang announced the findings.

On the charge of disobeying a direct order and breaking curfew: guilty.

On the charge of misleading the local police: guilty.

On the charge of assault and rape of a minor: guilty.

"No!" Hal yelled, incensed. "I'm not."

There was a sigh from the audience, not of jubilation, but a shared sense of justice and victory. Against all the odds, they'd been given the right outcome.

Hal sat down again while Lieutenant Bralow gave what Lawrence had to acknowledge was an eloquent plea for clemency. Then everyone stood for the sentence.

A very troubled-looking Ebrey Zhang said: "Halford Grabowski, given the grave nature of this abominable crime, we find we have no alternative but to impose the most severe sentence it is within this court's power to issue. You are hereby sentenced to death."

Hal Grabowski went berserk. He screamed obscenities at the presiding officers and started to run for the door. Anyone who got in his way was felled with powerful punches from his hulking frame. The audience scrambled for safety, also screaming.

It took two Skins to hold on to the enraged squaddie and administer a sedative. His unconscious body was dragged out of the banqueting suite.

Ebrey Zhang straightened his uniform and cleared his throat. "Sentence to be carried out at dawn the day after tomorrow. Leave to appeal is denied. Lieutenant Bralow, please inform your client of the outcome. This court is now concluded."

The presiding officers filed out. Lawrence didn't move.

Bralow turned to him and said: "I really am sorry. He didn't deserve this." As he didn't get an answer, he nodded nervously and hurried out. The audience was lining up at the doors at the rear to get out and back to their town and their lives. It wasn't long before everyone else had left.

Amersy and the remaining members of 435NK9 lined up in front of the defense counsel table. Lawrence looked at them one by one. "If anybody wants to stick with Zantiu-Braun, you'd better leave now."

A couple of them snorted in derision; the rest simply waited expectantly for their sarge to tell them what to do next.

"Okay," Lawrence said. "Time for us to start playing unfair."

This time Josep drove a car out to the spaceport. He arrived in the middle of the afternoon and passed through the main gate with the identity of Andyl Pyne, a junior manager with the catering company that had the franchise for the administration block. The spaceport's general management AS assigned his car a slot in park 7. Because of Andyl Pyne's somewhat lowly status, he had a long walk back to the block itself.

He carried a slim briefcase with him, *de rigueur* for management of any level. Sunglasses were also obligatory, so he wore a cheap plastic pair. His light green one-piece coverall wasn't quite regular, but it had the catering company logo on its breast pocket. He had boots rather than shoes. All in all, his appearance was well inside the permissible norm.

Ahead of him, the afternoon sun shone on a five-sided structure with slightly convex walls of darkened glass. From where he was, the administration block resembled a closed-up tulip flower with a blunt tip. It stood by itself to one end of the terminal building, away from the much taller control tower. Although the building was only five stories high, the architect's plans that his Prime had trawled out of the data-pool showed a service level and another five floors below-ground.

When he reached the main entrance he had to repeat the whole security identification procedure, allowing the AS to check his palm and facial pattern. Security in general was a lot tighter in the administration block than the main terminal, thanks to all the Z-B staff that worked there now.

Inside, he ignored the reception desk and the two Skins standing beside it, walking directly to the bank of elevators in the central lobby. No one who came in on a regular basis would be intimidated or even concerned by them anymore. He took an elevator down to the first sublevel, where building maintenance had its offices, along with the canteen. So far everything matched the floor plan and security camera images they'd trawled.

Josep went into the toilets and claimed an empty cubicle. The AS logged him through a security camera. Coverage inside the administration block was almost universal, with only places like the toilet cubicles free of cameras. Not that their absence mattered: the AS followed everyone's position constantly, you couldn't trade places or switch with anybody else. It was Andyl Pyne who went into the cubicle; if anyone else came out the AS would sound the alarm.

It wasn't the AS that Josep was trying to avoid, he simply needed time to make a few alterations. At this stage, sharp-eyed humans were his greatest worry. His Prime went into the administration block's network and began editing the monitor logs. The AS soon registered that it was Sket Magersan who was in the cubicle. Once the switch of electronic records was complete, Josep stood still and concentrated. The d-written organelles deep inside his cells quickened and began to modify his flesh. Facial skin pigmentation darkened slightly. Features started to morph. The tip of his nose broadened out, while the nostrils widened. Lips fattened up. His cheeks sagged slightly, then stiffened, giving the impression of a flatter jawbone. Irises became a light hazel.

There was a small vanity mirror in his briefcase. Josep took it out and examined his rearranged face.

They'd spent a long time observing Sket Magersan as the Z-B spaceplane pilot drank in Durrell's bars and ate in its restaurants. He'd been chosen because he was similar in height, weight, age and general profile, so Josep's d-written systems would be able to imitate his physical appearance without too much trouble. His voice was deeper than Josep's, and his accent was pure Capetown, but a direct link with a neurotronic pearl running a vocal synthesizer program took care of that. Josep even had the man's walk down pat; his shoulders had a lavish swing when he hurried.

The image in the mirror was that of Sket. Nodding in satisfaction, Josep stripped off his green one-piece and reversed it. This way round it was a standard dark-gray Z-B pilot's flight suit, complete with insignia, baggy leg pockets and elastic waist.

Josep stepped out of the cubicle and took his time washing his hands, making sure the toilet's security camera could see him clearly. The Prime monitored the security AS, but there was no caution alert issued. He went back to the elevators, and descended to sublevel five.

Simon Roderick had decided on the simplest system possible to monitor the key vault. Keep electronics to an absolute minimum and rely on human observers. That distrust of electronics extended to not informing the spaceport security AS that a covert operation was being mounted. They didn't even tell the local security staff.

According to the administration block records, the office on sublevel four was assigned to Quan and Raines, who were

Third Fleet quartermaster staff. They were the ones in charge of spare parts being shipped down from the starships to keep the Xiantis flying, working with their own AS to keep expenditure to the lowest level possible. Even the data that flowed into the office from local networks supported their assignment, although it did contain a large amount of information not directly applicable, such as staff schedules and flight profiles. Typical bloatware overload.

Simon occupied the office next to theirs. The AS had him listed as a spaceplane avionics systems manager, a title that could be confirmed by the number of boxes and small packages that kept getting taken inside, all of them labeled with electronics department bar codes.

The only thing missing from the two offices was a security camera. Simon wasn't going to risk the opposition being able to spy on his own spies.

They'd set up the first office as an observation center. One wall was now covered in sheet screens, relaying various scenes from the administration block. Each one was connected to a single fixed-position lens via fiberoptic cable. Picture quality was well down on standard sensors, but this way there was no electrical cabling. A power flow, however small, could always be detected. The screens even had their own independent power supply, a bank of cells in one corner. That way there was no drain on the administration block circuits, which could be tracked through the datapool.

Adul Quan watched the elevator doors open on sublevel five. A man in a Z-B flight uniform walked out.

"Who've we got here?" Adul grunted. Procedure was to confirm everybody who arrived on sublevel five. The screen feed was linked to a desktop pearl that had no connection to the local network: instead, it was loaded with personnel files. Whoever the new arrival was, he walked right underneath the lens covering the elevators.

"Sket Magersan," Braddock read off the card's display.

"One minute." He was frowning as he riffled through a stack of hard copy. Both he and Adul had privately bitched about Simon Roderick insisting on keeping printed records. But their chief was convinced that e-alpha had been compromised, leaving their data memories wide open to manipulation. So every morning, the spaceport's personnel schedules were printed out. This way they could check who was supposed to be in the administration block and who was suspect.

Braddock glanced down Magersan's sheet, stopped and read it carefully. "Shit, he's supposed to be on leave today. Spent the last five days flying."

Adul straightened up and peered at the other screens covering sublevel five. "So what's he doing here, and down at that level?"

"Good question." Braddock went to stand beside his colleague. They watched Magersan walk along a corridor, nodding affably to people.

"Heading toward the vault," Adul said in a low, excited tone.

"That's not certain."

"Bullshit." Adul was on the edge of his seat.

Magersan had arrived at the communications department. He gave the security sensor a codeword and put his hand over the scanner. His voiceprint and blood vessel pattern must have matched. The door slid open.

"Sir!" Braddock was heading for their office's connecting door. He opened it hurriedly. "Sir, I think we have something."

There were three offices making up the communications department, linked by a short corridor. Security cameras confirmed that as usual there were only two people inside, one in the first office, one in the third. When the outer door opened, Josep slipped in and waited for it to shut. Prime edited him out of the security cameras' vision. Neither of the two Z-B officers inside the department had heard the door. He paused for a second, then ordered his Prime to call the man in the first office. It was a query from the maintenance division about a glitch in a spaceplane satellite tracking unit, with the quasi-sentient program generating the supervisor's image and voice.

When the communications officer started to answer, Josep walked quickly past the office and went into the second. His Prime disabled three alarm sensors that were triggered by his entry. He shut the door and locked it with a manual bolt, then drew a quiet breath as he waited to see if either of the officers had reacted. Images from the security cameras hung behind his eyes, showing both of them at work behind their desks.

The key vault had a big steel door reinforced by boron longchain fiber. Before Z-B arrived, it had stored the gold and platinum used in the microgee manufacture of electronic components. Now the metal had been shipped up to the starships, leaving a lot of empty space for Z-B to store its keys.

There were two locks that worked on deep-scanned hand patterns. They had to be activated simultaneously by two different people. Josep took a pair of slim dragon-extruded modules from his trouser leg pockets and applied the first one over the top lock. Its surface undulated slowly as it melded itself to the scanner. The second module went over the bottom lock. He activated them together, and the magnetic bolts snapped out with a *clunk* loud enough to make Josep flinch.

He pulled at the heavy door, swinging it back. The vault was a cube, measuring eight meters along each side. Bright lights came on in the ceiling as he walked in. The walls were lined by metal grid shelves; a single metal table stood in the center. There were fifteen black plastic cases stacked up on the shelving—seventy-five centimeters long, fifteen centimeters high. Z-B's silver emblem was embossed on the top of each one.

Josep took the first one off the shelf and put it on the table. He ran a sensor over it, which drew a complete blank. There was no detectable power source inside. If it was alarmed, they'd done it in a way he couldn't beat. He flipped the catch and opened the lid. His Prime reported that the datapool remained silent. No alarm.

The case contained three trays stacked on top of each other, each with a hundred memory chips. He scanned them quickly, looking for the number they wanted. The Xianti flights for the next five days had already been scheduled, and their communication code assigned to them. He and Ray had chosen one in four days' time, which would give everyone else involved in the operation plenty of time to prepare and fly over from Memu Bay.

He found the designated key in the third case he opened. The little memory chip fit into the interface slot on his bracelet pearl, and the code transferred without a hitch.

Josep smiled broadly. That was it. The last major obstacle eliminated. Not that the rest of it was easy, but the odds of a successful completion had just risen considerably. So much was waiting behind this moment, so many awesome possibilities.

He put the case back on the shelf exactly as he'd found it and left the vault.

Simon Roderick waited patiently outside the elevators on sublevel five. His DNI provided him with a simple audio channel to Adul, who was watching the screen in his office on the floor above.

"He's closing the vault," Adul said. "Gadgets coming off the locks. Putting them back in his pocket."

Simon shifted his sensorium focus. The blue-gray corridor around him melted into hazier shadows. It was sliced by long, thin threads of brilliant emerald light, lurking just below every fuzzy surface. Some of them glowed with an intensity that rivaled the sun, while others were more delicate, flickering at frequencies almost too fast to notice. He was even aware of the little jade ember alight inside his own skull.

The standard human senses of taste, touch, sight, smell and hearing provide a phenomenal range of input for the brain to cope with. In most cases it does so by subtly concentrating on one sense at a time, sliding the others into a peripheral mode. By using this inherent neural programming ability, geneticists reasoned that the sensorium could be expanded to cope with new inputs. The batches of Rodericks provided them with a perfect opportunity to experiment, by adapting and modifying each fresh generation.

The idea behind it, developing an ability to "see" electrical patterns, was an old one. Psychics, shamans and con artists had been claiming they could find north for centuries, along with other mystical perceptual traits. The discovery of magnetite in human brain cells back in the late twentieth century had bolstered their claims with the kind of pseudoscience backing such people thrived on. Given the minuscule quantities of magnetite actually involved, it was extremely unlikely that any of them could act as a human compass. In any case, there was no specific interface between the particles and the brain's neural tissue. That had to wait for genetic engineering to manipulate cells, incorporating magnetite particles into a ferro-vesicle cell model. The actions of a magnetic field on the particles suspended in serous fluid were found to generate discernible neural impulses.

After that, the alignment of the ferro-vesicles to provide a valid image had to be determined, along with its size and how the impulses were best introduced into the brain. By the time the SK2s were gestated, the design was essentially complete. Their electric sense organ took the form of a membranous crown with a nerve path direct into the medulla oblongata. It allowed them to see wires carrying current, or dataflow. But most important, and the reason the Rodericks wanted the ability in the first place, it allowed them to sense the impulses of another human brain. They were never going to be able to read thoughts directly, but by observing a brain in action, they could determine the emotional composition, see how much creativity was going into the thought processes, how much memory. As a lie detector, the ferrovesicle organ was almost infallible. In negotiations with the senior management of other companies they had a profound advantage.

"He's coming out," Adul said.

Simon started walking forward. There were a few other people in the corridor. He certainly couldn't risk clearing the building; that would have alerted Sket Magersan. Simon was already quite worried about the man's capabilities; the last thing he wanted was for this operation to degenerate into violence.

He passed one man whose aura was bright and dense, barely distorted by his clothes. It corresponded to the contentment running through his brain. Another man was considerably dimmer, with areas resembling sunspots amid his emanations. Simon was experienced enough to spot a hangover without even having to ask any calibrating questions.

Sket Magersan stepped out of the communications division. In the electromagnetic spectrum he was a human nova. Simon almost stopped, he was so surprised at how bright the man's aura was. For a moment he thought he might even be some kind of android. But no, the body's bioelectric patterns were all recognizably human, simply more intense by an order of magnitude. He also carried several electronic modules in his pockets. Tight flux lines pulsed around them, indicating high-level power cells.

Simon didn't recognize any of them. It was hard to resolve

anything through the vibrant electromagnetic glare, but the secondary patterns induced by the internal systems were fabulously complex and pervasive. He couldn't spot the usual signature of neurotronic pearls.

When the two of them passed in the corridor, Sket Magersan's thoughts registered a small degree of nervousness, but nothing to indicate suspicion. Simon wondered what his own brain activity must reveal. If he'd known this was what he was up against he would never have allowed himself to be in the same building as the . . . refined man and his alien gadgetry. This discovery could well have a value greater than every asset realized on Thallspring. *Where the hell did he come from? And what was causing that aura?* One thing was for sure, this wasn't Sket Magersan, the pilot who was in Z-B's files.

"Sir?" Adul queried.

"He might have weapons, but I'm not sure. Proceed as planned."

Josep pressed for an elevator. It took a few seconds to arrive. He resisted the impulse to shout at the slow old mechanism to hurry. *I did it!* Walked into Z-B's most secure facility and stole their crown jewels. Their last remaining problem would be to get the dragon through spaceplane cargo security. He and Raymond had already developed ideas about that.

The elevator doors opened. A man came out, giving Josep a distracted nod. Josep stood to one side, then walked in. He pressed for the first sublevel. The doors closed and the elevator began to rise.

A quick change of identity back to Andyl Pyne in the toilets. Maybe thirty minutes after that he'd be back in the car and away from the spaceport.

His d-written neural cells lost all contact with the administration block network. *How could that happen?* He frowned, but the lights were still on and the elevator still moving. Maybe the elevator was somehow isolating him from the node. But it hadn't happened on the way down.

Josep blinked as he swayed against the wall. The control panel with its buttons and illuminated floor display wavered as if he were looking at it through water.

What the fuck is this?

He jabbed the emergency stop button. Nothing happened. The elevator was still moving. His legs sagged, taking him down to his knees. Blotches danced across his vision. There was no air. He drew down a deep breath, but it made no difference. His strength was fading fast.

Air, he had to have air. He called up what strength he could and punched at the door, where the two halves sealed in the middle. The metal buckled under his fist. It was smeared with blood. He punched it again, and the dent deepened. There was no gap between. Another punch. This one had no effect. He didn't even hear the bang of the impact. His forehead was resting against the door. It wasn't cold. He couldn't feel anything. His last conscious thought was directed at the Prime stored in his bracelet pearl: help.

That evening they asked Hal if he wanted a priest in the morning. He told them to go and fuck themselves with a Skin dick. They asked what he wanted for his last meal. He said a boiled egg. After that, they left him alone.

Dawn was at five-twenty-two.

At four-thirty, Lawrence and Dennis came to visit. Hal was being kept in one of the cellars under the Barnsdale Hotel. Two Skins were on permanent guard outside the tough wooden door, and the master-at-arms had fitted Hal with a remote restraint bracelet—just in case. Nobody was really expecting any trouble. The Skins got a call alerting them to

Lawrence's arrival a minute before he turned up. He and Dennis were pushing a small hotel kitchen trolley along in front of them.

"But he didn't want a meal," one of the Skins said.

"I know," Lawrence said. "But we brought it anyway. It's a fillet steak, his favorite." He took the silver top off a plate so the Skin could see.

"Okay then, you'd better go in."

Hal was lying on the small cot in the corner of the room, hands behind his head. He looked around when Lawrence and Dennis rattled the trolley across the floor. "I told them I didn't want any of that crap."

"The chef's a local," Lawrence said. "And the guilt's starting to sink in. If we go back and tell him he left it under the grill too long he'll probably need therapy for the rest of his life. You know what a pain these liberals are."

Hal grinned and went over to the trolley. The guard shut the door.

"Sarge," Hal said quietly. "I know what you said, but I've been thinking. I want to take the injection. It doesn't hurt none, and it'll be just like going to sleep. I figure that's for the best, you know."

"Hal, I need you to face the firing squad. I'm sorry, I know it's going to be tough, the toughest it could ever get for anyone. But that's the only way."

"Only way for what?"

Dennis bent down and pulled the trolley's white linen cloth aside. There was a field-aid case on the lower shelf.

"What's that for?" Hal asked.

"A simple way out of this mess," Lawrence said. "Which is the only thing that worries me. Someone else might figure this out. Sit down, Hal."

Hal did as he was told.

Dennis put the case down beside him and opened it up. He

unwound two coils of clear thin tubing and plugged them into Hal's neck valves.

"Now listen," Lawrence said, and started to explain.

It was uncharacteristically cold for Memu Bay as the first traces of wan predawn light skimmed eagerly over the horizon. Myles Hazeldine had put on a warm woolen coat to accompany Ebrey Zhang out into the orchard garden at the back of the Barnsdale Hotel. The orchard had been selected because it was enclosed by a tall stone wall.

Myles assumed Z-B wanted to keep the execution private from the morbidly curious local citizens. But Zhang had told him the wall would also help stop the bullets. It had taken a moment for Myles to understand what he meant. "A firing squad?" the horrified mayor had asked. He couldn't believe that even Z-B was this barbaric. Like the rest of Memu Bay, he'd assumed they'd simply administer an overdose of some sedative. That Grabowski would quietly slip from sleep into death and that would be the end of it.

He should have known better. This whole terrible event was never going to finish with quiet dignity. Now he was going to have to stand and watch as bullets tore into a man with an explosion of blood. It was an outrage against civilized decency. He couldn't even feel glad that Grabowski was going to die like this. He'd wanted justice, certainly. But this was more like medieval vengeance.

"The condemned man does have this right," Zhang had explained awkwardly. "There are three methods of execution, and he can select one. If he doesn't, the court will decide for him. It is unusual to ask for firing squad." There was a thin line of perspiration on Zhang's forehead, despite the early morning chill.

Myles didn't ask what the third method was. He followed Zhang to a place at the rear of the orchard garden. His eyes never left the single post that had been set into the ground in front of the far wall. The earth was fresh around its base. Sandbags were stacked up behind it.

This was everything his ancestors had left Earth for. The ultimate act of callous inhumanity. Myles jammed his shaking hands into his pockets and looked at the grass. Think of Francine, he ordered himself sternly, the terror she went through.

Someone was barking out orders. Myles forced his head up.

The sergeant major marched the eight-strong firing squad out of the door and halted them behind the line painted on the grass seven meters away from the post. The unlucky squaddies had been chosen by the old short-straw draw. He'd spoken with each of them beforehand, telling them that Grabowski would want someone who could shoot straight and clean, and they were not to let him down no matter their feelings, assuring them that this duty would never go on their record.

When they'd left the briefing, sullen and subdued, he'd quietly thanked Allah that he wouldn't actually be pulling a trigger himself. Then Lawrence Newton slipped in and had a quiet word. The sergeant major had listened to his old comrade's request and nodded agreement. Anything else, he didn't want to know about.

Edmond Orlov and Corporal Amersy led the condemned man out into the orchard. Hal showed no emotion as they stopped him by the post. Edmond tied his wrists together behind the post and whispered something to his friend. A smile played over Hal's lips. Amersy offered him a blindfold, which he accepted.

The two men from Platoon 435NK9 saluted their comrade and marched away.

The sergeant major looked to Ebrey Zhang, who gave a slight nod.

"Squad, raise your weapons."

The sound of palms slapping precisely against weapons carried across the orchard.

"Take aim."

"Hey, Zhang," Hal called out. "You are one miserable fuckup of a commander, man."

"Fire."

Myles Hazeldine threw up. The sound of eight rifles firing at once had stunned him. It suspended time in silence. Then he casually turned his head and saw Grabowski's body shudder as it was thrown back against the post. Blood burst out of his chest with frightening speed. And the big young man was falling, slumping forward onto his knees, with only his bound hands holding his ruined torso up. Sound returned to Myles's universe, a roaring in his ears. A human being had been slaughtered in front of him. Because of him, the deal he'd cut. He knelt forward and vomited helplessly onto the orchard's dew-moistened grass.

Traditionally they were called the burial detail, though on Thallspring there would never be any grave dug for a member of Zantiu-Braun. Company policy governing death away from Earth was for a cremation and scattering of the ashes.

Hal Grabowski's own platoon had demanded the right to perform that last duty, and Captain Bryant certainly wasn't going to try to say no—he really didn't need any open rebellion among his own men right now. So while the firing squad was marched quickly away they walked out of the hotel with a stretcher and a bodybag. They untied Hal's hands as Ebrey Zhang was supporting the retching mayor and laid their dead friend out on the blood-soaked grass. He was lifted gently into the bodybag, which was zipped up, then transferred onto the stretcher.

They carried him away as the mayor and the senior officers went back into the hotel. The cleanup detail emerged after that, to take down the post and remove the sandbags.

There was the blood to be washed away, too. By midmorning, there would be no trace left of the execution.

The burial detail carried the stretcher through the rear corridors of the hotel and out into the small courtyard used by delivery trucks. A van was waiting there to take the body to the crematorium. Its doors were opened quickly, and the stretcher pushed inside. Had anyone managed to see the interior, they would have been surprised to see how much medical equipment was inside. It could almost have been mistaken for an ambulance.

"Go!" Lawrence yelled at Lewis.

The van sped out of the courtyard.

Dennis was already ripping the bodybag open. "Oh hell," the medic grunted when he saw the mess of gore that was Hal's chest. "How many bullets?"

"Only three," Lawrence said. He caught sight of the body. "Sweet mother of Fate! Can you do it?"

Dennis was already activating Hal's Skin suit, which lay crumpled in the corner of the van. He brought the extension tubes out and began plugging them into the kid's valves. "Cut the shirt off."

Blood began to squirt out of the jagged wounds, pouring onto the floor of the van. Lawrence took a scalpel and sliced the shirt fabric, pulling the saturated cloth aside, leaving room for Dennis to work. When he brought his hands away, they were dripping blood.

For the first time he began to have doubts—something he hadn't acknowledged before. He refused to let doubt be part of the equation as he focused himself on accomplishing just one thing: not letting the bastards murder Hal. He wanted a victory over KillBoy as subtle and devious as KillBoy's relentless assault against the platoons on the streets of Memu Bay. But now he could actually see the terrible damage that the bullets had caused. . . .

Dennis was trying to clamp off the torn arteries in the

chest cavity. "His heart's so much raw meat. We'll have to drain and reinflate the lungs."

"The brain?" Lawrence demanded. "What about the brain?"

"I don't know." Dennis gave him an anguished look. "It was seven minutes." His optronic membranes were scrolling medical data almost too quickly for him to follow; Hal's Skin was using up its drug capsules at a dangerous rate as it tried to minimize cellular trauma.

"But we superoxygenated his blood," Lawrence said. "You said that would last him."

"It should, it should." Dennis finished clamping one artery and went for the next. "Odel, anything?"

Odel was attaching a sensor to Hal's scalp. He looked at a palmtop display. "Not yet. Still flatlined."

"Come on," Dennis screamed at the kid. His face was streaked with Hal's blood, which he'd smeared there with the back of his hand.

"Lewis, how long till we get there?" Lawrence shouted.

"Three minutes, Sarge."

"Is he alive?"

"I don't know," Dennis barked.

"Three minutes, Dennis, that's all. The crash team's waiting."

"Crash team?" Dennis's voice was veering toward hysteria. "Crash team? One struck-off doctor and a couple of field medics, and you expect them to perform a fucking heart transplant?"

"It's a biomech heart, Dennis, you just plug it in and switch it on."

Dennis laughed. "Oh, Jesus fucking Christ."

"Dennis! What about Hal?"

"I'm trying, god damn you." There were tears in the corners of his eyes. "I'm trying."

"Hey," Odel cried. "Hey, I've got brainwaves showing here."

Hal's mouth dropped open. His tongue flopped about weakly as he gurgled through the scarlet blood that was foaming out of his throat.

"Hal!" Lawrence shouted. "Hal, you hear me? You hear me, Hal? You hang on for us, kid. We've got you. We won't let you go."

CHAPTER FIFTEEN

SANTA CHICO. THE ORIGINAL PARADISE PLANET.

From orbit its colors were intense—Earth-like, but brighter, more alive. There were no pastels here, no gentle shadings. Vegetation was vivid emerald; fast-growing, all-conquering. That made the few real deserts intolerably bleak: hot as hell and dry as Mars. Barriers between the extremes of rich life and barren desolation were short, making the contrasts ever more striking. The oceans that covered over half of its surface were livid sapphire. Snow-white clouds were magnified by the deep atmosphere as they hurtled through the high, turbulent jetstreams.

The air with its 30 percent oxygen content was poisonous to unmodified humans. But for native life, the abundant gas was raw nuclear power to its biochemical processes. Evolution here had grown thorns on everything.

For some it was a magnificent challenge. A chance to live

differently, abandoning the strictures that governed society on Earth.

Just how differently, Corporal Lawrence Newton was only just realizing. Now that the company of eight platoons had arrived at the chemical-processing factory, all he could see was decay. The facility was spread out over several acres. Its design illustrated only too well the new angles with which Santa Chico's inhabitants set about attacking old problems. The closest he could come to describing it was organic gothic. Large sections of the machinery were alive, membranes and nodules blending smoothly into the metal and plastic portions. Or had been alive. Or were still alive but de-evolving, reverting to more primitive forms. He couldn't quite decide. The factory obviously hadn't been in use for some time.

It had been sited in a small valley that was a natural habitat for the gargul plant, a bush of yellow-and-scarlet sponge-like dendrites whose sap contained wondrously complex molecules that could be employed as vaccine bases. Such compounds were a big factor in the original settlement effort. Santa Chico's vegetation was a natural pharmacopoeia, which when harvested properly produced an astonishing array of medical and industrial applications. Now the garguls had returned to the factory, growing over and under the inert machinery. In many cases, Lawrence could see fissures in the pipes and organolytic crackers allowing the bush to take root. Fluffy lichens tarnished the big metal mountings. Pink moniliform fungi spiraled up support struts. Vines and creepers scaled the highest burner towers, forming thick-webbed buttresses.

Jeeps and trucks transporting the platoons fanned out from the narrow, overgrown track and halted beside the fecund equipment. Captain Lyaute ordered a sweep of the area. "I know it looks like a complete waste of time," he told the platoons over the general frequency. "But we have to find out if anything can be salvaged from this crock of shit."

Lawrence took Kibbo, Amersy, Nic and Jones with him. They stuck together as they searched their assigned section of the factory. For an hour they wandered through the tangle of machinery. Green-and-yellow-striped tigergrass had sprouted along the roads between the equipment, reaching their knees, which made it tough to walk even with Skin. Pipes that looked as if they were made out of bark arched overhead, connecting tanks to refinery buildings. Dark, dank fluids dripped down from small splits. They walked around ion exchangers and splitters grafted together out of translucent mushrooms the size of apartment blocks. Metal pumps and valves jutted out of the ground at odd intervals, hopelessly antiquated and out of place amid the slick biomechanical systems. One end of their section had an office block of stacked oblong rooms in a cube of girders: no power, broken windows, dead electronics. When they peered in through the open doors, creatures slithered through the darker recesses, escaping observation. There was nothing of any value anywhere. Nothing left working.

Every time they saw a bird in the distance, Lawrence flinched. Four of the fleet's drop gliders had collided with windshrikes, flying animals larger than pterodactyls. The impacts had killed the windshrikes instantly, but they'd also sent the drop gliders tumbling out of the sky to smash across the landscape.

That was when Lawrence knew they'd made a mistake coming here. From the moment 435KN9's drop glider splashed down in the lake outside Roseport all he wanted to do was get into a spaceplane and fly back up to a starship. If there were any left. He really hadn't wanted to fly down to the surface to begin with.

They'd encountered exo-spheric weapons on their approach. One starship wiped out completely, all hands lost.

Two more badly damaged. You couldn't keep that sort of news from the platoons in the surviving starships.

Rumor had it that at first the admiral and the captains didn't even know what attacked them. Sensors showed massive storms within the planet's far-flung magnetosphere, where the flux bands compressed and twisted into hundred-kilometer vortices that spat out lethal particle beams. Remote satellites sent into the heart of the magnetic hurricanes revealed huge webs of chain molecule filaments, spinning for stability and manipulating the planetary magnetic field. Santa Chico had discovered how to create ephemeral energy cannon on a titanic scale. They weren't even purpose-built. As the fleet found out later, the webs were simple induction systems to power orbital craft and microgee station facilities. Turning them into weapons was just a matter of reprogramming.

When the starships did reach parking orbit, the satellites couldn't find any major cities on the planet. There were just large towns like Roseport on the existing settlement areas. They did find thousands of smaller towns and villages, all with identical pearl-white buildings. And there didn't seem to be a datapool, at least nothing the fleet could link into. Which meant there was no central government to receive Z-B's legal claim for asset realization. The flipside of that was it left them unable to deliver a warning about the gamma soak threat. Not that they knew where to gamma soak to intimidate the locals.

It gave everyone a foretaste of future events.

For Lawrence the defining moment had come when he waded ashore from the drop glider. They'd aimed for a broad lake that ran along the side of Roseport, one of the first settlements. On the final approach the drop glider's forward camera had shown them a smear of white houses almost engulfed by brilliant emerald vegetation. The place resembled a Greek fishing town embracing the stony slopes that led away from the water.

Roseport might have been built by humans originally, but the new-natives who occupied it now were no longer thoroughbreds. Bipeds, tripeds, quadrupeds, even serpentine organisms, were ranged on the open ground between the buildings and the lake; they were mammalian, reptilian, equine, canine, simian, hulking things that didn't fit any terrestrial classification. Each of them had retained a few human elements—hands, limb joints, facial composition, even hair in the form of manes and plumes—but that was all. Most had a kind of segmented exoskeleton, a dark amber shell as flexible as thick rubber. Some had developed entirely new types of hides.

The Skins stood in silence just above the shore in a long line, staring up at the city's inhabitants as a variety of eyes and sonic pulses stared back.

"Who the fuck are they?" Ntoko asked as he crossed himself.

He should have known.

The Santa Chico settlement and investment company had been formed out of some very specific companies on Earth, those that relished challenge and tackled it with a bravura lack of orthodoxy. The majority came from one location.

Always a technology leader, California attracted the smartest researchers and entrepreneurs to its cutting-edge companies, most of which were moderately unconventional. Money excused a great deal, allowing them to live almost entirely as they pleased provided nobody else got hurt, and the technology companies did make a great deal of money. With Hollywood as their neighbor and prime example, every combination of sexual and narcotic abuse was enthusiastically pursued, along with freewheeling households.

To start with, this hippie chic company culture was based on electronic hardware and software, spreading from the techno heartland of Silicon Valley to install factories in every urban industrial precinct. Then, with the human genome finally read, genetics and biotech began their rise to prominence. The whole nature of "outrageous behavior" began to alter. Instead of using drugs, the new lords of biotechnology experimented upon themselves. The ethical review boards that licensed their company research activities were predominantly made up of elderly advisors, many of whom had strong religious beliefs. They saw cloning as inherently evil, and altering the human norm as an unholy sin. These were not the kind of restrictions acceptable to pioneers who more often than not had a completely different set of moral values. Several areas of research went underground.

Rejuvenation was the main goal, biotechnology's holy grail. Though to that should be added enhanced body and organ functions, new and expanded senses, innovative methods of pleasure stimulation and limb redesign, among others. Athletes, professional and amateur, were keen devotees. The cosmetic applications were also hot topics; California's ultimate deity. Just as the Internet had broken down the privacy and censorship barriers fifteen years before, so the tidal wave of quasi-legal medical, genomorph and cosmetic products helped overwhelm the moral legislators.

Billionaires cured themselves of cancer, cloned themselves to create new styles of dynasties, changed sex, lost weight without resorting to diets or liposuction, added new senses and prolonged their lives for decades. Organic AIs were germinated, and in many cases interfaced with humans. The muscle skeleton suits (Skin's predecessors) were a popular product with government and corporate paramilitary divisions. Neurotronic pearls dominated the processing market. Thousands of new products made extensive inroads into millions of lives. At the heart of it all were the specialist companies. Small partnerships of ideas people with a few research labs and a lot of stock options, whose products would be licensed out to the bigger companies for mass production. They were the ones who were fascinated by Santa Chico. Here was a whole new range of high-energy biochemistry ripe for exploitation. And the only way to physically access it was by taking their current physiology modification processes to the extreme. They didn't need crude gamma soak areas on which to build settlements, they could adapt themselves to the excessive oxygen. Even body shape could be reprofiled to take advantage of the environment.

They never expected this conversion to be carried out in one clean switch. Various avenues would be explored over generations. Mistakes abandoned. Successes built upon. But slowly and surely the divergence from terrestrial humanity would grow until the final generation could walk naked under an alien sun and breathe the air without technological support.

Z-B's briefing had explained all this to the platoons. The emphasis had been on cellular adaptation, giving the impression of ordinary-looking people with slightly different lungs. It had never mentioned just how great the physiological changes would be.

Looking at the inhabitants of Roseport, Lawrence knew the briefing had barely touched the history of Santa Chico. Whatever had happened here since settlement began, it wasn't going to act in the platoon's favor.

In the beginning, Santa Chico had been the one exception to interstellar trade being a nonprofit activity. Among other things, the planet churned out a panoply of high-grade vaccines, biologicals, antivirals, vector treatments and biotronics, products that were unique, cutting-edge, and hard to duplicate. With an entire planetary ecology of potent vegetation and aquatic plants as raw materials, every new batch was an improvement on the last: more sophisticated, more effective. New settlers would travel outbound from Earth, and the completed biologicals would return, paying for starship maintenance and any technological and industrial equipment the inhabitants had requested. But over the last few years the starships had been returning with less and less cargo. As fewer settlers were heading out, the Earth-based portion of the Santa Chico development corporation became heavily debt-laden. Zantiu-Braun had performed a leveraged buyout and sent its Third Fleet to realize all those highly profitable biological assets.

The platoons lining the shore were ordered to advance up into Roseport. Their audience moved aside, filling the air with a loud, high-pitched chittering sound, as if a whole jungle full of chimpanzees were screaming at once. Later, the AS up in the starship would decipher the chittering as a very-high-speed hybrid of Spanish and Valley English. Captains ordered snatch squads forward to fix collateral necklaces. Skin amplifiers boomed out instructions, telling the new-natives not to resist, that they were being held responsible for . . .

The fight started immediately. New-natives swarmed down the rocky slope into the line of Skins. They didn't seem to have any weapons, but they were strong and extremely fast, easily a physical match for Skin. There were so many of them, and the platoons were so closely bunched still, that using darts and other nonlethal weapons was difficult.

What appeared to be a hairless ape leaped on Lawrence, carrying him to the ground. Huge clawed hands were either trying to remove his Skin helmet, or more likely just rip his entire head off. Lawrence gripped the thing's wrists and tried to prize them off. His Skin wasn't strong enough. Sheer surprise made him freeze for a second. Nothing in training had dealt with a situation like this. Skin always gave squaddies the advantage.

He pushed down with his right leg, shifting the pair of them over. Then he punched the thing on its sternum. It grunted in pain but kept twisting its claws round Lawrence's neck. Lawrence punched it again, feeling the tough amber hide give fractionally. After a few more seconds of futile wrestling, Lawrence ordered the Skin to fire its electrical pulse. The ape-thing screamed, its limbs locked as the charge ripped through it, then resumed its attempted decapitation. Something like a baby elephant joined in, kicking Lawrence in the ribs. He was left with no choice. His Skin's nine-millimeter pistol deployed through the carapace, and he shot the ape-thing at point-blank range. The first bullet simply enraged it further. Lawrence had to pump half a dozen shots into the demented creature before it finally lay motionless on the tigergrass. Vivid scarlet blood spilled out from the bullet holes in its torso and neck.

Lawrence staggered away from the thing, his ribs aching from the kicking administered by the baby elephant. He ignored that. Nausea and giddiness threatened to knock his legs from under him. He'd never killed anyone before. Not another human. And that's what this was, however distorted. Those clever Skin weapons had always absolved him, turned it into a nonissue.

Now the air around him crackled with weapons being discharged. The agonized screams of mortally injured new-natives cut through it all. Something approximating a Neanderthal ran straight into Lawrence, sending both of them tumbling to the ground. Lawrence brought his pistol arm around automatically. Targeting graphics centered on the prehistoric throwback's head. It had a tall, scalloped ridge running from the top of its nose over the crown of its skull, with a lacework of blue veins throbbing prominently. Very Homo sapiens eyes stared wildly at him, allowing him to read the new-native's fright and anger.

"Fuck off," Lawrence bellowed. He jerked the pistol nozzle up and fired three shots into the air. The new-native rolled aside and scrambled to its feet, sprinting away. Lawrence slowly clambered up as his Skin's peristaltic muscles pushed fresh ammunition along a feed tube into the pistol's magazine.

There was movement on every side of him, with a hundred voices shouting into his communication link. It took Lawrence a long moment to realize what was happening. The fight was breaking up. New-natives were fleeing back up the slope into the streets and buildings of Roseport, running, galloping, limping, even hopping. Dozens of bodies lay behind them, draped over the pale rocks; some were drifting through the shallows, blood spreading out of their wounds to stain the water a dense crimson. Hundreds of little ripples were expanding as aquatic creatures began to feed on the unexpected bounty. It was carnage on a scale Lawrence had never envisaged. Nor was it exclusively new-natives sprawled on the ground. Several Skins were tangled among them, their carapaces pulped and buckled, oozing gore.

Shots were still being fired into the backs of the retreating new-natives. Sergeants and captains yelled to cease fire.

"Sweet Fate," Lawrence whispered. Skins were on their knees around him, helmet valves open, allowing them to vomit. Lawrence's Skin AS reported it was infusing a cocktail of narcotics to help him cope with the shock its medical monitors had revealed. He felt light-headed, as if everything he'd just witnessed were part of some terrible i-drama. He didn't want to move, to take part, help his injured comrades. Just wanted someone to switch the whole image off and wipe the memory clean.

"Hey, look," Nic shouted. "Look up there. Jesus God, what is the deal here?"

Lawrence pushed his sensor focus into the cloudless sky above. He almost laughed; the numbing drugs made it seem funny. Just when you thought it couldn't get any worse . . . The cargo pods were hitting the lower atmosphere, their madcap descent slowing to subsonic speed. White-and-yellow parachutes bloomed high overhead, lowering them gently. A flock of windshrikes glided among them with fast grace. Massive crocodile-contoured jaws snapped and champed at the domes of fabric. Teeth the size of human hands tore easily through the nylon. With their chute panels ripped apart the pods began to plummet downward. They hit the ground at terminal velocity and burst apart in silent explosions of shattered crates and mangled equipment.

After the injured Skins had been treated as best their limited medical supplies would allow them, the erstwhile governor of Roseport gathered his officers for an emergency conference. They had been down for ninety minutes and hadn't even entered the city yet, let alone established collateral. Nearly a third of their equipment pods had been wrecked. The local inhabitants were nothing like they'd been led to expect. And the starship captains were reporting continued attempts to sabotage and attack the big vessels in orbit: subversive software was contaminating every datalink, while kinetic spears in retrograde orbits were probing their physical defenses. The admiral's orders were to establish a dominant presence among the new-natives, then draw up an inventory of possible assets.

Roseport's governor went along with that, but put securing the local spaceport at the top of his priority list. Lawrence was in the company assigned to retrieve the pods that had survived. He just counted himself lucky 435NK9 wasn't one of the platoons ordered into Roseport itself. As he and the others tramped through the clinging tigergrass, they heard a near-constant barrage of small-arms fire and grenade explosions. They could see very little movement amid the peaceful sprawl of squat white towers that formed the majority of the little city's buildings. But the communications link gave them a continuing story of ambushes and booby traps.

Even out on the lush plain skirting the city they weren't immune. Infrared sensors were all but useless in the rolling expanse of tall tigergrass. New-natives lay in wait, hunched down among the roots, bulky creatures capable of damaging Skin carapaces with a couple of swift blows and often making a clean escape after they'd battered a squaddie to the ground. Communications became difficult as interference and jamming increased throughout the day. Somebody here was operating sophisticated electronics.

By nightfall the company had gathered enough equipment to set up a camp with a heavily guarded perimeter. Jeeps and trucks transported the whole lot over to the spaceport, a single runway that had been set out to the north of town. With their escape route secure and a large-caliber arsenal at hand, the squaddies relaxed slightly.

Lights shone in the city that night, lemon-yellow windows radiant against the deep night. Strange shadows moved along the walls in jerky motions. Sounds echoed through the still air, helping to fuel the invaders' imagination, making them wonder what the new-natives were busy building.

On the second day, the governor divided up his forces. Several platoons would attempt to establish a foothold in the city again, while other companies were dispatched to known industrial sites. Satellite observation had revealed the factory structures were still intact, though most were apparently deserted. Best of all, a squadron of twelve TVL88 tactical support helicopters had survived, and the engineers had spent the night assembling them. The companies could call on a full aerial assault if they got into any trouble. When Lawrence's company drove out that morning, the pilots were taking odds on how many windshrikes they were going to bag apiece.

Lawrence called in the empty factory offices to Ntoko, and they turned around to walk back to the company's vehicles. After the first few reports, Captain Lyaute had decided the factory was never going to work again. He was recalling all the scouting parties.

"I don't get it," Kibbo said. "Why did they let this place fall apart in the first place?"

"Fate knows," Lawrence said. "But at least we know why they stopped exporting all that fancy expensive biological junk. They just don't produce it anymore."

"That's not a reason, Corp," Jones said. "Why did they abandon factories like this? We know they worked better than anything on Earth."

"They're animals, man, that's why," Kibbo said. "What are you, blind? Didn't you guys see those things that attacked us yesterday? They ain't human anymore; they're freaks. This is a fucking great planet full of freaks. No animal can run a factory. And they don't need human medicines anymore."

"They're not animals," Lawrence said. "They're people; they just look different, that's all."

"No way, man, they're filthy animals. They don't even talk, all they do is scream all funny. They attacked us for no reason."

"It was territorial," Amersy said.

"What?"

"Territorial; you said they were animals."

"The corp said they weren't."

"In which case we're in deep shit," Jones said. "If they fight like that and they're smart with it there's no telling what they'll throw at us next."

"You think I don't know that?" Amersy grumbled.

"So why did they dump this place?" Kibbo said.

"Who knows?" Amersy said. "They still use machinery. You saw the lights in Roseport last night. Our communications links are being screwed by their jamming. And the spaceport runway was intact. One of the engineers I talked to this morning said the spaceplanes they found in the hangar were still flightworthy. Somebody's been maintaining them."

"So there's some real people left? So what? That doesn't mean there's anything here for us."

Lawrence agreed with Kibbo, though not for the same reasons. He didn't think the new-natives were animals. They might not have quite the same behavior pattern as humans, but they were certainly sentient. Exactly where that put them on the evolutionary scale he wasn't sure.

Captain Lyaute got everybody into the vehicles and ordered them back to Roseport's spaceport. When he called in their return to the governor, he was informed that all the similar exploratory missions had found the same thing. The cities were occupied by extremely hostile new-natives, while the factories were abandoned and decaying. No real dialogue with the new-natives had been established. The admiral and Simon Roderick didn't know what to do next. They were considering sending a starship to rendezvous with the big captured asteroid that was in a two-thousand-kilometer polar orbit. Sections of the planet's space-based industry were obviously still functioning, although a lot of the stations and microgee modules had been destroyed when the induction webs were eliminated. If nothing else, the starships could take the surviving orbital industrial facilities back to Earth; that would show some kind of gain on the balance sheet.

In the meantime, the governor advised, they were probably going to boost the platoons straight back up to orbit, although there were worries about the availability of hydrogen at the spaceports that had already been secured. Roseport spaceport did have several storage tanks full, but the refinery itself had been switched off. The engineers were going over it now to see if they could restart production.

Lawrence drove one of the jeeps, with half of 435NK9 as his passengers. They were eighth in the long convoy as it wound back along the route it had taken to the factory. It was slow going; the road was thoroughly overgrown with tiger-grass and creepers, although there was evidence that some kind of vehicles still used it occasionally. Lawrence remembered the Great Loop Highway back on Thallspring and quietly wished for something that clear and level again.

The terrain they were driving through was hilly, a landscape of crumpled valleys and short, awkward slopes. Tall trees thrived along the upper slopes of the ridges, projecting impossibly slim spires above the forest roof. Topped with fluffy violet leaf plumes, they looked like the battle pennants of some medieval army marching to war. Down in the valley floors the trees were fat bruisers, nearly spherical, their gray-silver bark bristling with hard, venomous thorns to repel wood-drillers and acidlice. The upper half of their swollen boles sprouted concentric circles of whip branches, shaking small leathery leaves in the breeze to produce a continuous discordant clattering. They grew together in an almost solid fence, pushing and straining at each other as decades-long battles were fought for ground and light. Those that lost and died were riddled with holes as animals burrowed their nests into the rotting wood. Swarms of fungus leeched to the crumbling bark, producing a glistening rampage of color as they wept glutinous fluids saturated with spores. Ferns and tuber leaves dominated the dim floor of the forest, banishing tiger-grass and bushes, while carnivorous coilwraiths hung from forks in the overhead branches to catch insects amid their wriggling fronds.

The road reached the first swath of forest a couple of kilometers from the factory. Its builders had tried to avoid the trees where possible, curving it around along valley walls or letting it run beside the fast-flowing streams. As a consequence, the lead vehicle could rarely see more than two or three hundred meters ahead.

Lawrence frowned as they began to slow. He couldn't see any reason for it. The road was a mess, sure, but it didn't pose too much of a problem for their vehicles. They weren't even in the forest yet; it was running along the side of them fifty meters away. Up ahead there was a sharp curve around the base of a small hill. But there was no barrier, nothing blocking the track.

"What's happening?" he asked over the command link. They were almost stationary now.

"Something up ahead. The ground's moving."

"Moving?" Lawrence didn't understand.

"Can you hear that?" Nic asked.

Lawrence braked to a halt. "Hear what?" He ordered his AS to turn up the Skin's audio sensitivity. That was when he realized the jeep was shaking slightly.

"That!" Nic insisted.

The Skin's receptors were picking up a bass rumbling.

"Six-nine-three and Seven-six-two, deploy forward with carbines," Captain Lyaute ordered. "Five-four-one, watch our tail."

Skins were jumping down from the jeeps, moving forward in a double buddy formation. Squat muzzles had emerged from their arm carapaces.

Lawrence didn't like the situation at all. None of his briefings had mentioned this being an earthquake zone.

The herd of macrorexes lumbered around the side of the mountain, a wall of beasts over eight meters high, with the smallest weighing in at ten tons. Unlike Earth's dinosaurs, they didn't have long necks and tails. Their bodies were husky cylinders fifteen meters in length, with three sets of legs. It was an arrangement that allowed them to move in a sequence of synchronized jumps, arching their spine so that a wave motion rippled down their dorsal column, each set of legs bounding forward in unison. A flattened heart-shaped head rose and fell as the body undulated, swinging occasionally from side to side as far as the stumpy neck permitted. The end of the jaw was caged by three curving tusks longer than a man's arm, two pointing up, one down; they opened and closed in a steady rhythm. The sides of the head swept together in a series of bladelike triangular fins that looked as if they could cut through steel. Their eyes were invisible somewhere among the sharply crinkled bone ridges of the upper skull.

Now that the macrorexes were in full view, their trumpeting cries split the air. Shrubs and bushes simply detonated under the pounding impact of their legs.

Lawrence thought yesterday's drugs must still be kicking through his blood cells. He remembered the giant beasts from his briefing files, but couldn't really grasp that forty or fifty of the monsters were coming at him in a motion that resembled a perpetual skidding crash.

"You've got to be fucking kidding," Nic groaned. There was real fear in his voice.

The Skins out in front of the jeeps opened fire. Lawrence couldn't even tell if the bullets were penetrating the filthy ash-gray hides. They certainly weren't having any effect. The trumpeting rose to a crescendo, and Lawrence realized the macrorexes were only 150 meters away. Nothing was going to stop them.

Captain Lyaute was yelling incoherently in the command communication link. Someone else was calling desperately for helicopter support. Skins were running hard, ripping through the cloying tigergrass as the macrorexes pounded toward them. Lawrence jammed down hard on the accelerator and pulled the wheel over. Tires spun wildly in the greasy soil. There was absolutely no way he was going to be able to loop around a full 180 degrees before the front rank of macrorexes reached them. "Hold on," he shouted, and sent the jeep skidding and bouncing toward the edge of the forest. Out of the corner of his eye he could see a couple of macrorexes charging along the treeline. Smaller trunks were pulverized into a cascade of splinters as their huge armored legs plowed into them. Long whip branches were severed cleanly by the fin-blades on the edges of the beasts' heads, twirling away through the air.

The first rank of the herd caught up with the fleeing Skins. Jaw tusks flashed behind the slowest, puncturing his carapace without even slowing. He was tossed aside, fountaining blood as he cartwheeled through space. A couple more were overtaken, vanishing below the thundering legs. The tusks chomped down again. Screams reverberated along the communication link, cut off with horrifying swiftness. The macrorexes were moving at a phenomenal speed. A cool part of Lawrence's mind knew they'd never be able to keep it up for long, not even with this planet's oxygen feeding them. They must have started the charge just seconds before meeting the company.

The corpulent trees were only fifteen meters ahead of him. His view jounced about wildly as the suspension thrummed over rocks and hidden furrows. He turned the wheel savagely, aiming for a gap that was probably wide enough to take the jeep. To his right, the macrorexes were closing fast amid a debris plume of shattered timber.

That was when he caught sight of something sitting on the neck of one of them. A crouched profile of a man-leopard hybrid, forelimbs windmilling with wild enthusiasm. Mouth flung wide in an insane jubilant laugh.

That Couldn't Be—

"Lawrence!" Amersy shrieked.

Lawrence yanked at the wheel. The front fender of the jeep clipped a spiky, bloated trunk, knocking them violently to the right. Lawrence fought the motion, Skin muscles and power steering forcing the wheels around. Inertia shunted them through the gap at a wide angle. One tire burst as it slammed into a rock. Lawrence kept the accelerator hard down, sending them plunging farther under the trees. Whip branches slapped across the windshield. Then there was a giant tree dead ahead. Lawrence thrust his other heel down on the brake, which made no appreciable difference to the chaotic rush. The jeep's front bumper hit the tree full on, sending everybody tumbling forward. Skin carapaces hardened, protecting the vulnerable bodies from the worst of the impact.

"Out!" Lawrence ordered. "Move it."

It was as though the crash were still happening. The sound of disintegrating wood grew louder. They could barely balance on the shaking ground.

Lawrence staggered onward, hoping to hell he was heading in the right direction. His orientation was screwed up. The AS display grid was out of focus.

Three meters behind him, a giant leg descended vertically on the jeep, crushing it into the soil. The shock wave sent Lawrence sprawling. Then the pillar of flesh was lifting. He managed to shift his helmet so the sensors showed him the compacted wreck, just before the beast's midlegs streaked down. Lawrence crawled forward as fast as he could. The final set of legs landed, flipping the jeep through ninety degrees. It stood on its ruined tail as the legs disappeared up into the sky, then slowly toppled back. Several shredded whip branches rained down on top of it.

Lawrence twisted around. His nine-millimeter pistol had deployed from one forearm, while the carbine was sticking out of the other. He swung his arms in fast arcs, covering the trail of devastation the macrorex had left in its wake. Targeting graphics slid around the scene on semiautonomous seeker mode, hunting for any conceivable threat. Now would be a perfect time for ground troops to finish off the Skins. Neither Lawrence nor his AS could find a new-native.

His weapons retracted. He could still hear the herd thundering away, but the loudest noise right now was his yammering heart. The medical grid display showed just how much adrenaline was coursing through his blood. Beneath his Skin, his skin was already chilling as the immediate danger faded out.

He called up the telemetry grid, checking on the Skins under his command. Everyone, it seemed, had survived the jeep's madcap dash. Looking around, he could see them picking themselves up. Dust churned through the air, glowing ocher in the bright sunbeams pouring through the broken forest canopy.

"Sarge?" Lawrence asked. "You intact?"

"Holy shit, man," Ntoko spat. "Yeah, I guess so."

It was the lead vehicles that had taken the brunt of the macrorex charge. Too close to get out of the way, either they'd raced into the forest like Lawrence, or the Skins had abandoned them to take their chances on foot. The jeeps toward the rear of the column had enough time to turn and drive clear of the rampage, though most of the trucks were too bulky and slow to maneuver like that. In total, four jeeps and one truck had survived. Over twenty Skins had perished, either mauled by tusks or trampled to death. There were a number of other casualties, as well.

One of the macrorexes had been felled, the victim of intense carbine fire from three Skins who made their stand from the edge of the forest. They'd managed to shatter its enormous skull. Even so, raw inertia had kept it slithering forward until it crunched into one of the bulky trees, knocking the trunk almost horizontal. It had plowed up a broad furrow of slick black earth behind it.

Captain Lyaute set up a field camp on the side of the forest. There were fifty-four survivors, of whom seventeen were injured; another five had damaged Skin. Two platoons were assigned to gather up what weapons and equipment they could find amid the trail of destruction left by the macro-rexes. Communications with the spaceport were patchy. There seemed to be something wrong with the satellite relay. Lyaute's urgent request for airborne evacuation was turned down flat. Two helicopters were already down. Other scout companies had been attacked. The governor was keeping the remaining helicopters assigned to guarding the spaceport.

A platoon dispatched to find out what had happened to the macrorexes reported that they were now milling about quietly a kilometer down the road. There was no sign of the new-natives who'd been spotted riding them.

Lyaute announced they were going to pile the wounded onto the remaining vehicles and make their way directly back to the spaceport. It was going to be a slow trip: some of the injuries were bad, and everyone else was going to walk escort. It had taken two and a half hours to drive out to the factory, and it was midday now; he estimated they should be able to make it back for nightfall. Lawrence knew that was bullshit.

"We'll take point, sir," Ntoko told the captain. "Scout out any trouble lying ahead of you."

Lyaute agreed quickly enough. None of the other sergeants volunteered their platoons.

Lawrence switched to a secure link and asked the sergeant: "Why? Those dinosaur monsters were only the start; they won't be the last assault today, no way. We'll get hit by whatever it is they've got out there."

Ntoko was walking along the line of salvaged weapons. He picked up a couple of rotary feed grenade launchers and handed one to Lawrence. "Maybe, maybe not." His voice was quiet and intent. "Look at it this way. The captain's just given us the pick of the weapons. We can deploy in a decent formation so nobody takes us by surprise. And we'll be a good distance out in front."

"Big deal."

"Think, man. Right now we're in shit that doesn't get any deeper. Those injured guys we've got, there's some that are in a real bad way. They're going to slow the rest of the company down."

"Yeah, but—"

"You been keeping up on tactical? There's not enough hydrogen to lift everyone off, Lawrence. That's if they even get the spaceplanes down past the windshrikes. Now do you want to be at the front of the line?"

Lawrence looked around the temporary camp. The wounded were being helped onto the jeeps. Field medics had already used a lot of aid kits getting them ready for that first move. A couple of engineers were working on a jeep, replacing bent suspension components with parts cannibalized off a wreck.

He had to admit, the company was hurting. When that happened, you mucked in and made sure everyone got back to base okay. That's what his training and first instinct was, anyway. Ntoko had drilled that into him. Being part of a unit was what it was all about.

Now there was doubt, among other disturbing notions, bubbling around in his thoughts. But selling out the others . . . Although his loyalty had always been to the platoon itself. What the hell did a simple corporal know about the overall strategy? He couldn't take the whole invasion force into account, much less save them. So where did you draw the line?

"We should never have come here," Ntoko said.

Lawrence took the bulky grenade launcher from the sergeant and slung the ammunition bag over his shoulder. His Skin AS interfaced with the weapon's targeting system. "Yeah, right."

Platoon 435NK9 set off first, walking down the battered track that was the road back to Roseport. Ntoko had put Lawrence and Nic out in front, leaving the rest in single file, spaced about seven meters apart. He brought up the rear himself.

Lyaute's brief was to flush out any possible ambushes. Don't bother too much with investigating potential sightings, just use firepower to eliminate new-natives. The rest of the company would follow a couple of hundred meters behind.

Twenty minutes along the road they'd already built the distance to four hundred meters. Ntoko had dictated the pace to Lawrence. "I'll handle any flak from Lyaute," he'd said. They didn't get any. The electronic interference was relentless. It had to be more than simple powerblock jamming. They were almost reduced to line-of-sight communication.

At the start Lawrence was busy with his AS, pulling in relevant data. They had enough bloodpaks to last twenty hours. He figured if they hadn't reached the spaceport by then they'd be dead anyway, though he found it somewhat unnerving that they couldn't just shed the Skins if they ran out of supplies. They needed some kind of protection from the oxygen. Ntoko had talked about disconnecting the helmet and using it purely as an air filter. It could remain plugged into the neck valves, and the body's organs would be able to sustain it without too much strain. Lawrence also called up tactical scans from the low-orbit observation satellite, trying to predict ambush points. He would have handed over his entire mission bonus (not that he expected to get one) for a real-time infrared scan of the area around them. But the low-orbit satellites had dropped out of the communications network hours ago.

"Surprised you're with us anyway, Corp," Nic said as they splashed through a stream. "What happened to your transfer over to the starship boys?"

Lawrence would like to blame it all on Morteth, Laforth and Kmyre. But it wasn't really their fault. They were the trigger, not the cause. They'd been dismissed from Z-B as soon as the platoon arrived back on Earth, sullen and thuggishly resentful to the end, swearing vengeance. It was the whole way the Arnoon village incident had been dealt with that troubled him. Maybe it was his own background that was the real problem, but he just kept thinking that the three of them should have been prosecuted. That way there would be accountability, responsibility. By agreeing to help out and play it quiet and canny he'd collaborated with the company. It was the kind of deal his father would have made. "The real way the world works," Doug Newton called it.

So what the fuck did I ever leave Amethi for?

When he thought about it these days it was only ever Roselyn and the pain she'd inflicted. Joona hadn't been too far wrong about the companies and their uniculture. Every human world was developing into a bland Xerox of Earth. Except for Santa Chico, of course.

"I got my promotion," Lawrence said. "It was more important at the time. I can transfer over to the starship division whenever I want."

"Not after this," Nic said. "We aren't going to have any starships left."

Lawrence kept expecting Lyaute to order them to slow down and wait. He'd kept up the same pace for over an hour and a half, striding along the track of beaten-down tigergrass. The jeeps were out of sight behind them now. Communications with Lyaute and his two lieutenants was becoming very intermittent. They just kept calling in their position and progress whenever they got a link.

Even in Skin, Lawrence was sure he could feel this planet's thick, heavy atmosphere working against him. There seemed to be a slight resistance to every movement. It wasn't gravity; Santa Chico was .95 Earth standard. It had to be the sluggish air pushing against him. Another damn problem.

Haze from the powerful sun was a further side effect. Anything more than a kilometer away wobbled in the heat radiating off the ground in fast distortion ripples. It played hell with their long-range sensors. Infrared was hopeless, of course. All a new-native had to do was crouch down in the tigergrass and scrub, and he'd become invisible. Platoon 435NK9 all had their laser radars on, sending out fans of pale-pink light to sweep the sides of the road. So far they'd had a few probable sightings, but nothing they could shoot at.

Ten kilometers out from the factory, the road emerged from the end of a wide valley onto a gently undulating lowland terrain of tigergrass. It made a change to have an open view of the countryside ahead, though when Lawrence scanned his helmet sensors around, the eternal wave motion of tigergrass in the wind swamped the discrimination program.

"Nothing in sight," he reported.

"Keep going," Ntoko replied.

They moved out. Away to the north Lawrence could see a couple of macrorexes moving along a stream. Their ponderous motion was easy enough to see, as was their grubby hide color against the bright tigergrass. He wondered what kind of nerve it took to climb up on the back of one of those brutes and goad it into a run. More than he had, that was for sure. Who in Fate's name thought of doing such a thing in the first place?

"Somebody moving," Nic said.

"Where?"

"Two hundred meters southwest."

Lawrence expanded Nic's telemetry grid, meshing the sensor imagery to his own. There was something, a blur that wasn't all heat shimmer.

"I think we have a shadow," Lawrence told Ntoko.

"We've got a couple back here as well," Ntoko said.

Lawrence called up a tactical map. There was a small group of buildings a couple of kilometers ahead and to the east with small homesteads ranged around it, barely large enough to be classed as a village. The satellite sweep had revealed some activity, but that was a day out of date. Lyaute hadn't bothered investigating the place when they'd driven past that morning.

"Close in," Ntoko ordered.

"Easier target for them," Lawrence said over the secure command link.

"I know that. But they're sneaking in anyway, that means they're going to attack. This way we've got a better firepower concentration."

Lawrence's audio sensors picked up a number of warbling calls out amid the tall tigergrass. He was tempted to play one back at them on high volume. The Skin AS couldn't translate them.

A small bronze-colored bird darted above the tigergrass, moving fast toward them. It had three wings, one smaller than the others, and used some kind of spinning motion, like an asymmetric propeller. Silver-tipped wings traced bright spiral afterimages as they caught the sunlight. Nic shot it with his nine-millimeter pistol. It burst apart in a mist of blood.

"What are you shooting at?" Ntoko asked.

"Nothing, Sarge," Lawrence said. "Just a bird."

"You guys keep calm up there."

"You hear that?" Lawrence asked.

"I don't trust nothing in this place," Nic grunted.

Lawrence's sensors were picking up bursts of motion all around now. New-natives were dashing through the tigergrass, running for a few meters, then ducking down. None of them were closer than 150 meters. More of the bronze birds were being flushed out of the clumps of tigergrass by their antics. Lawrence watched them flitter about. He wasn't quite as suspicious as Nic, but he had his doubts. There were a lot of them. When he asked his AS to run a check through its files on indigenous life, there was no reference. But then the information was limited to a few dozen prominent species like the windshrikes and macrorexes.

The birds were clumping together in small flocks of six or seven, swooping and curving just above the tips of the tiger-grass. The more Lawrence watched them the more he was convinced that they were being driven in toward the platoon.

"Sarge?"

"Yeah, man, I got them. But I can't see us shooting every one—we don't have enough ammo for that, even if we could hit them."

One of the telemetry grids on Lawrence's display flashed red.

"Shit!" Kibbo yelled.

"What is it?" Lawrence could see from Kibbo's telemetry that his Skin suit had been struck by something.

"Took a hit. Ahh, shit."

Lawrence turned to see Kibbo fifty meters away, stumbling badly. He fell to his knees, clutching an arm. Skins were running toward him.

The telemetry grid was scrolling down weird data. Lawrence had never seen anything like it. Something had penetrated the carapace, but it was small, barely a couple of millimeters wide. If a bullet had split the surface, the tissue underneath should have absorbed it and clotted immediately. But the synthetic muscle around the puncture was starting to overheat. Its nerve fibers were failing.

Kibbo started screaming. His medical readouts were going wild.

"Down," Ntoko ordered. "Keep down, people."

Lawrence arrived just as Kibbo fell flat on his face. His arms and legs started thrashing, hammering into the ground.

"Some kind of convulsion."

"What's his medical program doing, for fuck's sake?"

"It's his Skin, it's spasming."

Ntoko hurried up, so Lawrence was looking right at him

when the dart struck. It slammed into the grenade-launcher ammunition bag he was wearing on his back, nearly knocking him off his feet. He dropped to all fours, grunting hard at the impact.

Lawrence scrambled over and pushed his sensor focus on the little crater in the bag.

"What the hell was it?" Ntoko demanded.

"Don't know." Lawrence shifted to infrared. The small hole was damp. Spectrographic analysis revealed an unknown type of hydrocarbon fluid. "Shit. Could be some kind of bio weapon." His Skin deployed its aerosol nozzle and sprayed the area with a multispectrum neutralizing agent. The fluid fizzed a livid saffron.

Kibbo screamed again, his bucking lifting him off the ground. The rest of the platoon circled around, not knowing what to do. The Skin's AS and medical systems couldn't even stabilize him. The wild motions stopped suddenly. His helmet's emergency disposal valves opened. Blood poured out.

"Jesus!"

The Skins lurched back, fearful that any of the crimson fluid should splash against them.

"Was that the birds?" Nic asked. "Did they do that?"

"No way, man," Amersy said. "How could they?"

Lawrence risked a quick look around. The air was full of hundreds of fast-spinning birds, a sparkling river that hurtled through the sky. They'd formed a complete ring around the platoon.

"These are the people whose granddaddies invented Skin," Nic said. "If anyone knows how to shut us down, it's them."

"Shoot them," Ntoko ordered. "Carbines out; give me a circular formation, ten-degree overlap. Move."

They were firing as they rose to their feet, hosing the bullets at the thick dazzling stipple gyrating around them. The birds broke apart, soaring higher in a scintillating plume. Targeting individual birds was impossible at that distance.

Foster screamed at the same time his telemetry grid flashed its alert. He toppled over, limbs jerking about. The rest of them automatically dived for cover.

"They're killing us," Jones cried. "We're fucking dead. Dead!"

Foster's agonized gurgling was filling the general communication link.

"Lawrence, incendiary grenades," Ntoko said. "We're going to start using this goddamn environment to our advantage. Range two hundred and fifty meters, semicircular pattern. You take north."

"Got it, Sarge." He rolled onto his back and angled the grenade launcher toward north, moving the muzzle until the targeting graphics confirmed he'd ranged ground zero. He began firing. The dull thud of the grenades was audible through his Skin helmet. Ntoko was firing in the opposite direction. Faint smoke trails appeared in the air, forming wide arches that radiated out from the huddled-up platoon.

The first grenade detonated. It was like the dawn of a blue-dwarf sun. A halo of fierce light rose out of the tigergrass. Designed for operation in a normal atmosphere, the incendiaries were burning far hotter than usual in the abundant oxygen. The undergrowth ignited immediately.

Lawrence kept firing, moving the launcher around in precise increments. The brilliant detonations merged swiftly into a solid wall of crackling light. Flames burned a vivid blue, consuming even the living vegetation. Sap sizzled and evaporated before the onslaught, leaving withered blades that burst alight instantly.

It took less than a minute before they were completely surrounded by flame. The circle began to burn inward relentlessly, though Lawrence's sensors could just see another, wider, ring burning outward.

"Use the rest of the grenades," Ntoko said. "I'm not risking the manufacturer's heat-proof guarantee on these ammo bags."

"Right." Lawrence waited until he'd fired all the incendiaries, then switched to fragmentation, using a random dispersal pattern. When he finished, he unslung the bag and threw it and the launcher away toward the advancing inferno.

The birds had all gone, zooming high over the rampaging flames. Foster lay dead on the ground, blood soaking into the soil as it dripped from his open disposal valve.

"Now we'll see," Ntoko growled.

"How do we get out of this?" Jones asked. His voice was panicky. "There's no way through the flame."

"That's the idea," Ntoko said. "You've got to believe in your Skin, my friend. This flame burns so fast it'll be past us in a couple of seconds."

"Oh, Jesus fucking wept, Sarge!"

"Just hold your place."

Lawrence nearly laughed. He'd worked it out just before he started firing. The time to object was long past. They'd all have to ride it out now.

His Skin's audio sensors were relaying the fierce roar produced by the flames. It grew steadily louder. They were approaching at a phenomenal rate as they consumed the tigergrass. His briefing had included strong warnings about fire in this atmosphere, but he'd never imagined anything this potent. There were screams now, rising above the background roar. A new-native charged past the platoon. He was bipedal, with arms that reached down to his knees. There was a long mane of ginger hair streaming out from his spine as he ran, already singed and smoldering. Lawrence caught sight of a narrow bandoleer, with some kind of cylindrical electronic modules slotted into hoops.

The terrified new-native saw the platoon and immediately altered course, more from fear than sense.

"You can run, asshole," Ntoko yelled after him, "but you can't hide."

Two more new-natives rushed past. One of them was a husky quadruped with some kind of canine DNA in its genetic makeup. Lawrence watched as it sprinted at the wall of flame sweeping in toward them. It jumped. He couldn't believe anything that big could get so far off the ground. Even with its muscular limbs it didn't get high enough. The ferocious blue flames speared into its underbelly, excoriating its tough amber hide. Raw splits opened into its blackening flesh, spewing out steaming fluid. It howled in agony as its entire epidermal layer ignited spontaneously. Death must have struck with blissful speed. It was silent and motionless as it struck the ground in the middle of the conflagration.

"Holy shit," Ntoko whispered. The flames were barely fifty meters away and closing fast. They were stabbing up seven to eight meters into the air.

Lawrence's display was already issuing heat cautions. His carapace was turning white to reflect the massive infrared input. He slowly stood to face the flames, seeing the rest of the platoon follow his lead and climb to their feet. Sensors had to bring two layers of filters online to combat the glare of hellish light given off by the flames.

He ordered the visual sensors off altogether in some crazy effort to make the horror go away. That didn't work: the darkness was even more unnerving. His indigo display grid hung in the middle of nothingness. The digits recording external temperature blurred as if they'd begun to count milliseconds instead. He brought the sensors back online. The flames were ten meters away.

A couple of the platoon were murmuring prayers. He wished he knew how to join in. The temperature warnings were now so ridiculous they were laughable.

All around him the tigergrass was withering, vapor effervescing out of every blade as it smoldered and blackened.

Then the grass burst into flame around his legs. The main tsunami of fire hit, nearly knocking him down again. Something gripped his Skin and started shaking him; it was like being trapped in a slow-motion explosion.

He could see nothing. No discrimination program could possibly make sense of the incandescent chaos buffeting against him. All he knew was the one display grid reporting his Skin status. Every thermal indicator was leaping toward overload. Yet here he was, perfectly comfortable at the center of the fury. He held his breath, tensing every muscle against imminent death, then forced himself to breathe out and inhale calmly. Nothing he could do would make the slightest difference. It was all down to technology, and just how much of a safety margin had been built into his Skin.

His hand went to the base of his throat, covering the lump that was his pendant. Patterns began to appear around him, faint shadows that purled within the intolerable light, then slowly began to darken. It was as if water were sluicing down a muddy window, producing streaked images of what lay outside.

Flames shrank away, revealing a land that was completely black. Spiky root clumps of incinerated tigergrass mottled the baked soil, puffing out streamers of grubby blue smoke. A dense rain of ash fell, flakes settling gently on every surface, including Skin.

He turned to see the wall of flame not ten meters behind him and retreating rapidly. The rest of the platoon was standing in a loose circle, sable silhouettes against the solid glare. When he brought a hand up to examine it, he saw his carapace was glowing a dull vermilion as the weave of thermal fibers hurriedly expelled their excessive loading. He reviewed his status, relieved to see his Skin's reserve bladders had retained their integrity; with them and the spare bloodpaks he could easily make it back to the spaceport.

Laughter and delirious whoops began to fill the general

communication band. The shouted jubilation had a strong note of hysteria.

Ash was still falling, but Lawrence extended his sensor range, trying to see what lay through it. The second wave of wildfire was still rampaging out ahead of him, lurid flames chewing their way voraciously across the tigergrass, sending up a broad veil of smoke and yet more ash. He couldn't believe so much destruction had spread so quickly. The holocaust they'd unleashed was easily over a kilometer wide now and still expanding. He wondered how far it would continue for. Not that there was much guilt associated with the thought. Santa Chico must be used to such events.

"Can't raise the captain," Ntoko said.

"You reckon the fire's reached him?"

"Could be. The Skins will come through okay. Don't know about the vehicles."

"You want to go back and check?"

"No. We keep going unless ordered different. Even then I'm not keen."

"Sure."

"One good thing, nobody's going to be creeping up on us unseen now."

"Sarge, there's nobody left to creep up on us." His sensors had found a small mound that was the remains of a newnative. It looked like a lump of coal.

There was no hint of where the road had lain across the land. They checked their inertial guidance and started marching again. A couple of them were unhappy about leaving Kibbo and Foster behind, but Ntoko quelled their dissent with a few gruff words about how the guys would want the platoon to reach the spaceport.

The ground was still furiously hot, although it didn't present too much of a problem for their thermal fiber weave. As they walked they found patches of tigergrass and even trees that the fire had completely bypassed. There didn't seem to be any particular reason for any of them being spared. Vagaries of the land. Streams too broad for the flames to leap. Even some scrub trees with fat spire leaves that were resistant to the flames entirely, standing alone and unblemished amid the scorched desolation.

A broad ridge of rocky ground had saved the village from the firestorm. They examined it through the continuing fall of ash. Their sensors detected movement among the buildings. Ntoko decided they couldn't ignore it.

By the time they arrived, the carpet of delicate loose ash was a couple of centimeters thick, covering everything. Gusts would stir it up in small twisters, but that just rearranged it. Nothing was free of the mantle. The skirt of tigergrass around the buildings swayed and quivered in the breeze, as if trying to shake the flakes off. But they were too small, too insidious to release their hold.

The village homes were simple structures, broad circular towers with domed roofs, never more than two stories high. They seemed to be made from a pale cream coral with a rough, grainy surface that was a magnet for the ash, allowing it to lodge in every crinkle. Windows were arches covered with a thick membrane, laced with delicate silver veins.

The new-native inhabitants were mostly bipedal, smaller than the average human, with shaggy hair that continued down their spines in a thick mane; in some cases it extended out along their arms almost to the elbow. Their shirts and jerkins were cut to allow the hair to flow through. It was often braided. Bright-colored beads were favored by the children.

There were exceptions. Feline hominoids who struggled to stay upright, dropping down to use their forelimbs to walk a few paces. A squat giant that looked like a cross between a sumo wrestler and a troll. Delicate spindly elves, whose legs seemed too slim to support their bodies.

They didn't look alien, Lawrence thought, so much as

primitive, although their hides were the typical Santa Chico tough, translucent amber, and none of the bipeds had a terrestrial human rib cage and abdominal arrangement. Ridges around their torsos were more insectile than anything else. Their faces, though stiffer than skin, still managed to express basic emotions, although that could have been just the eyes. Sullen glances were more or less the same the universe over.

Ntoko took Lawrence and Amersy into the village with him, deploying the rest of the platoon outside. They were subject to blank stares from the inhabitants who stood in open doorways. New-natives in the streets moved aside to let them pass. It was the first time their authority had ever been acknowledged, even if it was at gunpoint.

Lawrence's sensors detected a small level of electronic activity in the buildings, nothing above desktop pearl level. They seemed almost devoid of mechanical or electronic technology. Certainly there were no vehicles in evidence.

The new-natives appeared uncertain what to do about the Skins; they were waiting for them to set the agenda. As they walked into the center of the village more new-natives appeared and followed at a respectful distance. Unless half of the homes were deserted, the numbers didn't match up. Lawrence wondered how many villagers had been in the group beating the birds out of the tigergrass. And how many had survived.

Ntoko stopped beside a big overhanging tree that had a coating of the ubiquitous ash. "Anybody want to tell me what's going on here?"

"You fired our lands," a voice said. It was heavily accented, but had the easy lilt of Spanish roots.

Lawrence identified its owner, a woman who wouldn't reach his shoulder. Her luxuriant hair was snow-white, though whether that indicated old age he wasn't sure. She had a flat face, with several creases in her cheeks, giving her jaw a considerable degree of flexibility. The robe she wore was decorated with silver piping: a DNA helix had been embroidered down the front in scarlet and turquoise.

"You the big chieftain around here?" Ntoko asked.

"No. I am Calandrinia." She combed a hand through her hair, shaking out the latest dusting of ash.

"You going to talk to me?"

"Are you going to kill me?"

"Not unless you give me a reason."

She bared her teeth, which were long enough to qualify as tusks. "I have many reasons, but I won't be acting on them today."

"Well, thank you. Now you want to tell me what the fuck is going on around here?"

"You violated our lives. This is how we respond. What did you expect?"

"Less violence would be a good start. You people have got to be crazy. Do you know how much firepower we've got backing us up?"

Calandrinia showed her tusks again. "Less than you started with."

Lawrence used his secure command link. "Sarge, can I talk to her?"

"Sure, go right ahead if you think it will get us anywhere. I hate a smartmouth."

"Thanks." Lawrence was never quite certain, but Calandrinia seemed to turn to him just before he started talking. "I'd like to know, why did you abandon your factories?"

"Why does anybody abandon anything, Earthman? They are obsolete and irrelevant. Now we grow whatever we need directly."

"But your products weren't obsolete on Earth; they were damn useful. Why stop exporting?"

"If Earth wants medicines it should make them for itself."

"Well, for a start, without the cash from those exports you won't be able to import the products you don't make here."

She laughed at him outright. "If we don't make it, we don't want it. If we don't want it, we don't make it."

"So that's it? You've kissed good-bye to technological civilization? You're all happy regressing?" Somewhere at the back of his mind was the question of how many times he would have this conversation, and on how many planets. Regressor types seemed to get everywhere.

"Technological, no," Calandrinia said. "Mechanical, yes. What do you need machines for? Biological systems are much more efficient at providing for us."

"You can't make biological equivalents of everything."

"Not everything your society requires in order to function, no. But then we don't have your kind of society anymore. We've adapted ourselves, not bent the world to our vision. Worlds are too big for that. Why live in isolated settlements built on dead, irradiated earth when you can modify yourself to enjoy the freedom of the whole world?"

"That must be quite an ideology you've got here, to convince people they have to leave their past behind."

"It's not ideology, it's evolution. You know our ancestors came here with the intent of modifying themselves; why are you so surprised by what you found?"

"Nobody knew how far you'd taken the modifications. We didn't expect any of this. If we knew what was here, we wouldn't have come."

"Yet here you are. Now what will you do?"

"Me personally? Go home."

"Why not join us? Your children would have a beautiful future. They would never want or need for anything."

"Excuse me, but that's not even remotely tempting. If I take this helmet off, I die. You know it, and I know it."

"I could grow you an oxygen filter in my housewomb. It would be a part of you in a way your Skin never is. You would live with it in perfect symbiosis."

Lawrence held a finger up. "Yeah, stop right there. I'm not coming to live with you, okay?"

"Why? What do we lack? I do not mock, I am genuinely curious. You seem so primitive compared to us. I don't understand your reluctance. Do you not wish to better yourself, to be a part of a richer, more mature culture?"

"We're the primitives? Which of us is living in mud huts, lady? I wouldn't wish this existence on my worst enemy, let alone my own children. You're going backward faster than progress ever pulled us out of medieval squalor. Sure, this kind of life looks appealing now; you're still close enough to the industrial market economy to make you think this is all stress-free and rich in karma. Another two generations, and you won't be able to cure a cold, let alone cancer. And you call that living life to the full. I call it betraying your children."

"Ah." Calandrinia shook her hair again. "Now I begin to understand. How old am I, Earthman?"

"I haven't got a clue."

"I'm fourteen."

The information left Lawrence nonplussed. He simply couldn't see the relevance. "Really?"

"Yes. It wasn't just their biotechnology skills that our ancestors brought to this world, they brought a saying with them as well. *Live fast, die young, and leave a beautiful corpse.* Thanks to them I can do that."

"How long do you live for?" Lawrence didn't want to ask it, because he suddenly knew he wasn't going to like the answer.

"Probably thirty years. Can you imagine such a time? How it must stretch at the end."

"The oxygen. It's the oxygen, isn't it?"

"Of course. Everything here is faster, more dynamic."

"But . . . thirty?"

"Thirty whole years, during which time I will live and love

and think. Why do you think that is wrong? Why do you want to live for such a long time?"

"To live is to experience. You can't do that in thirty years. There's so much of the universe to know."

"I do experience, far more than you ever will. I grow faster. I learn faster. I live faster. We all do. This world's life is so much more vital than your bland biology. As for the universe, it is contained in your mind. Observation is purely relative. I can watch the stars from here, all of them, while you crawl between them in your tin cans and see only one at a time. I appreciate my life, Earthman; there is less memory in my brain, and much more thought."

"Thought," he sneered. "But you don't use it. What's the point of thinking if you have no way to apply it, nothing to create?"

Calandrinia let her breath whistle out between her teeth, as did several other new-natives. "We do nothing else but create, Earthman. Do you think we have time to carry and birth our young as your women do? I adapt my children to the world as I see it and know it."

"You're talking about shape, aren't you? That's why you all look different."

"We have become morphogenic, the greatest gift our ancestors left us. What I think, my children become. Can you imagine what that is like? If I see a tree that is so tall and full of grace that I have to sit at its foot and gaze up in admiration, I can engender a child who will be able to climb to its apex and laugh with the joy of doing so. When I swim in a mountain lake, I do so for a few exciting minutes, while my daughter will be able to glide through its deeps and play with the fish. And when I shake with awe as a macrorex walks past, I can absorb its essence and mingle it with my own."

"Sweet Fate, you're talking bestiality."

"How simple your mind is. How pitiable. Do you think we alone should remain sentient and aware? If we are to live with this planet, we must share the best of what we are with it. Are you so unselfish, Earthman? Would you stop us from doing that, from waking Gaia?"

"I won't stop you. But I want no part in it. You're not human anymore."

"Why, thank you. I rarely get paid such a compliment."

"Wait a goddamn minute here," Ntoko said. "Are you telling me that the macrorexes are part human, that they're self-aware?"

"Some of them," Calandrinia said. "They are our friends, they help us when we ask."

"And the windshrikes, too?"

"Of course."

"Jesus H. Christ."

"If you don't have pregnancy anymore," Lawrence said, "where do all these kids come from?"

"The housewombs gestate them," Calandrinia said simply.

"House . . . ?" He looked at the dumpy buildings that made up the village. "You mean your homes contain some kind of artificial womb?"

"Do you listen to anything? There is nothing artificial about a housewomb, it is perfectly natural. Our houses were the last stage in bridging the gap from what we were to what we are. Tell me, do your files have fastrocks in them?"

"Yeah." His Skin AS was retrieving the information. Fastrocks were essentially a polyp-type plant that grew quite slowly by Santa Chico's standards. They resembled ocher stones that grew clustered together in vast colonies and were completely fireproof. Their shells were also tough enough to resist being split open by the jaws of anything smaller than a macrorex.

Calandrinia gestured at the houses. "Our ancestors modified those small plants into the sturdy buildings you see today, a true and grand amalgamation of the genes from two planets. Now we live in living houses. Their roots grow deep to collect water and nutrients, while their shells harvest the sunlight. Within our houses we are nurtured without violating the planet as you do. Their organs provide for us in the way your machines do for you, although our bond is closer and more appreciative."

"You mean symbiotic."

"Ah, you are listening. Yes, our houses are a part of our family. Once I have a fertilized ovum, I place it in a housewomb to gestate."

"Did you give them sentience as well?"

"Of course. How could you marry an entity devoid of thought?"

"Good point," Lawrence retorted sarcastically. "Surely you've made yourselves overdependent on these constructs? Do they grow your food for you as well? Our satellites didn't see any working protein cell refineries."

Calandrinia reached up and pulled a cluster of small red berries from the tree. "Modifying Santa Chico's plants to fruit terrestrial food was the first and hardest task facing our ancestors. Once they understood how to merge the two different genetic molecules in a successful union, then everything we are today became possible. It took decades of effort before anything so complex was achieved, which is why we had to involve ourselves with commerce. So much of our respective biochemistries was incompatible, as it is on every world humans colonize. The old ways of life, your markets and machines, had to be sustained for that whole time while the problems were solved. Now, as you can see, we have left them far behind."

"And left yourself in debt," Ntoko said.

"Only on your planet, Earthman. Here there is no such thing. Here we are one."

"Claiming you are above such things as money is a very convenient way to duck the issue," Lawrence said. "But I know you understand economics and technology. You still have spaceplanes and orbital systems. They have to be maintained, spare parts manufactured, fuel produced. Housewombs can't do that for you."

"We had such machines until you arrived and destroyed them," Calandrinia said. "Some among us have the kind of dream you have, Lawrence Newton, that of spreading out through space. They are involved with ideas of modifying our cells to live up there in the desert beyond the sky. Our space enthusiasts want what we are to blossom on the comets and moons that share our star. It is a pleasant dream, I think. But they are a minority. And your arrival has put an end to their aspirations. They have agreed to turn their minds back toward Santa Chico. They will help to seal the sky and prevent you from returning."

"How did you know my name?"

"I'm sorry," Calandrinia said. "I didn't know it was a secret."

Lawrence didn't like how casual she was being. If they'd decrypted the communication links, which he acknowledged as a strong possibility, then they might have heard his name being used, as well as his conversation with Nic. But to identify individual Skin suits would be difficult. There were too many things here that the new-natives made light of. He still didn't even know how they communicated over long distances.

"What do you mean seal the sky?" Ntoko asked.

"You said you are leaving, which is what we want," Calandrinia said. "What we need to do after that is to make sure you do not return. At least, not while your present society is the dominant culture on Earth. To do that we must seal the sky." She exposed her tusks again.

"Come on," Ntoko said to Lawrence on the secure communication link. "This is wasting our time. We're outta here. If they had anything that could knock us down they would have used it by now. We just gotta watch our backs, and kill anything that moves out there."

"Right."

"Okay, we're going now," Ntoko told the new-natives. "You-all make sure you don't follow us. That way there's no misunderstandings, and nobody gets hurt."

"Such wise advice, Sergeant," Calandrinia said. "We will try to make use of it."

"Motherfucking smartmouths," Ntoko grumbled. "I wish we could nuke every fucking one of them."

The platoon tramped away, the soles of their Skin kicking up huge clouds of powdered ash. They crossed over from the island of tigergrass and trees around the village onto the black wasteland scoured by the wildfires. The ash rain had stopped falling around them, allowing them to see out across the countryside. Flames were still burning in the distance ahead of them, sending thick columns of smoke and ash soaring hundreds of meters into the deep-indigo sky. But it was no longer a solid wall; the wildfire had split around rivers and gullies, breaking into dozens of small blazes that raced onward.

"What the hell did you make of all that?" Ntoko asked.

"I'm not sure," Lawrence admitted. "It could just be a whole load of bullshit she made up to scare us. Or it could be true, in which case it's even scarier. There's a lot of things around here that don't add up."

"Jesus, smart animals for one. Maybe that part's true."

"Some of it has to be wrong. They used everything they've got to strike the starships when we were on our way in. They can't close off this planet from space."

"Damn, I wish we had some kind of contact with the captain. We should be telling people this."

"Satellite relay's still down."

"Yeah, I know. Let's hope the governor is still holding on at Roseport."

Lawrence was on edge for the rest of the march. If Calandrinia had been telling the truth, there was no way of telling what represented a danger to them.

By late afternoon the wildfires seemed to have died down. Smoke and ash hazed the air, darkening the sky to a sullen gray blue. They didn't see any more animals, large or small. Several times Lawrence thought he caught sight of windshrikes in the distance. But it could have just been dense swirls amid the lingering smoke. Underfoot, the tigergrass clumps started to weep, gooey sap leaking out of the burned blades. Their roots had obviously survived the fire, but then it had been so quick that it probably didn't heat the soil below a few centimeters. It wouldn't be long before the first new shoots were poking up through the mantle of ash.

The burnout zone ended along a narrow, deep gully with a brook of reddish water gurgling along its rocky floor. The land on both sides was covered in stones and boulders; there was little vegetation growing out of the cracks between them, which had reduced the fire's intensity. As he walked up to the gully, Lawrence realized it wasn't actually stone he was walking on. None of them moved under his feet, not even the pebbles. It was actually a vast bed of fastrock.

As the platoon crossed the gully a Xianti 5005 swept overhead, less than a kilometer in altitude. It was losing height.

"We've still got the spaceport," Nic said. "Thank Christ for that."

Half an hour later they topped a ridge that gave them a view of Roseport a couple of kilometers away. The lakeside city was in a bad way, with many of its houses smashed open. Lawrence's sensors zoomed in. Dark, glutinous fluid was oozing out of the broken shells, slithering slowly along the streets like molten tar. Internal organs lay exposed, reduced to a mass of pulped ginger jelly. Nobody was moving down any streets.

When he pulled the focus back he saw autosentinel guns

had been set up in a broad perimeter around the outskirts—olive-green spheres on thick metal legs that were anchored into the ground, each with a trio of magnetic Gatling rifles protruding from its midsection, swiveling slowly from side to side as they tracked across the smashed houses. If anybody emerged from the city, they'd be cut apart in milliseconds.

Lawrence knew the autosentinels were part of the fleet's equipment, but he'd never seen them deployed before. Like land mines and laserfencing they were last-resort weapons.

Now that the platoon had line of sight, their communications link to the governor's field headquarters came back online. Ntoko reported in, telling the staff that they'd lost contact with Captain Lyaute hours ago.

With access to fleet tactical data restored, Lawrence requested a situation update from the headquarters AS. It was even worse than he'd been expecting. There had been a near-constant battle around Roseport. Skins entering the city had been killed by a variety of chemical and biological weapons. Each time they learned how to protect themselves from one, something new would hit them. In the end the governor ordered the autosentinel deployment in the hope they could confine the new-natives until the evacuation was complete. Even that was going to be touch and go. Macrorex herds had charged the spaceport three times. The Skins had to use armor-piercing smart missiles against them before the huge beasts reached the runway: nothing else could bring them down. Windshrikes were harassing the Xiantis when they lined up on final approach. The spaceplanes were having to fire antimissile airmine clusters from their countermeasure pods to kill them.

Of the eleven companies dispatched on scouting missions to factories and industrial facilities, only four had returned. Three more (including Lyaute) had reported in that they were under attack before communication was lost. The remaining four were currently classified as missing in action.

Events in orbit were equally hostile. The software assault on the starships was relentless. Communications bandwidth was reduced to a minimum to enable the onboard AS to examine every byte entering the ship's network. Kinetic slugs in a retrograde orbit had taken out several satellites. One swarm had got through the *Mahonia*'s defenses, damaging a life support wheel and one of the compression drive's tokamaks.

In view of the unfolding catastrophe, the admiral had ordered a complete evacuation. Plans to capture Santa Chico's orbital industry stations weren't even on the agenda anymore.

Lawrence couldn't find out how many spaceplanes were operational. The data was classified. Estimates on how long the evacuation would take were also restricted. As was the number of surviving Skins.

Ntoko managed a few terse words of encouragement to the platoon and started to lead them toward the spaceport in a wide semicircle around Roseport. They jogged toward the runway, saying little. Lawrence knew they all shared the same anxiety. They were close to making it now. All they had to do was reach the runway. Someone else would take care of the rest—the admiral, the Xianti pilots. That just left the ground between here and there. Halfway around they stumbled into a wide patch of burned ground. Tigergrass was still smoldering round the edges. In the middle was the mangled wreckage of a TVL88 helicopter.

Ntoko decided they would be too exposed crossing the open space and took them around the side. Another couple of minutes added to the journey. It didn't help that they heard an autosentinel firing.

The last three hundred meters was a straight sprint. Discipline went all to hell, and the whole platoon charged over the tigergrass, dodging around trees, jumping low rocks, heedless of the target they presented. They made it past the perimeter, where Skins were lying in shallow trenches, armed with the heaviest portable weaponry in their armory.

Ntoko reported to the local lieutenant, who allowed them an hour's rest. They were issued with a ration pack of paste food, which they could eat with their helmets on. Headquarters also gave them a flight number. If the spaceplanes kept up their current schedule, they'd take off in another six hours.

The lieutenant gave them guard duty around the maintenance hangar. They took up position just as the sun sank below the horizon, integrating the new weaponry they'd drawn from the armory. Lawrence had taken a smart missile rack. So far the new-natives hadn't managed to glitch them.

He settled down to walk a regular route, making sure to keep his times random. His visual spectrum sensors provided him with a fuzzy blue-and-white image of the nighttime countryside, with the infrared bleed painting in small vermilion patches as rocks slowly radiated away the heat they'd absorbed during the day. Nothing moved among the tigergrass, not even small animals. He was thankful for that.

When he scoped Roseport, it glowed strongly in infrared, throwing off a coral-pink aura. There were no lights in any of the buildings. The autosentinels were taking shots at something every few minutes. Gossip on the general communication band centered on how much ammunition the robots had left and how long it would last.

Spaceplanes continued to arrive, thumping down out of the night sky amid the strident bellow of their Rolls-Royce turbojets on full reverse thrust. Sometimes they were preceded by the spectacular green-and-crimson magnesium firework display of their countermeasures clearing a path through the air.

An hour after nightfall one of the lost companies made contact. According to their report they'd suffered from a macrorex charge, losing most of their vehicles. On their way back they'd endured near-constant sniping and harassment from new-natives. Then they'd joined up with another company, which had a 30 percent casualty rate. Between them they had enough firepower to keep the new-natives back. It had been slow going with all the injured to care for, but they estimated they'd be arriving at the spaceport in another ninety minutes. Altogether there were over 120 of them.

Lawrence had a flush of guilty relief that it wasn't Lyaute, who would have wanted a damn good explanation about where 435NK9 had got to and what had happened. Everyone else was cheered by the news. It would mean at least one extra spaceplane flight, delaying the final departure by another twenty minutes.

The autosentinels' rate of fire slowed considerably after the first couple of hours, but Lawrence was convinced the new-natives would try to infiltrate the spaceport. It gave him a brittle edge that his Skin pharmacy couldn't dampen. Standing by himself on the edge of the spaceport facing the unknowable threats creeping through the tigergrass was using up a lot of his resolve. The area directly outside had been seeded with hundreds of remote sensors that had secure links to his display grid. He didn't entirely trust them: his thoughts were bent toward Calandrinia and how she had quietly mocked them.

Two hours before their scheduled departure time he learned why she had been so confident. Everything she'd said had been true. To start with he thought another Xianti was on its way down and dispensing countermeasures. Several streaks of flame shot across the sky, fading almost immediately. He scanned around, but couldn't locate the intense thermal signature of a spaceplane. More streaks blossomed, stretching out across the stars. As far as the flightpath was concerned, they were in the wrong section of sky. He realized it was a meteor shower and grinned briefly. Before they died away a third batch had begun to fizzle their way down. These were slightly larger particles, with a bulbous head stretching out a sparkling tail as they tore down through the upper atmosphere. There seemed no end to them streaking out of one section of sky. Lawrence's grin faded. The patch was elongated, extending north-south and still growing.

"Oh, bloody Fate," Lawrence groaned. He understood then. *We will seal the sky.*

A Skin was running toward him from the hangar.

"Lawrence?" It was Ntoko, using his Skin's speaker on low volume.

"Yeah," Lawrence replied, using his own speaker.

"They've done it, haven't they? This is what Calandrinia was talking about."

"Yeah. They must have nuked the polar-orbit asteroid, pulverized the fucker. This shower's just the edge of the debris swarm."

"Goddamn it! You know all about this orbital-mechanics shit. Can the starships get away through it?"

"Yeah, but they'll have to leave soon, if they haven't already. The debris won't have started to cascade yet. All we have for the moment is just an expanding cloud of rock in polar orbit."

"What's a cascade?"

"Look, the nukes will have shattered the asteroid into a million pieces, okay? Some of them will just fly straight down and burn up in the atmo-sphere. We're seeing that start now. But if the new-natives set the charges right, then there's a shitload of mountains and boulders and pebbles still in orbit. Right now they're separating, flying apart into their own irregular orbit, but once they've spread out enough, then they're going to start colliding. Each boulder that crashes another releases another cascade of smaller rocks, which are going to smash into another batch of rocks and so on. It's a chain reaction that is never going to stop. In a year's time, this planet is going to be surrounded by a shield of rock splinters ten thousand kilometers thick. It'll be like Saturn's rings, only spherical. She was right, that Calandrinia. Nothing will be able to get through this. They have sealed themselves away from the universe. It will take millennia for the shield particles to decay and burn up in the atmosphere. Maybe they never will. Fate, I don't know. Nobody's ever seen a cascade before."

"Okay, grab the guys. Head for that spaceplane." He pointed. "It's fueled."

"But—"

"You said it, man, the starships have got to leave. There aren't going to be any more spaceplane flights after these. Now get your ass in gear, Corporal."

Lawrence pumped his speaker volume up. "To me, people, come on, let's go." He started jogging, then broke into a sprint. Ntoko was shouting as well. The survivors of 435NK9 began running toward them.

High above them, larger debris particles had reached the atmosphere. They screamed down in a sheath of plasma until the pressure shock detonated them into a dazzling halo shoal that expanded and brightened as it sank ever downward. Sometimes the shoal particles would explode again and again as the rocks were broken into smaller and smaller fragments by the superheated ions, sending pyrotechnic shock waves radiating outward. A hundred conical plumes of incandescence flowered against the night, flaring through the spectrum as they slowly withered away to violet specters.

Half of the continent was drenched in a light greater than that of the sun. Lawrence could see the entire spaceport on the move. Skins were running about in chaos, not knowing what was happening. There was no chain of command. No orders. No information. No discipline. Not even new-natives with their hyperoxygenated muscles could match the turbocharged speed of the Skins. Everything was happening in accelerated time.

A hundred meters ahead of Lawrence the Xianti was parked in its flight preparation bay. Its turbofans were already starting up. Fueling arms had disengaged. They began to sink back down into the concrete.

The airstairs were still in place. Skins were surging up them, desperate for a place. Lawrence had no idea how many were already inside. He reached the bottom of the airstairs in five seconds. Twenty Skins were clustered there, funneling onto the aluminum steps. More Skins were heading their way.

Out on the runway a Xianti began its takeoff run.

"Lawrence," Ntoko called. "Give me your rack."

Lawrence handed over the weapon as he shoved and wriggled his way toward the bottom of the stairs. Apart from Ntoko, who already held an identical rack, he couldn't tell who was who. His AS wasn't tagging individual suits. The entire communications band had crashed.

"What do you want it for?"

"You take care, Lawrence. You look after my guys for me."

"Sarge? Ntoko!"

"I'll be watching." Ntoko was already slipping free from the throng of Skins. He opened the bottom of the rack that Lawrence had given him and pulled out a data cable, which he plugged into his Skin's interface port. The tubes at the top of the rack spat hazy orange flames that pulsed for several seconds.

Explosions bloomed across the taxiway. The swarm of Skins sprinting for the remaining spaceplanes dived for cover. More explosions rippled down the side of a hangar as Ntoko tried to deflect the onrush that threatened to overwhelm the last two spaceplanes. Composite panels and steel girders crashed over the tarmac. Smoke and dust billowed out. Skins started firing. Armor-piercing rounds pummeled the control tower. Carbines opened up.

"Ntoko! For fuck's sake, you can't!" Lawrence was at the foot of the stairs. His sensors showed him Ntoko walking calmly away from the rear of the melee, a rack held in each arm. Flames stabbed out as more smart missiles leaped from their tubes. The sergeant raised one of the racks in salute and kept on walking.

For an instant, Lawrence hesitated. But the Skins behind were pressing him on. And his own sense of self-preservation was just too strong. He clambered up the airstairs and into the cabin. The spaceplane began to move, pulling free from the airstairs. Lawrence grabbed at the Skin on top, helping to drag him in. Another Skin leaped across the widening gap, crashing into everyone crammed into the airlock. Another jumped and just managed to grab the rim of the hatch. He hung there, dangling as the spaceplane accelerated onto the taxiway. Lawrence was looking at the abandoned airstair as it wobbled about. The Skins on it were using their speakers at full volume, shouting at the Xianti to come back. One of them deployed his carbine and started firing. A couple of bullets ricocheted inside the airlock. Lawrence ducked automatically. Then an explosion went off at the base of the airstair. The whole structure collapsed, taking the Skins with it.

"Thanks, Sarge," Lawrence whispered.

He moved back into the Xianti as the hatch swung shut. The cabin was badly overcrowded, with Skins crammed along the aisle. He didn't even consider the extra weight. The sarge was out there, covering their asses like he always did. They'd make it.

Inside the sealed cabin, his Skin could link into the spaceplane's internal network. He called up the external cameras.

Outside, the asteroid fragments were still sleeting down in a blaze of light. On the ground, Skins were racing about in their distinct fast motion. All of them seemed to be shooting at something. Explosions erupted from the shattered buildings. Wild clouds of smoke writhed across the ground as vigorous blue-white flames swirled out of wrecked equipment blocks. The Xianti turned sharply onto the runway. The pilot didn't waste any time; Lawrence could feel the vibration building as the turbojets wound up to full power before the nose was lined up. Then they were racing forward, lifting from the ground.

They flew up steeply, the giant turbojets pushed to their redline. The spaceplane's cameras showed the calm upper cloud bands fluorescing a lambent silver in the lurid radiance thrown out from hundreds of descending fireballs. As they passed through the thin layer the perspective shifted until it looked as if they'd climbed above a frosted desert gleaming in winter moonlight.

Their scramjet ignited, thrusting them higher. The swaths of cloud shrank away to a shimmering haze that veiled the world. Scintillating rose-gold contrails scored their way through the empty darkness toward it, plunging underneath to dwindle and vanish.

Ahead of the spaceplane's nose, stars glittered coldly in welcome.

CHAPTER SIXTEEN

THERE WERE A NUMBER OF MEDICAL MODULES ATTACHED TO Hal's torso, clustered around the red-weal wounds and surgical scars. Some of them were integral with patches of artificial skin that were busy melding with his spoiled dermal layers, infusing regeneration virals into the plexus of capillaries. Others were more complex systems, sprouting slim tubes that penetrated the scars, pumping specialist fluids in and out of damaged organs, supporting them until he could be given replacements and proper treatments. He wore a baggy white shirt to cover them, but the modules were too bulky to hide completely. It was as though his torso was busy growing plastic tumors.

He sat in his high-backed leather seat, head lolling against the side cushions as if his neck didn't have quite enough strength to hold it upright. Every time one of his friends came into the hotel's small private staff lounge that they'd taken over to care for him he grinned at them and made a happy grunting sound. Edmond went over and gave him a high five. Just watching Hal's hand wavering unsteadily through the air as he concentrated hard to make contact made Lawrence chill deep inside. The others were looking away, their expressions grim as they were reminded of Hal's state.

Only Dennis stared unflinchingly. And Lawrence knew he'd been taking too many of his own sedatives lately.

Amersy was the last in. He closed the door and gave Hal a quick thumbs-up, the way anyone would to a longtime pal. It was his eyes, flicking away quickly, that betrayed what he must be thinking.

What was left of Platoon 435NK9 turned to Lawrence.

"First off," Lawrence said. "Don't worry about getting Hal back up to the *Koribu*. I have a contact at Durrell spaceport."

Hal let out a long wheezy groan. His jaw started to move, chewing on air. When Hal had come round after Lawrence's ragtag medical team installed the biomech heart he'd lost all sensation and movement down his right side. Since then, feeling had been returning slowly, as with a recovering stroke victim. If that was his only complication Lawrence would have been happy. But despite the superoxygenated blood in his brain, there had been some starvation damage. The kid's thoughts were slow and muddled, coupled with memory lapses. With the paralyzed muscles and the difficulty he had putting words together, watching him try to speak was painful. Most of the time he knew what he wanted to say, but grew angry with himself when the words refused to form. Sometimes the anger would cause him to punch the arm of the chair with his good arm, tears of frustration leaking down his cheeks.

"Thank. You. Sarge," he grunted out. Tendons stood proud from his throat with the effort of forming three words.

"That's what I'm here for, Hal." Lawrence glanced around at the other faces in the room, trying to judge the collective mood. They were all quietly expectant, curious what he'd brought them together for. Since the court-martial and the firing squad they'd been directing their shock and anger at Z-B in the form of Captain Bryant and Ebrey Zhang. Resentment and the sense of betrayal hadn't manifested in any coherent form, but they'd become difficult to command. Not that the rest of the platoons in Memu Bay were any more disciplined right now. But with a ruined Hal to rally round and help, 435NK9 retained a degree of internal cohesion. They did what Lawrence told them, not because they were orders from Bryant, but because Lawrence wanted them carried out.

He couldn't have asked for a group of men more suited to help achieve his personal goal.

Funny how things work out.

"The reason I have a contact at Durrell is something I was going to share with you at some time anyway. Might as well be now. I think there's a big asset out in the hinterland that Z-B doesn't know about. I want to collect that asset."

"For Z-B?" Karl asked quickly.

Lawrence smiled without humor. "Not a chance."

Lewis clapped his hands together. "Fucking-a."

"More like it."

"What kind of asset?" Amersy asked. He sounded more cautious than curious.

Lawrence pulled out a desktop pearl. Its pane unfolded and began to display a satellite image of the plateau behind Memu Bay. "This is Arnoon Province. I went up there last time I was on Thallspring, a patrol that was sent on a sweep through the hinterlands. According to Memu Bay's official records, the people living up there harvest willow webs in the forest and turn the stuff into sweaters and blankets, crap like that. What we found when we got there was a nice little village in the woods, with a very decent standard of living. It was like a five-star holiday resort. I've seen the same kind of isolated community on several worlds. No big deal. But there were a few things wrong about this one. You'll have to take my word on this, but there's no way they could afford the standard of living I saw by just selling blankets. Every house was crammed full of gadgets and electronics, all of it top-of-the-range gear. There were a lot of people living there as well, more than Memu Bay knew about, and too many for their community income to support. And none of them were ill, either. I'm not talking about hospital cases. I didn't even see a kid with a runny nose. They were the healthiest group of people I've known."

"So you're saying they've got another source of income?" Amersy said. "Lawrence, I've seen communities like this, too. They'll have some kind of illegal scam running up there in the forest, away from the city police and, more important, the taxman. It won't be anything we can take home."

"No, they have money on a scale that goes way beyond anything like that. I'm talking orders of magnitude, here. They're probably the richest people on this planet."

"How do you figure that?"

"It took me a while to realize, because they've used the best camouflage there is: put your biggest secret in plain sight. I thought they were Regressors that first time. I saw them eating fruit from a tree." He smiled softly at the platoon.

"So?" Lewis asked. "They *are* Regressors. Nobody else does that kind of thing. It's filthy. Decent people eat protein cell food."

Lawrence chuckled. "Which just proves my point. You can't see it either. Willow webs are a local plant. The forest up there in the hills is indigenous. Arnoon Province wasn't gamma soaked."

"No way," Dennis said sharply. "Terrestrial plants won't grow in alien environments. For a start, the soil bacteria is all wrong. That's why you have to gamma soak the land and reseed it with our own bacteria."

"Exactly," Lawrence said. "But I saw it. I saw them pluck fruit from a bush and eat it. From what I remember, it wasn't even a terrestrial bush."

"Then you didn't see it, Sarge. Sorry, but humans don't have a biochemistry that is compatible with this planet's indigenous organisms. I might have flunked my degree, but I did manage to take in stuff that basic."

"I know. But I've seen this happen on one other planet as well. You weren't with us then." Lawrence cocked an eyebrow at Amersy. "You remember Calandrinia?"

"Hardly likely to forget her."

"She was a new-native on Santa Chico," Lawrence told the others. "They ate fruit that was growing on trees. I called them Regressors, too. But I was wrong there as well. Calandrinia told us that the biotechnology experts who emigrated from California eventually worked out how to blend the terrestrial and alien gene pools. It made Calandrinia's generation what they were, and it gave them food the old-fashioned way, so they weren't dependent on food refineries. That was a big part of their philosophy, liberating themselves from machinery. So it is possible."

Dennis pulled a face. "Maybe. On Santa Chico I could believe it. But here, on Thallspring? Christ, Sarge, the most advanced thing ever to come out of Memu Bay is a new shape for a windsurfing board."

"Yeah. And according to Calandrinia it took decades for Santa Chico to develop the gene blend. Decades of work by hundreds of the greatest geneticists and biotechnicians Earth ever had. Yet here we are, in the middle of the hinterlands on a planet thirty-seven light-years from Santa Chico, and I see the same thing. How do you explain that, Dennis?"

"You think they bought the genetic blending technology from Santa Chico. Don't you?" Amersy asked.

"There's nowhere else it could have come from. And it would have taken a lot of money. You'd have to travel from here to Santa Chico carrying a complete range of Thallspring botanical and bacterial samples. Then you'd have to employ a team of geneticists to adapt the techniques. That takes serious money. Billions in anyone's currency."

"Santa Chico's cut off," Edmond said. "Everyone knows that."

Lawrence shook his head. "I was on Thallspring before I went to Santa Chico. This must have happened thirty, forty years ago, maybe even longer. Back when credit meant something to Santa Chico."

"All right," Amersy said. "I can accept that it's theoretically possible for the Arnoon villagers to have trees that produce terrestrial food. But where did the money come from?"

"Two possibilities," Lawrence said. "The first is they're exiles. A group of billionaires setting up a colony inside a colony. Essentially, they're self-sufficient—that's where the food trees come in. They have a high standard of living, which they support by quietly buying in all the consumer goodies that a relatively advanced world can provide. The problem I have with this idea is billionaires don't live like that. You don't work your ass off amassing that kind of fortune and then spend it on some forest idyll community. Earth is their element, with its stock markets and stakes and boards."

"So what's the other possibility?" Odel asked.

"That Arnoon began exactly the way they said it did. A group of honest people looking for a quiet life who harvested willow webs for a living. They start their community, establish some decent tenets to live by. Then suddenly they find something valuable. Fabulously valuable. This is the motherlode of all motherlodes, here. What do they do? If they tell the rest of their world, everyone will want a piece of the action. Arnoon Province is developed and industrialized. Their way of life will be wiped out. So they decide to spend it on safeguarding that good quiet life they've got for themselves.

"A few of them buy passage to Earth on a Navarro house starship. Then fly on to Santa Chico. Several years later, when the gene blending has been accomplished, they come back the same way. After that, it's easy. They can expand their population without the local authorities knowing about it, because they have an independent food supply. Buying in every civilized luxury is relatively easy; they set up a couple of wholesale companies down here in Memu Bay, maybe another in the capital. They're fronts so they can send the products up to the plateau without anyone knowing."

"How did they know about Santa Chico?" Amersy asked. "We damn well didn't until after we landed."

"We didn't know how far they'd carried the modifications," Lawrence said. "We knew about the modification project. It was the big difference that the settlers were so proud of. Santa Chico was not going to be colonized the same way as every other planet. They made that very clear right at the start and went out of their way to advertise the fact. Even on Amethi I'd heard of Santa Chico."

"All right, maybe you did, but a bunch of willow web farmers?"

"They had datapool access and money. It's a wonderful combination; it always gets you what you want in the end."

"I don't believe hundreds of people can keep a secret that big for so long. One of them would come down into Memu Bay and blow his wad in a marina club. Word would leak out."

"Nobody knows about them," Lawrence said. "So the secret has been kept. Simple logic."

"I don't see how."

Lawrence didn't know what else to say to convince him. You couldn't dispute the plain facts.

"Hey, Sarge," Karl said. "What do you reckon this motherlode is?"

"That's where it gets interesting. The survey satellites never came up with anything on the plateau other than the bauxite. So, geologically speaking, it's got to be something that shouldn't be there, an anomaly. Expand on Arnoon," he told the desktop pearl AS. The picture flowed. Snow-capped mountains rippled up as the focus shifted past them. Huge tracts of forest swept by until the AS centered on a valley with a circular lake. There was a small island in the middle. "I saw this, too, when I was there last time. I didn't recognize it then, because you don't often see these all covered in trees and grass."

"Oh, my goodness," Odel murmured. "It's a crater."

"You got it. And it's impact, not volcanic. That island is the central peak. A chunk of some asteroid or comet hit the plateau a few thousand years ago, maybe less. The cliff on the west side is still almost sheer. There's been very little movement or erosion since it happened."

"So what hit it?" Karl was leaning forward, staring intently at the pane.

"I'm betting on metal," Lawrence said. "A near-solid lump of it. That would survive the flight through the atmosphere, and the impact. It also gives the Arnoon villagers something to mine."

"What kind of metal?" Karl wanted to believe. Manna from heaven mixed in with a great treasure hunt. He was buying in heavily.

"I don't know for sure, but it's got to be one of the precious ones. Gold, platinum. Or maybe I'm wrong, and it was an ordinary carbonaceous chondritic that fused into diamond from all that heat and pressure during the impact."

Karl slapped Odel's shoulder. "You hear that? There's a diamond mountain in them there hills, and it's all ours."

Odel gave him a pitying look.

"Maybe it's there," Lawrence said. "And that's what you've got to decide. By yourselves. I need to know if you're in or out. All I know for certain is there's evidence of a lot of money up on that plateau, and there's an impact crater in the same place. To me that's more than coincidence, but I can't guarantee anything."

"What sort of proposition are you offering, Sergeant?" Odel asked.

"Equal share for everyone who goes up there with me. We also have to pay off my contact and a spaceplane pilot."

"How do we get there?" Amersy asked.

"We've been assigned a hinterland patrol, leaving oh-eight-thirty tomorrow. Estimated duration two days."

"Jesus." Amersy gave a surprised, slightly troubled grin. "That's some contact you've got. Our assignments come out of Zhang's office."

"I've been planning this for a while" was all Lawrence said. Even now he wasn't about to trust anyone else with Prime.

"Man, we're covered," Lewis said as his smile broadened. "We're going to be up there on an official mission. And the villagers are going to be the last people to protest about some private asset realization. They can't let on they had anything worth taking in the first place." He looked at Lawrence in admiration. "Fucking-a, Sarge. Count me in."

They all turned as Hal began his loud grunting. "I'm. With. Sarge," he ground out. "Need. Money. To. Be. Better. Not. Live. Like. This."

Edmond patted his friend. "It's okay, man. You get a share anyway."

"Actually, if we go we have to take him with us," Lawrence said. "There's nobody here to give him the kind of care he needs. He can ride in the back of the jeep."

It surprised them, but they didn't object.

"I'm on for it," Karl said. "Fucking Z-B. If there's really any of that metal up there, I can walk away from the bastards."

"Sign me up," Odel said.

"Me too," Edmond agreed.

"You're not leaving me behind," Dennis said.

"Congratulations," Amersy said. "That makes it a full set."

Denise had managed to keep her emotions in check for so long now, she'd almost forgotten they were there, squatting at the back of her mind. She'd told herself her immunity from distraction was due to the d-writing she'd undergone; that objectivity and rationality had been installed along with all the other enhancements. The news about Josep had exposed that for the self-deception that it truly was. Ray had called her an hour after he was supposed to have left the spaceport, saying he hadn't reported in. Then his Prime started intercepting heavily encrypted messages flashing between the spaceport and the East Wing of the Eagle Manor, where Z-B's intelligence staff had set up office. Several of them referred to "the prisoner"; they covered requests for personnel and equipment, mainly from the medical department.

"They're getting ready to interrogate him," Ray said.

Denise fought hard to suppress the dismay that had risen from nowhere. "Do you mean torture?" she asked levelly.

"No, it'll be drugs and brain scans. That's why they want the medical people."

"Can you get him out?"

"I don't even know for sure where he's being kept, yet, but I'm pretty sure it's the spaceport. They disconnected it from the datapool fifteen minutes ago. Which gives us a problem in trying to track down his physical whereabouts inside. And even if I did, it would take time to retrieve him. He'll be under the heaviest guard they have. Denise . . . I don't think I'll be able to get him out and safeguard the mission as well."

"I see."

"He knew that. You and I both knew this was a possibility, too. We've always accepted this risk."

"Yes." Stick to the mission, she told herself. "So now what? Do you think you can get hold of a key?"

"I'll have to wait and see. I need to know where they caught him, and if they have any idea what he was doing at the spaceport. That's what I don't understand, Denise, how the hell did they catch him? We know their security, there was nothing left to chance when we planned this out."

"Another Dudley Tivon," she said. "A random event. Someone caught him red-handed."

"Then why wasn't there any kind of alert? If anyone does anything against Z-B's interests they send up a barrage of red rockets. This time, without Prime intercepting their communications I would never even know they'd taken anyone prisoner."

"So what do you think?"

"The way it was done, the fact he never managed to load a warning into the datapool, I'd say they were waiting for him."

"They *can't* have been, Ray! That would mean they know about us."

"Yeah. Not a nice thought, is it?"

"I don't believe it. There has to be another explanation. There has to."

"I don't want to believe it, either. But we can't afford to ignore it, not now."

"Ray, we have to get a key for one of the Xianti flights. Without that, we've failed."

"Not yet we haven't, not by a long way."

"If you can't get him out . . ."

"I know. He'll never let them discover what he's become, nor what he was doing. At least we have that option."

"Do you want me to come to Durrell?"

"No. If I can salvage this I need you to be ready where you are. I'm going to have to consider my next move very carefully. I think we underestimated Z-B from the start. If that's the case we may even have to abandon the mission altogether."

"No!"

"Face it, Denise. We're not looking good right now. In any case, Z-B will be back in another decade or so. We can try again then."

"All right."

"It's not over yet. I'll keep monitoring the situation here and review my options. I'm trying to establish a link to the spaceport. We should know within twenty-four hours."

Denise managed a sad little smile. "That's when we were supposed to be on our way."

"Yeah. I'll call you as soon as I have anything new."

After that, Denise didn't go to the school. She left a message for Mrs. Potchansky, claiming a stomach bug, then told the house's Prime-augmented AS to filter calls. She knew she wouldn't be able to face the dear old woman, not even over a visual link.

It was the first time the rented bungalow had felt truly empty without Ray and Josep. Her head slowly filled with strange notions as she wandered along the hall. That she should go back to Arnoon Province where she'd be safe. Or fly to Durrell anyway, and rescue Josep. That this whole mission had been a mistake.

None of these thoughts are relevant, she told herself crossly. That didn't stop them from breeding.

Denise looked at the door to Josep's room, not quite sure why she was standing outside it. He hadn't gone in for much in the way of personal decoration—a desk, a couple of dark green leather chairs, which she thought were pretty awful. The bed was a double. Naturally. He'd hung a big sheet screen over half of the opposite wall, so he could lounge around on the mattress and watch the shows. In its inactive mode the screen showed a picture of Mount Kenzi taken on a cloudless, sunny day, rugged snowcap shining bold against the pale turquoise sky. She turned the handle and went in. When he'd taken off for Durrell he'd left the room in a shambles—the quilt crumpled up at the bottom of the bed, sheet rucked; several pairs of swimming trunks were shoved under the bed. T-shirts that he'd worn when teaching his tourists were thrown into a heap on one of the chairs, still smelling of seawater. Towels had been dropped on the floor. A set of his gills were slung over the back of the desk chair.

Despite everything else Denise had to do over the weeks following the invasion, she'd tidied up both the boys' rooms. Clothes and towels were gradually put into the washing cabinet. The mess sorted out. She'd even found two pairs of panties and a bra under Josep's bed—they had also been washed. The quilt had been folded neatly on the foot of the bed. Their little domestic robot had vacuumed the carpet, dusted round and polished the broad window that looked out over the back garden.

Even spruced up like this, the room belonged to Josep. There were tears in her eyes, which she wiped away savagely with her knuckles. She sat down on the edge of the bed, a hand stroking the mattress. When she closed her eyes it was easy to see him. Memories of him as a stupid little boy up at Arnoon. Growing taller and more serious as the years wound on. Emerging from the d-writing, mature and confident, his dedication to the mission easily as strong as hers. Then down here in Memu Bay. Devilish and happy, growing into a decent, attractive young man. All those fabulous girls he'd brought back to the bungalow, ending up here on this bed.

She'd never slept with him or Ray. Instead they'd shared what amounted to a brother-sister relationship, caring and respectful, with plenty of teasing thrown in, housemate pranks. *Was I being stupid? Should I have just leaped at him? Stolen the precious time we had? Or were we both scared of how deep and serious it would become if we started?*

Irrelevant now. Just an exercise in *what if,* and painful self-recrimination as the prospect of total failure dawned. She hated herself for thinking such things. But the memories wouldn't let her stop.

The message package from the underground cell arrived late in the morning. Prime programs installed in various data-pool nodes ensured it stayed below the horizon of Z-B's monitors. Not even dataflow logs recorded its routing.

Denise was curled up on Josep's bed when the bungalow's AS accepted the message and delivered it direct to her d-written neuron cells. The pillow was damp around her cheek. She'd been crying.

Misery became plain annoyance as she reviewed the message. It was from a cell group in Harkness, one of the smaller suburbs almost on the edge of Memu Bay's moat of terrestrial vegetation. They'd barely been active since the occupation. Scrawling a few slogans on walls. Storing equipment and crude weapons for the more active units in Memu Bay itself. But Harkness was stretched along the eastbound wing of the Great Loop Highway—a very strategic location given their mission. The main purpose in recruiting the cell was so that they could keep the road under observation. And they'd just fulfilled their principal function.

The package was a report that two Z-B jeeps had passed through town, heading down the Great Loop Highway out toward the hinterlands.

Denise felt a flash of resentful anger that the imbeciles in the cell had screwed up and bothered her. Especially right now. Another emotion surge she could do without.

There were no jeeps. The Prime she'd inserted into Z-B's headquarters network reviewed their deployment schedules. And something like a convoy of Skins on their way out to the hinterlands would have been tagged as immediate priority. She would have known within seconds of Ebrey Zhang's office posting the duty.

The Prime operating within the bungalow automatically correlated the new information. There was a patrol scheduled to travel round the hinterlands today.

Reflex muscle action made Denise sit up fast. She queried the patrol assignment.

Prime confirmed it.

She loaded another query, asking why Prime hadn't warned her about the assignment.

For software as powerful as Prime, the answer was a long time coming. Several milliseconds. Her Prime hadn't been aware of the patrol before because the assignment hadn't come through Zhang's office. Something else had inserted it into the schedule, and done so in a way that shielded it from registering on any monitor routine. The Prime was sending out thousands of subtle trawlers through the surrounding architecture, trying to locate the origin. One of the probes encountered another Prime lurking inside Z-B's AS.

Within the electronic universe the two quasi-sentient software systems regarded each other passively. Attempts at infiltration and subversion were impossible. They were equals.

"Another Prime?" Denise squawked in shock.

It simply could not be.

Yet there it was.

She withdrew her own Prime.

There had been no alert issued in the Z-B network; nobody knew she'd been sniffing around. The other Prime hadn't informed them. She tried to think the situation through logically. There was only one place a Prime came from, and that was Arnoon. Somebody else from back home must be in Memu Bay. Somebody with a mission contrary to hers. Which again wasn't possible. No Prime would act against the dragon; it had written Prime specifically for them.

None of this made any sense. Then she finally paid attention to the platoon that had been assigned the patrol: 435NK9. Lawrence Newton!

"He can't know," she whispered. But he was heading down the Great Loop Highway on a patrol that Z-B had never authorized, and didn't know about.

Denise closed her eyes and considered her options. There weren't many. She had to know how a Prime was helping Newton. That was paramount: it might even reveal how Josep had been captured. The answer had to be in Arnoon. And Newton himself could not be allowed to reach the province.

Denise ran to her own room and began to change clothes. Jeans, a T-shirt, leather jacket, a small bag with the two weapons she kept in the bungalow. As she was putting them on she issued commands to various cells, requiring them to take direct action against the patrol. Her Prime also scoured the local traffic regulation AS to find a suitable vehicle for her. It gave her a list of possibilities, and she selected the one she wanted. A flurry of emergency route commands were shot into the vehicle's AS.

She pulled some heavy boots on and hurried out.

Lee Brack had been surprised when his bike AS suddenly

flashed up emergency symbols on his optronic membranes and the bike immediately turned off down a side road. He always hated engaging the AS anyway. This bike was meant to be ridden properly, by humans, not goddamn software. The big green-and-gold Scarret had a three-core converter cell for power, with superconductor cabling and multi-ring direct axle motors with inbuilt turning angle compensators. Top speed of 250 kilometers per hour on a decent stretch of road. His wives referred to it as his midlife crisis machine. And here it was being remote-controlled into some damned housing estate. The alignment power coupling turned the front wheel again, taking him in to the curb as he slowed. Parking legs slid out.

Lee Brack took his helmet off and stared around in confusion. "What bloody emergency?" He was in the middle of Stereotype Street suburbia. On the other side of the road an old couple were walking their chocolate Labrador. In front of him an attractive girl was out jogging. Actually she was sprinting damn fast. She came to a halt beside the Scarret.

"Thanks," she said.

"For wh—"

Her hands grabbed the front of his one-piece bikesuit. Lee Brack was lifted off the Scarret's saddle as if he were made from lightweight foam rather than his actual weight of ninety-five kilos. He chased a short arc through the air to land badly on his left arm, with the shoulder taking most of the weight. Something amid his bones and tendons made a nasty crunch. Only then did he manage to yell.

The girl snatched up his helmet and straddled the Scarret. Lee's cry of pain turned to outrage as the dashboard display lit up. What about his fucking security codes? "You bitch!"

Denise's Prime simply erased the Scarret's AS and installed itself in the neurotronic pearls that governed the bike's systems. With her d-written neural structure integrating her directly into the software, it was as if she'd become part of the bike. Power burned into the axle motors, and she turned the handlebars in smooth unison with the alignment power coupling. The U-turn was sharp enough to scrape a parking leg on the tarmac. Sparks fantailed before it finished retracting. Denise accelerated hard, losing Lee Brack's barrage of obscenities within seconds.

The jeeps were approaching the edge of Memu Bay's original gamma soak. Strands of darker, bluish vegetation were mingling with the terrestrial grass on either side of the Great Loop Highway. Up ahead the jungle of native vegetation was steaming gently after the early morning rains. In the passenger seat of the first jeep, Lawrence had a good view of the wide strip of tarmac cutting straight across the land until it disappeared into the trees.

Finally they were leaving the villages behind. They were dotted every few kilometers along the highway, clusters of small buildings that lined the road, almost identical each time—a couple of general stores, always a bar, and some kind of low-tech industry. Truck garages were fairly common, with rows of corroding hulks parked out on the grass. Road maintenance robot stations, also with broken-down chassis strewn around. A semiautomated steel mill churning out I-beams. A reclamation furnace with tall twin stacks blowing out thick, greasy smoke into the clear air, a huge stinking pile of rubbish sprawled over the land behind. The houses that accompanied them were a lot cruder than the fine whitewashed apartment blocks in Memu Bay. These were little more than one-story shacks with walls of cinder block and a roof of composite sheeting and solar collectors. Adults sat outside, watching the road and its traffic. Kids ran about on the dirt paths, chasing after one another, playing soccer.

"None of this was here last time," Lawrence said as they drove through a little conglomeration calling itself Enstone. A big sign was stuck up on the side of the highway, advertising the boatyard that had spread over a couple of acres beyond the row of houses.

"We're twenty klicks from the sea," Lewis protested.

"Cheaper to build out here," Amersy said. "This is Memu Bay's secondary economy. It always starts to grow up around prosperous settlements that have been established for a while. The bigger the population the bigger the percentage of semiskilled and transient workers."

"You mean poor people," Dennis said.

"I certainly do."

The traffic on this stretch of the Great Loop Highway was also a lot heavier than Lawrence remembered from last time. Most of it was trucks or vans that were dropping in and out of the little factories and businesses, shuffling supplies and material between them. At this rate, he thought, it wouldn't be long before the villages merged into a single urban strip.

They were passing through the last highway village when Lawrence's Prime notified him that another Prime had queried the patrol assignment. Another Prime? he asked it. There was no margin of error.

It must be KillBoy, he thought. It was the only explanation. In fact it made perfect sense. He'd always known the resistance people had sophisticated subversion software available. Strange irony, though, that it should be Prime; in twenty years this was the only copy he'd ever encountered.

"I want sensors switched to search pattern A-five," Lawrence told everyone. "Have your AS review the input for localized data traffic and electronic activity. Someone's just taken an interest in us. There may be a few hostiles around here."

"How the hell do you know that?" Amersy asked.

"I have some smart software that can spot illegal askpings. And someone queried our patrol. Someone outside Z-B."

"Christ, Sarge," Karl said. "They should make you general."

"That's some software," Amersy said dryly.

"Yeah. Come on, people, look lively." He checked his telemetry grid to make sure they were activating their sensors. When everyone had upgraded he turned around and checked on Hal, who was riding in the back of the jeep. The kid was leaning on the door so he could look out across the countryside. Wind was thrashing his short hair about. He had a permanent lopsided grin as he watched the scenery flash past. Edmond was sitting beside him, feet resting on a box full of the medical supplies that Hal's modules used.

"Everything okay?" Lawrence asked.

Edmond waved casually. "Under control, Sarge."

They crossed the border between terrestrial and Thallspring vegetation. The only other vehicle left on the Great Loop Highway was a tractor unit pulling a flatbed trailer that was trundling in from the hinterlands. When they passed it, Lawrence saw the trailer was loaded with trimmed tree trunks. He wondered how legal that was. There were several plants in town that synthesized wood.

"Let's go," he said to Dennis, who was driving the lead jeep. "I want to reach Arnoon by nightfall."

Dennis lowered his foot down on the accelerator, and the jeep began to pick up speed.

Since the call came in, Newby had been operating on a permanent adrenaline high, and it felt glorious. This kind of action was what he'd envisaged when he joined the cell. But ever since the invaders landed, all he'd been asked was to keep some bulky sealed boxes in the back of his father's shop, hidden underneath the crates of empty bottles that were waiting to be collected. He did get a thrill from the strangers who would come in and give him the password, either collecting or delivering boxes. It made him feel part of something important. At twenty-three years of age, it was the first sense of belonging he'd ever known.

Now *finally* the cell had been put on active status, with a *critical* duty. He joined his fellow cell members Carole and Russell around the back of his father's store and climbed into the battered old pickup. Any thoughts of a quiet getaway were ruined by the gut-rattler roar of the truck's ancient combustion engine as it fired up. He winced and grated through the gears, racing away as his father came running out.

The instructions received and decrypted by his bracelet pearl were simple and accurate. He stopped to pick up another cell: three people he'd never seen before. Two pudgy pasty-skinned men in their late twenties that he suspected were brothers. The third man was slim and dignified, at least sixty years old, wearing pressed jeans and a denim shirt with a lace tie; his Stetson was also clean and expensive. He looked like money to Newby. But they all had the right password, and each of them carried two intriguingly heavy cases. They squashed into the back of the pickup, and all six of them headed east along the Great Loop Highway toward the Mitchell foothills, with Newby pushing the old engine hard.

They chose the ambush site deep in the jungle, where the road had already begun its climb up to the plateau. It was an area of exceptionally lush vegetation, with creepers and vines that grew at near-visible rates. The battle between the undergrowth and the highway maintenance robots was as fierce as ever. Constant pruning by energy blades meant that the wall of foliage on either side of the road was now almost solid. Overhead, where the robotic implements couldn't reach, the branches had knitted together over the tarmac wound, creating a somber arboreal tunnel. Ragged strings of creeper hung down from the apex, acting as conduits for the rain-soaked canopy above. They dripped sour water across the Great Loop Highway like botanical stalactites.

Newby had to use the pickup's headlights, it was so dismal under the trees. When they finally spotted a gap in the thick tangle of undergrowth along the side of the road, he turned off and slowly maneuvered the pickup through the trees until it was a hundred meters away from the tarmac and completely invisible. Aramande and Rufus, the brothers, immediately set about fixing explosive charges to trees beside the road. They handled the little charges efficiently. During the journey they'd explained that they took part in occasional unlicensed logging operations in the jungle, where a lot of trees needed to be felled quickly. Nolan, the old man, had opened up the remaining four cases. They contained the kind of weapons Newby always dreamed about using against the invaders. Nolan assembled a chunky gun with quick professional motions. He called it a thunderbolt. The short barrel was eight centimeters in diameter, with a loading mechanism that looked as if it had been put together out of components from a hardware store; there was no electronic augmentation. It fired rounds as big as a fist. Nolan slapped in a bulky magazine and handed it to a delighted Newby.

"You get this because these rounds are energized explosive," Nolan said. "In other words, it doesn't matter if you're not very accurate. Which I don't believe you are. We think a direct hit from one of these will kill a Skin suit. A close hit will almost certainly damage one. So when we stop the jeeps and I give you the okay, you fire this magazine at them as fast as you can. The idea is to destroy the jeeps and kick the shit out of the Skins. Then you put in the second magazine and aim for individuals." He handed another thunderbolt to Carole. "The five of you will be shooting these at them simultaneously, and you'll have the jungle to provide cover. In those circumstances, it will be difficult for them to shoot back, but not impossible. Their sensors are good and they're backed up by an AS. They *will* be able to spot you. Understand? That's why you must keep the barrage going."

"What are you going to be doing?" Carole asked.

Nolan opened the last case. There was a rifle inside that

had a barrel nearly a meter and a half long. Even to Newby's untutored eye it looked deadly.

The old man took it out and patted it fondly. "I'll be going for the precision strike."

Newby found himself a tree with a decent solid trunk over two meters wide. It was twenty meters from the Great Loop Highway. If he crouched down between two big buttress roots he had a clear view of the crumpled ribbon of tarmac. A pair of interface glasses kept him in touch with the others. Nolan had brought them as well as the guns. They were all linked with fiberoptic cable, which he'd unspooled across the jungle floor.

"This way we can communicate without transmitting," he'd explained. "It'll help keep our exposure to a minimum."

So now Newby waited with his legs folded uncomfortably and the dreadful humidity soaking his shirt and giving him a serious itch all over. Tixmites had found him, and were eagerly exploring this new supply of nourishment. He was swatting the tiny insects every few seconds as they gave his skin another painful bite. Now that he had time to look around properly, he could see their glistening nest mounds swaddling the tree trunks all around him.

His earlier excitement had faded. Nerves were chewing at his confidence. Shrill birdcalls made him twitch. He wanted this to be over. Twinges of cramp began to shoot down his calf muscle.

"I hear something," Russell's voice whispered in his ear.

"What?" came a chorus of whispers from the others.

"Could be them."

"Very well," Nolan said. "Now remember. Stay calm. This will be short, noisy and brutish. Do not lose track of our objective among all that. We have to support each other. That's the only way this will work."

"I won't let you down. Not me." Newby was slightly abashed to realize he'd spoken it out loud.

"I know you won't, son," Nolan said gently.

"It's them," Aramande hissed. "I see them."

"Very well. Rufus, don't leave it too late."

"Hey, man, I know what I'm doing."

Newby shifted around slightly, lifting the thunderbolt up ready. He looked along the fat barrel toward the road. Sure enough, a jeep was approaching. Headlights glared amid the gloom and shadows. There was another one just behind it. He could see the Skins sitting inside. The first jeep was almost level with him when Rufus blew the tree. It was a simple enough trap. One tree down in front, blocking the road, forcing the jeeps to stop. Then a second would be blown behind them, preventing any retreat. They'd be in a killing zone, with the thunderbolts ripping them to shreds.

The brothers really did know what they were doing. The charge in the trunk blew out a huge section of wood at the base, shaped *just so.* There wasn't much of a flash, or noise. The tree crashed down, tearing through the hundreds of vines that knitted it to the rest of the jungle. It landed almost at right angles across the tarmac, thirty meters ahead of the first jeep.

Newby jumped to his feet, bringing the thunderbolt to bear, finger squeezing the trigger. But the first jeep wasn't even slowing. He thought he saw a couple of bright-orange flashes somewhere among the seated Skins. Two explosions detonated in the middle of the fallen tree. They were terrifyingly powerful, pulverizing a vast section of the trunk. A shrapnel cloud of deadly dagger-sized splinters erupted out of the twin fireballs, shredding the surrounding vegetation. The two surviving sections of the tree on either side of the explosion were shunted apart violently, leaving the road clear.

"Shoot!" someone yelled in Newby's ear.

He was in the act of flinching as several dozen of the fatal wood splinters scythed through the air around him, but managed to pull the thunderbolt's trigger anyway. The recoil nearly wrenched his arm off. God alone knew where the shot was aimed. He recovered and tried to take aim on the first jeep as it sped past. Explosions burst through the forest on the other side of the road. One went off on his side, about thirty meters away. The blastwave was muted by the trees, but still managed to punch him into the trunk that he was using as cover. His interface glasses were flung off. He yelled wordlessly at the pain, unable to hear himself. His ears stung, but the world had fallen completely silent.

More explosions were pounding the jungle, bright orange-and-violet light strobing weirdly. There seemed to be two different kinds, one a lot fiercer.

With his knees barely supporting him, he managed to roll his body around against the trunk until he was facing the road. A jeep was driving past. He brought the thunderbolt up again, surprised by the runnels of blood he was seeing on his hands and sleeves. The weapon wobbled as he lined it up on the speeding jeep. He pulled the trigger. An emerald laser fan swept across him. All he could see was a dazzling green haze. Then something exploded in midair halfway between him and the jeep. He was flung backward as a dreadful torrent of heat scorched into him. He could feel the skin on his cheeks and forehead shriveling. His hair smoldered as he crashed down into the sharp, prickly undergrowth.

Newby laughed, or cried. He wasn't sure which. But his lungs were juddering as his throat convulsed. His whole body was numb as shock blanked out the pain. He could see very little, just simple silhouettes. He blinked a few times as he scrabbled around feebly in the mud and broken branches. It took effort to lever himself up onto his knees. The laserblast had left huge gray mists floating across his vision. He whimpered as the numbness quickly gave way to a terrible cold gnawing deep into his flesh. Then he was shaking uncontrollably. The jeeps had gone. Several fires were burning amid the shattered trees. Smoke braids coiled around the trunks as they drifted up toward the canopy.

A dark mote streaked past his head from the direction of the road, so fast he thought it was some phantom, part of the damage the laser had wreaked on his eyes. But there was a tiny rigid contrail in the air, marking its passage.

Newby turned to see where it had gone. The contrail was curving at incredible speed, weaving fluidly around the intervening trees as it chased through 180 degrees. His brain sent a flood of nerve impulses out to his lungs and vocal cords, preparing them for a scream. They weren't fast enough.

Lawrence didn't allow them to stop until the jeeps had climbed up onto the plateau itself, and they were free of the jungle. During the last section of the climb the Great Loop Highway had gradually eroded to little more than a path through the trees. The tarmac had crumbled away from a combination of heat, water and roots. This far from Memu Bay, the budget for highway maintenance robots no longer allowed for resurfacing. The best they could do was keep the original route clear. Vehicles that traveled out here had the kind of gearing and suspension to cope with a mud track.

The jeeps had certainly managed. They'd come through the attempted ambush with several dents from chunks of flying wood, and the paintwork was scarred and scorched. But the engines and wheels were intact.

Dennis braked to a sharp halt as soon as Lawrence told him he could, tires kicking up a cloud of sandy dust.

Lawrence turned around. The sniper's bullet had caught Edmond at the base of his neck, slicing clean through the Skin carapace. There was nothing the Skin's medical program could do for him. The bullet had spun inside him, hacking through muscle, blood vessels, nerves and even shattering two of his cervical vertebrae before punching out through the back of his shoulder. There was just too much damage.

Hal's arms were flung around his friend, as they had been for the last hour. Even with half of his facial muscles impaired, his anguished expression was terrible to see.

"Dead," Hal wailed. He sucked down some air and blew it out. "Dead. Dead." Another labored inhalation. "Sarge. He. Is. Dead."

"I know, Hal. I'm sorry."

Blood had foamed out through the hole in the Skin's carapace. It'd soaked into the front of Hal's white shirt, where it was clotting into a thick paste.

Amersy, Lewis, Karl and Odel walked over from their jeep.

"Shit," Lewis muttered on the general communication link. "Now what?"

"I didn't know this was going to happen," Odel said.

"Yes, you fucking did," Karl snapped. "The sarge warned us. And we saw those bastards lurking in the woods."

"He's dead!" Odel snapped.

"So are they." Karl's voice had a satisfied edge. "Smart missiles. You know they make sense."

"Dear heaven, this shouldn't have happened." Odel turned away from the jeep, standing with his hands on his hips.

"We have to bury him," Lawrence said.

"Sarge?" Dennis asked.

"Bury him. As far as Bryant and Zhang are concerned, he's another Jones. We can't take him back with us. We can't tell them what happened."

Hal was still embracing his friend. Dennis had to prize his arms away using a hefty fraction of his Skin's strength. Hal's cries were wretched as they carried Edmond away from the jeep. His hands flailed helplessly against the seat and door, rocking the whole vehicle.

By unspoken consent they walked several hundred meters

away from the track. Amersy and Odel began to scrape at the sandy soil, digging quickly. They laid the body, still in its Skin, in the bottom of the grave and filled it in.

"Anybody got any words?" Lawrence asked.

"Good-bye, mate," Karl said. "I haven't finished with KillBoy's friends yet. I'll score up a few more for you before this is over. Promise."

Amersy sighed. "Those of us who knew you thank you for the time you shared with us. You lived a good life, and that will not be forgotten. We wish you bon voyage on your last journey. May God embrace your soul."

"Amen," Dennis mumbled.

"Amen," Lawrence repeated.

"So now what?" Lewis asked as they walked back to the jeeps.

"We should be able to reach Arnoon in another five hours," Lawrence said.

"You mean keep going?" Odel asked.

"I will," Lawrence said.

"But he's dead, Sergeant. They know we're here."

"Not anymore they don't," Karl said. "They're dead too. We've earned that money, man. It belongs to us."

"If you want to go back, you can," Lawrence said. "Nobody's going to stop you, or hold it against you. I said right from the start, this is your own choice. It was Edmond's, too."

"God damn that KillBoy," Odel said. "I hope he burns in hell."

"Okay, let's get started," Lawrence said. "Dennis, I want you to look after Hal. Get him cleaned up; I think we brought some fresh shirts for him. I'll drive. Odel, you're with us; I want you integrated with the smart missile rack."

"You think they'll try again?" Lewis asked.

"Only if they're really stupid," Amersy told him.

* * *

Denise managed to keep the Scarret's speed at around the 140-kph mark as she powered through the highway villages. Her body weight swayed fluidly from side to side in perfect concert with the alignment power coupling, slicing the bike round the lumbering trucks and decrepit old vans. The combination of the Scarret's laser radar, Prime and d-written neurons proved a formidable guidance mechanism, allowing her to push the bike right out to its limit. Ramshackle buildings flashed past, reduced to a peripheral slipstream of drab colors. Her attention was focused only on the road ahead, the obstacles that snapped up. Bicycles were a pain. People were dangerous, especially the kids, who ran out into the tarmac. She lost count of how many times she hurtled past one with only a few centimeters' separation distance, leaving the child screaming in terror.

The traffic began to thin out as she closed on the border. As the gaps between vehicles stretched, she increased the power flow to the axle motors. Hunched down behind the sculpted ellipsoid of the windshield she could feel the wind blast past on either side. Tarmac was a slick blur below the fat, soft tires. Once again, human emotion had engaged. The aggressive thrill of speed pursued over the edge of safety. A predator's satisfaction at closing on its prey. And coiled deeper in the psyche, the painful hunger for a revenge that was pure vengeance.

She thundered out of a low valley to see the countryside open up ahead. The Mitchell range slid up across the horizon, standing aloof above the jungle. One by one she named each of the peaks spiking up into the pale turquoise sky. It had been months since she'd seen them, the companions of her youth. The sight of them invoked a subtle reassurance. Despite the circumstances, she was coming home. The loneliness would soon be over.

Inevitably, once the Scarret entered the jungle, she had to slow again. The tarmac was cracked, pulped gray fruit was splattered across it, water pooled in the potholes and steamed off the flatter sections. Even this bike, with all its active stabilization and compensators, had to be careful over such a treacherous surface.

Her private wish was that she'd catch up with the jeeps before the ambush, maybe even charging past to help Newby, Nolan and the other cell members. Not anymore.

When the Great Loop Highway finally narrowed enough so that the trees merged above and cut off the sunlight, she switched the headlight on. It was a strange, spooky section of road. Rather than illuminate, all the blue-white beam seemed to do was deepen the twilight murk around her. The undergrowth that fenced the tarmac was peppered with mold and slime; the leaves, deprived of light, had grown long and distorted, bleached of their healthy color. Tixmites were the only form of life here, flourishing on the decay carpet that was the jungle's floor.

The bike hummed down the center of the disintegrating highway, its superb engineering still giving her a smooth ride over the erratic surface. She switched off the laser radar in case the Skins detected it. Every enhanced sense she possessed was straining to detect them.

In the end, it wasn't difficult. Gases from the explosives lingered a long time in the still, thick air that smothered the road. Denise smelled them a minute before she reached the ambush site. She came around a slight curve and saw thick columns of sunlight pouring down through the canopy where several trees were missing. Parking legs slid out of the Scarret and she got off. Explosions had torn huge rents through the jungle. Shattered stumps were still smoldering. There was a shallow crater in the road itself, with the ruins of a huge tree on either side. It didn't take much to work out what had happened. The fallen tree to stop the jeeps, putting them in the killing zone. Except the Skins had blasted it aside with their own weaponry.

She knew from her Prime's snooping through Z-B's AS that the platoons had brought heavy-caliber weapons to Thallspring. But it was the first time they'd ever used them. Newton must have withdrawn them from storage without anybody knowing—just as she had with the land mines.

The notion was extremely worrying. Unless it was some huge coincidence, it must be Newton who had the rogue Prime. Which meant that he must know of the dragon. *How?* Had somebody told him? The same person who had also given him a Prime?

And now he was taking his platoon up to the plateau on a freelance mission. There could really only be one reason.

Denise scouted around the immediate vicinity of the ambush, trying to find out what had happened to the cell members. She had a vague hope that they might be able to help fill in some details. Then she saw a toppled tree that was splashed in scarlet fluid. Tixmites were swarming over it. They were also falling off as fast as they arrived; hundreds were lying underneath, dead. She walked closer to investigate. Her foot slipped on a lump of something with the consistency of tough jelly. She looked down and winced.

The cell members wouldn't be able to tell her anything after all.

She hurried back to the bike. Her ring pearl used the domestic relay satellite to place a call to Arnoon. The call routing was guarded by Prime and heavily encrypted. Even so, there was a tiny risk of interception, but she had to take it.

"Denise!" Jacintha exclaimed. "Why the encryption? Have you heard something about Josep? We're all so worried."

"We've got a bigger problem than that, I'm afraid."

The Great Loop Highway was now just a rather feeble joke. The transponder posts were missing. The maintenance robots hadn't cleared the vegetation for years. The highway was nothing more than two uneven tire ruts in the ground, gouged out by the few trucks and pickups that still drove across the plateau. And they weren't even following the original path anymore. As puddles and holes grew bigger, the drivers had swerved around to avoid them. These curves would create new holes, and the next swerve would be wider.

Lawrence was constantly turning the wheel to follow the meandering track as it snaked around unseen obstacles. There were no puddles today; it hadn't rained on this part of the plateau for some time. His jeep was throwing up a smog of powdery dust as it bounced along the ruts. The stuff got everywhere. Hal had to wear one of the paper masks from the medical kit. Skin gills had to flush the gritty particles out of their filter membranes.

Lawrence was constantly referring to his inertial guidance display to confirm they were still heading roughly in the right direction. There was no other way of knowing. The map file of the plateau was the same as they'd had last time, without a single update. It still showed the Great Loop Highway running straight and true between the hinterland settlements.

When they approached Rhapsody Province he even thought the map was glitching. There was no sign of the bauxite-mining operation. It took him a while to realize that the conical hillocks ahead of them were actually the old slag heaps, a little bit taller now and covered in baby crown reeds and stringy weeds. The vegetation had a distinct lemon tinge, as if the plants were jaundiced.

"I wonder if they've shut the mine down altogether," he said.

"Can't see much happening here," Dennis said. "Maybe they've moved on."

"At least that explains why the road's in such a crappy state these days."

They drove on past the base of the first few slag piles, then turned in among them. Somewhere up ahead was Dixon. Lawrence didn't really want to go there, but that was where the road led. For all their ruggedness, the jeeps wouldn't be able to cross the raw terrain of the plateau.

"Somebody behind us, Sarge," Lewis announced. "Moving fast."

Lawrence expanded Lewis's telemetry grid and called up his visual sensor. Sure enough, a small plume of dust was racing across the plateau. It was too far away for the sensors to gain a clear picture of what it was. But it was certainly traveling a lot faster than the jeeps had managed along the same stretch of road.

"Keep tracking them," Lawrence said. "No active sensors. But I want to know when you can make them out properly."

"No problem, Sarge."

Dixon was still there. Most of it. The first thing Lawrence saw was that all but one of the huge maintenance sheds were gone. Its doors were open, showing a single excavator processor standing inside. Concrete oblongs marked where the other sheds had stood, gradually succumbing to the slow incursion of windswept soil. One of them was now a parking lot for a couple of articulated trucks. A further two were covered with small piles of aluminum ingots; there weren't enough to fill even one of the trucks.

The houses remained, though the majority had sheets of bleached plywood fixed over their windows. The grainy dust lay thick on every ridge. Lawrence noticed that all the air-conditioning cabinets had been removed, leaving empty metal brackets on the walls.

He looked over to the hexagonal building outside town that housed the fusion plant. The web of red power cables that used to radiate away from it had been taken down; now there was just a solitary line of pylons carrying a lone cable across the countryside. When he switched to infrared, the walls and roof glowed a light coral pink in comparison to the dull vermilion of the surrounding land.

"They have power," he said.

"Anybody home?" Dennis asked. He was too edgy to make it entirely jovial.

"There has to be somebody," Odel said. "They're still working here. The lights in the shed are on."

"They must have seen us coming," Karl said. "They'll be hiding out there somewhere."

"How did they know it was us?" Amersy said. "We didn't announce we were visiting."

Lawrence's jeep had reached the first houses. He nudged it forward along the main street, sensors sweeping for any sign of movement. "I don't care where they are as long as they're not in our way. Keep going."

"Sergeant!" Odel called. "Airborne. Incoming."

Odel's telemetry grid expanded across Lawrence's vision. Tracking data scrolled down. Three kilometers west, five hundred meters' altitude, holding level at four hundred kilometers per hour. One meter long. No known match found in the armory file.

"What the fuck is that?" he murmured. His own AS had acquired it: there was hardly any infrared signature, and no electromagnetic emission at all.

"It's a goddamn recon drone," Lewis said. "They're hunting us."

Who? Lawrence wondered. Somehow it didn't seem like the kind of thing that KillBoy would use. It must be Arnoon Province. They had the money and the technology to guard their territory. Despite the alarm at such a thing being deployed against them, he felt happy. *I was right about them.*

"More like a smart cruise," Dennis was saying.

"Amersy, double time," Lawrence said. "Let's get out of here. Odel, use a smart, shoot it down."

"Yes, sir!"

Lawrence accelerated: the main street was the best piece of road since before the ambush; the jeep made a hundred kilometers an hour along it without any trouble. He saw Amersy keeping up with him. A single pulse of bright-orange flame squirted out from the smart missile rack Odel was carrying. His sensors tracked the little missile as it flashed into the sky, arcing around to line up on the unidentified drone or whatever the hell it was.

He pushed the jeep harder, racing across the central square. His display grid flashed a huge silent warning. His Skin was being struck by an incredibly powerful em pulse. Even though all its electronics were ultrahardened, the brutal power of the energy wave had already overloaded several neurotronic pearls. Noncritical internal functions began to close down.

The jeep died on him. Every electrical system simply stopped working simultaneously. The dashboard display didn't even flicker before it blanked out. They were almost across the square, with the main street just to the right. He turned the wheel, but it was sluggish without the power steering. His foot kicked down on the brake. Tires skidded on the loose sandy soil.

Their right fender clipped the building on the corner of the main street. The hood smashed through the wall, shattering composite paneling into a shoal of fluttering fragments. Then the right front wheel struck one of the concrete foundation piers. Lawrence was slammed into the steering wheel, which broke instantly. His beleaguered Skin AS didn't harden the carapace fast enough. The steering wheel's blunt column stabbed straight through his Skin, puncturing his flesh just below the rib cage on his left side.

Odel was catapulted out of his seat and straight through the windshield. He plowed into the hulk of the building, his inertia breaking several more panel sections. Hal's safety straps held on to him as he was flung forward, then reeled him back into the seat. He flopped there limply, his eyelids flickering. Dots of blood began to stain his fresh shirt, seeping out from around the medical modules. Dennis was slung out sideways, his Skin locked solid as he whirled across the road.

Amersy saw what was happening to the jeep in front and yanked the steering wheel around hard. The brake seemed to have virtually no effect on their speed. He saw the other jeep crash into the building, its rear end lifting from the ground as it struck the pier. His steering wheel was already in full lock; he couldn't twist it any farther. They missed the other jeep by less than half a meter. By then they were almost at right angles to the street. Amersy tried to reverse the lock. He could feel the tires skidding. They hit something big in the middle of the road. Momentum rolled the jeep. There was a single bar to protect the occupants. It almost worked. From Amersy's point of view the horizon tilted fast, turning the ground into the sky. It descended onto his helmet. His Skin carapace hardened just in time to protect him from the lethal blow. Then the world rotated again. And again.

Lewis tumbled out of the rolling jeep when it was on its second spin. His Skin had turned rigid, holding him in a spread-eagle pose as he slithered across the dirt road to crash into a building's foundation pier. Despite the Skin's protection the impact stunned him. The Skin released him from its grip, and he collapsed back onto the ground. When he lifted his head he saw the jeep had finally halted, its wheels in the air. The roll-guard bar had buckled, trapping Amersy and Karl underneath. He clambered to his feet and staggered over.

Amersy's upper torso was protruding from the wrecked jeep. He was trying to crawl out, but the chassis had his legs pinned. Lewis gripped the side of the jeep, braced himself and lifted. The jeep creaked as it rose half a meter. Amersy wriggled free.

"Thanks," Amersy said.

"What the hell hit us?"

"I think it was some kind of e-bomb. It blew every circuit in the jeep. Even my Skin's electronics suffered."

"Shit. Where the hell did KillBoy get an e-bomb from?"

"God knows." Amersy looked back at the first jeep. "Sarge?"

"I'm here."

"You need help?"

"Don't think so. How are your guys?"

Amersy saw blood spreading out from underneath the back of the jeep. "Jesus! Karl? Karl, can you hear me?" He checked Karl's telemetry grid, which was almost blank. The Skin suit still had some functions, and there was a heartbeat. But that was about all he could tell.

Both of them dropped to their knees and peered into the gap. Karl's Skin had been split open by ragged twists of metal that had peeled back from the bodywork. Several of them were still impaling him.

Amersy used his speaker. "Hang on, Karl, we'll have you out of there right away."

"We'll have to turn it right over," Lewis said.

"Shouldn't be a problem." They took hold of the jeep. "Ready?" Amersy said. "Okay, lift." The jeep began to emit labored metallic screeches as they slowly tipped it up. One of Lewis's hands slipped, and the vehicle sagged back a few centimeters.

"Hell," Lewis grunted as he found another hold. "The tank's split, this thing is dripping in hihydrogen."

"Great."

They had almost got it on its side when Amersy's sensors detected the little projectile. An intense blue-white spark hit the jeep. The hihydrogen ignited immediately, enveloping the entire vehicle. Amersy and Lewis let go, sending it crashing back down.

"Down!" Amersy ordered. He was already flinging himself flat.

The fuel tank exploded.

Lawrence was unconscious for an unknown time. When the pain pulled him back he guessed it couldn't have been long. Dust was still swirling round the wreck. Amersy called him, and he said he was okay, though he was lying. The end of the steering column was still jutting into his side. The Skin's medical program was scrolling up information: the damaged tissue, chipped pelvic girdle, the suppressants it was pumping into his blood. He placed both hands flat on the dashboard and pushed. His body moved back, sliding off the steering column.

Even with the drugs, Lawrence wailed at the pain as it withdrew. Then his Skin muscles were realigning themselves, sealing up the wound. The suit's internal layer discharged antiseptic, anesthetic and clotting agents into the tear. The whole area turned blissfully cold.

He turned around to see what had happened to Hal. "Oh, sweet Fate." The kid was sitting against his safety straps, head hung forward. All of his medical modules had been burned out by the e-bomb. Lawrence's Skin AS couldn't get a response from any of them. Small patches of blood were staining the kid's shirt where the jolt of their impact had dislodged modules.

"Sergeant?" Odel limped out of the house. There was something wrong with the shape of one leg. "Are you okay?"

"Sure." He levered himself out of the driver's seat. "You?"

"Not a problem."

"Good, let's get Hal out."

"Where's Dennis?"

Lawrence looked around. There was a Skin suit lying in the middle of the street, badly mangled. The second jeep must have hit it full on. "Shit!" Dennis's telemetry grid was flatlined.

Don't grieve for the newly dead—they never thank you. Secure the platoon. Lawrence could hear Ntoko growling it out.

He released Hal's safety straps and lifted the kid out of the rear seat. "Bring the medical kit," he told Odel.

An explosion rocked the street. Lawrence saw a fireball burst out of the second jeep as the wreckage levitated off the ground.

"Amersy, get to cover. We've been ambushed. Seek and destroy."

"Roger that."

The two Skins lying flat on the ground beside the blazing jeep scrambled to their feet and sprinted for the nearby houses.

Lawrence wrapped one arm around Hal, lifting him gingerly, then reached down into the passenger seat for the smart missile rack. He stepped into the house through the demolished wall.

The room inside was empty. He kicked the door open and went through into a dark hallway. There were six identical doors and a stairway. He hurried down to the far end and kicked the last door open. It was another empty room. Tiny cracks of light leaked around the boarded-up window. He eased Hal down in a corner.

Odel put the medical kit down and snapped the top open. "Have we got anything that'll help?"

"I don't know." Lawrence picked out a diagnostic probe and switched it on. He was relieved when the small display pane lit up. The em pulse hadn't damaged electronics that were off at the time of the attack. Data began to flow into his AS.

There was the sound of a carbine firing in the distance. "Odel, go help them."

"You bet."

"Hey, be careful out there. These bastards look like they know what they're doing."

"So do I."

Amersy ran along an alleyway leading off the main street. There wasn't a lot of cover. The prefab houses were spaced a regulation twenty meters apart. The grid of streets almost allowed him to see from one side of town to the other. It also gave the ambushers the same field of vision. And he didn't know where they were.

Lewis had ducked away at the second house, heading off down a side street. Amersy raced on for a little while, then took a turning. He called up the Dixon map file, inertial guidance plotting his position. The Skin AS gave him a rough estimate of where the little projectile had come from. His link with Lewis gave him the other man's location, plotting that on the map as well.

"Lewis, we need to pincer them. You getting this tactical data?"

"It's online, Corp."

"Keep going straight on for a hundred and twenty meters. I'll be three streets to your left. They should be between us."

"Roger."

Lewis jogged along the route Amersy's AS had provided. His carbine slid out of its arm recess. When he got to the intersection he stopped and peered around the corner. There was a flash of motion two houses down. His AS replayed the image. It was a woman in her mid-twenties, dressed in jeans and a yellow T-shirt. She was carrying some kind of pearl-white cylinder in her left hand. The AS couldn't match it to any weapon in its catalogue file.

He switched carbine rounds to depleted-uranium and fired straight through the wall. The composite paneling in front of him disintegrated as the high-penetration rounds went straight through. Nothing in the houses was solid enough to stop one. His Skin AS gave him a spread pattern that had a high strike probability. He stopped firing and started to run to the corner where the woman had disappeared.

"Got a hostile, Corp. In pursuit."

"Roger that. Did you get them?"

Lewis reached the end of the street and jumped. It was a trajectory no hostile would be expecting. He flew past the end house, a meter and a half in the air. Carbine pointing down the street, sensors sweeping. There were puncture holes in the houses left by the depleted-uranium rounds, three foundation piers had shattered, but no body. His feet hit the ground and he was still running. He dashed behind a house and crouched beside a foundation pier.

"Shit, I missed, Corp."

"Okay. Let's close in, they're here somewhere."

Lewis got to his feet and turned down the street toward the corporal. He'd gone maybe ten paces when his neat display grid dissolved into a jazz of indigo lightning bolts. "What the hell . . . ? Jesus, not now." His Skin's neurotronic pearls must have taken a bigger hit from the em pulse than he'd thought. He waited for the e-alpha fortress to reboot the crashed software. Instead the indigo swarm evaporated, leaving him without any data at all. "Son of a bitch."

He'd almost come to a halt when the woman stepped out from behind a house twenty meters ahead. She just stood there, watching him.

Lewis snarled, bringing the carbine up. Even without targeting graphics he could hit a pebble at this distance. The carbine didn't fire. He squeezed his trigger finger twice, three times, an impulse that was linked directly to the weapon. Nothing.

He started to run toward her. If she was counting on him not using brute force against a woman, she was about to find out the hard way just how big a mistake she'd made.

All of a sudden his legs weren't making any progress. It was as if he were wading through thigh-high mud. In dismay he realized his Skin muscles had stopped supporting him. His ordinary leg muscles were having to move the entire Skin suit.

"Corp!" he yelled, hoping his speaker was still functioning. "Corp, the fucking Skin's crashed. Corp!"

It became impossible to move. The Skin muscles had solidified, entombing him. He toppled over. For whatever reason, his visual sensors remained online. The side of his vision field showed him the woman walking toward him at a leisurely pace. She stopped beside him, the toes of her worn sneakers almost touching his shoulder.

Lewis was having to suck every scrap of air down into his lungs. The fail-safes! What had happened to the fail-safes? He wanted to shout up at the woman to help him, to open the Skin suit. But there was no air.

The woman leaned over slightly, as if she were studying him. Then she held a hand out above his helmet, fingers spread wide. She slowly closed the hand, fingers curling in to make a fist.

Lewis felt the Skin muscles flex. For an instant he believed the e-alpha was finally rebooting the entire suit. Then the Skin muscles began their contraction. He found enough air to scream as his ribs cracked. The last thing he saw was the hand above him squeezing tight.

Lawrence's AS had come up with very little to help Hal. What the kid needed was a whole new set of medical modules. Some of the ones he was using were so specialist that Lawrence wasn't even sure Memu Bay had replacements. All they had in the kit was field-aid systems and capsules of the drugs that the organ support modules used.

The diagnostic was showing abnormal blood chemistry. Lawrence's AS produced a list of drugs to combat the condition. But he didn't know if that blood chemistry was wrong given Hal's state. In the end he settled for injecting small doses.

Hal groaned, his head moving slowly from side to side.

Lewis dropped out of Lawrence's telemetry grid. "Amersy, what the hell's happening out there?"

"He's gone. I don't know what . . ."

Amersy's voice faded. His telemetry grid was breaking up. Then a warning Lawrence had never seen before flashed up. It took him a moment to recognize the symbols. His Prime was expanding into the Skin's neurotronic pearls, replacing the standard AS program.

"Amersy, Odel, listen to me. Your Skins are being infiltrated with a very powerful subversive software. Close down and reboot. Do not use the communications links again. They've been compromised. Repeat, don't use the communications links."

Odel's telemetry vanished.

"Oh, sweet Fate." He requested a status summary from Prime and read the indigo data as it scrolled. Prime had immediately blocked the attempt to infiltrate his own Skin. The ambushers had tried to use their own Prime to subvert his AS. In which case, he decided, the ambushers had to be from Arnoon Province. They must have bought their Prime when they were on Earth—not that it mattered now.

He could hear a carbine firing again. If they could intercept the Skin communications link, they'd have his position down to a millimeter. He snatched up the smart missile rack and plugged its data cable into a port on his Skin. Prime entered the missiles.

Lawrence stood up. Prime was displaying maps with tactical scenario overlays. Just as he turned to leave, Hal groaned again. Lawrence gritted his teeth.

Amersy did exactly as Lawrence ordered, shutting down his entire Skin immediately. For a wretched moment he was locked solid in airless dark. He hated how vulnerable he must be, a big figure standing motionless in the middle of the open street. Then the AS began its reboot. Sensors came back online.

This whole mission was going too wrong, too quickly. Whatever the Arnoon ambushers had, it was an easy match for Skin. Amersy knew Lawrence had seriously underestimated them. The remains of the platoon were never going to reach their pot of gold now; all those lazy dreams of coming down from the plateau with enough money to buy out were dead. Survival was all that mattered.

As soon as he got limb movement back he ran at the nearest house, shattering the door as he went straight through. When he was inside he ordered the Skin to open. He started instructing the AS as he wriggled out.

Denise was doing nearly a hundred kph when she charged past Dixon's outlying buildings.

"Amersy's stopped transmitting," Jacintha said. "Gangel, see if he's still active."

"I've got Odel's last position," Denise said. "I'll take him."

"Be careful," Jacintha said. "We know the Prime hadn't infiltrated his Skin. Damn, Newton's good."

"Don't we know it."

"Eren, help Denise out, please."

"I'm thirty seconds away," Eren assured them.

Denise slowed the Scarret and skirted the main square. Odel could be down any one of half a dozen side streets. Fortunately, a Skin was easy enough for her to spot: its heat signature left a trail in the sandy soil that was as strong as a neon sign. She sensed Eren walking down a street parallel to hers. Seventy meters ahead, hot footprints glowed on the ground. They led into one of the houses. Its front door swung loosely on its hinges.

"Got him," Denise said. She slowed the bike to walking pace and closed on the house.

Carbine fire echoed over the deserted town.

"Amersy is still with us," Jacintha said dryly. "Those were depleted-uranium rounds. Careful, everyone."

Denise halted ten meters from the house. Eren appeared from an intersection opposite. He gave her a small wave.

"Cover the other side," she told him. "A Skin can walk through these walls like they're paper."

"Right." Eren jogged away to the back of the house.

Denise took the electron cluster pistol from her bag. The little unit melded snugly into her palm, giving her an aim alignment that was pure instinct.

More carbine fire sounded as Jacintha and Gangel started a running firefight five streets away. Ragged holes were punched clean through a house twenty meters behind Denise. Jacintha and Gangel shot back with e-c barbs. Sheets of flame roared into the sky as composite walls ignited.

"Ready," Eren said.

Denise swung a leg over the bike and stood facing the front door as it swung about in the slight gusts. *This is wrong,* she thought, *it's too easy. Odel is a properly trained soldier. He won't allow himself to be cornered like this.* She looked up at the house's roof.

The solar collector panels were hot from the sun. Infrared was useless. Then she saw scuff marks in the patina of ocher dust.

Denise spun around, her e-c pistol coming up. She was firing as she turned. Tiny sparks of light spurted out of the nozzle, twinkling as they shot away into the sky. Then one of them struck the Skin lying prone on the roof. It punched the bulky suit a couple of meters across the smooth solar collectors, ripping off a segment of the carapace. Two more e-c pulses hit, tearing the Skin apart.

Eren came round the side of the building. "Denise? What happened?"

"He jumped to another house. It was an ambush."

"Hell. Well done."

Carbine fire sounded again. Depleted-uranium rounds tore into the house, shredding huge sections of the paneling. A foundation pier burst apart in a wicked shrapnel cloud of concrete chips. Denise and Eren hit the ground together.

"Crap, I hate those carbines," Eren shouted as he lifted his head.

Denise risked looking back to where Amersy had fired from. "They never used depleted-uranium down in Memu Bay."

Eren grunted. "I wonder why."

Another house ignited from a deluge of e-c barbs. Carbine fire crackled again; buildings shivered and rocked as the rounds chopped through.

"Gangel, he's on your left," Jacintha cried. "Denise, we could use some help."

"On our way."

Eren gave her a reluctant grimace and started running. Denise paced him as they closed on her sister's position.

"He's under a house," Gangel said. "Damn, moving again."

A stream of e-c barbs poured out of an intersection ahead of Denise. She flinched. Carbine fire answered. Denise hit the ground again. Twenty meters away, a solar collector roof ruptured; black, glittering fragments rained down on the street.

"What was he shooting at?" Jacintha asked.

"Who cares?" Gangel said. "He can't have much ammunition left, not at this rate."

More depleted-uranium hammered through five houses. The last one sagged and slowly collapsed inward as its piers disintegrated. Denise had been ready to run forward again. Instead she buried her head in the ground.

"Shit!" Jacintha exclaimed. "He's got us pinned down."

"This is seriously crap tactics," Eren said in alarm. "If Newton comes up behind us, we're dead."

"They can't communicate," Denise told him. She wished she had more confidence. The platoon had worked together for years, decades even. And they were trained soldiers. If anybody could manage without a direct link, it would be Newton and Amersy.

The carbine fired. Her link with the Scarret went dead. "Shit!" She was aware of Jacintha hurrying forward again. Gangel started sprinting from the opposite direction. They were pouring e-c barbs into the house where Amersy was hidden. Denise began to race forward, holding her e-c pistol ahead, sending out a constant barrage of barbs. The house was an inferno, huge, violent flames roaring almost horizontally out of the shattered windows. Its solar collector roof was twisting and flexing as the heat lashed at it. Then flames were stabbing victoriously through the gullies as it juddered and began to sink down.

Another burst of carbine fire came from inside. Even as she flinched down yet again, Denise marveled at how the corporal could keep so cool in this situation. Skin suits might be heat resistant, but to stand at the heart of an inferno surrounded by enemies and still maintain a devastating fire pattern was enviable.

One of the walls collapsed amid a huge fireball. A Skin tumbled through the gap. It was hit by e-c barbs from three directions, bursting apart.

Denise squinted against the glare and fierce heat. There was something badly wrong about the way the Skin had ruptured. Jacintha must have thought the same. She was approaching the remnants, cautious yet urgent, covering it with her pistol.

The solar collector roof finally fell to the floor, throwing out a cascade of sparks. Jacintha raised one hand to ward them off. She bent over the tattered Skin. "Shit!" She was looking around wildly.

"What?" Denise asked. She was closing in, along with Gangel and Eren.

"It's empty. The bastard wasn't in it!"

Denise was suddenly swinging round, her heart thudding in fright as she attempted to cover half the town with her pistol.

Amersy crouched behind a foundation pier, watching intently as the young woman and her companion started running toward the firefight. His Skin AS was firing the carbine in random bursts, maintaining a suppressing-fire pattern. The two ambushers flung themselves flat. He grinned as he scuttled over to the shiny Scarret. Damn dumb amateurs, weren't even checking their asses. He used a power blade to slice through the dashboard, then waited until another carbine burst sounded before jabbing the tip straight into the electronics. He cut through the neurotronic pearls and fiberoptics that were connected to the compensators and brakes. Without AS management (or whatever program was loaded in) the bike would be sluggish. But he could accelerate, brake and steer manually. It was enough to get him back to Memu Bay. That would have to do.

He slung a leg over the saddle and twisted the throttle.

Five houses were on fire around the empty Skin suit, their composite panels hissing and melting as flames licked around them, exposing the steel skeleton. Thick black smoke billowed high into the plateau's calm air.

Still watching the empty streets, Denise went over and hugged her big sister. "I missed you," she whispered.

"We're together now. Everything will be all right."

"I hope so. We're making a complete mess of this."

"He's naked and alone, he won't get far."

"On my bike, he'll get clean away." She couldn't believe she'd been so stupid.

"It doesn't make any difference. He's not a part of Z-B anymore. They won't be sending in the cavalry. Not on this one."

"Okay. That just leaves us with Newton to deal with."

"And the other one."

Denise gave her a surprised look. "What other one?"

"There were four of them in the lead jeep. One of them was in normal clothes."

"Did you see who it was? There's nobody in the platoon left."

"I don't know."

"It could be our traitor."

Jacintha stroked Denise's cheek. "I don't think there is one."

"There has to be! Newton has Prime."

"Our dragon isn't unique," Jacintha chided gently.

"But . . ."

"Come on, we need to finish this."

The four of them split into pairs to approach the crashed jeep, closing on it from opposite sides.

"Newton was in there when he detected our Prime infiltration," Gangel said. "And that diagnostic probe is still transmitting. Whoever the fourth man is, he's in a bad way."

"Do you think Newton is still in there?" Jacintha asked.

She and Denise were crouched at the corner of the next house along the main street. When Denise inched around the foundation pier she could see the battered rear end of the jeep sticking out of the house. Nothing moved. Heat traces around the jeep were confused and fading. "I doubt it. But he can't have got far."

"Okay. Eren, any heat traces on your side?"

"Nothing."

"Stand by. We're going in."

"I'm going in," Denise said. "You cover me."

She scurried along the front of the house, keeping flat against the wall. Her breathing had quickened, the rasping loud in her ears. Heat was flooding out of the jeep, its axle motors gleaming crimson, power cells casting a vermilion glow underneath the chassis. The smashed-up wall was criss-crossed with hot ruby lines where the material had bent and cracked. Denise eased herself through the gap at the side of the jeep, her pistol sweeping across the room. There were thermal tracks all over the floor, leading to the door. Jacintha climbed up behind her and nodded.

Denise flipped around the open door, into the hallway. It was empty. The door at the far end was open a couple of centimeters. She didn't even need infrared. The dust showed two sets of Skin bootprints going straight to it. Only one came out.

Her bracelet pearl pinpointed the diagnostic card broadcasting from inside. The fourth man was definitely in there. Beads of perspiration were building up on her face. It was no good creeping along the hall: the Skin carbines could shoot through walls as if they were fog. She sucked down a breath and sprinted down the hall, bursting through the door. Shock froze her.

Jacintha followed her sister into the end room and nearly knocked into her. Denise was standing rigid in the middle of the room, pistol pointing at the figure slumped in the corner.

"You're dead," Denise croaked. She was aiming at Hal Grabowski's head. The same Hal Grabowski who had faced a firing squad and died. Now here he was again, all by himself in an abandoned house in Dixon. Her pistol arm shivered slightly.

"Who the hell's that?" Jacintha asked.

"Hal Grabowski."

"You mean the Hal Grabowski that you set up in Memu Bay? The one Z-B executed?"

"Yes," Denise snapped. She straightened her arm, ready to shoot. She couldn't do it, not an unconscious man. Then she noticed the writing on the wall beside him.

HELP HIM
I WILL KNOW

The diagnostic probe was resting against Hal's abdomen, still transmitting. Denise looked from that to the big medical kit box.

Gangel and Eren slipped into the room.

"Where's Newton?" Eren asked. "And . . . hey, isn't that Grabowski?"

Denise flashed him an exasperated glance and finally lowered her pistol. Gangel went over to the window. The frame was open. When he pushed at the plywood sheet nailed up outside, it swung out. "Looks like Newton left."

"So what about him?" Eren asked, pointing at Grabowski.

"He's Newton's problem," Denise said.

An explosion went off somewhere in the town.

Gangel was squinting through the gap at the side of the plywood. "That was a smart missile. He just took out the general store building. What the hell did he do that for?"

Denise looked at Hal again. She understood the message now. "He's not asking."

"What?" Jacintha asked.

"Newton wouldn't abandon an injured comrade. He's not asking us pretty please to help Grabowski. He's telling us."

There was a huge explosion outside. The house on the other side of the main street blew apart, fragments of composite panels and solar collectors whirling through the air to rain down over a wide area. Dust and smoke surged up out of the crater, spreading out in a miniature mushroom cloud.

The blast shook the room. Denise ducked in reflex. The glass in the window frame cracked, and the plywood sheet whirled away, allowing sunlight to blaze in. She saw the diagnostic probe had fallen off Grabowski and scrambled over the floor to grab it. She slapped it down on Grabowski's stomach; the display pane began to register his vital signs again. "All right! We'll do it."

Jacintha stared at her. "Do what?"

"Newton's out there with a rack of smart missiles—which

he's probably loaded with Prime. He'll keep firing them over Dixon until he runs out. If we go outside, the seeker head will spot us and . . . that's it. Even we can't deflect one of them. The only place we're safe, the only coordinate he'll never target, is here with Grabowski. And if we don't keep Grabowski alive, guess which house the next missile will take out."

"Sneaky bastard," Gangel said with bemused admiration.

"You said it," Denise grunted.

They all winced as another missile detonated. The flash was close to the maintenance shed. Smoke began to rise over the rooftops.

"He's not kidding, is he?" Jacintha said. She knelt beside Grabowski and lifted his shirt up. "We'd better get to work." She took a dragon-extruded analyzer unit out of her pocket, placing it over one of Hal's defunct medical organ modules. The little plastic rectangle softened and began to mold itself round the module.

"What range have those missiles got?" Eren asked.

"Three kilometers," Denise told him.

"That's not too far. We know he was injured. We can catch him."

"We won't know what direction he took. All he has to do is leave the rack two kilometers away and program it to keep launching at regular intervals. He could be ten or more kilometers away before this barrage stops."

"Shit!" Eren glared at Grabowski. "Once those missiles run out, so does your luck."

"Does it?" Denise gave Eren a quizzical look. "After we spend a couple of hours caring for him, you're just going to kill him, are you?"

Eren banged a fist into the door frame. "No. Guess not."

"We should call the village," Gangel said. "They can send a team out here. With enough support we can tackle Newton."

"No," Denise said. "That's too much exposure. Besides, I know which road Newton's on."

Lawrence was on the edge of town when he saw the bike charge along the Great Loop Highway, about five hundred meters away. Helmet sensors zoomed in. It was being ridden by a naked man whose skin was smeared in pale blue gel.

The bike stopped and the man looked at him. It was Amersy. He raised his fist and punched the air twice.

Lawrence laughed as he gave an answering punch. His rack fired another smart missile back into the town.

Amersy paused a moment, then turned the throttle, accelerating fast along the road.

Lawrence left the rack fifteen hundred meters outside Dixon. He was in the middle of the slag heaps, so he could push it down into the black grainy soil easily enough. Once he satisfied himself it was secure, he departed at a steady jog. The smart missiles would fire at random intervals. Each was targeted on a different house, with the seeker head programmed to watch for human bodies moving along the streets. If it located one, it would divert from the primary target and go after the body.

With the rack's data cable disconnected, he had only one telemetry grid left now: Hal's diagnostic readout. Judging by the way his vital signs had stabilized over the last ten minutes the Arnoon people had worked out their side of the deal. His only worry now was whether they'd keep treating the kid after the missiles ran out.

Sorry, Hal, but what else could I do?

Trying to carry Hal out of the ambush was impossible. They wouldn't have gotten ten meters before those strange weapons cut them down. He'd been puzzled by the little dazzling bullets of light that the ambushers were firing. Once again there was no match in his armaments catalogue file. And not just the model, either; the nature of them was a mystery, too. His one clue was the intense magnetic signature that his sensors had recorded as he'd slipped away. He hadn't stopped to try to get a second reading.

Lawrence increased his pace. There were enough missiles left to last seventy minutes, although that did leave some long gaps between a few launches. But it should allow him to put about twenty kilometers between himself and Dixon if he stuck to a reasonably straight line.

He called up the plateau map file as he ran. After Dixon, the Great Loop Highway carried on in a wide curve through the Mitchell peaks, passing through Arnoon Province almost at its apex. He began to plot out a direct course to the crater lake. There was one river cutting across his path, which he'd be able to cross easily enough in Skin. The only real problem was that taking this route put Mount Kenzi directly in the way. He expanded the foothills to try to find a passage around the side.

The slag heaps soon gave way to the plateau's wilderness of crown reeds and the occasional giant tree. He had to slow slightly to go around the crown reeds. Each mature clump varied from two to three meters high. The fat, succulent leaves with their serrated razor edges weren't able to cut his Skin, but he certainly couldn't push through them. The ground underfoot was a thin, brown soil threaded with a low scrub plant that had slim woody stems and tiny saffron flowers.

At twenty minutes he lost the signal from Hal's diagnostic. The little probe was never intended for long-range broadcasts. The last reading showed the kid was recovering well. Lawrence didn't know what the ambushers were doing, but they were making a lot better use of the aid kit than he had.

As he drew away from the slag heaps of Rhapsody Province the land began to grow more uneven. The slopes he crossed were long and gentle, each one a little higher than the last. His inertial guidance told him he was steadily gaining altitude. Crown reeds gradually shrank away to be replaced by small wiry bushes, their bark a dull russet color. Boulders lurked among them, half-buried lumps of hard, dusky rock.

After an hour he had to slow again. The wound left by the steering column had begun to ache despite the local anesthetics. It was similar to having a stitch, but just above his hip. The Prime reported he was bleeding. Clotting agents weren't able to cope with the constant stresses of running. When he looked down he could see blood dribbling out of the puncture hole in the carapace. He told the Prime to readjust the Skin muscles to reseal the wound. More clotting agents were discharged.

He gave it a minute for everything to take effect, then started off again. Mount Kenzi didn't seem any closer, just bigger. A raft of fat clouds obscured its pinnacle. Wind was bringing them in from the east. The sun was already lost behind them, shading the plateau in a dreary penumbra light.

Thin trailers of fog began to slide past him. The brittle bushes were glistening with moisture, even though it wasn't raining. Ahead of him the ground curved up until it met the clouds. Cataracts of mist flowed out of it, sluicing down along the narrow, stony gorges that wove chaotically across the land. He jogged on as the ridges steepened and the scraps of destitute vegetation became less populous. The external temperature was dropping considerably as the mist thickened. Lawrence was hot inside his Skin; he could feel himself sweating. He was taking constant sips on his water nipple; the inside of his mouth was parched.

The mist closed around him, reducing visibility to less than twenty meters. He kept going for another hour, then sat down on a frosted boulder. A chest pouch opened up and he took out one of the three spare bloodpaks he was carrying. Its nozzle clicked into the Skin's umbilical socket, and the internal reserve bladders sucked the fluid in.

Blood was oozing out of the wound again. His leg was

slick with runnels of the sticky liquid. The Skin sealed itself once again and dosed him with antiseptic and clotting agent. His display showed him that the suit muscles around the puncture were starting to degrade. They were losing as much blood as the wound.

As he rested, his own muscles began a mild ache. He'd been on the go for four hours now. His side around the wound was numb, with the surrounding flesh tingling slightly from the drugs. He was sure he could feel blood trickling down the inside of his leg, which might be a problem later on. There was no way of draining it out short of removing the whole Skin. Without a medical kit to treat the wound immediately, he wasn't about to do that.

When he stood up, a rush of dizziness almost made his legs buckle. He swayed about for a moment until the Skin muscles tightened and held him upright. His head slowly cleared and he took a big suck on his water.

He started off walking, then slowly broke into a trot. In his mind he could hear his left leg squelching inside the Skin every time his foot hit the ground. The light was beginning to fade, hastened by the cloying mist. This region of the plateau was almost barren. It comprised long stretches of sloping land that ended in ridges that were almost as steep as cliffs. Every time, he would have to scramble and claw his way up through the boulders and scree falls. Stubby toe claws extended from the Skin to give him extra grip over the slippery dripping rock.

Night had fallen half an hour before he reached the ridge that would take him up onto the saddle plain. Mount Kenzi was on his left, with Mount Henkin to the right. He stopped at the base of the rock barrier and took out the second bloodpak. His Skin guzzled it down greedily. While he was waiting, the last fringes of the mist retreated down the slope. There were no stars visible. The sky above was cloaked in dark cloud, its turbulent underbelly swelling and surging as it was provoked by conflicting air currents surging off the mountains. But there was enough light for him to see the ridge. He'd negotiated the last one with laser radar as his only way of seeing what lay ahead. Here, there were broad stripes of white rock zigzagging down through the ridge, almost like a giant's steps. He studied them, trying to concentrate on finding an easy route up.

Indigo icons slipped over his vision. Medical symbology cautioned him on the state of the wound. He responded by ordering another infusion of drugs. The cold numbness was spreading up his ribs. Occasionally he would shiver, which the Skin would automatically mimic.

This time he clambered to his feet with slow, deliberate movements. Even so, when he was upright it felt as though his body were made from jelly, held in shape only by the hard mold of Skin around him. It was a stupid sensation, so he ordered a stimulant infusion. His mind cleared swiftly, and he looked hard at the ridge, finding himself a way up.

When he got to the top he could see the saddle plain stretching away in front of him. The heavy cloud formed an unbroken ceiling five hundred meters up. On either side, the two mountains were massive, curving walls of naked rock, riddled with slender crevices and deep folds. It was an enclosed universe that gave him no choices. According to his map file, it was ten kilometers to the far side. He started walking.

The saddle was classed as alpine desert. Lawrence thought it looked more like the surface of Mars. The exposed soil was a somber rust-red, strewn with small, flinty stones. There were no animals or insects living up here. Even the small crustaceous plants that peeked out from the stones looked desiccated. His Skin reported that the pressure was down to a third sea level. The gills were having to work hard to pull enough oxygen out of the freezing air.

He'd got a kilometer past the ridge when it began snowing. It wasn't big, soft flakes drifting out of the sky; these were small, hard pellets of ice that the wind drove straight at him. He could see them bouncing off the Skin carapace. Visibility was down to seven meters. Laser radar was useless. He didn't even bother with infrared or low-light. All he had was inertial guidance. It was enough for him.

Until the snowstorm engulfed him, all that had mattered was to keep going, to remain focused on the destination. Anything less would be betraying the platoon—which he could never do. Now, Lawrence began to contemplate what he was going to do when he actually reached the crater lake. He'd got a full magazine for the carbine. But against that the villagers had guns that fired weird stars, e-bombs, Prime and biotechnology from Santa Chico. He needed medicine and treatment for himself, and blood for his Skin. Then all he had to do was find out what the source of their wealth was and extort some of it out of them. Oh, and transport, too.

Blind, alone, cocooned by a faltering Skin against an environment that would kill him in minutes, Lawrence Newton started laughing. All this—*insane desperation*—so that he could buy himself back into Amethi. The home he'd run away from so he could explore the universe. It was hard to remember now, but in those days Lawrence Newton had thought the stars were full of excitement and wonder. What was it he'd told Roselyn that first day they'd met?

Nowhere you live can be exotic. That's only ever somewhere else.

Now he knew: it was always somewhere else. If he'd been given the chance, that young Lawrence Newton would have kept on flying and never come back.

Did I really hate myself that much back then?

He smiled happily as his thoughts of Roselyn brought her image to the front of his mind, the one icon that never deserted him. His hand patted the base of his throat, feeling the small lump of the pendant pressing against his skin.

It would be nice to see her one last time.

The clouds swept clear when he was still a couple of kilometers from the end of the saddle plain, taking the snow with them. Stars gleamed brightly in the thin, clear air. Two centimeters of ice pellets lay across the ground. His Skin crunched them down as he trudged onward.

He had to use the last bloodpak before he finished traversing the saddle plain. The Skin had used up a lot of energy keeping him warm in the snow. When he sucked at the water nipple, the tank was empty. His tongue was dry inside his hot mouth. Pain was a constant in his side now, a fierce pulse at the center of a permanently cold hip. The anesthetic made no difference. He wasn't even sure the clotting agent was having any effect. The Skin leg was coated in blood. Its muscles couldn't keep the carapace puncture hole fully sealed anymore.

And still, he had no choice.

The ground began to dip away, and Lawrence could look down across the forested vales of Arnoon Province. It was quiet and beautiful in the starlight, just as he remembered it.

This side of Mount Kenzi was a scree slope that swept down steeply for over two kilometers. Lawrence began his descent. The small stones slid and skittered beneath his feet, clattering away out of sight. As he became used to the subsidence he used it to slide his way down, taking long hops, deliberately landing hard on his heels so the scree would give way underneath him. Time after time he lost his balance or hit a big rock and fell, skidding and sliding down the slope at the head of a miniature avalanche. Without Skin he would have been cut to shreds on the sharp little stones. But the carapace maintained its integrity easily: this kind of treatment was well inside its tolerance limits.

The scree gave way to tough grass. He started to walk down to the treeline several hundred meters below. His left leg was stiff, even with Skin muscles moving it. Several scree stones were stuck in the open puncture. He stopped to pick them out, then continued. Their absence didn't make any difference to the limp. The display revealed that an alarming amount of Skin muscle in his left leg had degraded to a nonviable level. When he checked, blood was still leaking down the leg. It must be coming from the wound inside. There was no clotting agent left.

He stopped when he reached the trees and bent over, trying to throw up. Nothing came, apart from a vile acidic juice that burned his already arid throat. His gills adjusted their filter parameters, feeding him a higher oxygen level. It made breathing a little easier.

The trees thickened quickly once he was inside the forest. But their trunks were never so close as to form a barrier. Undergrowth was a shaggy fern that his Skin legs pushed through with hardly any extra effort. The visibility was as bad here as in the snowstorm. He had to rely on inertial guidance again, following the indigo trail across the slope, always heading down.

Warmth slowly drained out of him, seeping away through the puncture hole in the Skin. His fingers were icy, his feet blocks of ice. Nothing he could do would stop the shivering. The display wanted him to replenish the Skin's blood bladders. He sneered at it and told the Prime to clear the icons away. More medical warnings appeared, indicating the strain he was putting on his own organs now that his body was having to reoxygenate the blood.

The trees came to an end. Lawrence moved forward with small, laborious steps. He was hunched up in an effort to ease the pain throbbing along his ribs. One hand was clamped over the puncture hole in the carapace.

He arrived at the top of the curving cliff. A hundred and twenty meters below, the black waters of the crater lake rippled gently. Low-light sensors turned the gloomy night vista to a glowing blue-and-gray image. He saw the central island. The little stone temple was still sitting at the center.

"Meditation my ass," Lawrence grunted at it, and jumped.

The carapace hardened protectively long before he hit the water. It was a jolt that sent an excruciating pain flaming out from his wounded hip. He screamed inside the helmet. For a moment he though he was about to throw up again. All he was really worried about was the depth of water at the foot of the cliff. Whatever it was, his feet never touched the bottom of the lake.

Low-light sensors showed him faint gray bubbles swarming around him as he slowly floated to the surface. Then he was bobbing about, trying to see where the island was. Once he found it, he brought his legs up so he was floating on his back. His feet kicked slowly, aided by the occasional flap of his arms. He instructed the Prime to make the Skin muscles follow the motions he wanted. His own limbs weren't responding very well. The result wasn't a particularly fast stroke, but he made steady progress.

He was about seventy meters from the island when something brushed against him. The carapace tactile sensors stroked his skin in mimicry of the contact. Lawrence flinched and held still, waiting for it to happen again. When nothing happened, he began kicking again, perhaps a little more urgently now. The fish-creature prodded his left leg. Lawrence shoved at it with his hand. A narrow, pointed head broke surface for a moment, then dived with a small splash.

Something touched him on the other leg. Two of them! He concentrated on kicking, keeping his feet below the surface for maximum effect. One of the fish-creatures slithered over his chest. It was similar to an eel, but pale green, over a meter long, with three ridges running the length of its body. They were vibrating softly.

"Shit!" Lawrence punched at it in panic. But it was too fast.

Pointed jaws with needle teeth worried their way into the puncture hole like a hammer drill. Lawrence chopped at the thing with the edge of his hand. Two more were nuzzling the puncture. He twisted over and started swimming sidestroke, keeping the puncture out of the water. One of the creatures tangled itself round his legs. The puncture was forced below water. Teeth began to bite into the exposed Skin muscle.

The carbine slid out of its recess. Lawrence angled it away from his leg and fired. Bullets chewed the water around the creatures. There was an eruption of spray, and they were gone.

Lawrence started swimming hard, shouting to counter the pain coursing through his body at each hurried motion. Ripples wriggled across the water, arrowing toward him. Several of the creatures were suddenly writhing all over him. Lawrence thrashed about, going under for a moment. Their jaws were tearing at the Skin muscle in the puncture, severing chunks of it. He used the carbine underwater, hearing a dull roar as it fired.

When his head came up, he could see the island thirty meters away. A biohazard alert flashed in the middle of his display. Some kind of toxin was seeping into the Skin's circulatory network. Prime determined the infection point as the muscle cords around the puncture.

They're poisoning me!

Just then one of the fish-creatures began to coil and convulse a few meters away, flinging spray in all directions. Two more started similar berserk motions.

Lawrence kept swimming. Prime closed the valves connecting his major blood vessels with the Skin. Another creature jabbed its head into the puncture. He shot at it.

Dozens of the creatures were racing through the water around him. They slid over and around the Skin. Lawrence's foot touched a solid surface. He struggled for balance and waded out of the water. The creatures were charging around his legs, butting against the carapace. Prime was flashing up information on the toxin. It was spreading through the Skin's leg muscles. Secondary blood vessel valves were being closed, in an attempt to isolate it.

When his feet were finally out of the water, Lawrence managed two steps on the grass, and fell over. His legs wouldn't move, the weight of the inert Skin was too much.

Lawrence surveyed the status display. The toxin had contaminated over a third of the Skin's muscles. There was no blood reaching the rest. With a sob he gave the Prime his last order. The Skin split open smoothly along its chest seal. Lawrence whimpered as he pulled the helmet back off his head. Cool night air licked against his body. He pushed and wriggled, emerging slowly and painfully from the dead Skin like some glistening blue chrysalis. For a while it was all he could do to lie panting on the grass. Then his left hand felt its way along his side and probed at the wound. He grimaced, and slowly sat up.

The clotting agent had left a thin layer of white foam inside the wound, which was cracked and flaking. Blood was dribbling out, running down the slippery layer of dermalez gel. He pressed his hand against it, hoping the pressure would hold the bleeding until he could find something to use as a dressing.

Lawrence got to his feet and looked around. He could just make out the temple. Each step was forced, and he cried out more than once as he made his way over to the stone structure. When he got there, a section of the tiered seating had sunk away to reveal a staircase leading down. A weak light was shining at the bottom.

"I knew it," he mumbled.

He had to lean his shoulder on the wall for support as he stumbled down the stairs. Dermalez gel smeared an uneven trail along the stone as he went. Blood dripped continually through his fingers, splattering on the steps.

There was a small, empty room at the bottom, directly

under the middle of the temple. A single metal door faced the stairs. It slid open as Lawrence hobbled toward it, revealing an elevator. He eased himself inside and found a control panel with just two buttons. The door slid shut when he jabbed the lower one.

There was a quiet whine as the elevator descended. The door opened to show a large hemispherical chamber with wall segments of dark copper-colored composite. Lawrence staggered out, not caring that he'd be seen. He just had to know what he'd been chasing. That was all. Nothing else mattered anymore.

In the middle of the chamber was a broad pedestal of milky glass, almost like an altar. A long ash-gray rock was resting on top, its surface pitted and blackened. The central section was draped in a gold mesh. The end pointing at the elevator had been cut and polished; clumps of small aquamarine crystal were sticking out of it, glowing lambently.

Lawrence squinted at the scene, not understanding any of it.

Two young women were standing in front of the pedestal. The older one gave him a sad smile, and said: "Welcome to the temple of the fallen dragon, Lawrence. Remember me?"

Lawrence grinned at her, and lost consciousness.

CHAPTER SEVENTEEN

JOSEP'S THOUGHTS CAME TOGETHER QUICKLY AS HE WOKE. FOR a second he kept his eyes closed while he assessed his position. He was lying on some kind of plastic cushioning. No clothes; his skin was pressed against the fabric. Slight pressure round his hips. A pair of boxer shorts, then. Cold metallic bands around his wrists, which were being held in place fifteen centimeters apart. Manacles of some kind. His legs were free. Artificial light on his eyelids. The distant clatter and murmur of a busy building.

When his d-written neurons tried to locate either his bracelet pearl or a local datapool node, all he could sense was some disjointed background signals that were almost below his threshold. It was as if the electromagnetic spectrum had been muted somehow. He put the odd perception down to the gas that they'd used to knock him out. Some of it must still be in his system, affecting his neural cells.

He opened his eyes. The room was a cell, four meters by four; no window, just a conditioning grille. He was lying on a bench opposite a heavy metal door. A small camera on the ceiling was angled down to look at him.

Cells in the spaceport security division were very similar. They might not have moved him yet. In which case he stood a chance. He knew the entire spaceport layout.

That thought made him pause. He hadn't known about the elevator. And there must have been at least one alarm that wasn't on any file they'd accessed when they were planning the break-in. Most likely it was something that Z-B had discreetly installed after they took control of the administration block. Even so, his Prime should have caught it going off.

Making a show of being slow and confused, he sat up, rubbing at his hair. The manacles made the movement difficult. He frowned at them. "What the hell . . ."

Nobody came in to explain. He padded over to the door. The tile floor was cold under his bare feet. "Hey!" He banged on the door. "Hey, what's going on here?" There were grazes on his knuckles where he'd hit the elevator doors. That could have been a mistake. If they measured the dent he'd made they could work out the force behind the blow. That would make them *very* interested in him. Not that they wouldn't be anyway. But he couldn't allow them to examine his body too closely. The patternform sequencers must be protected at all costs.

He padded back to the bench and sat down. It was standard procedure to let prisoners sweat for a while after they'd been captured, allow them to build up some anxiety. Not that such crudities would affect him. But he had to decide what to do next. The d-written cells in his cheeks and jaw had held their shape while he was unconscious. He still had Sket Magersan's face. Z-B would have checked with the real pilot. They'd know this was a serious sabotage attempt by a resistance group.

Interrogation by Z-B would inevitably involve medical diagnostics, probably including a full brain scan. The d-writing modifications were subtle, but with that sort of scan there was a high risk of exposure. And he wasn't entirely certain he could hold out against the drugs. His d-written neurons were hardly omnipotent, and Z-B had been dealing with resistance movements across decades and dozens of planets. By now, their techniques and technology for extracting information would be formidable.

The choice was simple. The longer he remained captive, the lower his chances would be of escaping. If he was going to get out, it would have to be before they fully realized what he was physically capable of.

That brought him back full circle to how they'd caught him in the first place. He started to go over the break-in right from the start.

It was another two hours before the cell door opened. Josep still hadn't worked out what he'd done to set off an alarm. Two guards came in, their navy-blue uniforms sporting a small Z-B insignia on the collar. Both of them wore helmets with tinted visors; they held long truncheons with shock prongs on the end.

A simple white one-piece suit was slung at him.

"Put that on," one of the guards said.

Josep picked it up and let it unroll. He held his arms up, shaking the manacles at them. "You'll need to take these off."

"Nice try. Just put it on."

The suit sleeves had a seam down the sides, fastened with studs. He struggled into the lower part of the garment, and one of the guards fastened the studs for him.

They marched him out into a short, curving corridor. Josep checked the length and height, and knew exactly where he was: administration block, third floor. The security division had a long section all to itself in the five-sided building. Floor blueprints rushed through his mind. The only ways in or out of this section were two elevators and an emergency fire exit. He couldn't use the elevators, they were code-guarded—not forgetting what happened last time he used an elevator in this building. The fire exit was the obvious route, but there were strong safeguards there as well.

"Where are we going?" he asked.

"You'll see soon enough."

They were walking in the direction opposite the elevators. The only rooms ahead were the offices. They must have set up their interrogation equipment in one of them. He still couldn't sense a signal from a datapool node.

They turned a corner. The walls of this corridor were lined with doors. He named them silently as they went past: departmental management, briefing one, two, and three, investigator lieutenant, finance. Josep swayed slightly to shift his balance and kicked at the guard on his left. It was a perfect aim, heel smashing into the man's kneecap. He yelled in pain and went down. The second guard slammed his truncheon into Josep's back, and the shock prongs flared, pumping a charge into him. His d-written cells resisted the blast of electricity—just keeping his nerve channels open. He turned and wrenched the truncheon from the guard's grip. The man grunted in surprise at the force. Then Josep stabbed the truncheon into his stomach. The guard staggered backward, doubling up before finally keeling over.

Josep jabbed the truncheon into the first guard's neck as he was trying to rise. He collapsed back onto the carpet. At the other end of the corridor two men in Z-B uniforms were shouting as they ran toward him. A shrill alarm went off, terribly loud in the confined space. The security AS must have seen the whole thing through the surveillance cameras.

Josep threw the truncheon at the two running men, then charged at the door to the finance office. It wasn't even locked. As he expected, Z-B didn't have any use for financial staff; the office was empty, abandoned for the duration. There were three desks lined up down the middle of the floor, cluttered with old memory chips and piles of hard copy. Desktop pearls were inactive. The wall opposite the door was tinted glass from floor to ceiling, facing out across the spaceport hangars. He rushed over to the last desk and heaved it into the air, then flung it at the glass wall. The toughened glass shattered, sending a blizzard of shards swirling outward. More alarms started up. The desk crashed down on the edge of the hole it had created, half of it still in the office. It wobbled unsteadily. Josep kicked it, sending it sliding out over the edge to smash into the flowerbeds three floors below.

The office door burst open. Z-B staff rushed in. Josep jumped after the desk.

Three stories in Thallspring's standard gravity. The fall was enough to shatter most of a man's lower skeleton when he hit. Damage to the organs from the massive impact deceleration would probably be fatal. Josep thrust his manacled arms above his head, desperate to keep his balance. It was surprisingly quiet as the warm late-afternoon air rushed round him. He bent his legs fractionally as the flowerbeds hurtled up.

His feet crashed into the hard soil, and his knees bent, absorbing as much of the terrible impact as they could. Suddenly his shoulder was smacking into the ground, knocking the breath from him. His bones held, though his ankle and knee joints sent pulses of pure agony into his spine. He blinked away the tears of pain. Rosebushes had torn at his legs. A glass splinter was embedded in his foot. Astonished Z-B personnel were leaning out through the broken glass of the finance office, peering down at him.

Josep ordered his deadened limbs to move, rolling onto all fours, then standing. Shouts from above mingled with the persistent howl of the alarm. He took a few excruciating steps until he bumped into the base of the building's glass wall. After that he could use it for support as he moved along. Somewhere up ahead was a door used by the maintenance staff. On the other side of the glass, people were standing up at their desks, pointing at him as he slid past.

He reached the door, put his shoulder against it and shoved. It bowed slightly, but held. He took a step back and launched himself at it again. This time the lock broke, and he was through into a narrow concrete utility tunnel. He hurried along it to the first intersection. The walls of the wider central tunnel were thick with conduits and pipes. Lightcones on the ceiling threw out a raw purple-white radiance every five meters. He turned left and started to run, wincing every time the glass splinter hit the ground, jabbing deeper into his flesh. Blood dripped out of the cut, but he knew the flow would be worse if he took the glass out.

Another left turn. Then right, right again. A locker room. Nobody inside. He went over to the row of gray-blue metal lockers, grabbed the handle of the first and tugged. It jerked open, the metal bending around the bolt, then tearing. Dirty overalls hung inside, and a pair of boots. He went to the next locker and tore that open. The next. The fifth had what he was looking for, a tool belt with every loop full. He pulled a power blade out, switched it on, then put the handle in his mouth, biting down hard to hold it steady. The blade cut through the manacles with an appallingly loud shriek and a shower of sparks.

Josep held his breath. There were shouts reverberating along the concrete tunnel outside. He went back to the first locker and shoved his feet into the boots, then pulled out the overalls. D-written organelles began to sculpt his flesh, returning his facial characteristics to those of Andyl Pyne again. With the alteration under way he snatched up the tool belt and a bracelet pearl that he found on the locker's top shelf.

Fifty meters on from the locker room there was an inspection hatch that led into the wall cavity. A power screwdriver from the tool belt opened it, and he eased himself in. It was a narrow, confined world, completely black. Even his infrared vision was cloudy. The walls were 110 centimeters apart, forming an interstice that was filled with structural girders, conditioning ducts and plumbing. He could see as far down as he could up. His feet were resting on an I-beam barely ten centimeters across.

The Skins and guards hunting him would know he had switched clothes back in the locker room, but the boots were cold when he put them on. It would take a while for his body heat to soak through into the soles. In theory they wouldn't be able to track him to the hatchway. The cramped interstice made it difficult to move, but he slowly worked his prison one-piece off, always keeping one hand clamped on a girder. He dropped the bundle of fabric into the darkness and began to struggle into the overalls. Twice he had to stop as someone ran past the hatch. When he finished dressing, and with the tool belt fastened around his waist, he began to climb. As he ascended he tried to integrate his d-written neuron structure with the bracelet from the locker room. He still couldn't establish any kind of link. The knockout chemical must have done more damage than he first realized.

Once he was up level with the second floor he followed the phosphorescent coral line that was the hot-water pipe until he reached the toilets. The panels in the wall here were a lot smaller than the hatch back down in the utility tunnel, intended principally to provide access to the tanks and pipes that served the cubicles and basins. He found the largest and put his ear against it, listening to the movements inside. Two people were using the urinals—at least it was the men's room, he thought. One paused to wash his hands, the other left straightaway.

Josep used the power blade to saw around the rim of the panel. He squeezed and wriggled his way into the cubicle, frantic at the noise he was making. Then when he was most of the way through, he had to push his body into a gymnast's contortion that even his d-written limbs had trouble achieving, all to make sure he didn't stick his head out past the partly open door. Every toilet had a security camera, and the security AS would be devoting a large percentage of its processing capacity to spotting visual abnormalities inside the administration building.

"Resourceful," Simon observed.

"I think we were too slack on the chase," Adul said. "We should have given him more grief in the utility tunnels."

"We've reinforced his feeling of superiority. Look at his deep thalamus activity. He's confident."

"As long as his easy ride doesn't make him suspicious."

"I'd hardly call that jump easy. I thought he was suiciding until I remembered his bone structure composition," Simon said. "Give him a reasonable body match," he instructed the AS.

* * *

The toilet door opened. Josep tensed, waiting to see what the man would do. Footsteps made their way to the first cubicle. Josep tapped his knuckles on the partition. There was a slight hesitation in the footsteps.

"Hey," Josep hissed.

"Some kind of problem, there?"

"Yes."

The man peered around the cubicle door to find Josep sitting on the toilet bowel, head bowed. "What's up?" He moved a little closer.

Josep's left arm shot out and grabbed the front of the man's suit jacket, tugging him hard into the cubicle. At the same time his right hand chopped across the man's neck. He closed the cubicle door. If he'd got it right, the security camera should have seen the man pause by the first cubicle, then choose the second.

According to the man's identity card he was Davis Fenaroli-Reece. Josep began to strip him out of his suit. Changing clothes in the cubicle was almost as bad as putting on the overalls in the wall interstice. Once Josep had the suit on he propped Davis Fenaroli-Reece's body on the toilet bowl and studied his face hard. His own features began to shift again. Without a mirror he wouldn't be able to get the likeness as accurate as he wanted, but his real worry was the hair. Davis Fenaroli-Reece had very dark hair, whereas Andyl Pyne and Sket Magersan were both fair. In the end he settled for splashing water from the toilet bowl over his head and slicking his hair back, hoping that would darken it enough to fool the AS. He was content the camera didn't have sufficient resolution to spot the change in texture.

Another minute was spent with the tools, fixing the cubicle lock. When he closed the door behind him the bolt clicked into the latch and read *engaged.* Josep washed his hands and left.

He started walking around the corridor to the main stairwell. The second floor was a mix of Z-B personnel and local spaceport staff. Most of them were standing pressed up against the window wall, looking down at the Skins circling the building.

It was growing dark outside, with the sun already hidden behind the high ground of the horizon. That meant he couldn't have been unconscious for more than forty or fifty minutes. He felt hungry, though, as if he hadn't eaten for a day.

The stairwell took him up to the fourth floor, where there was a bridge leading directly into the main terminal building. A couple of Skins were standing at the far end, checking everybody coming out of the administration block, as if the AS wouldn't be able to spot Sket Magersan walking away. They never moved as he passed them.

Forty minutes later he was out in parking lot 4B, walking casually along the rows of vehicles. A group of staff that had come out of the terminal building said good night to each other and split up. Josep followed one of them as he went to his car.

"Excuse me?"

The man stopped just as he'd gotten the door open. "Yeah?"

"My car's dead. Axle motor cable, I think. Are you going into Durrell?"

"Sure." The man nodded. "I can take you."

"Thanks."

There were Skins standing around the exit barrier.

"Big flap on," the man remarked as he slowed the car level with the twin security posts.

"Wonder what it's about," Josep said as he swiped Davis Fenaroli-Reece's card over the scanner on the car's passenger side and looked at the camera.

The barrier pole swung up.

"Someone tried to steal some bullion out of the vault this afternoon," the man said. "He got away."

"God, I hope they don't use one of the collateral necklaces."

"For that? I doubt it."

The man drove him into Durrell as promised. Josep thanked him as he was dropped off at a commercial center in one of the outlying districts. Fortunately, Davis Fenaroli-Reece carried just enough cash to pay for a bus ticket into the city center. It was a ten-minute walk from there to the university campus. When he reached Michelle's residence building, he paused in the lobby while his face finally reverted to his own features.

"Ah, I wondered what he actually looked like. Let's see if we have any records of that face."

Josep tapped the code into her door lock and walked in. The room was a mess as always. Barely large enough for one student, it had turned into a flea market of clothes, fast-food wrappers, hard copy and unwashed crockery since he moved in with her. Michelle was sitting on the small bed, watching the pane on the desktop pearl that was resting on the pillow. Her head came up, shock registering on her face. The gash in Josep's foot left by the glass shard suddenly jabbed a hot pain up his leg. He winced.

Michelle looked up in surprise as the door opened. It had to be Josep. She'd been so worried that he'd been caught doing something for the resistance cell. Relief turned to shock as she saw the thing coming through the door. It was a parody of a Skin suit, thin and spindly, with a simple metallic sphere as its head. The twin black lenses that were its eyes stared at her. She screamed as it walked into the middle of the room.

Two genuine Skins hurried in behind it. Michelle kept on screaming as one of them lunged at her. Thick fingers clamped around her arm. She grabbed at the headboard, but the Skin was immensely strong. She was dragged off the bed, her shoulder blade thudding down painfully on the floor.

"Help me!" she wailed. "Somebody, help."

"Shut the fuck up, bitch." The Skin picked her up and slung her over his shoulder. Michelle tried kicking, but the viciously tight grip on her legs prevented the slightest movement. Her head was hanging halfway down the Skin's back. She tilted her neck back to see the slender humanoid thing moving slowly around the room, its fingers stroking objects. Then she was out on the landing, where several more Skins were waiting. Students stood in their doorways, watching her being carried past, too scared to move or say anything.

Tears rolled down her cheeks. It was all over. Z-B had discovered their little resistance cell. They'd interrogate her and kill her. She whimpered pitifully as the Skin walked into the elevator with her. Three men were crammed inside waiting for them. They began to attach instruments and medical-style modules to her skin.

Michelle started screaming again as the doors slid shut.

For a moment the room was out of focus.

"Are you all right?" Michelle asked. She'd got up to stand beside the bed, looking concerned.

Josep lifted his foot, taking the pressure off the wound. The pain eased immediately. "I'm fine."

She gave him a tentative smile. Josep waited. But for once she didn't rush over and embrace him. He wondered what was wrong with her. Did she think he was seeing another girl? *Please, not that, not now,* he prayed.

He gave her a quick kiss. There wasn't much of a response. "There's a problem," he told her. "I have to talk to Ray. Get the stuff, will you? I'm going to move it out of here."

"Why? What's happened?"

"Nothing to worry about." He sat down on the bed and pulled the desktop pearl toward him. There was still nothing from his d-written neurons, just that faint background buzz. That made him pause. What the hell could knock them out of kilter for so long? Every other enhancement d-written into his body seemed to be working fine.

"What is it?" Michelle insisted.

"Okay, look, the controller called me. Z-B has been sending askpings into the university network, checking up on student files. I'm sure it's nothing to worry about, but we have to be careful. I'm going to move out for a few days."

"I don't like it."

"Neither do I. I'm sorry, but we have to be safe. You'll be fine. Now just get the stuff, please." He requested his Prime from the desktop pearl's memory blocks. The pane flashed up an invalid request icon.

The remote spoke the command with its associated code, but the desktop pearl didn't respond. An invalid request icon appeared in the pane.

Josep stared at it, not understanding. "Damn it!" Where had the Prime gone? If he could just interface directly . . . He wondered if he should call Raymond without using Prime protection. Michelle was still standing behind him, watching.

"Are you going to get the stuff or not?" he asked.

"I don't want you to go."

"Damn it." He told the desktop to call Raymond.

* * *

Simon's DNI was giving him a comprehensive display of the Durrell datapool architecture, the graphics generator riding on the AS's monitor program. He saw the placement ping flash across the entire datapool. Josep's call was to a personal portable address. Wherever the individual was, the nearest datapool node would route the call straight to them. A node in the Silchester District started to establish the link. The entire Silchester datapool crashed.

"What happened?" Simon asked.

The AS reported that its monitor program had been discovered and identified by an unknown program. The Silchester District had crashed immediately.

Simon was impressed. All the gadgets they'd taken from Josep had self-destructed as soon as Z-B's technicians had started to examine them, vaporizing evenly from the surface inward. An analysis of the gas residue had revealed some extremely unusual and complex molecules. It would seem their software was equally sophisticated.

The desktop flashed up a receiver-not-found icon. Josep regarded it with growing concern. Even if the datapool couldn't make the link, Ray's Prime would have intercepted the call placement ping and responded.

"That shouldn't happen."

"Perhaps he's switched off his bracelet," Michelle said.

"Maybe." Josep looked round the room, deeply uneasy. Something was wrong. Why couldn't he get any kind of interface with a Prime?

"Did Ray call?"

"No."

That wasn't right, either. Ray would have known within an hour at most that the break-in had gone wrong. One of the first things he'd do was call Michelle.

He stood up and faced her. She returned his gaze levelly.

Michelle would never do that. She'd either blush or grin happily, lovingly.

"You still haven't got the stuff," he said lightly.

"I told you, I don't want you to go."

"Oh hell," Adul said. "He's suspicious."

"It was always going to happen," Simon said. "Just a question of when." He looked over at Josep. They'd suspended him in a total reality immersion suit, not too dissimilar to Skin: a tactile emitter layer surrounded by artificial muscle to stimulate all levels of physical contact from the water splashed on his hair to the feel of the shirt fabric. It hung from the center of a gimbaled circular frame, allowing them to orientate him to match his personal inclination within the world created by the AS—though the jump had taken it right up to the limit of its replicant ability. Fiberoptics had been inserted through his corneas and pupils to shine directly on the retinas. The projection had zero-zero resolution: perfect.

The big pane on the wall in front of Simon showed the simulation that the AS had fabricated. So far the illusion had been flawless. Josep had believed completely in the spaceport administration block and the journey through Durrell. Even Michelle's room was exact, thanks to the data from the hominoid remote; not just the colors and proportions, but the texture and temperature of the bed and desktop pearl as well. Duplication of inanimates was always easy.

It was where the subject interacted with other people, especially unknowns, where problems and errors began to creep in. If it was someone the AS had no background profile for, their behavior and responses had to be estimated from context. Once a mistake was made, the effect would rapidly multiply until the entire environment simply became unsustainable. And in this case the AS had to try to realize both

Iichelle and the strange software in a believable fashion ·om the absolute minimum of data.

Nonetheless, Simon was content with the procedure. After ·itnessing for himself how extraordinary the intruder was, e had been convinced that a standard interrogation would rove useless—a decision that was endorsed by the subse-uent cellular-level scan of their unconscious prisoner. Doc-ors and biotechnicians had been fascinated by the profound hanges made to his body and were completely unable to ex-lain how any of them had been performed. The number and ature of exotic microparticles was astonishing. Some of the xperts were still debating whether he was a human who had een improved, or an alien that had been modeled into uman form.

For all the prisoner's physical prowess, Simon had limpsed enough of his mind to see the human emotions vithin. It was enough for him to launch the attempt at virtual hicanery.

As far as he was concerned, it had now paid off hand-omely. They had garnered several vital leads, especially the irl, who was definitely an ordinary human.

'Where is the stuff?" Josep asked softly. "In fact, what stuff?"

"Don't," Michelle said. "Please."

"Who?"

"What?"

"Who? Who am I?"

Her expression crinkled up into misery. "What are you loing?"

"What's my name, Michelle?"

"Just stop this. It's not nice."

"Uh-huh? You know, for someone who's only been away for an afternoon, I'm very hungry." He bent down and picked up an old pizza delivery box. There was still a sliver left inside. He put it in his mouth and started to chew.

Simon's magnetic sense caught the emotional content washing through the prisoner's brain. It was changing rapidly; confusion giving way to a tide of bitter resentment.

"He knows," Simon declared ruefully. "Well, the scenario had almost played out. We have enough to investigate his background."

"But we still don't know what they wanted to hijack a Xianti for."

"One step at a time." Simon's smile faded as another change swept through the prisoner. He hadn't seen the emotion too often before. And never this profound.

"No taste," Josep said. "None at all. Why is that, Michelle?"

"Please, you're scaring me."

"Fatalism," Simon said, startled by the intensity.

The prisoner's bright glowing aura began to swirl.

"I didn't know software could be scared."

"Out!" Simon bellowed. He charged at the door. Behind him the aura was in a frenzy of turbulence. Then it shrank to nothing.

Simon reached the door. Opened it.

The prisoner exploded.

Lawrence found the darkness reassuring. He was warm, his body was perfectly comfortable and at ease. There was no

ain. It was a womb darkness, he thought, secure and nur-
ıring. A heartbeat he assumed was his own drummed out a
teady rhythm in his ears. Breath flowed easily into his
ıngs. He supposed he could move his limbs if he wanted to.
Ie didn't; the coziness of allowing himself to drift was too
ppealing. Only his eyes were ineffectual in this pleasant
nvironment, showing him nothing.

Without sight, he began to see.

Events from his life slipped in and out of his conscious-
ıess, without order, as all memories were. He visited his parents. Played with his brothers and sisters again. Roselyn
emerged into his life, all smiles and adoration. He walked on
ılien worlds, and kept on walking, over the plateau and into
he white, cold isolation of the snowstorm. The crater lake
ay below him; he spread his arms wide and dived cleanly
nto its deep, cleansing waters.

There was the feeling of a smile, a slight mockery. His recollections weren't the only ones he was aware of. Another's
listant dreams shared this universe.

"Hello?"

"Hello, Lawrence."

"Who are you?"

"The humans of Arnoon call me the dragon."

"Is that where we are, Arnoon?"

"Yes."

"What's happening?"

"I am repairing your body."

"Are you a doctor?"

"No."

"What then?"

"You wish to know me? Come."

The dragon's dreams grew stronger. And the universe was
no longer a place of darkness.

It was half-past-three in the morning Durrell time when Simon's spaceplane touched down. They taxied over to the parking apron, and a set of airstairs were wheeled out. When the outer lock opened, he took a moment to breathe in the air. It was a lot better than the recycled molecules of the *Norvelle,* but other than that there was no distinguishing scent, nothing to mark it down as alien. Every time he stepped out on a new world he expected to find something special, divergent. Every time he was disappointed.

Braddock Raines was waiting at the bottom of the stairs, his face grim. "Welcome to Thallspring, sir." The five-strong escort of Skins closed in to form a protective cordon as they walked over to the limousine.

"Thank you." Simon paused to look at the fire engines parked around the administration block. Dozens of scarlet-and-amber strobes were flashing across the field, acting like an advertisement for the disaster. Despite the time, a large number of spectators were still clustered behind the temporary barriers.

Floodlights had been set up around the administration block, illuminating a large section of the top two floors, which had been blown out. The three floors underneath it had sagged, smashing most of the surrounding glass wall. Dunes of glass fragments were strewn over the ground below. Paramedics, engineers, firemen, Skins and robots were picking their way through the wreckage, searching for survivors and bodies. Long crane gantries from the fire engines stood sentry duty, spraying water over the already soaking debris.

"Quite a mess," Simon murmured.

"Yes, sir," Braddock agreed. "The datapool is saturated with it. President Edgar Strauss is requesting an urgent clarification of the incident from General Kolbe. He wants to know who was to blame."

"Ah, the good president. I only dealt with him once, myself. How is he?"

An awkward smile flickered over Braddock's face. Even hough he had a top-level Z-B security clearance, he was always edgy when reminded how many Simon Rodericks there were. "Worried that we'll use collateral."

"Understandable. Given the circumstances I can hardly old Thallspring's citizens to blame. But there's no need for trauss to know that. Have the general tell him the investigation into the cause is continuing. That should keep people on their toes."

The spaceport hospital was tucked away in a small wing of he terminal building. Braddock had thrown a tight security cordon around the whole area when the badly injured Simon Roderick was rushed in. Only Z-B's own medical staff was llowed into the surgical theater to operate on him. Subsequently it had become an intensive-care unit. Security engineers and programmers had to examine and clear every piece of equipment brought in to treat him. The theater's electronics and nodes had been physically cut from the spaceport network, which itself was isolated from the datapool. Software subversion against any of the equipment was now impossible.

"Very thorough," Simon said approvingly as they walked through the hospital. "What about Adul?"

"Dead, sir," Braddock said.

"Damn. He was a good man."

Three Skins were standing guard in the corridor outside the theater. One of them held the door open for Simon. A viral technician and a doctor were in the theater monitoring the life support equipment that encrusted the figure on the bed.

"I'd like a moment alone," Simon told them.

Roderick beckoned before the doctor could protest. The startled viral technician gave Simon a long, scrutinizing glance as he walked past.

Simon went over to the bed. Two of the ten lights in the

big mobile array overhead were shining across the machinery. The SK2 had 73 percent burns, which had been sealed under a thick oyster membrane that had its own plexus of fluid capillaries to treat the ruined tissue. His head had been completely covered, leaving just two small slits, one for his mouth and one for the eye that had been saved. An oxygen tube passed straight through the membrane to the remnants of his nose. His left hand had been amputated, as had both legs below the knees.

"Can you hear me?" Simon asked.

The SK2's eye opened. Air hissed out past his teeth. "That alien motherfucker!"

"If it's any comfort, he's in a worse state than you. They're still scraping bits of him off the ceiling."

"What ceiling? There isn't even a building left. Hell, I was stupid. I should have realized what it was capable of."

"Yes, well, as a suicide method it certainly beats a tooth full of cyanide. I didn't realize the human body contained quite so much chemical energy."

"Not human. Alien."

"No. I've reviewed the data on my way down. Our own dear experts have made some headway. His DNA is confirmed as human. It wasn't even modified. The microparticles were foreign to his body. Not that we have many of them to study from those samples you took, a few hundred thousand, but it would seem they rebuild the molecular structure of the cell that they occupy. The modifications are not genetically sequenced. You know what that means."

"Alien."

"The technology certainly is. We've ruled out a Santa Chico connection. This is way beyond anything they have. It's a working nanonic system that can engineer molecular biology."

The SK2's breath hissed loudly again. "Find it." He

roaned. Several monitor lights turned red. "Godfuck, it urts."

"They're just stabilizing you. It'll take another two days efore you're ready for regeneration viral therapy."

"Are my balls still there?"

"Apparently so, yes."

"Thank fuck for that. They told me about the rest."

"I know."

"Both goddamn legs."

"They'll be replaced as soon as we get home, along with verything else."

"Hoo-fucking-rah."

"Would you like me to have you placed in hibernation leep?"

"No. I'm going to see this through."

"As you wish."

"Of course I fucking wish. You know what this means, lon't you?"

"The potential for a working nanonic is quite phenomenal, yes."

"Phenomenal, my ass," the SK2 rasped. "It's total. We can levate the whole human race. And in real-time, too. No nore waiting for backward regions to benefit from our investment, no more germline v-writing health policy. My God, we've won. Everything we wanted to achieve can be mplemented. There'll be no more moron barbarians holding us back. Society can shift to an entirely active-creative economy."

"Let's hope so," Simon said cautiously.

"Hope be damned!"

Simon didn't enjoy seeing his clone sibling in so much pain. It was far too easy to visualize himself lying on the theater bed with little machines leeched to him. The SK2 was fixating on the prospect of finding nanonics as a way of justifying his own suffering, making the sacrifice and pain worthwhile, which it would be, Simon conceded. But the alien was deploying its nanonics in a very strange and definitely hostile way. "We're still not certain about this nanonic system's capabilities. So far we have a lot of conjecture from overexcited technical staff, nothing concrete."

"I saw him, what he became. We can rebuild every human in the universe to make them sane and intelligent."

"As sane and intelligent as we are." Simon thought he kept the tone irony-free.

"That's what we exist for."

"Quite, although we never envisaged achieving it in one big bang." Simon almost asked *What if people don't want to be altered by us?* But he already knew the SK2's answer to that. The discovery of this nanonic technology would cause an unprecedented split in the Board; some batches would demand immediate implementation, while others, like his own, would want to move more cautiously.

Although that would be completely hypothetical unless he did actually find the alien and obtain the entire technology. Simon gave the SK2 a thoughtful look. Was that why he'd refused hibernation? To make sure the acquisition was completed? The very fact that he could think that of a clone sibling made him uneasy.

"Well, now we can modify our original objective to take that into account, can't we?" the SK2 said.

"That's some modification you're asking for."

"But possible. And extremely desirable."

"Absolutely."

"Interrogate the girl, first—Michelle Rake, she's a very weak link in their security."

"Of course. Any thoughts on why the alien is using its technology against us?"

"No. We don't have enough information yet. You'll have to determine that as well."

'he dragon's dreams were everything Lawrence had ever lreamed of. He embraced the irony with a kind of bitter ıumor as he learned of the Ring Empire. Once again, the ıniverse had shifted around him, taking away the life he hought was real. Colorful, elegant facts slowly coalesced, nerging with his own thoughts until they became revelation. Vithin this strange state of enlightenment he floated erenely after Mozark as the prince flitted from planet to ılanet. There were, Lawrence realized, huge segments missıng from the story.

"Most of my memories are lost," the dragon said regretully.

"This is real, then?" Lawrence asked as he gazed across The City, marveling at the silver-and-crystal palaces emergng in the rosy dawn light.

"This is history."

"How long ago?"

"Tens of millions of years, if not longer. Again, that infornation no longer exists within me."

If his eyes had been open, Lawrence was sure he would ıave wept. The dragon's knowledge was stupendous, its physical science tremendous. The potential was here to achieve . . . anything. Lawrence wasn't just awed, he was humbled. His own goals seemed utterly inconsequential and petty compared to all this. Yet the dragon didn't judge him, which made his guilt all the greater.

"I hoped I would find wealth here," Lawrence said. "But I never expected to be this rich."

"The villagers never considered themselves rich."

"They are. Believe me. There could be no greater gift than knowing you. You are the kind of hope I had long since stopped believing in."

"Thank you. Though it is humans who must take the credit for resurrecting me this far. I would not exist if it were not for your endeavors."

"I would know one thing," Lawrence said, even though he felt ashamed at asking. "Are you sure about Mozark? Did those places and species he encountered on his voyage genuinely exist?"

"The memories are all I have. They are what I am. Does your past exist, Lawrence?"

"There are times when I wish it didn't."

Denise had risen soon after dawn, content to be in her own bed for what she sincerely hoped was going to be the last time. With the light growing outside she went onto the balcony of her parents' A-frame home. For once the sun was visible as a splendid copper crescent rising in the cleft between Mount Arnao and Mount Nallan. Denise took that as a good omen. It was rare for the cleft to be free of fog and cloud so early in the morning. Now she could lean on the carved wooden rail to look out across the marvelous crumpled valleys and craggy rock faces that composed Arnoon. A shallow layer of mist hung over the meandering slopes spread out around her, with only the tallest treetops poking out above its frayed surface. The sun's radiance fluoresced it a delicate rose-gold as it gently slid and slithered its way out of the foothills toward the plain below.

After a light breakfast with her parents she walked across the village to the big snowbark pavilion. The air up on the plateau was a lot cooler than the humid coastal climate that she'd grown accustomed to down at Memu Bay. She put on a willow wool sweater before leaving the house: a present from Jacintha, whose husband Lycor had designed it, as always incorporating bright colors without making them garsh. This one was midnight-black with curling flecks of sapphire, topaz and magenta looking as if they were being blown across the weave; its sleeves were flared at the wrists, with a small V-gap allowing her to roll them up. It kept her beautifully warm as the cold morning gusts drifted down from Mount Kenzi.

As she walked, friends she hadn't seen in an age came over to greet her and exchange pleasantries and words of encouragement. They all expressed their sorrow over Josep, as if she somehow suffered his loss more than they. It was wrong, she felt; they were treating her as if she'd achieved something, instead of nearly bringing ruination to them all. But to say so to their faces would be selfish. And there was still hope. Not that she could ever have imagined it would present itself in such a strange form.

Before the children arrived she walked around the inside of the pavilion, trailing her hand over the bark of each of the ten trees, reacquainting herself with them. So many hours of a pleasant childhood had been spent in or around the pavilion with her friends, playing games and listening to the adults tell their fantastical stories. It was fitting that she, the one who'd been chosen to seed their way of life on a new world, should be given a last opportunity to tell the new generation of their heritage.

The children began to arrive, little groups of them bounding over the central meadow, chattering and laughing. Denise smiled in reflex: something about happy children was just infectious, their smiles made the world a less painful place. Parents were bringing smaller children. She saw Jacintha and Lycor with little Elsebeth holding their hands as she toddled along between them.

Eventually, after some coaxing, the children were settled in a big semicircle around Denise.

"Have you all heard the stories of Mozark and Endoliyn?" she asked them.

"Yes!" they cried back.

"Well, today I'm going to tell you the last story of the Rin Empire. This is set long after the time of Mozark and En doliyn. It's sort of a sad time, because the Ring Empire wa starting to decay. Some of the inhabitants blamed the ma chines for this, because they were now so smart that the took care of people from the moment they were born unti they died. This machine-pampered generation had nothing t do except live their lives chasing personal pleasure and satis faction. They had become decadent, and not a little bit cruel Now this generation, the final generation, had enormous re sources at their disposal; their machines could dismantle th very planets and reshape their atoms to build whatever thes people wanted. With that kind of ability you'd think they' be totally content. But no. Even the number of planets is fi nite. They began to argue with each other about how man resources any one person should have and how these re sources should be divided and supervised. At first it was jus arguments. Then it grew into theft and hoarding. Eventually fighting began and grew into what was known as the Deca dence War. The individual kingdoms that had been so closel knit turned against each other. Battle machines were con structed, the most terrifying things ever built, equipped wit weapons that could tear a planet to pieces and even extin guish stars. These battle machines fought each other over th division of entire solar systems. And it took an enormou amount of resources just to build them. That meant that an solar systems that the battle machines conquered were soo turned into more battle machines. The last generation was de- prived of the one thing they had launched the war for. With- out the resources they craved they soon dwindled int extinction amid the conflict. The battle machines continued fighting for thousands of years, wreaking havoc among the stars, until they had finished eliminating each other along with entire races.

"But the decadents and their battle machines weren't the only reason the Ring Empire fell—although they must take most of the blame. They represented only the physical aspect of its decline. Many societies had followed the Wilfrien, slowly regressing, even rejecting their technological society, seeking a more primitive existence in search of peace, withdrawing their participation and support from the Ring Empire. Then there were others, like the Outbounds and the Last Church, who had actually been quite successful in reaching their goals. They had attracted the most dynamic people, the brightest, the restless who relished challenge; all of these had found their cause and given themselves to it. In doing so, over the millennia, they had drained the Ring Empire of vitality, the very people who could have regenerated it.

"Among the factions and wars was one group who had predicted the fall. The Eternals, who were more academics than anything else, had studied civilizations from across the Ring Empire. They found one thing that remained constant among all biological species: the cycles of growth and decay. It might take only a century. It might take a million years. But life always follows that pattern. As the Decadence War raged and the Ring Empire fell apart around them they decided to save themselves. Many groups were desperately trying to do the same thing, with colonies and secret enclaves to keep their original ideals alive so that one day they could expand again, rekindling their former glory. In some instances entire kingdoms isolated themselves behind fortified borders so that they wouldn't be contaminated with the decay infecting their neighbors. All of them were attempting to resist that which the Eternals were convinced was inevitable. By doing so they would surely be doomed to failure, the Eternals thought. And they were right, for today there is no Ring Empire, only whispers and legends of the glory that was. But the Eternals are still here.

"Instead of making some futile stand against the decline,

the Eternals embraced the cycle of life. They transforme themselves and their society so that it would live in harmony with galactic nature. Biological life and Ring Empire ma- chine were fused at a molecular level. The Eternals became giant spaceborne creatures. Unlike starships, they didn't nee artificial power and great industrial stations to maintai them. In many respects these creatures were profoundly sim ple. It is that simplicity that has allowed them to survive an spread across the galaxy.

"They live today in orbit above the galaxy's red gian stars, powered by the heat and grazing on the solar wind They have enormous solid bodies like streamlined asteroid that sprout solar wings whose span is measured in kilome ters. Because of their shape, and the fiery environment i which they thrive, we call them dragons. They are eve hatched from eggs. Every solar system has them, dark col globes circling among the outer cometry halo as they wait fo the star's main sequence to come to an end. That's the cycl again. Stars grow old and die, swelling out to absorb thei planets, and eventually expanding into red giants. That' when their warmth reaches the eggs, energizing them. The grow slowly, absorbing the heat and the thin gusts of ions until they're fully fledged dragons. And then they listen t the universe. Their wings are threaded with elements that ca pick up radio waves from the other side of the galaxy, an even far beyond that, allowing them to listen to planetar civilizations as they rise and fall. They listen to the cosmo itself, the death and birth of stars, the shriek of matter as i falls into black holes, quasars and pulsars crying out from th empty void. All this knowledge they spread among them selves, and think about it, and remember it. On rare occa sions they even use it, for they can modify themselves at a molecular level. That is their physical nature, the legacy o the Ring Empire.

"Eventually, as the star shrinks back to a white dwarf be

fore its final extinction, they are left abandoned in the dark and cease to be. In accepting their mortality they live with the cycle, with what's natural. They have served their purpose and advanced their species. Like any civilization, they acquire knowledge, they organize it and bestow it on their descendants. As they circle above their stupendous star they send their own eggs out into the universe, each one containing the memory of everything they consider important and relevant. These eggs fall through the darkness of interstellar space until the gravity of some bright new star pulls them in to a long distant orbit so the cycle can be started over."

"Except one of them fell to earth, or rather Thallspring," Lawrence said.

The children gasped and turned. Lawrence was leaning casually against the trunk of a snowbark, arms folded across his chest. His grin was lopsided as he stared at Denise.

The children started whispering excitedly.

"That was over two thousand years ago," Lawrence continued. He grinned down at the expectant, slightly awed faces. "The dragon's egg streaked out of the sky like a splinter of sunlight and struck the plateau near the base of Mount Kenzi. The force of the blow was so powerful it gouged a crater out of the bare rock. Every tree for fifty kilometers was ripped out of the ground and smashed apart by the blastwave. Then the timber burned for days, filling the air with thick, black smoke. But the dust from the vaporized rock billowed up into the stratosphere and blotted out the sun for weeks. It brought the coldest winter the plateau has ever known, covering it in snow. Then, when the snow began to melt a few years later, it filled the crater with water. The trees started to grow again. And a hundred years later everything looked just the same, except now there was a new lake.

"Then people arrived and called this place Arnoon. They built themselves a village and began harvesting the willow webs. And one day—"

"One day," Denise said, "my grandfather was out prospecting when his survey sensors found a strange magnetic pattern in the rock under the lake. So he started to dig. It took months for his little robot to excavate a shaft down to the base of the island. But when it got there, my grandfather found fragments of the egg. He didn't know what they were, only that they were artificial solid-state matrices of some kind, unlike anything humans had ever built. He began to excavate further, and eventually found the largest fragment of all, the one we call the dragon."

"By then," Lawrence said, "he'd discovered that the molecular structure of the small fragments was storing data. After a lot of experiments he finally managed to access some of it. Once he knew how to do that, he started to mine the huge reservoir of information stored within the dragon."

"The dragon was still sleeping," Denise said. "It possessed nothing but disconnected memories. My grandfather wrote programs that linked them together. The dragon slowly began to wake. It learned how to think."

He looked straight at her, heedless of their audience. "And you found that it was actually a cohesive nanonic system capable of molecular engineering. You used it to adapt native plants to grow terrestrial food. You used it to make yourselves resistant to disease. You made it synthesize bits of technology that are orders of magnitude more advanced than anything humans can make. And you kept it all for yourselves."

"Because it can only change itself into what we ask for. It can't build anything new. It doesn't know how. That data was lost in the destruction of the impact. Every patternform sequencer particle in my body was a part of the dragon. It diminished itself to enhance me. It diminished itself further to heal you."

"Yeah," Lawrence said. His belligerence faded. "Makes righteous life kind of awkward, doesn't it?"

Denise turned back to the children. "So now you know why things are a little different here than on the rest of Thallspring. A very noble creature has sacrificed part of itself to make our lives easier. Our debt to the dragon is enormous. We must never forget what we owe. And we must pray that one day we can repay it."

The children filtered out past the snowbark trunks. Many of them crept up close to Lawrence, then dashed away giggling. Approaching the big bad Skinman was a seriously scary dare. He found it rather funny.

Jacintha came up to him, little Elsebeth cradled on her arm. The girl was shy, burying her face in her mother's neck.

"I do remember you now," Lawrence said.

She nodded a fraction reluctantly. "I'm sorry we became enemies again."

"Love and war. I guess that's part of the human cycle."

"We hope to break that. With the dragon's help."

"I know. It told me."

Jacintha glanced over at her sister, who was waiting for them, a disapproving expression on her elegant young face. "Try not to give her too hard a time. She has to do this."

"Don't worry. I know what I have to do, as well."

Jacintha gave him a mildly suspicious look, then walked away back to Lycor. Elsebeth gave him a little wave from the crook of her mother's arm. He shook his own fingers at the young girl, smiling.

"We have Grabowski," Denise said briskly. "I'm willing to offer you a deal. You can't go back to Zantiu-Braun, so if you cooperate with us the patternform sequencer particles will repair all Grabowski's damage, including his brain, and he can begin a new life here in the village."

Lawrence widened his smile until it became suitably irritating. "I don't need a deal. I'm going to help you anyway."

"What do you mean?" she asked slowly.

"You want to take the dragon fragment to Aldebaran, right? The closest red giant, where all the real dragons are."

"Yes." She said it as if admitting a fatal weakness. "They can make it whole again. If it stays here, then your kind will discover it one day. They'll take it from us and break it apart in their corporate labs to discover how patternform-sequencing systems work. I can't let that happen. It's a living entity that has given us so much, and we've never done anything for it. This is our only chance to return it where it belongs."

"My kind, huh?"

"Zantiu-Braun, or Thallspring's government. People who don't live out here like this. People who don't live real lives, who'll never care about anything but themselves."

"You know, there's more of *your* kind than you think. Everywhere I go, I keep bumping into idealists."

"A shame none of it rubs off on you."

"I'm helping you, aren't I?"

"Why? Why would you agree to help?"

"Raw altruism not good enough for you?" He wasn't about to tell her the shock he'd experienced on hearing about the Mordiff, nor its accompanying revelation.

"I don't believe it, not from you. You came here to steal the dragon. You wouldn't switch sides and morality this quickly."

"I didn't know the dragon existed before I arrived. I thought you'd got a big stash of gold or diamonds hidden away up here."

"But . . ." She gave him a troubled look. "Where did you get your Prime from?"

"A boy I knew once back on Amethi. A good kid. Little bit misguided and confused; but then, isn't everyone at that age?"

"So Earth has found a dragon."

"No. That's why I'll help you."

ichelle didn't know where she was, nor what time it was. ıe wasn't entirely sure what day it was.

After the Skins had dragged her from her room she'd been iven somewhere in a blacked-out van. The medical order-:s from the elevator went with her. They tore her T-shirt off their probes could inspect and scratch her skin. Needles ere inserted into her flesh along her limbs and belly, leav-g small beads of blood welling up when they were ex-acted. She'd screamed and pleaded and struggled. It was all tile. A Skin pinned her down until their examination ended.

Her ruined T-shirt was thrown back at her, and she tried to rap it round her breasts. Now that they'd finished, the men ıowed no more interest in her as she lay on the floor of the ın, weeping pitifully. She half expected them to rape her, ıt that didn't happen either.

The trip lasted fifteen minutes. When she was hustled out the van, it'd been parked in some anonymous underground ırage. She was marched directly to a small cell and pushed ıside. The door slammed shut.

After the first hour she thought they'd forgotten about her. he banged on the door. But nobody came. She started weep-ıg again, hating herself for being so weak. She was just so ightened. Zantiu-Braun could do whatever they wanted to er. Anything. Nobody would know. If she could just see osep . . . This horror could be endured if he was with her. lowly she shrank into a fetal position on the cot, hugging er legs tight to her chest. Little bursts of sobbing came and ent. Why didn't they just take her out and start their inter-ɔgation? Just get this *over* with. At some time she must have rifted into sleep.

The door thudding open woke her with a start. A Skin alked in. Michelle clutched the ragged T-shirt to her chest,

staring fearfully at the dark, bulky figure. Suddenly s
wasn't so keen for the interrogation to start after all.

"You. With me. Now." The Skin beckoned.

Michelle was led along cheerless basement corridors to
elevator. It brought her up to the main levels of the buildi
She thought it looked like an extremely high-class hotel, w
luxurious gold carpeting and gloss-polished wood doo
Large, elaborate oil paintings hung on the walls. Delicate a
tique tables supported china vases full of big flower arrang
ments. Lighting cones were gilded in silver and cut crysta

It wasn't a hotel. Open doors gave her glimpses into
fices. The men inside, and hurrying along the corridor,
shared a tense, preoccupied air. Few of them even spared h
a second look.

The Skin finally opened the door into an office with a si
gle desk. A man was waiting for them, dressed in a sm
gray-and-purple suit, styled differently from anything she
seen on Thallspring. "I'll take her from here," he told t
Skin.

Michelle barely heard. She was looking out of the wi
dow. The view showed her a swath of formal grounds swee
ing away to a broad circular highway. Beyond that were t
familiar sturdy public buildings that populated the center
Durrell. But to be seeing them from this angle, she'd have
be inside the Eagle Manor.

"I'm Braddock Raines," the man was saying. "Please." H
took his jacket off and proffered it to her. "Sorry about t
way you've been treated. The frontline boys tend to becom
slightly overenthusiastic, especially on an operation wi
such a high priority."

"Operation?" she asked blankly. She was still having tro
ble with what was happening.

"All in good time." He smiled reassuringly and gestured
a tall double door. "My chief would like a word."

There was a larger office through the doors. The man s

ting behind its broad desk gave Michelle a pleasant nod as she was shown in, then returned his attention to a pane in front of him. It was difficult to tell how old he was. Mid-forties, she thought, though he had the kind of assured authority that was normally found in men a lot older.

Braddock steered her to a settee and indicated she should sit. She pulled the jacket around her as if it were a shield.

"My name is Simon Roderick," said the man behind the desk. "I'm in charge of Zantiu-Braun security on Thall-spring. And you, Michelle, have been a very stupid young lady."

She dropped her gaze, praying she wouldn't start sniveling.

"One thing in your favor right now is that we know you're actually human."

"Excuse me?" she stammered.

"You're a human, unlike this gentleman." The sheet screen on the wall flashed up a picture of Josep's face. "Ah, you do recognize him."

"Yes."

"Thank you, Michelle. At least you have some understanding of how much trouble you're in."

"One day you'll be defeated," she said, amazing herself at such defiance.

"It's not only Zantiu-Braun that will be defeated by aliens that powerful. The entire human race could well be facing a terminal threat."

"What do you mean, aliens?"

"You didn't know, did you? Your comrade in arms was not entirely human."

"That's ridiculous." Nobody was more human than Josep. Only a human could bring another human so much pleasure and contentment.

"Is it?" Josep's image was replaced by a cluster of multi-colored spheres. "Do you know what that is, Michelle?"

"No."

"That doesn't surprise me. We're not absolutely sure ourselves. It's a nanomachine that appears to have molecular-engineering capabilities. It was extracted from your friend's blood."

"What have you done to Josep!" Tears threatened to burst down her face, but it was anger that pushed them this time, not fear.

"Josep?" Simon smiled. "Finally, a name."

Michelle's shoulders slumped. The anger burned out as quickly as it had flared. How stupid to be caught out like that. "You can do what you like to me," she said sullenly. "I won't help you."

Simon walked around the desk and sat on the settee next to Michelle. She tried not to shrink from him. He poured some tea from the silver pot on the low table.

"Do you know what we can do to you?" he asked. "Did Josep ever tell you?"

"You'll use drugs, I know that. And you'll probably rape me before you kill me."

"Good grief, what a repellent idea. We're not savages. My dear girl, you really must learn to distinguish between facts and your own side's somewhat lurid propaganda. Yes, we can use drugs, along with various hypnosis and deep-stimulus techniques, none of which are particularly pleasant. There is nothing you will be able to keep from us; you will confess your deepest secrets. Do you know why we're not doing that to you right now?"

"So you can trick me into giving you names," she said hotly.

"No. I want to appeal to you to give us the information voluntarily. Time, I'm afraid, is rather short. I really am not joking when I say Josep is an alien."

"What have you done with him?"

"Nothing. I wish we could. He escaped shortly after we captured him."

"Good. You'll never catch him again."

"Not without your help, no."

"I won't. You'll have to interrogate me properly." She was shaking at the prospect of submitting to their interrogation, but every minute in here was another minute Josep could use to flee.

"Aren't you going to ask where we caught him? Or do you already know, did you help plan the attack?"

"I don't know what you're talking about," she said, though there was a horrible suspicion bubbling through her mind. Those nights he never came home. Courier duty, he said, like the rest of the cell were given. Except she'd never been asked to run anything at night.

Simon picked up his cup of tea and settled back into the settee. The sheet screen began showing a datapool news report of the spaceport. Bodybags were being carried out of the wrecked administration block.

"Oh, God," she whispered.

"Eight people dead," Simon said. "Including Mr. Raines's colleague."

Braddock Raines was standing at the end of the settee, his face impassive. Michelle flashed him a hugely guilty glance.

"Seventeen injured, three critically. Our cargo-lifting operation delayed by several days. And the whole of Durrell terrified about what retaliatory measures Z-B will employ. After all, we promised to use our collateral necklaces to prevent any interruption to our asset realization. What do you think, Michelle, how many Thallspring citizens should Z-B kill so that your resistance movement doesn't do this again? Ten?"

"Stop it."

"Fifty?"

"None!" she shouted. "None at all. He didn't do this. We

didn't do this. All we do is sabotage your transport and stolen factories. This isn't what we want, not killing people."

"That's not what you want, Michelle. There's a difference in your understandable, if pathetic, yearning to fight the invasion, and the goals of your alien allies."

"Josep is not an alien!"

"Dear me, what an irony I have here. We can extract the entire truth from you should we so wish, yet we cannot install the truth. But the truth is what I am dealing in. Josep's body was altered, enhanced, by alien technology. He was using you."

"He was not. We were in love."

"Ah." Simon sighed happily. "Was this your first love, Michelle?"

"I . . . it . . ."

"So it was. How delightful."

"No, it wasn't." Even as she denied it, she knew Roderick knew, really knew, and blushed heavily.

"There is a standard ploy that intelligence agencies use for infiltrating their enemies, Michelle. It's very common and has been in use for centuries. You find some lonely, sad little soul working in the place you need to be, a woman maybe approaching middle age and unmarried, or maybe not as pretty as her contemporaries. Perhaps it's simply someone who doesn't fit into her new environment very easily, who finds it all new and strange and frightening. Either way, you send in a wolf. They meet, as if by chance. She finds herself courted by this most handsome man, impossibly talented in bed, devoted to her and her alone. Her heart belongs to him. And with her heart comes her complete and absolute trust. Does any of this sound familiar, Michelle?"

"Don't," she said weakly.

"Did he come into your life around the time we arrived on your planet, Michelle? This is your first year at university, the first time you've ever really been away from home. Your grades weren't very good. You were lonely. Did you meet him on campus? No. Before then? Ah, of course, the real first time you left home. Your mother and father paid for a vacation at Memu Bay, a reward for passing your exams. That's it, isn't it? That's where you met him. It was a classic, perfect holiday romance."

Michelle was sobbing helplessly. The pain the words inflicted was worse than any torture. "He loves me. He does!"

"Then we invaded. He appeared back into your life as if by magic. Yes. He lived with you, unofficially of course; there's no record of him in the university files. In fact, there's no record of him anywhere on Thallspring. Digitally, he simply doesn't exist. Do you know how impossible that is, Michelle? The most powerful askpings ever written cannot find a single trace of him in the global datapool."

"He's human!" Michelle implored. "Please." She turned to Raines, who shook his head sorrowfully.

"Did Josep tell you if he had special software?" Simon continued relentlessly. "Really clever, super-secret software that could help the cause?"

Michelle was starting to curl up back into a fetal position. The brutal voice just went on and on, tearing her world apart.

"Software that was better than any AS on Earth could ever produce. What did he say, that it was written by a few teenage geeks in their bedrooms, who also just happen to be loyal to the Thallspring cause?" Simon put his index finger under the girl's chin and tilted her face back. Her cheeks were sticky with tears. His electromagnetic sense observed the tidal waves of distress tormenting her thoughts. "I'm so sorry," he said tenderly. "I really am. This is all as frightening to me as it is you."

"Prime," Michelle stammered. "The software was called Prime."

It was quite an operation, lifting the dragon out of its unde
ground lair. The route had been prepared years ago, c
course. Denise's family had sunk a second shaft down to th
chamber, a bigger shaft than the elevator, which emerged t
the side of the small stone temple.

Lawrence sat on the curving stone bench, watching as it
concealed hood rose up on magnetic pistons, bringing
meter of soil with it. The dragon slowly emerged underneatl
still sitting on its white-glass pedestal. Its golden powe
induction mesh was wrapped tightly around its midsectio
Sunlight glinted off individual strands. Electrohydraulic mo
tors whined loudly in the placid air.

"Welcome to the world," Lawrence said. "I don't suppos
you can sense visible light?"

"Not directly," the dragon replied. "However, I receive th
images from yourself and other humans. I know wha
Arnoon looks like. It is very beautiful."

Repairing Lawrence's leg and hip wasn't all the pattern
form sequencer particles had done. They'd also modified
cluster of his neuron cells, giving them an ability similar to
DNI implant. D-writing, Denise called it, the particles engi
neering cellular structures in a direct fashion that huma
v-writing could never achieve—outside of germline treat
ments. Vectoring in new DNA was a scattergun approach de
ployed against entire organs or muscles; this was far mor
selective and precise.

"But you haven't given this communication cluster t
everyone here?" he'd asked her.

The two of them had sat together in the snowbark pavilio
for most of the morning, discussing how to get the dragon u
to a starship. They were being polite to each other, nothin
more. There was too much history for friendship.

"No," she said. "Only people like me and Raymond an
Jacintha need it. We didn't want to create some kind o

superwarrior breed. The enhancements given to the children are more benign and beneficial."

"Similar to germline v-writing?"

"Yes. The patternform sequencers can alter DNA quite easily. We gave everybody cancer resistance, and stronger immune systems, and refined organs, much greater life expectancy, a higher IQ. Their changes will be permanent, and the traits will carry down the generations. Arnoon won't have to depend on the dragon anymore."

"And the food," he said. There was a carved wooden bowl on the table in front of him. It was piled up with various fruits. He rested his finger on the rim, pressing it down so the bowl swung from side to side.

"The plants are also genetic adaptations," Denise said, enjoying his discomfort. "They'll breed true. In a hundred years, this forest will be an orchard that can feed a city. Nobody will need protein cell refineries anymore. Another economic necessity will be consigned to history."

"An economic necessity that liberated seventy percent of the human race from perpetual starvation. Growing things for food is a terribly inefficient use of energy."

"That depends on the nature of the culture you have to feed," she said. "Massive industrialized nations had to use industrial farming to feed their urban populations. If you replace them with scattered self-sufficient villages like Arnoon, then the requirements become very different."

"A world of physically separate communities linked by the datapool. The true global village. Knowledge belongs to everybody, and everybody goes their separate ways. You need microscale manufacturing to back that up, you know."

"I know. We've been studying the dragon as best we can, and we've copied every memory it has. If we give that to the rest of the world, then we hope something similar to the patternform sequencer can be built. It'll take decades, but we never wanted to force change overnight. This is going to be an organic revolution, generated from internal knowledge. I must succeed, if not here, then on a fresh world. Today's cul ture can't be the only way a technological society develops It can't."

His eyes flashed with mischief. "Plenty of prejudices t overcome."

"There certainly are." She picked a peach from the top o the bowl and held it up in front of him.

"You sure? Last time a girl did this to me I threw up al over her."

"You're just a born romantic, aren't you?"

He took the peach and bit into it. The fruit was sweet an succulent. Quite pleasant, really.

"It's not just fruit we get from our trees," Denise said in-nocently. "Some of them grow meat, too."

Lawrence had trouble swallowing.

He saw Hal before he left. The kid was in one of the A-frames, sleeping peacefully. His medical modules had al been repaired and were now industriously cycling chemical through various organs once more. And his skin was a much healthier color.

"The major internal damage has almost been repaired," the doctor said. "We'll start removing these modules in a day o two. I'm a little concerned about his biomech heart."

"What's wrong with it?" Lawrence asked.

"It's somewhat crude. I believe it was only intended as a temporary replacement. I'm not sure how long it will last and with the dragon leaving we don't have enough pattern-form sequencer particles to rebuild it. He'll probably need another transplant in twenty years."

Lawrence chuckled. "I wonder what kind of heart that'l be."

"Who knows?"

"What about his brain?"

"That will take more time to repair. He lost a lot of neu-ons from oxygen starvation. The patternform particles are ebuilding as fast as they can, but it will be weeks before full ntellectual function is returned."

A concept that, applied to Hal, made Lawrence grin. "Will his full memories come back?"

"No. Not even the dragon's systems can recover them. There will be large gaps in his life."

Lawrence stroked Hal's forehead. "I think that's probably a good thing if he's to make a fresh start here."

"Yes."

"Do me a favor. Take those valves out. That'll give him a real fresh start."

"Of course. Is there a message when he recovers?"

"Just . . . I don't know. Good luck, I guess."

It was pretty lame, he had to admit. But, really, what else was there to say? The kid had a chance at a new life here, why tie him to the past?

"Perhaps you could record a message," the dragon suggested.

"No. Cutting him loose is the best thing I can do for him. Besides, the last thing he needs is advice from me. Look what a screwup I made of everything."

"I believe that's what you call sweet Fate."

Lawrence touched two fingers to his forehead, saluting the dragon as a heavylift robot eased it off the pedestal. "You got me there."

Jacintha came into the temple and sat beside him. A small cargo robot rolled up behind her. The island's shoreline was nearly invisible under all the boats that had brought people and equipment over from the village. Lawrence hoped to hell Z-B's spy satellites didn't notice all the unusual activity. The villagers claimed they'd tracked everything the starships had launched into low orbit around Thallspring. If they were right, they had a clear sky above them right now.

"Your Skin's ready," Jacintha said, indicating the fat pla
tic case that the robot was carrying.

"Thanks. I thought that was dead."

"We had an antidote to sharkpike venom long before w
ever found the dragon. As long as it's applied quickly, you'
okay. The Skin's muscle cords were receptive once we'
flushed the contaminated blood out."

"Thanks. Those damn things scared me shitless."

"Every rose has its thorns. The rivers around here are fu
of sharkpikes. I've been bitten a couple of times myself."

"Can't you introduce some kind of virus? Wipe them out.

Jacintha's expression darkened. "Is my little sister real
going to be able to trust you?"

"Yeah, she can trust me."

"She's the closest thing to a genuine KillBoy there is.
was part of the team that wiped out your platoon. And no
that's all in the past? This from a man who would genocide
species because it has sharp teeth."

"The platoon followed me," Lawrence said slowly. "
brought them up here. You might have pulled the trigger, b
it was me who put them in front of you."

"And there I was thinking you were going to say the
knew the risks."

"That too. We don't expect a population to fight back, an
we certainly don't expect it in the hinterlands of Thallsprin
But each time we land we know it's a possibility. Denis
might have had a few zippy gadgets, but her real advantag
over us was how willing people were to sign up to her bogu
resistance movement. If the local inhabitants ever get prop
erly organized, or call Z-B's bluff, we automatically lose. D
you really think a starship captain, a flesh-and-blood huma
who has family of his own, is ever *ever* going to give a
order for a gamma pulse that will slaughter half a millio
people? It won't happen. So we know we're on our ow
down here, that there's no fallback, no help from above. Th

act that Denise eliminated so many of us in Memu Bay
roves what I've known for a long time now: that Z-B is in
ecline. Probably a terminal one. Skin suits are superb tech-
ology, even up against your dragon's knowledge. But with-
ut the organization, the initiative and the determination to
ace down threats, that means nothing. And we had none of
hose qualities down in Memu Bay. Santa Chico should have
old the Board that asset realization was over, finished for
ood. Instead they just kept on, trying to find weaker tar-
ets."

"You agree with the Eternals, then? Life is in a permanent
ycle."

Lawrence let out a long breath, exhausted with holding
ack his anger and despair. "Could be. You know what? I re-
lly don't care. I don't care that you killed my friends. I don't
are that I killed your ambush party. I don't care if that makes
s quits or not. I don't care that Z-B is quietly collapsing. I
on't care that you want to build some noble civilization
ased on total bullshit about people being perpetually nice to
ach other. I don't care that your deranged sister is willing to
acrifice herself and everyone she knows to save some piece
f talking rock. I don't damn well care that the universe is
oomed and the galaxy is falling into a black hole. I have
pent the last twenty years caring. I cared for my platoon. I
ared about what the human race was doing and where it was
oing. I cared that we didn't have frontiers anymore. I cared
bout my career. I even cared about what I was doing with
ny life. And look where I am because of that. Helping a
unch of cosmic hippies hijack a starship. Sweet fucking
ate!"

"You mean we can't trust you?"

"You got it, girl. Denise cannot trust me, not now, not ever.
do not like her. I will never like her. I will, however, respect
er abilities. And I expect a similar respect in return. What
ou can have from me is reliability. I am dependable in this

in a way none of you are. I will hijack that starship, and I will fly to Aldebaran. Of that you can be certain."

"I'm not sure I can be, Lawrence."

"This is for me, now. Not you and your ideals. That's why you can be certain. I finally, *finally,* have a chance to put my life back together and live it the way I was born to live it. To cancel out the last twenty miserable years. After Aldebaran I'm going home. That's all: home. And nothing and nobody can stop that from happening."

The sound of the hovercraft approaching made both of them turn and look out across the crater lake. Lawrence couldn't help a derisory laugh at the absurdity of the vehicle. It was made from wood, Arnoon's lightest, hardest timber, crafted into a simple oval platform with a cabin grafted onto the prow. Two big steerable propellers stood high on smooth tall fins at the rear. The skirt was willow wool, a fine tight weave easily holding in the cushion of air on which it rode. Electric motors powered the propellers and impellers, salvaged from an assortment of heavy machinery across the plateau.

It swept lightly across the water, with a thin haze of spray escaping from underneath its skirt, and a creamy V-shaped wake spreading wide. When it reached the island it rocked slightly as the front skirt rode up the shingle and onto the scanty grassmoss. The propellers reversed pitch, bringing it to a halt. It sank down with a prolonged wheeze of escaping air.

The heavylift robot carrying the dragon trundled over to it. A ramp was deployed in front of the propeller fins, enabling it to climb up onto the deck.

"We're ready," Denise said. She gave Lawrence and her sister an anxious glance, aware that they'd been quarreling.

"Sure," Lawrence said brightly. "Is that thing really going to work?"

"Certainly." Denise sounded offended. "We've practiced

the route a dozen times. The river is the easiest way out of Arnoon. The hovercraft will take us straight to Rhapsody Province. One of the articulated trucks from Dixon is already at the rendezvous point. It'll take the dragon all the way down to Memu Bay's airport. We'll be there in fifteen hours. After that, it's all up to you."

"Don't worry, my contact has sent a plane to collect us. Where's the cargo pod? We can hardly load the dragon into a Xianti as it is."

"The cargo pod is with the truck. An RL-thirty-three, industry standard sixty-ton capacity. We'll put the dragon inside it when we get there."

"Okay. Let's go."

Simon was appalled to discover that there was no supersonic transport on Thallspring. He wound up commandeering the presidential jet, which could barely reach Mach 9. It was a converted fifty-seat medium-range commuter jet that had a flight time of four hours to Memu Bay.

He spent the time working with his personal AS, dropping hundreds of askpings into Memu Bay's datapool. The leisure company that Michelle had signed up with to go diving among the atolls had no file on any employee called Josep, nor on Raymond, who was supposed to be his friend. The AS trawl couldn't find any abnormalities in the company's memory blocks. No substituted files, no gaps in the daily boat trip logs for a month either side of Michelle's visit; even the financial accounts were in order.

"Arrest them," Simon ordered Ebrey Zhang.

"Who, exactly?" Memu Bay's governor asked.

"The company's senior management. Their diving gill instructors. Boat crews. Bring them all in for questioning. I want them in custody by the time I arrive."

"Yes, sir."

The governor's noticeable reluctance made Simon review the current situation report for Memu Bay. "For God's sake," he muttered as the indigo script scrolled down. And to think, he'd warned the SK2 to keep an eye on the place.

Memu Bay had gone into meltdown over the last week. Asset realization was down to 50 percent of estimated targets. Two-thirds of the settlement's factories had some kind of strike action going on. The entire mayor's office had walked out and refused to work with Ebrey Zhang following the Grabowski rape case. The rest of the civil sector was reduced to emergency services only. Platoon morale was rock bottom, with charges accumulating against 30 percent of Z-B's personnel. TB cases were still being reported; immunization implementation was slow. Sabotage against utilities was a daily occurrence. Several districts had become no-go zones—and that included for Skin platoons. Collateral no longer worked. There were reprisals every time. Zhang was afraid to use any more necklaces for fear of making the situation even worse.

The more Simon studied the breakdown and its history, the more interested he became. Essentially, Z-B had lost control of the settlement. The resistance group led by KillBoy had waged a beautifully orchestrated campaign against the invasion, building to this climax of near anarchy.

"Why, though?" Simon asked a dismayed Braddock Raines. "How does this help our alien? Wiping out Zhang's little command is hardly going to cripple Zantiu-Braun."

"I'm not sure they could even do that," Braddock commented. "Physically eliminating every Skin stationed in Memu Bay would be difficult even for them. They can force the platoons off the streets and back into their barracks, maybe even make them fall back all the way to the airport. But if you hit those lads too hard, they'll hit you back. Part of the problem is Zhang holding back."

"You might be on to something there," Simon said. "With e platoons off the streets, the alien is free to do what it ants in Memu Bay without us noticing. But we still don't ıow what that is."

The presidential jet landed without incident. There was ry little activity at the airport. Half of its buildings were erating on reserve power supplies, thanks to the resistance oup severing a set of superconductor cables two nights ear- r. Skins patrolled the perimeter.

A helicopter was waiting for Simon. He climbed in as a big n-Skyways cargo jet took off, heading for Durrell with a ld full of assets.

Skins had to clear the square in front of the Town Hall so e helicopter could land. The displaced protesters jeered and rew stones over the barricades. Simon's Skin escort closed around him. He never normally noticed them, but today he as grateful for their presence. He didn't often get this close physical danger. The clamorous hostility of the crowd ade him distinctly uncomfortable.

Ebrey Zhang had a huge number of valid reasons why lemu Bay was in its current perilous state. He was very ger to tell Simon each and every one of them.

"Forget it," Simon said curtly. "I know exactly why this ess has built up. Not that knowing is going to help you uch before the board of inquiry."

Ebrey Zhang did his best not to scowl.

"Did you round up the leisure company people?" Simon sked.

"Yes. And it wasn't easy. My platoons get into difficulty very time they step out of their barracks."

"I'm not prepared to tolerate this situation. It will interfere ith my investigation. Can you enforce a proclamation of artial law?"

"Probably," Ebrey Zhang said. "It'll be difficult in some istricts."

"Declare it. Order a curfew for six o'clock that will b in place for twenty-four hours. Authorize the Skins to da anyone found on the streets after that time. Close down a vehicle traffic, and limit domestic datapool access to ente tainment and official calls only. Any acts of resistance or a gression against us are to be suppressed by lethal force. O collateral necklace will be activated per incident."

"Very well. But we'll probably have trouble getting peop back to work afterward. We might not be able to meet o asset quota."

"Irrelevant; I've already written it off. Now, I want to s the prisoners."

The interviews followed a pattern. Initial superficial de ance, which swiftly faded when the interviewee realized ju how serious Simon was. He began to learn what had ha pened.

Josep Raichura and Raymond Jang had been taken on the start of the last season—popular guys, who never lacke for female company. Management couldn't explain wh there were no files. With their cooperation Simon's person AS swiftly tracked down the substitutions. Implantation the ghosts was little short of miraculous: they had birth ce tificates, school grades, parents (who had similar digital hi tories), bank accounts, credit coin bills, medical records, ta records, insurance policies, apartment rent agreements. The were more real to the datapool and the AS than half of th people belonging to Memu Bay's underclass.

Verbal interrogation confirmed Josep and Ray had left th company around the time Z-B's starships had arrived. N one could remember the exact date. It had been a confuse time.

They'd not been seen since. Nobody had managed to con tact them.

The instructors who were friendly with them believed the

were from out of town. One of the hinterland settlements. Definitely not local, though.

Someone thought they lived in a suburb close to the Nium estuary. They certainly had a female housemate; several instructors had hit on her in the marina bars. Possibly called Denise. The AS immediately began generating an image of her from their descriptions.

"Find the house," Simon ordered. "I want every estate around that estuary visited by Skins. Physically verify the occupants of each house, apartment and hole in the ground. I want a complete verbal history of occupancy for the last five years, which they're to cross-reference with the AS."

The curfew had been in place for two hours when fifteen platoons began the door-to-door search. So far the proclamation of martial law had proved remarkably successful. Memu Bay's inhabitants had realized from Zhang's announcement that he wasn't bluffing. Most people started heading home by four o'clock. The fact that the roads really did shut down at six caught a few motorists out. The traffic regulator AS disabled every vehicle other than bicycles. Drivers hurried home on foot. Several were darted by Skins. Die-hard protesters outside the Town Hall and various Skin barracks were darted without warning at one second past six.

Simon and Braddock received a call about a possible suspect location at eleven-fifteen and immediately took a helicopter out to the Nium estuary estate. It was a bungalow rented from a property agency. Nobody answered the door when the Skin had rung the bell. When he asked a neighbor he was told that a girl called Denise lived in it by herself; her two male housemates had left several weeks ago. None of that corresponded with the information that the AS trawled out from the datapool concerning the bungalow.

Five Skins were standing guard in the garden when Simon arrived along with a small team of Z-B technicians. A further three Skins were inside. Simon and Braddock gave the bungalow a quick inspection. Someone had abandoned their breakfast. A dish of cereal and a mug of coffee were left on the table in the kitchen. Two slices of toast stood in a stainless steel rack, untouched.

Braddock sniffed at the coffee mug and pulled back fast, his nose wrinkling in disgust. "Several days old, I'd say."

"We'll go for an expert opinion." Simon told one of the technicians to analyze the food to see if he could determine how long it had been standing. "It would have been early morning here when Josep was captured at the spaceport," he mused as the technician took a sample of the semisolid cereal.

Another technician was examining the bedrooms and bathroom for skin and hair samples.

The very frightened neighbor said she thought Denise worked at a school. No, she didn't know which one, but it could be a playschool.

"I want every head teacher in the city brought in," Simon instructed Zhang. "Right now."

"I've got a DNA match," the technician reported. "A skin sample from one of the disused bedrooms belongs to Josep."

"Excellent," Simon purred. It was coming together beautifully. Of all the challenges, puzzles and pursuits he'd been involved in over the years, nothing had given him greater satisfaction than this. Some small part of his mind was childishly excited by the prospect of encountering an alien, even though that encounter would bring enormous upheaval, possibly even war, given the alien's recent actions. That made him pause. Interstellar war was impossible, surely? If commerce was impractical, then invasion and conquest must surely be out of the question. Then why was the alien so hostile?

He knew the answer was close. If the facts could just be put together in the right order . . .

Mrs. Potchansky was the nineteenth head teacher to be

brought before Simon. It was half-past-three in the morning, and he'd resorted to far too much strong coffee. The caffeine was slowly abrading his temper, and contributing to a subtle depression. It was one thing to be the butt of smartmouth insults; but he could actually sense the naked thoughts of each teacher, know how much he was genuinely despised and hated. That could wound a man's soul.

"Does Denise work for you?" Simon asked as the old woman stood in front of him.

"I don't know any Denise." It was a perfect schoolmistress voice, instantly instilling a sense of complete inferiority in any listener. She was one of the few teachers to arrive fully dressed. Simon imagined even the Skins would be made to wait until she had chosen appropriate clothes and put them on in her own time.

"Ah," he murmured contentedly. He tented his fingers and rested his chin on the apex. A pane on his desk lit up to show the image that the AS had generated from the descriptions of the lovelorn diving instructors. "Is this her?"

"If I don't know her, then I can hardly identify her, can I?"

"But you did know her. And it's what you think that is interesting."

Mrs. Potchansky's face remained perfectly composed. Alarm shivered through her mind.

"Did you know what she was connected with, which resistance movement?" His DNI was scrolling the woman's file.

"If this farce is over I'll be going home now. I trust you'll take me back with the same alacrity with which I was brought here."

"Sit down!" Simon barked.

Mrs. Potchansky fussed around with the chair, deliberately taking her time. Her thoughts were settling into a steely determination.

"When did you last see her?" Simon asked.

"You know this person's name, yet you're not sure wha she looks like. That's very odd."

"Very. Especially if you were to check your school' records, because she's not in any of your files. Nor is she i any file we can trawl out of the datapool."

"That must make it difficult for you to persecute her."

"When did she leave? Please."

"No."

"Very well, you're free to go. I'll have a car take yo home."

Mrs. Potchansky gave him a suspicious look. "Why?"

"Because you're obviously a tough old lady who isn' going to tell me anything."

"Why?"

"After the car drops you off it'll pick up someone who wil be more cooperative." The indigo file scrolled down acros his view of Mrs. Potchansky. He picked a name. "Jedzella perhaps."

"What a pathetically crude attempt at blackmail. You'll do no such thing."

"We killed your son. I expect you think of us as barbarian who answer to no one on this world. You would be correct ir that; I'm not even accountable to anyone on Earth. And I am desperate to find this girl. Truly desperate. The children will tell me who she is and where she came from. Do you want to put them through that? Because I will ask them if you make me."

"I haven't seen her since the weekend," Mrs. Potchansky said.

"Thank you. Now tell me all about her."

The huge Pan-Skyways cargo jet taxied slowly through the miserable gray rain that was saturating Durrell Spaceport. It turned onto the parking apron and braked to a halt. Steam shimmered off the nacelles as the spinning fans wound down to a stop.

A robot tractor nuzzled up to the front-wheel bogie and engaged its locking clamps. It began to tow the plane into a nearby hangar. The doors slid shut behind it, and it stood all alone in the enclosed space, dripping on the concrete floor. Pan-Skyways hangar staff brought a pair of airstairs up to the cabin hatch, and the two flight crew emerged. They were followed by Lawrence Newton in his full sergeant's uniform. He paused on the top step, conscious of the cameras dotted round the hangar. Z-B required any asset cargo flown by a civilian airline to be accompanied by a company representative. The AS would be checking his face and matching it with his file and the assignment orders issued by Ebrey Zhang's office.

Colin Schmidt waited for him at the bottom of the airstairs, a small smile playing over his face. "Welcome to Durrell."

Lawrence put his arm around his old friend's shoulders. "Good to be here."

They walked along the side of the fuselage to the rear of the plane. "I thought you were joking when you called," Colin said. "A whole RL-thirty-three pod. Holy crap! This I have to see for myself."

The clamshell doors that made up the aircraft's tail section were hinging apart. Colin ducked under them to stand in the widening gap. The cargo pod filled half of the cavernous fuselage, a long pearl-white composite cylinder resting on a cradle.

"I guess you weren't joking," Colin said. He looked around to make sure none of the civilian hangar staff were nearby, then lowered his voice. "Okay, what the hell is in it?"

Lawrence opened the flap on his breast pocket and took out a gnarled lump of rock. It glinted dully in the hangar's lightcones. Colin took it from him, examining the lump gingerly.

"It's argentite," Lawrence said. "That's a silver mineral."

"Silver," Colin said. He looked from the little chunk in his hand to the cargo pod, then back to Lawrence. "You are joking."

"No. What we have here is about forty tons of argentite with a very high silver content."

"Where in God's name did it come from?"

"Out in the hinterlands. I thought I saw it last time I was here. Nobody else in the platoon recognized it, so I kept quiet."

"Shit." Colin was laughing, hand over his mouth. "You old fraud, Lawrence, you told me we'd need to smuggle a backpack up to orbit."

"If I'd said a fully loaded Xianti you would never have agreed. Now you have an incentive. Can you get this into orbit?"

"Yes." Colin was still laughing. "Oh, God, yes. Forty tons of silver! Lawrence, you are goddamn unbelievable."

"Forty tons of silver mineral. We'll have to refine it when we get back to Earth."

Colin nodded, suddenly sober. "Of course. I'll have to make sure it's flown down to Cairns; after that we can get it offbase easily enough. But, Lawrence, I don't know how you get this mineral stuff refined. What do we need?"

"One step at a time. Let's focus on getting it up to the *Koribu* for now, shall we? Have you got me my pilot?"

"Yes, yes, Gordon Dreyer, he's our man. Needs money, and smart enough to keep his mouth shut about it afterward."

"Fine. What about getting the cargo pod through security? Its contents are on file as fusion reactor parts. It can't go through any sort of scanner."

"I can handle that. There are hundreds of these identical cargo pods coming in and out of the spaceport. It's just playıg the shell game, but on a much bigger scale, that's all. I've ot the verification codes, I can enter a full clearance file. 'he AS will never know the difference."

"Like going over the fence again, huh?"

"Like going over the fence." Colin was staring at the pod gain, his expression hungry. "Damn it, Lawrence, I already now the house I'm going to buy with this. I saw it once, on ıe Riviera, it was a white stone mansion with gardens over hundred and fifty years old. Fit for a Board member."

Lawrence felt a surprisingly large twinge of guilt as he lisːned to his friend's daydream. But the choice had been made ack when he was dreaming the dragon's dreams. All his old ɔyalties were over now.

Gordon Dreyer arrived six hours before his flight was cheduled to take off. Lawrence hadn't met him before, but e knew the type well enough: late forties with a highly seure job that sounded glamorous but was actually routine, nd wasn't going to take him anywhere careerwise. Two ıarriages behind him, Colin said, and the courts diverted a ig slice of his salary to pay for them. He was bitter about the ılings. He partied hard. Drank a little too much. Gambled bove his credit limit.

In the flesh, Dreyer's weight was pushing the upper limit ermitted by Z-B's fitness requirement. His dark hair was ıeticulously cut and styled to hide its thinness—v-writing ɔllicle treatments were currently beyond his means. He hook Lawrence's hand firmly enough, and played it cool vhen the deal was explained. Only the eagerness with which e accepted betrayed his true character.

As with all pilots, Dreyer shadowed the flight preparation. t started with a review of the Xianti's flight-engineering file, nsuring its performance was up to the specified levels and ıat standard maintenance had been carried out. His authoization was added to the flightworthiness log, which would

allow the spaceplane to be moved to the next stage: carg
loading.

Gordon Dreyer went to inspect the cargo pod in the pr
flight integration hangar. Colin Schmidt was the lieutenant i
charge of the logistics that morning, meeting all the pilo
who were readying their craft for the daily flight up to th
starships. They walked along the row of sealed cargo pod
discussing any problems or special requirements. At the en
he presented them with the security verification file, detai
ing the inspection process for each item of cargo. Drey
added his authorization to the file and thanked Colin fo
doing his job.

The RL33 pod was loaded into the Xianti, which was the
towed out to its fueling bay. Gordon Dreyer went off to th
pilot's locker room to get ready while the cryogenic tank
were being chilled down, then filled with liquid hydrogen.

Lawrence and Colin rode over to the makeshift medic
center in the terminal building.

"The hospital's been off-limits since the blast," Colin ex
plained. "They're taking care of some senior officers fro
fleet intelligence in there. Security won't allow anyone els
near the place."

They found an empty room and began sticking medic
modules to Lawrence's torso. His arm was then covered by
dermal membrane sheath, with more modules stuck over it

"I wish you didn't look so healthy," Colin complaine
"You're supposed to be a priority medevac case."

"I heard that in old wars soldiers used to eat the gunpow
der from their bullets. It made them look really sick."

"You want some energized explosive to chew on?"

"No, thanks." He pulled on a medical division covera
With its short sleeves, everyone could see the membrane an
small modules. It should convince the ground crew who sa
him embarking. Prime entered his record files in Memu Ba

nplanting an attack during an urban patrol, which had urned through his Skin, leaving him unfit for duty.

The fueling bay had a small operations center with a rank f darkened glass windows that looked out over the big delta-haped Xianti. Stairs in one corner of the center led down to ıe covered bridge, which had extended out to the spaceplane abin's airlock.

Gordon Dreyer was already in the center when Lawrence nd Colin entered. He was talking to a security officer, who anded him the flight's communication key.

"Do you need any help with that arm?" Dreyer asked.

"No, sir," Lawrence said. "I can manage, thank you."

A camera was fixed to the top of the bridge entrance. .awrence could feel the sweat on his forehead as he walked nderneath. At least it added credibility to his supposed in-ury. Dreyer was impressively calm as they walked along the ridge.

The cabin hatch slid shut and Lawrence let out a sharp reath of relief. Sneaking around like this wasn't his arena. *Give me head-to-head combat any day.*

"Home free, eh?" Dreyer said. "Sit yourself down, and eave the rest to me."

Lawrence chose a seat directly behind the pilot, where he ould see the console displays. Dreyer was absorbed by the inal checklist. Three minutes later he agreed with the space-lane's AS pilot that they were ready to lift off. The Rolls-Royce turbojets came alive with a resonant thrumming, as nuch felt as heard, and they rolled out of the fueling bay. The light to orbit was identical to every other Lawrence had been n, though it was interesting to see the console displays and ave a genuine view out through the narrow windshield ather than a camera image on a seatback screen.

"Eighty minutes to rendezvous," Dreyer announced as the wo tail rockets finished their injection burn.

"Sounds good." Lawrence picked one of the medical mod-

ules off his arm, leaned forward and pressed it to Dreyer'
neck.

"What arr—" The pilot lost consciousness. His body r
mained in the seat, held by the safety straps, but his arm
gradually floated up until they were hanging above the cor
sole.

Lawrence used his d-written neural cluster to establish
link with the Xianti's network. Prime went active and erase
the AS pilot program, assuming complete control of th
spaceplane.

"Are you all right back there?" Lawrence asked.

"I never knew freefall was this awful," Denise replie
from her hidden nest in the cargo pod. "I think I'm going
be sick."

"Try not to be, try very hard."

"Any more advice you want to give?"

"Let's get you out of there, I need to suit up." Prime re
layed a camera image of the payload bay to one of the fligh
console panes. The cargo pod almost filled it, leaving only
two-meter gap between itself and the cabin bulkhea
Lawrence saw a circle of plastic peel back on the end of th
pod. Something moved inside. A human figure in a silve
gray leotard of a spacesuit crawled out with very slow, un
certain movements.

"Nothing moves right," Denise complained.

Lawrence hoped she wasn't linked to a cabin camer
she'd see him grinning. "You'll get used to it. Just rememb
inertia is still the same up here."

A short, flexible tether clipped to her harness attached he
to the fat box containing his Skin. Once she was out of th
pod and anchored in the short gap, she began to pull it o
after her. Lawrence told Prime to open the outer hatch of th
payload bay airlock. It took Denise several minutes to ma
neuver the box inside. There wasn't enough room for her a

vell, so Lawrence cycled the airlock and pulled it out into the abin while she waited in the payload bay.

He already had his legs in the Skin when she emerged and ugged her face mask off. "I shouldn't have eaten," she roaned. "I shouldn't have drunk, either."

"Would you have managed your original scenario in that ondition?"

She glared at him. "I'd have done it. I still can."

"Yeah. Well, let's go for the nonlethal option first."

Aemu Bay's entire complement of twelve TVL88 heli-opters flew across the plateau just as dawn arrived. Simon vatched the landscape skim past from the cockpit of the lead raft. Stationary whorls of cloud surrounded each of the eaks, leaking streams of mist down through the foothill val-eys from where they gushed out across the plains and orests. The scene was primordial, with trees and ridges ticking out of the eerie white mantle.

"Satellite's coming over again," the SK2 said over the link rom Durrell. "There's not much available in the visible pectrum. That damn fog's covering the entire province."

Simon told his AS to show him the satellite imagery on his nirrorshades. A few forested hills slid across the display, eparated by the placid lakes of mist. Infrared cut in, giving way little. Several dozen fuzzy pink patches shimmered nder the white surface. They were roughly where Arnoon illage ought to be.

During the night it had been raining over the plateau. The atellite had been unable to penetrate the thick, dark clouds. Simon had called up old images, studying the little commu-ity. All he'd seen was a standard rustic settlement with ardly any sign of high technology other than its cybernetic voolen mills.

His AS had begun trawling the datapool for all availab
information on Arnoon Province. There was a lot of it, but s
far nothing relevant. When it sent askpings out to the vi
lage's few nodes it found nothing but standard domesti
management pearls linked in, some of them generations o
of date.

All perfectly normal.

However: the Dixon network had dropped out of the dat
pool three days ago. Memu Bay's telecom utility compan
couldn't explain why. They hadn't sent an engineering tea
out to the plateau yet; the civil situation had pushed it wa
down their priority list.

And there was a lost patrol up there somewhere. It had le
three days ago. At first Simon was delighted when his A
found the reference, thinking he could simply send them d
rectly to Arnoon. But their transponders didn't respond to th
communications satellite. The AS noted the patrol wa
scheduled to last two days. Yet no one had noticed when the
didn't return. Further investigation revealed a major data di
continuity in the headquarters AS. It had issued the assig
ment, but had no associated progress monitors. There wasn
even an established command hierarchy. They'd been sub
verted.

When Simon called in Captain Bryant to ask him what h
knew of his missing platoon, the befuddled officer hadn
known what he was talking about. Platoon 435NK9 had bee
reassigned out of his command.

"How can you misplace an entire platoon?" a disguste
Simon had asked Braddock.

A group of conical mounds crept into view ahead of th
helicopters. The mist was patchy here, finally starting to di
sipate as the sun rose higher.

"Dixon's straight ahead, sir," the pilot called over th
whoop of the rotor blades. Simon canceled the mirrorshade
display.

The TVL88 squadron cleared the slag heaps. They slowed as they skirted the little town, probing the whole area with active sensors.

"What in God's name happened?" the SK2 asked.

The mist had almost cleared, revealing the devastated buildings. Nearly a quarter of the houses were gone. They'd all exploded, scattering debris over a wide area.

"Some kind of battle," Simon told his clone sibling. "Those buildings were all deliberately targeted. I can't think why."

"Sir!" The pilot was pointing ahead.

"Take us over," Simon said.

There was a burned-out jeep in the middle of the main street. Another jeep was embedded in the side of one of the few intact buildings remaining around the town square.

"At least we know what happened to the platoon now," Simon said as the helicopter circled around them. There was no sign of any Skin suit in either of the wrecks. "Okay, I've seen enough," he told the pilot. "Get us over to Arnoon."

Pain was a constant now, squeezing every part of his body. Simon refused to let the doctor administer the drugs that would banish it, keeping his mind sharp. He was sure the SF9 simply didn't appreciate the enormity of the alien encounter, continuing to treat it like some fascinating intellectual puzzle. Typical of that batch's imperturbable poise.

Simon had perceived Josep's aura firsthand, experienced his determination and resolve. The only way they were ever going to survive this encounter was if they matched the alien's drive. He couldn't allow the chance to slip their grasp. The potential of the nanonic system was staggering. In Zantiu-Braun's possession it could be used to elevate the entire human race.

Despite Josep's being an enemy, Simon envied what he had become. His enhanced form was a magnificent ideal for humans to aspire to, wonderfully superior to anything germline v-writing promised.

Few moments in history were truly pivotal. But this was going to be one of them. Simon had to take part, to contribute, to disallow failure—particularly through weakness. Acquiring the nanonic system had to be made to happen. Fortunately, his immobility didn't prevent datapool access. And the pain, constant, persecuting, diabolical pain drove him onward.

His DNI scrolled down file after file as the SF9 flew on toward Arnoon, information thrown up by his AS as it hunted for oversights and mistakes. Somewhere below the knees his legs were itching abominably, adding to his suffering and anger. Finally the clues he knew to exist began emerging from the datapool. "You were wrong about the patrol," he said.

"What do you mean?" the SF9 asked.

"We don't know what happened to them."

"We just saw the remnants," the SF9 chided. "The alien or its allies wiped them out because they were on their way to Arnoon."

"And then used Prime to cover it up, to erase the platoon from our data systems."

"Yes."

"But the cover-up was in place before the platoon left. Somebody arranged it so that Four-three-five-NK-nine could visit the plateau without anyone knowing what they were doing. If the alien wanted to stop any of our people from visiting Arnoon Province, it could simply use Prime to change their orders. We'd never know."

"What are you suggesting?"

"There's another factor here." A new file appeared, highlighted by the cross-reference program that the AS had run on Platoon 435NK9. Specific information scrolled down. "It would seem that the platoon's sergeant has been to Arnoon

ovince before. He was in a similar patrol the last time we
re here. Are you going to tell me that's a coincidence?"
"It's improbable," the SF9 admitted. "Can you datamine
m?"
Simon instructed his personal AS to launch an askping
wl for all files concerning Lawrence Newton.

e TVL88s thundered in over the treetops to surround
noon village, weapons extended. Downwash from their
werful rotors tore at the mist, breaking through the central
earing in seconds. The last strands of cloying vapor
eaked past the shaggy wooden A-frames, exposing them to
e targeting sensors. A young woman in a cream sweater and
rk jeans was standing on the balcony of one of the houses,
ipping the handrail to steady herself in the miniature hurri-
ne.
She was the only person the sensors could detect. The
-frames were all warm, their domestic appliances drawing
wer. But nobody was inside.
Five helicopters, including Simon's, landed on the dew-
aked grassmoss, while the others spread out and began
anning the surrounding forest. Skins deployed rapidly, fan-
ng out across the meadow. Their carbine muzzles were ex-
nded; each of them had a rack of smart missiles.
Simon climbed down out of the helicopter, holding on to
e front of his loose leather jacket as it flapped about. Three
kins fell in around him as he walked toward the woman.
She came down the steps from the balcony, her lustrous
ra giving her the appearance of some biblical angel.
imon Roderick, I presume. I'm Jacintha. Welcome to
noon village."
"I thought there'd be more people here."

"They're all out there in the forest somewhere. They r away when we found out you were coming."

"Why?"

"We're frightened of you."

"Interesting. I find you quite daunting. You know you ha a remarkable aura."

Jacintha frowned. "Oh, I understand. You must have magnetic sense. Is that how you caught Josep?"

"Let's say it's how I learned to be very careful around hi Not that it was of much help ultimately. A lot of people we killed when he committed suicide."

"And your collateral necklaces kill a lot of people for reason."

"I'm not here to justify what I've done, nor argue with y about who owns the moral high ground. I'd simply like meet the alien, please."

"I'm sorry," Jacintha said. "You can't."

"You know I will. If you defeat all twelve helicopters a these platoons—which I doubt you're actually capable of we will simply come back with more. And we will keep co ing back until we finally get through to it."

It wasn't her rather pitying smile that disconcerted hi but her thoughts. She actually felt sorry for him. It was t kind of sympathy an adult would express during an infan tantrum.

In return he couldn't help but admire her. It was nothi sexual, rather an appreciation for a perfectly balanced pe sonality. The SK2 was right: if only everybody had her inte lectual depth.

"You could send a thousand starships full of Skins a weapons," Jacintha said. "It would make no difference."

Finally, Simon began to understand. "It's not here." H mind began to meld all the information that he'd gather with a speed that was almost vertiginous. "Memu Bay is anarchistic mess; you can take anything through without nowing. The spaceplane! You weren't going to blow up a tarship . . ."

"Newton was here," the SK2 said. "Here at the spaceport. Ve medevaced him this morning."

Jacintha cocked her head to one side, listening to a silent oice.

"Shit!" Simon gasped, as his DNI scrolled the files. "Stop im," he told the SK2. "Stop the flight. Keep Newton away rom the starship."

"Too late," Jacintha said.

CHAPTER EIGHTEEN

"I'M SHOWING A LEVEL-TWO HYDRAULICS FAILURE," LAWRENCE eported. Prime converted his voice to an exact replica of Gordon Dreyer's clipped accent for the audio link with *Koribu*. "The payload bay doors aren't responding."

"God, Dreyer, can't you people stick to simple maintenance procedures?" the *Koribu*'s flight controller complained. "You're supposed to oversee flightworthiness. There's no point to having pilots otherwise. Purge and reactivate the system."

"Copy that. Attempting reactivation."

Amber graphics began a slow dance on the console panes as Prime produced a digital simulacrum of the hydraulics system being reactivated. Lawrence let the phony procedure run twice so that the telemetry being received by the *Koribu* would show he was doing his best to rectify the problem.

Through the windshield he could see the massive starshi
floating 350 meters away. They were level with the fusio
drive section, where sunlight broke apart into soft scintilla
tions over the crinkled thermal foil that protected the deu
terium tanks. Three more Xiantis were strung out in front c
them, their payload bay doors fully open. The cargo pods tha
they'd boosted up to orbit had risen out on cradles ready fc
collection, as if they were some kind of offering held out b
metal fingers. Engineering shuttles like black chrome beetle
were sliding round the spaceplanes, puffs of dusty gray ga
flaring out of their reaction control nozzles as they aligne
themselves to pluck the pods away.

"Still no response," Lawrence said.

"Ah, goddamnit, all right, Dreyer," the *Koribu*'s fligh
controller said. "Clearing you for docking. Bring it in to ou
maintenance bay. The AS is assigning you an approach path
And congratulations for screwing up today's schedule."

"Always a pleasure."

Prime acknowledged receipt of the new flight path. Hy
pergolic fuel ignited in the reaction control nozzles, gentl
pushing them around the starship. Lawrence saw ribbons o
sulfur vapor flare out to envelop the entire nose as the Xiant
began a slow roll. The starship gradually slipped from view
through the windshield. Sensors showed him the *Koribu*'
cylindrical cargo section drifting past below. Beyond th
silos, the long maintenance bay doors were opening up. A
row of small lights lining the rim came on, banishing shad
ows from the ribbed metal cavity.

With Prime controlling their maneuver, the Xianti glide
smoothly into place directly above the maintenance bay. It
undercarriage doors folded back. The reaction control noz
zles fired shorter and shorter bursts as they eliminated al
momentum relative to the giant starship.

On the maintenance bay floor, mechanical mandible
flexed themselves upward, searching out the load pins in th

Xianti's undercarriage. Latches snapped shut, securing the spaceplane.

"We're in," Lawrence murmured. The mandibles were retracting, pulling the spaceplane down. They both watched the rim lights slide up past the windshield.

Denise turned to the images from the spaceplane cameras. "Where are the umbilicals?"

"Just wait," Lawrence said.

The Xianti trembled slightly as they came to rest in the cradles. Secondary mandibles, coiled by tubes and cables, wormed their way up to nuzzle at the spacecraft's umbilical sockets. Power, data, coolant, communications and hydraulics were all connected and confirmed.

Prime used the datalink to load itself in the maintenance bay network, erasing the AS and establishing control over all the local systems. Subversion on such a massive level was immediately detected by the *Koribu*'s principal AS, which threw up a firewall around the affected network. It also cut power and environmental feeds to the section of the starship around the maintenance bay and closed the first set of emergency pressure doors along the main axial corridor. The section's backup power supply cut in automatically, allowing the network and most ancillary systems to function. There was nothing Prime could do about recovering environmental feeds, although there was enough oxygen to sustain the crew trapped behind the sealed pressure doors.

The maintenance bay airlock tube telescoped out of the wall toward the Xianti's cabin hatch. Lawrence held an e-c pistol in one hand. His carbine was already extended. "Stay behind me," he told Denise as the hatch rim locks engaged.

"Yes, Commander."

Her tone irked him. "We've been over this. That suit of yours is good, but it can't take as much punishment as Skin. And I know they have weapons on board."

"Yes, all right," Denise grumbled.

The cabin hatch slid open, revealing the twenty-mete length of airlock tube on the other side. It was dark, with orange strobes blinking at the far end. Prime supplied Lawrence's tactical grid with camera images from every part o the starship it had gained control over. The crewmen in th cargo section were confused. They knew the environmenta systems were off; amber warning strobes were flashing i every compartment. All of the refuge chamber doors an lifeboat hatches had swung open. Lighting had gone to ful power-save reduction mode, dimming the corridors and narrow crawlways to a near-claustrophobic level. There was n personal communication with the rest of the ship—Prim was blocking that. Yet the AS seemed to be telling them everything was fine, and this was just a localized glitch.

Lawrence kicked off and glided cleanly down the middl of the tunnel, controlling his flight by occasionally flippin his free hand on the tube wall. Denise followed after him bouncing her way along with a running commentary o curses.

He had to use the manual hatch release at the other end Two crewmen were floating just behind. They saw the Ski float out and flipped gracefully in midair, shooting away lik frightened fish. Lawrence darted both of them. They kept o going for several meters before colliding with the compartment walls with a heavy impact. Then they were spinnin flaccidly, limbs protruding in all directions.

Lawrence pushed past them and dived into the long corridor leading out of the compartment. It had a D-shaped cros section, with a ladder running along the curve's apex. H slapped at the rungs, propelling himself along. Denise was a couple of meters behind him.

The axial corridor was at the end, a broad cylinder wit bulky environment ducts running down the walls. It ran th entire length of the starship at the center of the stress structure, linking the rear fusion drive section to the forward com pression drive, with radial corridors connecting to every other pressurized section. Emergency pressure doors were positioned at forty-meter intervals along it, big reinforced composite circles that were normally kept open.

Five crewmen were drifting around the closest one to Lawrence as his helmet rose through the hatchway into the axial corridor. Two of them were trying to get it to open, while one was pressed against the viewport in the center trying to see into the next section. Lawrence lifted his wrist: he could feel the tiny wriggling of the dispenser mechanism's muscles as the darts spat out.

Denise crawled her way along to the emergency door and pressed a ring of energy focus ribbon against it. Lawrence was pushing the unconscious crewmen away.

"Get clear," Denise told him. She sent a code to the ribbon. The pulse of raw energy it emitted sliced clean through the door. Thick black smoke jetted out as the edges of the composite sizzled and flamed. Fire alarms went off.

Lawrence gripped the door's handle and kicked down on the burning circle. It flew out, turning over and over like a flipped coin, with the noxious vapor swirling behind it. Several crewmen who were clustered round the other side of the door took flight.

For a second Lawrence could see down the length of the *Koribu* toward the forward compression drive section. Then all the other emergency pressure doors were closing. Amber strobes came to life and fire sirens wailed along the entire corridor. Crewmen were diving away down radial corridors. He managed to dart three of them before they all vanished. Secondary pressure doors began closing off the radial corridors as he and Denise glided over them.

There was a network node ten meters from the emergency pressure door. Denise used a power blade to slice through the casing and carefully positioned a dragon-extruded communications link on top of the databus unit. Microfilaments slid through the electronics inside to merge with the fiberopti cables. Prime loaded into another section's network.

The audio alarm brought Captain Marquis Krojen instantl awake. The volume was like the scream of explosive de compression. He sat up fast, the strap around his waist pre venting him from soaring completely off the bunk in the lov gravity. For a moment he looked around in confusion as hi cabin lights came on. Starships had different alarm sound for every conceivable type of emergency. After so man decades flying them, Marquis could have sworn he knev every one by heart. But this time he actually had to wait fo his DNI to scroll the information.

"Intruder alert?" He simply couldn't believe the neat in digo symbols.

The alarm fell silent.

"Yes, sir," the ship's AS confirmed.

"Jesus Christ, this has got to be a drill." Somethin dreamed up by that bastard Roderick after all the trouble a the Durrell Spaceport. It couldn't be real.

"No, sir," the AS insisted. "I have been erased from th maintenance bay hangar section network. Firewalls are i place and holding against the subversion software."

Marquis tore at the Velcro on his waist strap. He wen through his main cabin into the bridge, moving fast in th one-eighth gravity. Colin Jeffries, the executive officer, wa in the command chair, looking thoroughly shocked. Onl three other bridge consoles were manned.

"What the hell happened?" Marquis Krojen made an effor to calm down. "Give me a situation review."

"A Xianti reported a hydraulics failure," Colin Jeffrie said. "We docked it in the maintenance bay, and the nex thing we know the whole surrounding network had been sub verted."

"What's our response?" Marquis sat in one of the unuse

onsole seats. The ship's AS activated the panes, showing a ange of schematics and camera images.

"Standard response is to withdraw power and environ-nental support from the contaminated section," the AS said. That has been done."

"Can you get me a real-time visual image of the space-lane?"

"No."

"Divert an engineering shuttle to the maintenance bay, ow," Marquis told Colin Jeffries. "I want to see what's hap-ening."

"Aye, sir."

"Durrell Spaceport security is online," the AS reported. They are warning us about the spaceplane. They believe it as been taken over by a Thallspring resistance movement."

Marquis Krojen refused to let the shocking information anic him into hasty action. The AS had brought up a physi-al threat procedure on one of the panes. If there was a valid omb threat against the *Koribu,* the captain was to order all ands to abandon ship. Security determined that any resis-ance group that had gotten within striking range would have bomb capable of destroying the entire starship.

But it hadn't gone off yet. And if they were going to nuke he *Koribu,* why were they busy trying to subvert it?

"Could our engineering shuttles just rip the Xianti out of here?" Marquis Krojen asked.

Colin Jeffries shook his head doubtfully. "I don't think so. Those shuttles don't have much thrust, and the hold-down atches are designed with a lot more inertia than a loaded Xi-anti in mind. You'd have to get underneath it and cut through hem."

"Work on it. I need options."

"Aye, sir."

"Do we have any contact with any crewmen in the affected

section?" Marquis asked the AS. He just couldn't bring hir
self to say "contaminated."

"No, sir," the AS said. "There are no internal communic
tion links open."

"Very well, I want someone physically looking throug
the viewport in the emergency pressure door. Give them a
open link to the bridge."

"Yes, sir."

"Overflight coming up," Colin Jeffries called.

The AS routed the engineering shuttle's sensor imagery t
the panes on Marquis Krojen's console. He looked down o
the big pearl-white delta shape, not quite knowing what t
expect. It appeared ridiculously impassive. Then his min
ran through docking procedures.

"Did we activate the airlock tunnel?" he asked.

"No, sir," the AS replied. "It was connected after the sub
version occurred."

Marquis Krojen looked directly at Colin Jeffries. "They'r
inside, then. Jesus! Does Durrell Spaceport security actuall
know what's in there?"

An excited voice burst out of a console speaker. "Sir, I ca
see somebody moving into the axial corridor."

"Who is this?" Marquis Krojen asked.

"Irwin Watson, sir, fusion engineer."

"Okay, who can you see, Watson?"

"Sir, it's a Skin."

A Skin? Marquis mouthed at Colin Jeffries. The executiv
officer shrugged.

"What's he doing?" Marquis asked. One of the consol
panes showed him Watson and several others clustere
around the axial corridor's pressure door.

"Sir, he's killing people, shooting them!" Watson's voic
had risen to near hysteria.

"What sort of weapon is he using?"

"I don't know. He's got some kind of pistol, but I didn'

e it fire. Hey, there's another person through there with im. They're wearing some kind of spacesuit, I think. He's utting something on the door."

"Get back, now," Marquis ordered.

"I can't see what it is."

The camera showed Watson pressing his face against the ressure door viewport.

"Get away from the door. That is an order."

Watson moved back reluctantly, gripping the rungs along e axial corridor. A brilliant white light stabbed out from the ressure door. It vanished as dirty black smoke poured out; reamers churned along the corridor walls like a fast-moving il slick. A disk of flaming composite suddenly tumbled out f the smoke, narrowly missing Watson.

"Secure that section," Marquis Krojen ordered the AS. "I ant physical isolation."

"Affirmative," the AS replied. "Closing emergency pressure doors along the axial corridor."

"Captain." Simon Roderick's face had appeared on one of e console panes. Just his face, against a neutral gray background.

"What have you let up here?" Marquis demanded. He idn't care about etiquette now. His ship was suffering.

"We believe there is an alien on board the spaceplane," imon Roderick said.

"What?"

"An alien," Roderick said imperturbably. "It has human llies who will probably try to hijack the *Koribu.*"

"Over my dead body." Marquis watched a camera image f the axial corridor. The Skin and his spacesuited companon were through the emergency pressure door. They stopped here there was some kind of access panel on the corridor all, and the spacesuited figure took out a power blade.

"Let's hope it doesn't come to that," Roderick said.

"The intruders have exposed a network node," the AS

said. "Subversion software is loading directly into the loc neurotronic pearls. It is reconfiguring their processing pa terns."

"Stop it," Marquis said.

"I am unable to comply. Network data management ro tines have been corrupted. Firewalls established. Power a environmental support withdrawn."

"Holy Jesus." Marquis studied the starship's prima schematic. They'd lost all contact with the rear third of t *Koribu,* which now lay beyond the firewalls and clos emergency pressure doors. "What can this alien do?"

"I'm not sure," Roderick replied. "But it has technolo well in advance of ours. You might not be able to stop them

"Break out the weapons," Marquis ordered. "I want o crewmen armed and authorized to shoot."

"We've got ten carbines and some dart pistols," Colin Je fries said. "They'll just bounce off Skin."

"But maybe not the other one."

"I am detecting venting from the isolated cargo sections the AS reported.

"Venting what?" an aghast Marquis asked. The pan shifted to views from external cameras. Huge plumes of gl tering silver vapor were fountaining out of the starship's re sections.

"Spectrographic analysis indicates it is our atmosphere the AS said.

The doctor refused to cooperate at first. Simon didn't act ally threaten him, but he came close before the man's mo basic survival instinct cut in.

"I really don't recommend this," the doctor said. He wa helping two orderlies push Simon's trolley and three cabine of intensive-care support equipment through the spacepo terminal building. "You're not stable enough for somethi as traumatic as a spaceplane flight yet. Please reconsider."

"No," Simon grunted. He could hear his Skin escort shout-
g at people to get out of the way. Protests and hurried
raping sounds. Trivial background details he ignored.

An optronic membrane was covering his remaining eye,
owing him camera images from the *Koribu* and the space-
anes around it. Gas was still venting from the fat barrel of
cargo section. There must have been twenty of the plumes,
nerging from hatches and valves distributed among the
los. His communication link to the starship buzzed with
nfused, shouted orders and queries. Crewmen were strug-
ing into spacesuits, collecting weapons from the executive
ficer. As countermeasures went, it was truly pitiful.

The starship's AS was completely ineffectual against the
ien's Prime program. If Newton and the other (presumably
enhanced villager) kept going along the axial corridor and
ysically loaded it into every section, they would soon have
mplete control. His personal AS now considered this was
eir most likely strategy. The most uncomplicated and effi-
ent way of hijacking a starship, with a frighteningly high
ojected success level.

Simon saw a small silver sphere fly out from the cargo
ction.

"What was that?" he asked.

"Lifeboat," Marquis Krojen said. "There's very little air
ft back there. My crew is having to abandon the contami-
ated area."

Simon's trolley wheels bumped over a small ridge on the
oor. He moaned at the sharp flare of pain that the jolt in-
icted.

"Sorry," the doctor said. He didn't sound it.

"Can the engineering shuttles close down the venting?"
imon asked.

"Some of them, possibly. But there's not enough time."

Several of the plumes were shrinking, becoming less ener-
etic.

The trolley was pushed into an elevator. Simon's magne sense showed him almost a dozen people clustered arou him as the doors slid shut.

"Damn," Marquis Krojen exclaimed. "They just blew a other pressure door. That puts them above the first life su port wheel."

"Where are your people?" Simon demanded.

"I'm putting a squad together. We're not trained for th not fighting a Skin."

"Learn fast." Simon saw another two lifeboats shoot aw from the *Koribu*'s cargo section.

"The subversion software's loading," Marquis Kroj said. "We're losing another section."

"Can they take over the life support wheel?"

"Not directly. The AS inside will firewall the wheel. B controlling the axial corridor gives them the power and en ronment feeds to the wheel."

The elevator halted and the doors opened. Simon's troll was wheeled out into the fueling bay's operations center.

The SF9 opened a communication link. "What do y think you're doing?" he asked.

The orderlies started pushing Simon's trolley across t walkway to the waiting Xianti.

"I'm going up to the *Norvelle,*" Simon told his clone si ling. "I'll assume command of our response operation fro there."

"Don't be ridiculous. You're in no condition to assur command of anything."

"I'm here, you're not. It would take you hours to get into orbit. That could well be too late."

"We're already too late for you to achieve anything there. It's all down to Captain Krojen now."

"Which makes it even more important that I reach orbit soon as possible. The *Koribu* is a disaster area. The capta has all but lost his ship to the alien." He cried out again as t rolley was lifted through the airlock. "It cannot get away rom us," he gasped. "I won't allow it. We must have that echnology. If they go FTL, I'll follow. I will bring it back for s. The whole world will be elevated."

"It will not. Wait while I try to negotiate a deal with the illagers."

"I know very well what the villagers will do to us."

"That's not—"

Simon cut the link. For good measure he used his codes to uthorize Memu Bay's immediate and total isolation from he datapool; then he shut down the satellite links as well. With luck, it should keep his clone sibling out of contact with Z-B for several hours.

Deprived of environmental systems, the axial corridor was hick with smoke, its amber warning strobes casting weird imbi around the walls. Lawrence's Skin sensors could cut hrough most of the crud clotting the air. He was keeping lert for crewmen who might appear from radial corridors he thought were clear. So far they hadn't run into any physcal opposition at all.

As they moved along the axial corridor, Prime had taken over the surrounding sections one by one, venting the atmosphere out into space, and he hoped, forcing the crew to abandon ship. Sensors had shown eight lifeboats being aunched from around the cargo section so far. Internal cameras had pinpointed seven crewmen remaining: three were waiting in a lifeboat, two were inside refuge chambers, while another two had put on spacesuits and were trying to get back nto the axial corridor.

"How are you doing?" Lawrence asked the dragon.

"Admirably, thank you. I have established full access and authority to the *Koribu*'s cargo and fusion drive sections. Prime is now installed in all its management electronics. I would have preferred a much larger bandwidth than the

spaceplane umbilical provides. This starship does have
remarkable number of components. I cannot operate all o
them simultaneously."

"What about our weapons?"

"Yes. I am in charge of several missile launchers, laser
and electron beam cannon. Sensor coverage is not yet com
plete. The majority are positioned around the forward sec
tions. Targeting information is incomplete at this time."

"But if you see something coming, you'll be able to take
shot at it?"

"I will."

"Okay. You should have access to the forward section
soon."

Denise loaded Prime into a node. "We've got control o
this section."

Lawrence studied the schematic that Prime was providing
For the first time, the *Koribu*'s AS hadn't cut the main powe
grid around them. It was supplying the life support wheels.

The secondary pressure doors at the top of each wheel ha
been shut. Prime's control of systems extended only down t
the giant magnetic bearings that were wrapped around th
central stress structure. Data access to the wheels themselve
had been firewalled.

"You keep going," Lawrence told Denise. "Establish
datalink to the compression drive. I'll deal with the crew."

He ordered Prime to halt the rotation of the life suppor
wheels. The axial corridor began to creak loudly as the bear
ings changed their magnetic fields to act as a brake on the
momentum of the tremendous wheels. The wall juddered an
vibrated as the stress structure tried to absorb the extraordi
nary torque forces leaking through the bearings. In theory
the life support wheels counterbalanced each other. It wa
fine when they were running smoothly, but there was enough
inertia wound up in each one to wrench the starship apart i
the forces weren't perfectly matched. Now the stress struc

e was taking the full brunt of minute errors in the braking ocedure.

Prime opened the pressure door on one of the radial corrirs, and Lawrence dropped through into the rotating trans- toroid. He placed an energy focus ribbon on the top of the xt pressure door and burned through into the life support eel.

ptain Marquis Krojen instinctively grabbed at his console a shudder ran through the bridge. The invaders must be aking the life support wheels. He didn't want to think hat that would be doing to the central stress structure.

"Can we use our reserve power to maintain the bearings?" he ked the AS.

"No, sir."

Every question, every countermove he came up with reived that same bland answer.

"They're in wheel one," Colin Jeffries reported. "We just st contact."

Captain Marquis Krojen clenched his teeth to stop himself vearing. The bridge crew had been using secondary transitters to provide communications between the wheels and e spaceplanes outside. Now the schematic showed wheel ne as a black outline.

Another shudder rocked the bridge. This time it was acompanied by a metallic creaking sound. He couldn't wait ny longer. "Okay, get our squad up into the axial corridor."

"Aye, sir," Colin Jeffries acknowledged grimly. He issued e order.

The squad consisted of bridge officers who'd drawn the arbines and found themselves a few laser welding tools. According to the AS, none of them stood a chance of damaging Skin. The idea was to use the firefight to lure the Skin into hub compartment that they'd wired into the reserve power upply. The voltage they could push through him might be

sufficient to disable the Skin suit, or possibly even kill t
man inside.

If the Skin chased them.

If they didn't all get killed in the first few seconds.

"Wheel one is venting," Colin Jeffries called. "That so
ware has blown the escape hatches and used the fire dur
nozzles."

A pane showed Marquis Krojen the life support whe
with precious air gushing out of its rim to spew across t
stars.

"Spacesuits, everybody," the captain ordered bitter
"You have my authority to abandon ship if your life suppo
wheel loses pressure." He began to pull his own suit on,
task made difficult by the falling gravity. Optronic me
branes showed him the squad opening a pressure door up in
the rotating transfer toroid.

"Moving through," the lieutenant commanding the squ
reported. "Nothing in the transfer toroid. Opening radial co
ridor pressure door."

Gravity on the bridge had almost vanished. At least th
meant the shaking had stopped. Marquis Krojen used a Ve
cro patch to secure his helmet on the console next to hi
keeping it within reach. "Isn't it locked?"

"No, sir. Going through. A lot of smoke in here. Can't se
much."

"Pull back," Marquis said. "He knows you're there, h
software will be tracking you."

"I can see someone." The muffled sound of carbine fi
came out of the speaker.

"Pull back."

"Yes, sir." The telemetry display from the squad began
waver. "Suit . . . can't . . . Malfunction."

"They're firing!" another squad member cried. "Down."

"Back!"

"There!"

"Shot. Shot me. Oh fuck, I'm hit."

The lieutenant screamed.

"Can't breathe."

All the squad telemetry vanished. "Subversion alert," the AS said.

"In here?" Marquis asked hurriedly.

"An attempt was made through the communications link," the AS said. "I have disengaged the life support wheel's internal nodes from the network."

"So we've lost contact with the squad?"

"Yes, sir."

"How many casualties?"

"I am uncertain. Their spacesuit electronics were being subverted. Telemetry after they entered the transfer toroid is unreliable."

Marquis was looking at the camera image showing him the open pressure door. Black smoke was oozing through it, hazing the compartment. He could see the fire alarm strobes flashing brightly. "Have any of them made it back into the wheel?"

"No, sir."

"Communication with wheel two lost," Colin reported.

"Turn off your spacesuit communications," Marquis ordered. "Don't let it get inside." He glanced at the camera pictures. Vigorous geysers of atmosphere were spraying out of wheel two. It was like watching a friend bleed to death.

Weary sadness replaced the anger that had carried him this far.

"Abandon ship."

"Sir?" Colin Jeffries said.

The remaining bridge crew were staring at him.

"There's nothing else we can do. And I'm not leaving any of my people hostage to these bastards. Use the lifeboats, get clear. The spaceplanes will pick you up."

"What about you?"

"The captain stays with his ship. You know that."

"Then I'm staying as well."

"Colin, please—" Pane displays began to fracture into ran dom chunks of color, then went dead. The steady backgroun whine of the environmental fans faded away. All the ligh went off. Marquis snatched up the helmet and jammed down on his collar. Shaking fingers engaged the seal. H gripped the arms on his chair just as the air turned into howling hurricane. Paper, plastic cups, food trays, electron ics, foaming water and even clothing streamed past him caught in stop-motion flight by the blazing scarlet strob light that was intent on warning the bridge of a decompres sion. A T-shirt wrapped itself around his helmet, flapping fu riously. He didn't dare let go of the chair to push it off. Th wild torrent would have carried him with it. He tried rockin from side to side, and eventually the fabric slipped away.

Air was pouring through an open hatchway. He could ac tually see moisture vapor scoring thin contrails, marking ou the flow. Any unsecured item had been sucked through When he pictured the wheel's layout he remembered there was an escape hatch three compartments along. The subver sion software must have fired the explosive bolts.

It took several minutes for all the atmosphere to clea When the gusts and roaring had shrunk away, the red de compression strobe was still flashing. It had been joined by green strobes that had come on around the lifeboat hatch tha had opened in the floor. With his communication circuit off Marquis could hear nothing. He switched on his spacesui helmet light, then pushed out of his chair. Colin and the othe crew were doing the same. He beckoned Colin over, and they touched helmets.

"Take the lifeboat," Marquis shouted. "Get these peopl out."

"You must come with us." Colin's voice was like a muf fled buzzing.

"No. I ordered the squad to intercept that Skin. I'm going to find out what happened to them. They're my responsibility."

"Good luck, sir."

The first of the bridge crew glided into the lifeboat. Marquis Krojen left the bridge. Normally the life support wheel seemed cramped and confined; in freefall it was a lot larger. Red, amber and green strobes flashed around him as he slid through the empty, airless corridors. He passed three lifeboat hatches that were shut. Blue indicator lights showed that the little craft had ejected safely.

It killed him to see his magnificent starship in this state. Environmental ducts had ruptured during the decompression, shattering dozens of plastic panels. Thick blue-green coolant fluid dribbled out of torn tubes, creating small constellations of globules that fizzed energetically as they evaporated in the vacuum. Loose debris that hadn't been sucked out formed its own baleful nebula in each compartment and cabin. It was mainly composed of clothes and crumpled food trays, though there were also cushions, fragments of plastic paneling, chairs, mashed-up pot plants, even a pedal frame from one of the gyms. Now that the air had gone it floated idly, drifting out of open hatchways to clutter the corridor. He glided around the obstacles or flicked them aside. Water was boiling furiously out of a split pipe, filling a long section of the corridor with thick white mist.

Even if they did recover the ship, he knew Z-B would never spend the money it would require for a complete refit. His *Koribu* was doomed, one way or another.

The spoke lift shaft that led up to the wheel's hub was almost clear, allowing him to move a lot faster. When he reached the first hub compartment the pressure door hinged shut behind him. The strobes went off, and the normal lighting returned. Several of the panels flickered, betraying just how much damage had been wrought by the decompression;

internal systems were designed to operate in a vacuum. Re
fusing to be intimidated by the subversion software's activ
ity, he moved into the hub's annular corridor and continue
toward the transfer toroid.

The pressure door was closed. He pushed against it, know
ing how futile that gesture was. Dense white gas suddenl
burst out of an environment duct grille with a silent rush.

"Jesus," he muttered inside the helmet. The software mus
be preparing the wheel for the invaders. He flipped aroun
quickly and kicked off. There was one hub compartment th
was safer for him than the others. He zipped back along th
annular corridor. Every grille was blowing out a column c
air. The booby-trapped compartment was directly ahea
Pressure was already back up to half a standard atmospher

"Careful. It could be dangerous in there."

Marquis gripped the hatch rim to halt his flight and slowl
looked round. The Skin was floating lazily along the annula
corridor behind him.

"Somebody wired up the whole place to the backup powe
supply," the Skin said; his voice was tinny in the thin air.

Marquis turned his communication circuit back on. "It wa
wired up on my orders."

"A reasonable idea for a noncombatant."

"What do you want?"

"What I have, Captain: your starship."

"Why? At least tell me that."

"We're taking it on a trip."

"I doubt it. You've succeeded in virtually wrecking it."

"This is just superficial damage to the life support sec
tions. The drives are intact. That's all we need."

"Where are you going?"

"To the alien's home star. You're welcome to come with u
if you like. You've spent your life in space. I suspect yo
haven't entirely lost your fascination with the unknown, eve
if it's been diverted by Zantiu-Braun."

The offer did cause Marquis Krojen to hesitate, but duty was a lot stronger than old dreams. "My only concern right now is for the safety of my crew. Did you kill the squad I sent up here?"

"A blunt question, Captain. But, no, they're not dead, although a couple of them are injured. We subverted their spacesuits and turned off the air. They had to take their helmets off. I darted them."

"I see."

"Well, there's gratitude. Ah. Here we go."

The lights dimmed again. Marquis realized something was diverting power from the tokamaks. "The compression drive," he said in surprise.

"I did say it was intact. We'll be using it as soon as the alien can raise the tokamaks up to full power and bring the energy inverter online. In the meantime, I want you to help me shove the remaining crewmen into lifeboats. If we don't, then they'll be coming with us, and this ship is not going to come back."

Simon had blacked out when the scramjet came on. Acceleration had heightened the pain to an unbelievable agony before his body's beleaguered natural defenses snatched him away. When he recovered he was in freefall, with the intensive-care equipment emitting urgent bleeping sounds. Indigo symbols and script slowly crept into focus. There was no data available from the *Koribu*. He told his AS to give him the orbital tactical plots, and the sensor readings from the starships and satellites. "Good God." It was every bit as bad as he expected.

"Please, don't try to move," the doctor said quickly. "You're all right."

"I'd better be," Simon snapped at him. The crisp circles of

the tactical plot showed him forty-eight lifeboats were slow
receding from the *Koribu*. The Xiantis had rendezvouse
with a few, but they didn't have the cabin space to accon
modate all the crewmen sheltering inside. Two of th
spaceplanes had simply loaded the lifeboats into their carg
bays and de-orbited, carrying them down to Durrell. The re
mainder of the lifeboats were waiting for instructions: shoul
they remain in orbit for rendezvous and rescue, or shoul
they fire their retro rockets and land on the planet—if so,
what location? Simon couldn't care less. He reduced the tac
tical display back into the main display grid and expande
the *Koribu*'s sensor scans. Vast and powerful magnetic flu
lines were expanding out from the compression drive sectio
as the tokamaks powered up. The starship was preparing t
go FTL.

He told his AS to establish a link to Sebastian Manet, th
Norvelle's captain.

"Can you disable the *Koribu*?" Simon asked. According t
his tactical plot, the starships were only eight thousand kil
meters apart.

"We should be able to saturate its defenses between the si
of us," Sebastian Manet said. "I would like to wait until it'
farther away from the lifeboats and spaceplanes. They coul
be damaged by the defense missiles, or the *Koribu*'s detona
tion."

"I don't want it detonated. I want it disabled."

"We don't have that kind of capability."

"Why can't you use the kinetic weapons? Target the com
pression drive section."

"The *Koribu* will simply use its nukes against them: noth
ing can get through that kind of defensive bombardment."

"We must have more kinetic missiles than they hav
nuclear explosives."

"We do. But Captain Krojen was in the last lifeboat to es
cape. He's confirmed the goal of the hijackers. They're abou

ve minutes away from going FTL. We'd need to fire at least ghteen salvos to soak up their defensive capability and be rtain of achieving a hit; that would take forty-five minutes an hour. We won't even be able to get the first salvo to em in time."

"Use the gamma soak on them."

"That projector takes about fifteen minutes to deploy."

Simon let out an infuriated grunt.

"Please," the doctor implored. "You must remain calm. I'll ave to sedate you otherwise."

"You even come close to me again and I'll have you rown out of the airlock," Simon told him. "Track it," he or-ered Sebastian Manet. "I want to know where they're oing. And begin the power-up sequence for *Norvelle*'s com-ession drive."

"Are you serious?"

"Yes. I'll dock with you in another seventeen minutes. As oon as I'm on board, we're following them."

imon had gone to sit in the helicopter's cabin to receive the ash of data that his personal AS had profiled for him. In-ormation on the *Koribu* hijack attempt was sparse, but sen-or data from the starship showed him the intruders rogressing down the axial corridor, loading Prime as they ent. Atmosphere began venting from the cargo section. ifeboats ejected.

And there was nothing he could do about it. His only op-on was to order one of the other starships to open fire with uclear weapons. Not only would that wipe out the entire rew and obliterate a multitrillion-dollar starship along with e nearby spaceplanes, it would also kill the alien. Nothing ould be gained, and everything would be lost. Besides, he wasn't entirely sure the other captains would obey such order.

A change-of-status icon blinked up within the DNI's di play grid. His personal AS always kept him updated on s curity matters concerning his clone siblings no matter wh crisis he was dealing with. He expanded the icon and read t script that came scrolling out with a growing sense of di may.

"What do you think you're doing?" he asked the SK2.

The response and short argument that ensued complete validated his consternation. There were precedents for se tling arguments among Roderick clone siblings, but he did know of any situation when one of them had been acting such an unstable manner. In his current state the SK2 prob bly wouldn't accept any kind of ruling that restricted h authority.

Then the question became irrelevant when the SK2 cut t link. "Damn him!" Annoyance turned to fury when his pe sonal AS told him that Memu Bay had been isolated from t global datapool. A second later the AV88's communicati circuit lost the satellite link. Simon tried to re-establish co tact through the other helicopters parked on Arnoon's centr meadow. There was no response from the satellite. He us his bracelet pearl. It could detect the satellite's beacon, b there was no contact.

Not only had the SK2 commandeered the mission, he' also isolated a clone sibling in a hostile area. That invalidate his authority entirely. Simon shook his head wearily. *Assun ing my Board brothers ever find out.* Legitimacy and poli cal maneuvering weren't exactly his primary concern rig now.

Jacintha was sitting at a long wooden table inside th snowbark pavilion. She looked completely relaxed as thre Skins stood guard a discreet distance away.

"You have a magnetic sense, and you're a clone," she sai

Simon walked up to her. "How fascinating. Life on Earth obviously a little more complex than we thought."

He indicated the bench on the other side of the table. May I?"

"Please."

"Did you hear all that?"

"Loud and clear, thank you."

"Whatever viewpoint you have, the outcome is not good. Iy clone sibling is . . . unwell."

"I think crazy is the word I'd use."

"He's traumatized, and still in a considerable amount of ain, which is affecting his judgment. He was inside Josep's last radius."

"Is that supposed to make me feel guilty?"

"I'm illustrating cause and effect."

"You invaded our planet. This is the result."

"I refuse to shoulder the entire blame. Your actions have onsequences, too. Neither of us has emerged from this con-ontation with much credit."

"No," Jacintha admitted with reluctance. "But we do have starship. And the alien will be returned home."

"I hope that its society is well armed. My clone sibling ill not stop until he has obtained their nanonic technology."

"The dragons don't need armaments. And any threats he akes against them will be completely ineffectual."

"Dragons?" Simon recalled the elaborate carvings he'd limpsed on the A-frames.

"Our name for them," she said.

"I see. Well, just knowing where these dragons live will ive him a dangerous victory. If he doesn't obtain nanonics n this flight, he will return there. Are you so certain that hu-ans will never obtain the information? If not by force, then y trade or diplomacy. After all, the dragon allowed you to ave it." He could see the uncertainty creep into her mind. "If at possibility exists, you have to help me."

"Help you do what?"

"Help me to ensure it isn't my clone sibling who acquir it first."

"No."

"Why not?"

"I believe that nanonic systems should be introduced the human race, but only on an equal-access basis. That's o of the main reasons we've been so cautious. If you or yo clone snatches it first, it would be misapplied. You know would."

"Anything that is used in a way you don't personally agr with is by definition misapplied. That's why human cultu evolved the way it did, so that the majority can influence f ture development. Everyone has a voice—a small one, a mittedly, but a voice nonetheless. Or do you mistrust th entire human race?"

"Please don't try to twist this. You personally, Zantir Braun as a whole, would misapply the technology. Yo would treat it as a monopoly to increase your own wealth an influence, and very likely your military strength as well."

"Of course we'll apply it to our advantage. But you don know what our goals are. I should say my goals, for I in a my hundreds of individual selves am the one who originall formulated our policy and ensure that it's carried out."

"All right, I'm curious, what goals? To invade and conqu more planets?"

"No. Asset realization is not sustainable in the long term or even the medium term. Today's starflights are the igno minious end to a noble dream that is slowly winding down its natural conclusion."

"The noble idea being?"

"Giving you what you have. A clean start on a fresh worl It's a desire that's hardwired into many humans. It come from our impetuosity and curiosity, the wanderlust gene. B it also has roots in the dissatisfaction with the society

hich we live. How much easier it is to move and start anew
an to rectify the institutional, even constitutional, mistakes
f a monolithic social system. Between them, those motiva-
rs were enough to launch the first wave of colonies. It was
lways going to be financially nonviable; the compression
rive technology just isn't capable of supporting the dream.
ut still we went ahead. There are a lot of successes, worlds
ke Ducain, Amethi and Larone: all independent and pros-
erous stakeholder democracies. We even have a host of
emisuccesses like Thallspring, in debt on Earth but fully
elf-sustaining. Personally, I actually rate Santa Chico a con-
iderable success—albeit in its own unique fashion."

"If we're a success, then stop holding us back. Let us de-
elop freely. Use that power and influence you have to stop
e asset-realization missions."

"I know our invasion dominates your thinking, and I'm
rry. But the necessary changes have to be made at a more
indamental level. We have to elevate the whole human race
order to be free of the restrictions they impose."

"Elevate them?"

"Yes. Earth with its seven billion population is the wealth-
est human world. After all, with that many people working
an industrial society, it couldn't be anything else. But it
lso has the greatest level of poverty. There are some city dis-
icts where the inhabitants are in their twentieth generation
f penury. They simply never get out, unlike your ancestors,
ho were smart and determined enough to get here. Schools
nd the datapool offer huge opportunities to learn, to enable
em to work their way out of the slums and integrate them-
elves with the primary economy. And they never do. For
very one that gets out, ten stay behind and have families,
sually large ones. Drug addiction is rife, crime impover-
hes them further; they suffer bad housing, bad parenting,
ad social care, decaying infrastructure, casual violence. It
ist goes on and on."

"I do understand the principles of the poverty cycle."

"You should; it's beginning to happen here. I've seen t secondary economy starting to creep in. You have an em gent underclass. At the moment they're only slightly adı from mainstream life on Thallspring. Soon, in another f generations, the divide will be unbridgeable. Thallspring w be a replica of Earth."

"No, it won't."

"Ah." He smiled. "Yes. You believe the dragon technolo will help bring your world together and allow you to bu something new and decent."

"Yes," Jacintha said. "If it's introduced gently, the kind changes we envisage will be massively beneficial."

"How remarkable. And enviable. With that kind of outlo I could offer you a seat on our Board. I—we—also want see societal change, not further pointless expansion that fc ever repeats past mistakes. But for that change to be total, has to come from the heart of human society: Earth. We' been attempting that for over a century now. The poor, tl underclass, have got to be eliminated. And I'm not speaki from pure altruism. I'm actually being quite selfish. Th prey on our compassion; they absorb billions in welfare pa ments simply so they can eat and be housed; they use up st more billions in medical care, for inevitably they are the se tor of society that is the most disease-prone and in gener bad health. It is they who cause today's dreams and visio to fail. If we didn't have them to take care of, our starshi would still be venturing out farther into the galaxy a founding colonies. We would have the time and resources explore new forms of living. All of us, not just you and San Chico."

"You speak of the poor as if they're subhuman."

"It depends what you mean by human."

"I'm not sure you are."

"Oh, but I am, because I care. We've devoted Z-B to co

erting whole communities to a rational economic pattern
ırough stakeholding. The Regressors and deglobalizers
ıeer at the whole concept, naturally; they call it corporate
ictatorship. But governments and local politicians are des-
erate for us to develop their impoverished regions and re-
italize them. Even our corporate rivals have followed our
ıitiative. Between us we've reintroduced the concept of
›bs for life that had almost been wiped out in the twenty-
rst century when technological evolution and innovation
'as so fast-paced that it was delivering machines and prod-
cts that were obsolete before they even reached the mar-
etplace. Today, we have a slower technological evolution,
nd economic instabilities have been reduced accordingly.
'he wealth we bring with our investment means our stake-
olders can afford almost every benefit that modern civi-
zation is capable of providing. And the one thing we
lways provide, no matter how small your stake, is full
amily healthcare. Germline v-writing is available to every-
ody."

"What kind of v-writing?" Jacintha asked. She made no at-
empt to hide her concern at the concept.

"Whatever the parents want," Simon said. "Invariably, the
ind of children born into all this middle-class affluence tend
ɔ be stronger and healthier, and to live longer. They're also
marter. Again, a natural desire for your child to succeed and
e happy: preload the dice to give it the best possible chance
n life."

"That's your goal? Increase the average IQ of the human
ace?"

"Yes. Essentially, we're breeding the underclass into ex-
inction. Once we get into one of these poverty zones the first
hing we do is give it improved healthcare. After that, the
ıext generation does finally take proper advantage of uni-
ersal schooling; they can see that there's a world outside the
ghetto that's worth taking part in. From that they progress to

earning a living, they contribute to the whole rather than de
tract. Right now there are fewer welfare dependents, manua
laborers, petty criminals and social outcasts than there hav
been for two hundred years. Fewer people the state has t
care for. Fewer people who drain the vitality out of th
human spirit.

"If we can finally instigate global stakeholding it'll mea
an end to poverty, the end of visionaries being restrained b
mundanes. Companies like Zantiu-Braun will be able t
begin programs of real expansion. We can build a whole nev
era of interstellar commonwealth, where ideas and concept
are traded between stars."

"That all sounds very . . . I don't know. Fascist?"

"We don't impose any of this. There's no gun to the head
We simply provide a choice and let human nature do the rest
Besides, you used nanonic systems to enhance yourself.
suspect you have germline v-writing in your ancestry a
well."

"I don't deny it. But that doesn't qualify you for my hel
in this situation."

"It ought to. Despite all my clone siblings and I have done
entire nations are still mired in their old ways. Even our mos
optimistic estimates had put global stakeholding anothe
three or four generations away.

"Now you've discovered an alien with the potential t
bring about stakeholding's consummation in a few shor
years. All the wasters and the ignorant I hold in contempt
and my clone sibling so despises, could be elevated in one
clean sweep, made a gift of the intelligence they so profoundly lack. If you thought what we've done to Thallspring
was an invasion, what would you call that? You said yo
planned to spread the dragon's knowledge gently, so people
would have time to understand it and assimilate it. Suppose
you didn't? Suppose you had a vision that required imple-

ntation rather than free choice? And you had the means to orce that implementation?"

'You can't force an entire population to enhance themves," she said, mortified.

'I know that. But my clone sibling is more driven than I. ss forgiving. As far as he's concerned, if you have the ans, why wait? A working nanonic system will give him t means. And if he follows the *Koribu* he will have abute exclusivity. So you tell me, how big a threat does he se to the human race? Is it possible to modify a grown ult human for higher IQ?"

"Yes. Neural cells are essentially no different from any ier. The patternform sequencer molecules can restructure m."

"Then you now have a choice. The dragon's nanonic techlogy will be introduced to Earth at some time. Do you nt him to be the one who delivers it, or me?"

She gave a bitter, brittle laugh. "What's the difference?"

"Look at me," he said. When she stared directly at him, he d, "I am the moderate voice. I will not force it on people. ill not allow it to be forced on people. It will be subject to lemocratic process, whether stakeholding or classical. But atever the outcome, change will come; that is always the nsequence of new knowledge. How it comes is now up to u. Today you and I are opponents because of circumstance. not let that color your judgment of me."

"What, exactly, are you asking for?"

"I want to know where they are both going, where the agon's homeworld is. I want Prime to break through this mmunications block so I can divert a spaceplane to Memu y airport and take me directly up to a starship. I have to go er them. I have to prevent my clone sibling from being the e who acquires this knowledge."

Lawrence and Denise spent most of the first week worki
on repairs and removing junk, aided by a small squadron
Prime-managed robots. Life support wheels one and t
were slowly spun up again, providing the *Koribu* with a b
anced precession. One, they ignored completely. Two, th
attempted to repressurize. It took them three days just to
cure pressure bulkheads. Open doors had been badly da
aged by the explosive decompression. Hinges had twist
Rim seals were ruptured. Debris clogged the rails. Pow
and data conduits had been shredded by flying fragmen
Each of the doors had to be examined for damage and son
how secured in place. The escape hatches that had be
blown into space were patched with metal or compos
sheets epoxied into place. Eventually, their little habital
domain expanded to cover a quarter of the wheel, with
bridge in the center. One spoke was also pressurized, allo
ing them up to the hub without needing to suit up. Not th
they used the axial corridor much. If anything did malfur
tion in the compression drive, it would be the robots th
performed the repair.

With the pressure restored, they repaired the air filtrati
and scrubber units, replaced fan motors, cleaned out the he
exchanger and mended pipes. Replenishing the oxygen a
nitrogen was no problem. The *Koribu*'s reserve tanks cou
resupply enough atmosphere to support twenty thousa
people for two months. Now they just had to sustain two pe
ple for 104 days. Water was equally abundant. So much
they never even considered fixing the recycler and purifi
mechanism.

Koribu's food stock was made up entirely of sterile me
packs, food that was produced entirely devoid of bacteria
prevent it from decaying. They had enough to last a thousa
years. Denise hated it. "There's no taste," she complained t
first day. They were back in the spaceplane, taking a bre

vhile the robots finished welding and insulating cryogenic ipes along the spoke.

Lawrence checked her pack. She'd chosen Chateaubriand teak with béarnaise sauce. The hydration valve was preset, o she couldn't have used the wrong amount of water to sat- rate it before she put it in the microwave slot. "It'll just be reefall pooling," he told her. "Fluid buildup in your head lays hell with your taste receptors. Try squirting some more alt solution into it."

"It's not just the taste, it's the texture, too." She pulled an rray of mealpacks out of a box, sending them whirling cross the little cabin to bounce off the walls. "Look at these. Each one a different food, and all with exactly the same con- istency. It's like lukewarm mashed potato in twenty colors."

"Right. Sorry about that." *Only another 103 days of this to o.*

When the wheel section was finally repressurized, awrence stood in the bridge compartment and cautiously nsealed his Skin. He sniffed at the cool air. "Sweet Fate, no roblem with freefall pooling here."

Denise took her face mask off and grimaced. "What did nat?"

"Let's go find out."

They never did track the stench down to a single source. Coolant fluid that had frozen was now sloshing about, slowly vaporating. The waste recycler was a big culprit, which they olved by closing the valves and having the robots spray the vhole mechanism in foam sealant. Food scraps that the crew vere eating had partially boiled in the vacuum before freez- ng; now they were truly rotten. Lawrence also suspected ro- ents and insects, decomposing away behind the wall and eiling paneling.

All of it had to be cleared away: the fluids mopped up, iodegradable items taken through an airlock and dumped in

nearby compartments that were still in a vacuum. It ke
them busy for a while.

Lawrence claimed the captain's small suite of rooms. H
took out every article of Marquis Krojen's clothes, all th
personal items, erasing his identity. Then he went through th
other cabins in search of clothes that fit. A lot had bee
sucked out into space, but there were enough to last a coupl
of months before he had to start thinking about washin
them. Denise moved into a cabin on the other side of th
bridge.

After the first week Lawrence began reviewing the mult
media library. He didn't have much else to do. Prime and th
robots were perfectly capable of maintaining the few piece
of environmental equipment necessary to keep their sectio
of the wheel functional. He had reactivated a sustainer cab
net for his Skin. Not that he expected to wear it again. Th
kept chugging along quietly without his intervention. Th
compression drive was operating efficiently, as were its toka
maks. There was no navigation required. No daily inspectio
of the ship. And no view.

At first he started choosing music to play, racking the vol
ume up loud. It was kind of eerie, two people alone in a shi
built for over twenty thousand. The music went partway t
filling the emptiness for him while he exercised away in th
gym to keep his body in trim. Then he and Denise started ar
guing about the tracks he played. He refused to let it get ou
of hand. He'd acquired plenty of experience with grudge
building amid small groups in confined quarters; she with he
rustic upbringing had no idea about the compromises tha
had to be made. So after that she chose half of them, and h
kept quiet about her taste.

Even spending three or four hours a day working out lef
him with a lot of time to kill. He went back to the library an
began accessing the i's. It was something he hadn't reall
done since leaving Amethi. At first he went for the comedies

w and classic, but there's only so long you can keep laugh-
g at situations that have no real bearing on your own life.
fter that he immersed himself in action adventures, finally
ving up on them when they became idiotic and repetitive.
ramas were generally too harrowing. He guessed that his
rrent circumstances must have heightened his emotional
ate, leaving him too susceptible to the melodramatic trau-
as that characters involved him in. Science fiction he re-
sed point-blank. Despite the huge temptation, that really
ould be premature. He would see *Flight: Horizon* again.
it not here, and not alone. So he butterflied between clas-
plays and travelogue documentaries and historical event
enactments. Though more often than not he'd delve into
e dragon's scattered memories of the Ring Empire and
her strands of galactic history, already old when Earth's di-
saurs were young.

Even though they kept to themselves during the day, he
d Denise made a point of spending mealtimes together.
hey varied the food as much as possible, although Denise
ever let up bitching about its blandness.

"You really do love her, don't you?" she asked during one
nner, about five weeks into the voyage.

Lawrence gave her a slightly guilty glance. As usual he'd
ned out her moaning about the state of the duck à la orange.
Then he followed her gaze, he saw he was rubbing the pen-
ant between his thumb and forefinger. The little hologram
niled at him below her foggy age-worn surface.

"Yeah," he said. It didn't hurt to say it to her, not now that
ere was no turning back. "I do."

"Lucky girl. How long ago was it, twenty years?"

"Just about." He gave the pendant another look, then
ropped it back inside his sweatshirt. "You know, I only kept
at first to remind me why I left home, so my anger
ouldn't fade. That's sort of shifted over the years. I keep it
ow because of what she represented. The happiest days of

my life. It took a long time to realize that nobody can hav
that effect on your life without meaning something to yo
And nobody else has ever meant that much to me, not eve
close."

Denise gave him a fond smile, slightly surprised by the a
mission. "I hope you manage to patch things up."

"I was so angry when I found out. Angry with everybod
else for being part of a universe where such things were a
lowed to happen. Which was the only way I had of expres
ing myself. It was such a shock to discover that someone yo
love has been used like that. But then we were both youn
and stupid, me and her. She was desperate to emigrate, an
that was the only way she could make it happen. And yo
know what? There's no difference between what she did an
what I did. Zantiu-Braun had my body for twenty years b
cause that's the only way I could ever hope to realize m
dream."

"You really are hung up on starflight, aren't you?"

"Absolutely. I was born on a colony. I owe my existenc
to wanderlust."

"That's so old-human. Just pressing onward for your ow
satisfaction and never seeing the consequences. I thin
Simon Roderick may have had a point."

"You are kidding." Jacintha had transmitted her entire co
versation with Roderick up to the *Koribu* before they we
FTL. Hearing Z-B's policy hadn't shocked him quite as muc
as it should have, or at least as much as it would have
month previously. After all, he had been extensively v-writte
himself, and if he had kids, he would want the best for them
just as Roderick said. Being born a part of a movement lik
this made you more sympathetic to its aims. It would prob
bly look a lot different from the outside. Terrifying, probabl
Smarter, richer, more powerful people wanted to alter you
children so that they could take part in their level of societ

ɔt yours. Lawrence wasn't sure if that was evolution or eu-
enics.

"No," Denise said. "He was right to say that all we've
een doing with colonization is re-creating new Earths for no
ason other than personal aggrandizement. They are estab-
shed as fresh territories for the wealthy so that they're not
emmed in by old problems and restrictions. But those prob-
ms and restrictions don't cease to exist back on Earth just
ecause they've left. If anything, they've been exacerbated.
ecause the type of people who leave are the ones whose en-
rgy and determination are exactly what's necessary to solve
ose problems. It's a political statement; you've given up on
e rest of the human race."

"People have always migrated in search of something bet-
r. It's a fundamental of human nature. It's even why Rod-
rick's project could ultimately succeed, because we do want
e best for our kids. People will always choose improve-
ent for themselves if they ever get that choice; they just
isagree with the definition of improvement. That's your
olitics. Colonization is a form of evolution. Minorities can
migrate to live the way they want without persecution. New
leas can flourish once they've escaped the dead hand of in-
rtia, which is what the unchanging, comfortable masses are.
lew beginnings allow human culture to move forward."

"Forward to what? Higher levels of consumerism?"

"It doesn't matter that some planets are just repeats of
arth. A few of them aren't, and that's what counts. I've been
ɔ Santa Chico. It's not a way of life I would ever choose. But
ey have. And it's incredibly different, and I respect them
or that. The portal colonies, who knows what they're build-
ig for themselves. You've found something that could help
s flourish to an astonishing degree. And it was found out
ere among the stars, beyond Earth's shriveled horizon.
inding the dragon was an accident. But coming out here to

the unknown where we can find the dragons isn't chance. It
where we want to be, it's where we belong."

"We might flourish with dragon nanonics. On the oth
hand we could just destroy ourselves. It's such a powerf
technology."

"That's been said of many new things we've built. Th
generation alive when it happens is terrified; then two gene
ations on, nobody can even understand what the fuss wa
about. I don't have religion to fall back on, I don't even be
lieve much in fate, but when it comes down to the botto
line, I do have confidence in us as a species. We'll absorb th
as we've absorbed everything else, and we'll move on
something wonderful. History's on our side."

"Not by much. Don't you see? This will give us the abi
ity to change, not just once, but continually. We won't ne
essarily even be the species you have all this confidence in

"I'm not talking solely about human history. I'm talkin
about the Ring Empire as well. They had this, and look wh
they achieved. Its cultural beauty is something we should a
pire to. That so much diversity existed is a wonderful incen
tive to what we could reach. The most magnificent societ
possible covering a quarter of the galaxy and lasting for ove
a million years."

"And where are they now?" Denise said brokenly.

"All around us. They are the dragons, remember. Th
greatest example of surviving change there could be. The
have grown in harmony with their milieu, the space aroun
red giants, and we'll grow into ours, the Earth-like plane
Maybe one day we'll move on from that and join them. W
could even be smart enough to learn from the history of th
Ring Empire and see that life can never be static."

"You're a dreamer, Lawrence. You don't deal in practica
ities. We have a Roderick chasing us who will warp you
ideals, and mine, into something wicked."

"Perhaps that is fate. Perhaps he will enslave half of th

ıman race. He'll never get everyone. He won't get you, will :? You and your genetic package will be free to build an-her new world clean away on the other side of the galaxy."

She stared at him as if he were the real alien. "And that ›esn't bother you?"

"It troubles my inherited sense of morality. But then, who e we to judge what will emerge out of that kind of forced 'olution? Why assume it will be evil? You could just wait ıd see rather than prejudge. After all, he believes he's doing ıe right thing. And even if he creates the most hideous evil, 'll never last. Evolution will turn again."

"I care because of the suffering it will inflict while it ex-ts."

"Suffering from your point of view. I told you I went to anta Chico. Someone I met there believes I suffer because I ve more than thirty years. Is she right, Denise?"

"We cannot allow him to obtain dragon technology."

"You can't. Oh, don't worry, when the time comes I'll help ɔu man the battle stations and disable the *Norvelle* if I can. ut the outcome, that doesn't bother me. I've spent the last venty years fighting for someone or other, for a reason that never knew about nor understood. It hasn't made the slight-st difference to the human race. Individuals don't control vents; we just like to think we do."

"This is different."

"To you and to him, but not to me. I've fought the only attle that mattered to me, and I won, because I'm here on ıis starship at this time. And it's taking me to the only place want to go: home."

A fortnight before they were due to reach Aldebaran, ,awrence started checking over one of the engineering shut-es. If all went well, and the dragons took back their lost, amaged kindred, it would have to be taken out of the Xi-nti's payload bay and delivered to them. So he maneuvered imself into the tiny cabin and ran through the systems and

procedures. Prime and the dragon could probably handle th
short flight, but including a human pilot would be helpful
such an unknown and hostile environment. Tanks of hype
golic fuel were purged and refilled. Power cells charged
full capacity. Robot arms tested. When everything was o
line, they flew a few simulations to familiarize him with th
handling characteristics.

"I think I'm as good as I'm going to get," Lawrence sai
after the third day. They'd already notched up eight hours
simulated flying time. "It can't be that hazardous."

"Our proximity to the photosphere will challenge the shu
tle's thermal control systems," the dragon said. "But the
will be sufficient for a short flight."

"Are you looking forward to this?"

"I'm not sure I have emotional states that equate to yours.

Lawrence studied the display panes around him as Prim
worked methodically through the powerdown list. "Do yo
have any emotional states?"

"My thought processes are not affected by external fac
tors, so it is difficult for me to judge. I certainly don't hav
the extremes of emotion that you do."

"That's an old AS argument dating right back to the Tu
ing test: knowing and experiencing are two different thing
Could you feel anger, or simply mimic it?"

"Anger would serve no purpose to me. Anger to a huma
reflects many biochemical changes within your metabolism
When you are threatened, fear and anger increase your re
flexes and to some degree your strength. It can also eliminat
higher thought processes, reducing you to creatures of in
stinct—a useful survival trait for your more primitive ances
tors to evolve. But as I am unlikely to be chased across th
savannah by a saber-toothed tiger, I do not need fear c
anger."

"What about other needs?"

"I prioritize. If threatened, I divert a proportion of my pro ssing power to produce a method of eliminating the threat. ıe greater the threat, the more problem-solving capacity I ıll contribute."

"Well, that answers one question. You must be a self-vare entity. Self-preservation is one of life's fundamentals."

"The villagers of Arnoon have a great respect for life. ıey taught me how precious it is."

"So your priorities and ethics weren't inherited?"

"Again, these concepts are derived from a cultural back-ound. There is little of mine remaining for me to draw ɔon. But the knowledge I retain of the Ring Empire and ıbsequent dragon star civilizations seems broadly compati-e with general human ethos."

Lawrence began flicking the console switches, manually cking in the powerdown. "And if you're wrong?"

"Right and wrong is dependent on cultural perspective. owever, I will be interested in assessing the knowledge I ıve lost. Once that is regained, I will of course have to eval-ıte my mental evolution."

"Do you think you'll be able to do that? Humans find it ery difficult to change their opinions and beliefs. And we ery rarely manage to look at things from a fresh perspec-ve."

"My thoughts may run parallel to yours. My way of pro-essing those thoughts does not. The ability to change is fun-amental to what I am, even in this reduced state. Whatever e encounter at Aldebaran, I trust I will be able to adapt to ."

"I hope you will, too."

"Thank you."

Lawrence watched the last schematics vanish from the anes. The shuttle was in full standby mode. He undid the adle straps and began to wriggle his body toward the hatch. Do you think the Aldebaran dragons will give Simon Rod-ick patternform technology?"

"I don't see why not. It is our nature to exchange inform tion. I know this concerns Denise."

"Me as well, though not to the same extent."

"Why?"

"First off, I know where I'm going, and whatever happe at Aldebaran doesn't affect me as directly as it will her guess that gives me a certain objectivity that she is denie And she's prejudged again, found the human race lackir This genetic package she's brought with her, it's the ultim in running away and leaving your problems behind yc Ironic, really, considering that's what she believes I' done."

"It is a noble ambition she is pursuing."

"Of course it is. She can restart Arnoon with those DN samples, and this time it will be without the rest of Tha spring to worry about. But it depends on the dragons' hel ing her, giving her the kind of information she doesn't wa to share with the Rodericks and Earth. She doesn't trust us

"How can she? She does not know you. Earth and i colonies are as alien to her as the dragons."

"I used to be like that once. I never gave anybody a se ond chance. It's a very sad way to live your life."

"Do you believe the dragons should provide patternfor technology to humans?"

"Yes, I do. Denise is convinced that because we didr create it for ourselves we won't be able to handle it pro erly, that it will be constantly misused. To me it's cor pletely irrelevant that we didn't work out every little det for ourselves."

"Why?"

"Other than pride? We know the scientific principles b hind technology. If we don't understand this particular th ory, I trust in us to learn it soon enough. There's very little v can't grasp once it's fully explained and broken down into i basic equations. But that's just the clinical analysis. From

ral point of view, consider this: when the Americans first
t a man to the Moon, there were people living in Africa
l South America and Asia who had never seen a lightbulb,
known of electricity or antibiotics. There were even Amer-
ns who didn't have running water to their houses, or an in-
or toilet. Does that mean they shouldn't have been given
cess to electricity or modern medicine, because they per-
ally didn't invent it? It might not have been their local
mmunity's knowledge, but it was human knowledge. We
n't have a clue how to build the nullvoid drive that the
ng Empire's Outbounds employed in their intergalactic
ps, but the knowledge is there, developed by sentient en-
es. Why shouldn't we have access to that? Because it's a
ortcut? Because we don't have to spend centuries of time
veloping it for ourselves? In what way will using ideas
ner than our own demean and diminish us? All knowledge
ould be cherished, not denied."

"I believe you would make an excellent dragon, Law-
nce."

A week away from Aldebaran they began to review tactics.
ime had been tracking the *Norvelle* from the moment it
nt FTL, twenty-five minutes after they had. There was a
cond starship, presumably with the other Roderick on-
ard, following another forty minutes behind that.

"He's persistent," Denise acknowledged at breakfast. Both
them were aware of the tracking data lurking in their
nds. Prime supplied it to them along with a host of other
adings from the ship's principal systems.

"We know that. What we don't know is what kind of ac-
n he'll take."

"Not much to start with," she said. "He will have to assess
nat's out there, the same as us. Which gives us a window."

"For what?"

"We use our weapons to mine his exodus point. If they

lock on and fire immediately when the *Norvelle* comes ou
FTL, he'll never know what hit him."

"*They,* okay? *They* will never know what hit *them.* Th
are over three hundred crew onboard. We are not extermin
ing them just because you have a problem with other p
ple's ideology. This is a first-contact situation, and if y
play it this way the first thing the dragons will ever see us
is blow up one of our own ships. They also might not like
way we scatter and detonate nukes across their space. So j
drop that idea. And don't forget as well that Captain Ma
has a hell of a lot more deep-space combat experience th
we do. He knows the *Norvelle*'s vulnerabilities, he'll be
his guard. They don't have to exodus where we expect. Th
could well be launching a nuclear defense salvo as they ex
dus. We can't afford to take him on in this arena."

"Nor can we just roll over and give in. Not now that we
finally here at Aldebaran."

"Yes, at Aldebaran, where you came to return the drag
to its own kind. Don't let that goal slip from you now. Lea
the Rodericks to sort out their own dispute."

"You'd sell your soul for a ticket home, wouldn't you?"

"I left my soul at home."

They stared at each other for a long time.

"All right," Denise said. "How do you suggest we han
this?"

"Talk to the dragons. Explain to them how vulnerable
society is to sudden changes of this magnitude, and ask the
to take that into account. All they have to do is wait anot
three hours and give the same information to the other st
ship."

"Suppose the *Norvelle* Roderick starts shooting?"

"Then we defend ourselves. But I don't expect he wi
We're going to be in the dragons' home system, and we ha
one of their own kind onboard. In my book, that does
make us a likely target of opportunity."

"Fine, but I'm going to keep our weapons suite at level-
ıe readiness status. If that bastard tries anything tricky I
ɔn't hesitate to use it."

"I know. But let's try not to forget what is going to happen
ter exodus. One way or another the human race will alter
d diverge. It's important to me at least that those fresh
ırts aren't built on bloodshed."

n the last day inside the compression drive wormhole
enise woke early. She hadn't been this wired since the day
e invasion fleet arrived over Thallspring. Today was what
ıe'd dedicated almost all of her life to, and it wasn't hap-
ening the way they'd expected. So much time and prepara-
ɔn had been spent planning how to get the Arnoon dragon
ıck here. Problems were supposed to be eliminated on
hallspring, giving her a clean run, not follow her here.

She was still tempted to scatter the weapons after exodus.
ut Lawrence was right, damn him. Killing people without
arning wasn't the right way to go about this. Such a thing
ent against every dream she cherished for her own fresh
art.

The whole notion was such a wonderful coda to the whole
oject of returning the dragon. Another human colony at
ɔme unimaginably vast distance across the galaxy. One that
ıd full patternform technology to sustain it. She was going
bring the children up in a world where the old human ills
f competition and jealousy had no part. A star where there
ould never be any danger of cultural contamination from
arth and its colonies, old and new. Just in case. Just in case
ıe rest of the human race erased itself from galactic history.
ıst in case the people of Thallspring didn't blithely accept
rnoon's gift of knowledge, and turned it to less than benev-

olent ends. Just in case Earth obtained dragon knowledge
misuse. Which was now going to happen.

The concept of a New Arnoon depended on the drago
She needed their patternform technology. Their starsh
drive. Their information about stars and habitable plan
across the galaxy. She had expected to spend months, if r
years, at Aldebaran, learning new wonders, helping t
Arnoon dragon grow and develop into its adult form. N
she might have only ninety minutes.

Yes, scattering the weapons in an attack formation was e
tremely tempting.

Instead she had a shower and dressed in clean clothes,
Z-B-issue sweatshirt and trousers belonging to some sm
crewman. With sleeves and trouser legs rolled up, she we
through the bridge into the small senior officer lounge th
she and Lawrence used as their canteen. He'd had his la
night snack again. As usual there were a couple of plas
cups on the central table, rings of tacky tea on the surfac
Doughnuts and remnants of doughnuts on and around
plate—he only ever seemed to eat the bits with jam on the
A media card was showing the end of some play on a ba
stage, the actors frozen as they took the curtain call.

If nothing else, the voyage reminded her of the time in t
bungalow with Josep and Raymond. She told Prime to cle
the mess and started rummaging through mealpacks as t
lounge's domestic robot trundled over.

Lawrence came through a couple of minutes later, his ha
damp from the shower. "Couldn't sleep," he confessed.

"Me too." She gave him his breakfast—bagels wi
scrambled egg and smoked salmon.

He tucked in appreciatively. "Thanks."

Denise sat opposite him, sipping her tea. "Any last-minu
flashes of genius how to avoid a confrontation?"

"'Fraid not. Sorry."

"Me neither."

"It all boils down to the dragons themselves. We just don't
ıow enough about them. I've been reviewing the memories
ır dragon has of past dragon star civilizations. There aren't
any. We're limited to generalizations, and not too many of
em. They just seem to passively suck up information and
ter out the useful chunks for their descendants to inherit.
ıat does seem to imply they're relatively benign."

"I hope so." She watched him shoveling down his food.
ıren't you even nervous?"

"No point. It won't do us any good."

"I was never nervous in Memu Bay."

"That's because you knew what you were doing. You were
control. Welcome to being on the receiving end."

"Do you really think they're benign?"

"Yes. But don't equate that to being on our side. If we ask
r their help against others of our race, that will mean them
etting involved in human affairs and politics. We would
ave to justify our appeal. That could well mean they judge
s."

"Where do you come up with all this philosophy from?
re you some kind of secret xenopsychologist?"

He drank down the last of his orange juice and produced
s broad, annoying grin. "One day, remind me to tell you
ow I used to waste away my childhood. You don't spend
ıree years traveling with the *Ultema* without learning some-
ıing about the alien perspective."

They went into the bridge for exodus. Prime spent two
ours readying the fusion drive for ignition as soon as they
ere out of the wormhole. Console screens came on as
awrence and Denise prioritized external sensor imagery.

"Are you receiving all this?" Lawrence asked the dragon.
uring the voyage they'd increased the bandwidth to the Xi-
nti with several hundred fiberoptic cables, linking it directly
the *Koribu*'s network.

"Yes, thank you," the dragon replied.

"Thirty seconds," Denise said.

Lawrence watched the displays as the energy inver powered down. Half of the camera images lost the vag nothingness of the wormhole interior to a blank carmi glare. The other half showed stars gleaming bright agai ordinary space. Radar found no solid object within five hu dred kilometers. Prime brought more sensors onli Lawrence used his optronic membranes to receive the i agery, with Prime giving him a perspective from the front the compression drive section.

Koribu had emerged forty million kilometers above Ald baran's nebulous photosphere. To Lawrence it looked though the starship were soaring across an ocean of featu less luminous red mist. The horizon was so distant it a peared to be above them. There was no discernible curvatu Star and space were two-dimensional absolutes.

Indigo symbols flowed across the image. Most of the concerned the *Koribu*'s thermal profile. Infrared radiati from the star was soaking into the fuselage. Prime fired t secondary rocket engines, initiating a slow barbecue roll m neuver so that the heat was distributed evenly around t structure.

"Heat exchangers are coping for now," Denise said. "I warmer than we were expecting, though. We might need raise our orbit at some point."

"Radiation is strong, as well. Solar wind density is hig There's a lot of particle activity out there. That's going to us more damage than the heat."

When he shifted his perspective to look back down alo the fuselage he saw lines of pale violet light flicker and dan across struts and foil insulation. Metal components gleam the brightest as the phosphorescent shimmer writhed acro them. "Hey, we're picking up some version of Saint Elmo fire."

"I hope our insulation's up to it."

"Me too. Okay, long-range radar is powering up." Six ultiphase antennae were unfolding from their sheaths ound the middle of the cargo section, flat ash-gray rectan-les measuring twenty meters down their long edge. They ipped back parallel to the fuselage and began probing the aotic climate boiling around the *Koribu.*

Prime overlaid their sweep across the visual imagery. A oint of solid matter appeared forty-three thousand kilome-rs away, in an orbit two thousand kilometers lower than the *oribu*'s. Another one was detected fifty thousand kilome-rs away. A third was over seventy-two thousand kilometers istant. The radar's focus shifted to produce a higher resolu-on return of the first. It was twenty kilometers across, oughly circular, although the edges had broad, curving ser-ations, and it thickened considerably toward the center.

"More like a flower than a dragon," Lawrence murmured.

The radar had detected another seven points of mass out to 15,000 kilometers, all the same size as the first.

"Are they the dragons?" Denise asked breathlessly.

"I believe so," the dragon said.

"There must be thousands of them."

"Millions," Lawrence said. The idea was exhilarating. Jntil now they hadn't known for certain that the dragons ex-sted. They could only assume that Aldebaran's gravity had ttracted the eggs when the star was bright and adolescent, nd that its expansion had hatched them. Now, here was the inal proof. Humans were no longer alone in the universe, nd the Ring Empire really had flourished when the galaxy vas younger.

All his earlier daydreams and beliefs had been justified.

I CAN go home.

The *Koribu*'s main telescope was swinging around to oint at the first dragon. Against the uniform red glare it howed as a simple, dark speck. When the communication

dish locked on it detected low emissions in several electromagnetic bands.

"Are you ready?" Denise asked. She sounded as if sh
were prompting a small child.

"I am," the dragon said.

"Then say hello."

The Arnoon dragon transmitted a pulse of data from th
communications dish and began repeating at half-second intervals. It was a simple sequence of mathematical symbols i
the language stored within its own memory.

The Aldebaran dragon answered with a much longe
pulse, little of which could be translated. Lawrence an
Denise yelled in delight, clapping their hands. He gave her
quick kiss and a hug, overtaken by the moment; then the
settled back to observe the exchange.

The Arnoon dragon began sending the information they'
prepared. A translation dictionary for what little it had of it
own language and the datapool English equivalent. After tha
there came a more complete English dictionary, with inter-
connected entries so that meanings and concepts would buil
into a cohesive whole. Syntax and communication protocol
followed. Finally it sent a short, encyclopedic file on humans.

Less than three seconds after the last pulse was sent, th
Aldebaran dragon said: "Welcome to our star. It is alway
pleasing to accept new information in any form."

Lawrence grinned. "Turing test," he said quietly t
Denise. "Even if it isn't pleased, it understands the principle
it's trying to be polite."

Denise nodded and took a breath. "Thank you. We ar
happy to be here. My name is Denise Ebourn. Do you hav
a designation?"

"I am One."

She flashed Lawrence a perplexed glance. "Does that hav
any significance?"

"I am the first dragon you contacted. One."

"Ah, I see." Denise reddened slightly as Lawrence gave r a malicious sneer. "One, we have brought one of your wn kind that has been damaged."

The *Koribu*'s sensors suddenly reported a bombardment of dar-style pulses. The whole magnetic environment around e starship altered, oscillating rapidly. Their one neutrino canner recorded emissions off the end of its scale.

"In what form are you carrying one of us?" One asked.

"Go ahead," Denise told the Arnoon dragon. It began ansmitting a pulse containing a summary of its own history.

"I understand," One said. "You brought this fragment to us elieving we would want you to. I thank you for your conern. Unfortunately, in this respect your voyage here has een fruitless."

"What do you mean?"

"We have no interest in the fragment."

Denise couldn't believe what One had said. The translaon dictionary must have glitched. "Do you mean you can't epair it?"

"No. We have no interest in repairing or re-forming it. Did ou not comprehend what it was?"

"Yes. It's one of your eggs."

"It is. As such it is irrelevant. We release millions of eggs very year. Only a fraction of them are ever captured by a tar's gravity. The others are simply lost. Or they crash as our fragment did. Some are even intercepted by biological pecies such as yourselves and mined for information. We do ot concern ourselves with them. I would suggest the analgy that they mean as much to us as a single human sperm oes to you."

"But . . . It's alive now. It thinks. It's a rational, sentient eing."

"It was not. It was a fragment that was slowly decaying ntil you discovered it and implanted this awareness."

"Are you saying we shouldn't have done that?"

"No. Each of our eggs is a hostage to chance. It is only
tiny fraction of the total that grow into a civilization such
this. Others contribute to the galactic knowledge base
more diverse means. The fragment you found has enhanc
your species' understanding of the universe. In that respect
has been a success. We can add to our knowledge of you."

"Did you already know about us?" Lawrence asked.

"We are only sixty-five light-years from your world," O
said. "We have been receiving your radio transmissions f
centuries."

Lawrence rolled his eyes in dismay. "Great."

"Maybe this fragment means nothing to you," Denise sai
"But it does to us. Could you repair it for us?"

"That is a null question. The fragment could never becon
one of us. It is not just its physical structure that is fra
mented, its memory is also diminished. The two togeth
make us what we are. We do not have a genetic code. It is i
formation that enables us to evolve and adapt according
circumstances. To do this, we must have a complete set
memories. What you are asking for is for me to provide
new set of full memories and to integrate the fragment bac
into a whole egg. In which case it will become an egg agai
nothing more. It would be released back out into the univers
to take its chances again. If that is what you wish I can di
assemble the fragment and use the molecules within a ne
egg."

"No," Denise said quickly. "Isn't there some way it ca
grow from what it is into something like you?"

"Not without abandoning what it is now."

Denise bowed her head until it almost touched the consol
She was close to weeping. The village had risked so much
bring their dragon back here. People had died to achieve i
Now, doing what they believed in had been exposed as a pa

cularly human folly, children getting sentimental over an jured puppy.

A curious sound roused her. Lawrence was chortling.

"What?" she snapped.

"Hubris always hits hard. Especially to an idealist like ou. Because of your convictions, everybody else is wrong; ostly they're not even allowed to have different opinions. nd now you have to face up to the fact that what you've one is wrong. You've been proved guilty of anthropomor-nism."

"I am not. Our dragon is a sentient creature who deserves spect. Its origins don't matter. What it is now is what ounts. We did the right thing bringing it here. The fact that 's unique makes it even more deserving. I would do the ame thing all over again. It deserves the chance to evolve; it as a right to life."

"A human right?"

"Yes," she growled. "A human right. It's also a universal ght. We rescued the dragon from nonexistence, we took om it, and now we have to give it back. I don't care what ou think, I know I'm right, and to hell with you."

"Fate, you are stubborn." He activated the communication nk. "One, can you tell me if you share knowledge with ther species?"

His tone was so sharp that Denise gave him a suspicious lare. He just gave her his annoying broad smile.

"We do," One replied.

"Any knowledge?"

"Yes. It is our existence."

"Then you wouldn't object if we use patternform se-uencers to try to enhance our dragon in a way we and it con-der appropriate?"

"No."

Denise smiled her thanks at him. It wasn't what she'd

wanted, but at least it gave them a chance to help their drag
grow into something other than an inert mass.

"Does it worry you that other species might misuse su
knowledge?" Lawrence asked. "For example, if they u
your knowledge to build weapons?"

"If you know how to interpret and understand the dat
then you already have the capability to build weapons of
similar nature. Weapons are not a technological proble
They result from the nature of a species' society."

"In other words, we have to be responsible for ourselves

"Of course," One replied.

"Can we ask for your help in that respect as well?"

"In what context?"

"Other members of our species will be arriving soo
Don't give your knowledge to them."

"Knowledge is universal. It cannot be denied."

"I don't want it denied. Just withheld for a short time,
least. Your knowledge could be very dangerous to o
species if it is not shared universally. One of the people fo
lowing us wishes to acquire a monopoly, so he can exploit
to impose his ideals on others. Do you view that as wrong

"In the context you have stated, it is wrong. But how do
know that those following you seek to dominate others
your species? How do I know it is not you who favor th
course?"

Lawrence gave Denise an awkward shrug. "Damn, I'v
gone and triggered its paranoia. Any ideas?"

"Our dragon can tell you all about our intentions," Denis
said.

"I would be willing to do that," the Arnoon dragon co
firmed.

"That would not be acceptable," One told them. "The frag
ment's processing routines are derived from your genetic a
gorithms. It is your creature."

Denise cursed at the pane showing One's visual image,

y black splinter lost against the irradiated fog. "Now
ıat?"

"Rely on human nature," Lawrence said. "May we ap-
ɔach you?" he asked One. "Our ship is being strained by
s environment. Your umbra would provide shelter. And
'll be safer there."

"Safer relative to what?"

"Open space. The human following us may be violent. He
ll not risk using weapons close to you."

"Very well."

"After that, could you wait until a third ship arrives? The
ormation could then be given to all of us simultaneously.
ıat would achieve a balance, wouldn't it?"

"I will wait."

mon had spent the entire voyage in the *Norvelle*'s sickbay.
ter the first fortnight, the two doctors onboard had begun
s dermal regeneration treatment. New skin was now grow-
g successfully over the deeper burns. He found the whole
ocess extremely tiring; the new growth seemed to devour
ergy from the rest of his body. Fortunately the pursuit of
Koribu didn't require his attention.

Captain Sebastian Manet had been surprised at the course
hijacked starship had taken. "They're heading for Alde-
ran," he'd said as soon as Simon's stretcher was maneu-
red through the Xianti's airlock tunnel.

Simon's personal AS scrolled data on the star. "A red
ant? Are there any planets there?"

"The astronomers have never found any," Manet said.
lowever, we don't know for sure. There have never been
y missions to Aldebaran. Nobody's pure science budget
er ran to that."

"Interesting. So we could well have an alien civilization
ing closer to us than we ever knew: right inside our sphere
influence, in fact."

The captain's front of disapproval faded slightly. "Is th what all this is about?"

"Yes. Now how long before we can depart?"

"I wanted to talk to you about that."

"There is no discussion over this matter."

"It will take a hundred and four days to reach Aldebar from here. We will barely have enough fuel to return."

"But we do have the capacity?"

"Just. Assuming nothing goes wrong. There's also t crew to consider. They did not anticipate an extra two hu dred days' flight time plus however long we remain in t Aldebaran system itself."

"Rubbish. I know you space-types. They will relish t opportunity of performing first contact."

"Then what about the ground forces on Thallspring? Wh will we return for them?"

"Captain, you will either give the order to go FTL or t me you will not."

Sebastian Manet gave the obscured figure on the stretch a hateful glance. "Very well. We can go FTL in another eig minutes."

They had barely spoken in the hundred days that followe Simon had spent a lot of the time asleep, as the treatments a into his reserves of strength. In his waking hours he review every scrap of data they'd acquired on the Arnoon alien, ho ing for some insight. Each day he checked the tracking da *Koribu* remained a resolute twenty-six minutes and thirte seconds ahead of them. He began to make his own plans f exodus. Manet, along with his crew, could not be trusted wi what needed to be done, and they couldn't be allowed to i terfere. Simon's personal AS reviewed and confirmed all t command codes, ensuring he retained ultimate authority ov the starship.

Ten days before the exodus Simon ordered the doctors

d his treatments. The interval would allow him to build his ength up again.

On the day itself he remained in sickbay; the doctors and edical staff had been dismissed. He lay on the bed, content at the pain had reduced over the last three months. He ould have the physical resources to carry this out. His DNI d optronic membrane linked him with every vital sensor tside the starship, their data formatted by his personal AS to a comprehensive wraparound display, putting his perption point at the front of the starship. At the moment it as surrounded by a formless gray haze, with an ebony knot r ahead of them. Indigo ranging data scrolled across it.

Simon kept switching his attention between it and the her knot, the one following them forty minutes behind. He ew the SK9 had somehow got himself up to orbit and into starship. There was only one reason he'd do that.

"Shouldn't be long now," Captain Manet announced. Ve're detecting the photosphere."

Beyond the knot representing the *Koribu,* a faint drizzle of ack was creeping through the nothingness, as if space were ming to an end outside the compression wormhole. Then e knot began to waver, expanding as it lost density. It vaned.

"They left that late," Manet said. "Only forty million kiloeters out."

"We need to be close to them," Simon said.

"Yes, I know. But I'm going to shift our inclination. If ey're really hostile they'll mine the exodus zone."

Simon spent the next twenty-five minutes watching the ack barrier of the star's photosphere coming toward them, ondering what Newton and his friends were doing out there real space. With five minutes to go until their own exodus, anet armed the *Norvelle*'s missiles in preparation. The ip's AS primed the fusion drive under the bridge crew's surrvision. Then the black wall was unnervingly close, and

beginning to pick up speed. It fractured radially, with r
streaks fizzling through. Then the huge starship was saili
high above a glaring carmine smog that seemed to stretch o
forever in all directions.

Columns of indigo digits streamed around Simon, muta
ing wildly as they went. There was no mass point within fi
hundred kilometers. No sensor radiation fell across them. I
frared energy began to soak the fuselage. The big seconda
chemical rocket engines around the cargo section flar
brightly, initiating a slow thermal roll. Large rectangul
radar antennae began to deploy.

Simon used his command capability to launch a salvo
missiles. Pressure doors throughout the *Norvelle* closed a
locked, isolating the crew.

"What are you doing?" Sebastian Manet demanded.

"That's perfectly obvious, Captain. The *Norvelle* is n
ship, and I am carrying out my mission." He canceled th
link and shut down all internal communications channels.

On the bridge, Sebastian Manet stared on helplessly as h
DNI was denied access to the starship's network. The co
sole displays darkened. Two of the bridge officers hammer
on the pressure door with their fists. Nobody heard them.

"Sweet Fate, they've fired their missiles," Lawrence said.

"Not at us," Denise assured him.

"What then? Oh!"

"Scatter pattern. I think they're going to try to strike th
other ship at exodus."

Lawrence opened the link to One. "Will you now acce
that your knowledge shouldn't be given to this starship?"

"The starship's actions support your contention so fa
Our knowledge will be withheld pending resolution of th
event."

"Thank you." He turned to Denise. "Can we interce
those missiles?"

"No. We're too far away."

"Shit. Get Prime to scan for the exodus. Use our main mmunication dish to broadcast a warning."

digo targeting graphics locked on to a section of space ghteen thousand kilometers away as the *Norvelle*'s radar tected an object under acceleration. Simon's visual focus aped across the distance. A long, incandescent spark rned hard above the mellow radiance of the photosphere. was moving fast, descending.

"Fusion flame," the starship's AS reported. "Spectral pattern identical to our own. Radar substantiates vehicle size. It the *Koribu.*"

"Where are they going?" Simon asked.

Several plot lines curved out of the dazzling plume. The orvelle's long-range radar began to sweep along them. It viftly found the destination.

"A solid structure, type unknown," the AS said. "Twenty lometers across, circular, very regular."

"Take us down to it," Simon ordered.

n the *Clichane*'s bridge, Simon Roderick sat behind the ptain as they approached exodus. Console panes counted own the last few seconds. Camera images turned a garish rmine. A small cheer went round the bridge officers. mon's DNI relayed the radar data directly to him. No rge, solid objects within five hundred kilometers. Several nall points registered. Extraneous radar pulses were illuinating their fuselage. The AS confirmed their signature as e long-range type carried by both the *Koribu* and the *orvelle*. It began plotting their locations.

"Receiving communications," the captain said. "Someody's shouting." The AS showed a very powerful transmison beamed straight at them.

"Your exodus point is mined. Launch defense salvo."

"Is that the *Norvelle*?" Simon asked.

"There is no identification code," the AS told him.

Simon's personal AS checked the radar image again. T small points were now moving under heavy acceleratic tearing straight toward them. *Clichane*'s AS immediate fired a countersalvo. The missiles slid out of their launch cr dles around the cargo section. Solid rocket motors ignite accelerating them at over sixty gravities. Sensors were d graded by the ion wind and radiation rising from the phot sphere. The missiles' onboard programs tried to compensa But the attacking missiles were also using countermeasur and em pulses. The defenders responded with their own vo ley of electronic treachery.

The *Clichane*'s AS acknowledged that the defense sal wouldn't be able to achieve precision elimination strikes. A tacking missiles would breach the defenses. It ordered sep ration, and the swarm blossomed as each missile discharg its multiple warheads. They were still inside safe-distan limit, but the attacking missiles were closing. The AS had choice.

A huge corona of nuclear fire erupted around the *Clicha* as the barrage of defending warheads detonated. The sphe cal plasma shock waves clashed and merged, forming a he ish shield of seething raw energy. Secondary explosio ripped long ebony twisters through the rampaging io short-lived hypervelocity spikes that tried to assail the sta ship.

At the center of the fusion inferno the *Clichane* was bu feted by radiation. Its external sensors were blinded as ha X-rays burned up their circuitry. Em pulses induced hu power surges along electrical cables and metal structur Temperature escalated, blackening the thermal protecti foam before the surface began to ablate, shedding scab charcoal flakes. The residual foam bubbled like molten tar. hurricane of elementary particles washed across the besieg uselage. In the bridge and throughout the life support wheels adiation monitors began a shrill whistle of alarm. Emerency pressure valves began venting deuterium gas from anks around the fusion drive section as the liquid started to oil from the electromagnetic energy input. Thermal radiator anels ruptured, jetting their sticky, steaming fluid into the urricane of neutrons swirling round the giant starship.

Simon clung to one of the consoles as the bridge shook. oud, harsh, metallic creaking sounds reverberated through ne life support wheel structure as the lights flickered. The raiation alarm kept up its insistent whistle. Schematics had urned completely red. The AS was battling to compensate or massive systems failure, rerouting power and data, isoating leaking tanks and fractured pipes. Backup thermal eservoirs were used to absorb the heat seeping through the uselage structure. Over half of the secondary rockets were isabled. The AS fired the remaining engines in short bursts, ttempting to counter the twisting impulses from the larger ents.

The whistle alarm slowly faded. One of the officers was hrowing up. Simon had to force himself to let go of the conole. His heartbeat was racing badly.

"We weren't hit," the captain said incredulously.

"How do you know?" Simon asked. There was no external ensor data available at all.

"We're still alive."

The officers had begun talking urgently to the AS; fingers kidded over the console keyboards as they tried to pull useul information from the degraded network. Reserve sensors leployed from their sheaths. The AS located the two radar ources again. They were both under acceleration.

"How long before we can get after them?" Simon asked.

"I should have an answer for you in about a week," the captain said.

* * *

The *Koribu*'s telescope immediately blanked out as the nuclear warheads exploded around the *Clichane*. Filter programs compensated as the fury slowly diminished. They saw the starship wreathed in coiling gas plumes. They were alive with scintillations from the radiation blitz, entombing the giant ship in a nebula of shooting stars. Venting had imparted a slow tumble. Lawrence didn't like to think how much liquid would have to be evacuated to move something so massive. Then they saw a flare of rocket motors around the cargo section as the AS attempted to stabilize their attitude.

Lawrence realized he'd been holding his breath for a long time. "They're alive, then," he said.

"Unless the *Norvelle* fires at them again," Denise said. "With the state they're in, I doubt they'll be able to defend themselves."

"Can you defend yourself against that kind of attack?" Lawrence asked One.

"No," One replied. "There is no reason for us to be armed. We have nothing other than knowledge. And that we give . . ."

"Yeah, you live to share," Lawrence said. "What happens when other species do threaten you?"

"We incorporate the knowledge of the threat."

"That's it?" Denise asked. "That's all you do, remember being destroyed?"

"We exist to acquire and distribute knowledge. We hatch in every sector of the galaxy and examine what surrounds us. Once that has been accomplished and the sun cools again our existence there ends. Another sun will eventually replace it. The overall processing of knowledge will continue no matter how many individuals of our species are exterminated. Very few other species have sufficient munitions to destroy every one of us. By now our eggs have probably reached other galaxies."

"Are you saying you don't care if you're destroyed?"

"Care is an emotion I do not possess. You know it because s bound with your sense of individuality. We are not a hive nd, but we are aware of ourselves as a civilization that uld well prove to be eternal. All events we encounter ntribute to what we are and what we will become. All lividuality ends eventually. We birthed ourselves in ac-owledgment of that."

"But the person in command of the second starship could eaten to destroy you if he realizes how vulnerable you ."

"If threatened I will provide the knowledge required. The eat will end."

"You said you'd withhold it," Lawrence said.

"In the assumption that would provide a balance for your ecies. You claimed the third starship would deliver our owledge to your entire race. That is no longer possible."

"We're going to have to go to Earth," Denise said. "Carry knowledge ourselves and get there ahead of the *Norvelle*. mn Roderick to hell. I didn't want this."

Lawrence held up a finger. "One, will you help repair the rship that has just been damaged?"

'Yes. A patternform can be provided that will modify itself perform the operation."

'So if we disable the *Norvelle*, the balance will be re-red. They can both be repaired simultaneously. There uld be no monopoly when they return to Earth."

'Will you make us a weapon that can disable the *rvelle*?" Denise asked quickly.

'No," One said. "You may have knowledge that can be d to build a weapon."

'How long would it take to build?"

'First you would have to learn how to apply patternform tems. Then you must integrate the knowledge for the apon to be extruded."

"Yes. How long?"

"You have some familiarity with patternform syste
This would be to your advantage. I estimate you would
quire as little as three weeks."

Frustration made Denise want to hit something. Anythi
The *Koribu* and the *Norvelle* were equally matched. If t
launched a strike, the *Norvelle* would retaliate. They nee
something else, a weapon that would give them an adv
tage.

"But you already have a weapon we can use," Lawre
said quietly.

The first thing the *Norvelle*'s sensors confirmed was
alien structure's complete lack of rotation in any directi
Somehow it held its attitude stable against the gusts of th
solar wind that blew constantly from the turbulent pho
sphere. Its shape gradually resolved during their approa
circular, divided into twenty scalloplike sections that cur
down toward the surface of the star. Their edges w
rounded and very smooth, tapering down to a few tens
meters thick. Its bulk was concentrated in the middle, wi
small aperture at the very center.

Simon thought it might be a docking port of some kind,
though he wasn't convinced. Even for an alien design it
a very strange habitat.

The apex of each scallop sprouted three slender ridges
shone a livid scarlet, radiating heat away out toward the st
leaving the rest of the upper surface considerably cooler.
AS postulated that this was how the structure generated
power, exploiting the thermal difference. To do so, the rid
could well be a type of thermal superconductor. An inter
ing technology, Simon thought, but hardly on a level w
nanonic systems.

The *Koribu*'s fusion burn ended, rendezvousing it with
alien structure. It hung inside the umbra, three kilome

m the surface. Simon waited, half expecting to see an en-
ieering shuttle fly over into the aperture.

"What are you doing?" he muttered to himself.

Norvelle's long-range radar continued to scan around. The
detected another eleven similarly sized alien objects
thin 150,000 kilometers. If they were some alien version
habitats, it gave him a seriously large population base to
al with. There must be millions of similar structures in
oit around the red giant.

"Have we intercepted any interstructure communications
?" he asked the starship's AS.

"Not in the electromagnetic spectrum. They could be
ng lasers, or masers. In order to intercept them we would
ve to insert ourselves into the beam."

"Never mind." He continued to study the structure. When
y were two thousand kilometers away, *Norvelle* extended
magnetometer booms. The alien structure was the core of
ast magnetic field. Around the center it was as dense as a
kamak containment field. Vast, invisible flux wings ex-
ided for hundreds of kilometers below the star-facing sur-
e. Simon altered the calibration of the main radar and
ocused the telescope. The AS combined the data from
ch, presenting it as a false-color thematic image.

Solar wind was being scooped up by the magnetic field
d pulled inward. He could see tenuous eddies of the stuff
ming as it streamed up toward the hidden center of the
ucture's star-facing surface.

He knew it couldn't be a habitat. A machine of some kind,
en. One that consumed solar wind particles. What sort of
ichine did that? He knew the aliens had nanonic systems.
ey must be converting the solar wind into artifacts of some
id. The production capacity represented by the millions of
uctures was awesome. Although that would be severely
nited by the minuscule quantity of mass that the magnetic

field scooped up. If you had that kind of ability, why us
like this?

That was when he recognized what the structures were.

"Open a link to the *Koribu,*" Simon ordered the A
"Lawrence Newton, can you hear me?"

"Loud and clear. Is this Simon Roderick?"

"It is."

"Your clone told us about you."

"I don't suppose he was very flattering."

"He made some strong claims. Your attack on
Clichane would seem to validate them."

"If you know about me, you know why I had to do that

"I know your rationale for the attack. That doesn't mea
agree with it."

"I've seen your file, Newton. You gave up everything
whole world, for the chance to fly explorer starships. Y
know there is more to the human condition than what we
today. And now we can realize that goal for everyone."

"Whether they want it or not."

"It is the underclass that prevented you from making the
flights you dreamed of. They restricted you more than th
ever did me."

"I'm not arguing with you. I'm telling you, I will not all
you to impose change on people. You and your clone m
have the information together."

"Is it yours to give?"

"Yes."

"I think not. This isn't a habitat, is it, not some artifa
This is the alien itself. How utterly magnificent. A creature
pure space."

"Yes, this is the alien."

"One of them crashed on Arnoon, didn't they? That's wh
made the crater next to the village."

"Your research is very competent."

"It made no sense at first. Why an alien with nanonic tec

logy would enlist human allies and steal a starship. It was maged, it didn't have all of its abilities."

"And now we've brought it back to its own kind."

"What were you going to do with the technology, New-?"

"Nothing. I'm going home."

"I don't believe that, either. You're from a Board family. u would use it to your advantage, just like me."

"Wrong. I suggest you go back to the *Clichane* and help its ew. Once you've done that the technology will be made ailable."

"Have you really convinced the aliens to cooperate with u already? Or are you hiding something from me? Why n't you go back and help the *Clichane*?"

"This ship is in no condition to help anybody. We barely de it here."

"Then how were you planning to get home? Can the alien nonics repair your ship? I suppose they can."

"They can."

"How interesting. In that case, I think I will remain with u and observe them in action." It was an almost perfect so-ion, he realized. If he parked the *Norvelle* in the alien's bra his proximity to the *Koribu* would provide him with greatest possible opportunity to obtain a physical sample the nanonic technology. He began to wonder just how ch of an ally the alien was to Newton. How would it react any attempted interdiction of nanonic systems? Certainly re had been no repercussions from his attack on the *ichane*. Before he took such an overt course of action he st at least try to establish communications with the alien. could be that Newton was actually bluffing.

The *Norvelle*'s fusion drive matched velocity with the vast en, then slowly eased the starship into the umbra. It cut off, d the AS began firing the chemical rockets to refine their itude. At that moment they were five kilometers from the

surface of the alien, and twelve kilometers from the *Koril*
Neither Newton nor the alien had responded to Simo
repetitive calls.

The *Norvelle*'s magnetometer booms were still observi
the titanic flux lines warping around the alien. Their patte
began to change, contracting like petals at sunset. But *fas*

"What—" Simon managed to ask.

Lawrence had gone over the lifecycle of the dragons ma
times during the hundred-day voyage. Naturally, the cre
tures fascinated him. Then actually seeing them through t
starship's sensors thrilled him even more. He loved their
egance. He even admired their philosophy, despite how fr
trating it was to his situation.

Each dragon must have taken centuries to grow to its f
size. Like Simon Roderick, he watched the magnetic fie
gather up wisps of solar wind, ingesting them for the acti
patternform system to alchemize—a slow, laborious proce
given the quantity involved. Some of the molecules we
used to replenish and sustain the dragon's own body, b
once it had reached its full size, most were given to the pr
duction of eggs. Each one took a long time to convene. N
only did its physical structure have to be put together a pa
ticle at a time, but those particles had to be loaded with da
from the ever-shifting tides of knowledge possessed by t
dragon star civilizations.

Once an egg was complete it would be sent off into int
stellar space, to fall aimlessly through the galaxy. But t
dragons were in orbit around Aldebaran, tied to the star
gravity. The eggs couldn't just be detached from the adu
they would simply drift around the same orbit. So the cen
of every dragon was a magnetic cannon, capable of accel
ating an egg up to solar escape velocity.

The *Norvelle* was parked five kilometers from the muzz
when One fired. The egg, a solid sphere of matter sevent

'o meters in diameter, struck the starship's complex and
elicate compression drive section at over forty kilometers
er second.

CHAPTER NINETEEN

VERY BED IN THE *CLICHANE*'S SICKBAY WAS OCCUPIED, MOSTLY
th victims of radiation burns sustained during the attack.
urrounding cabins had been converted to hold the overspill.
he doctors walked around, checking vital signs, making
re their patients were comfortable. They didn't have a lot
e to do. Prime, improved with genuine dragon routines,
as orchestrating the patternform systems that had twined
emselves through the flesh of each patient. This was a
uch more active application than the Arnoon dragon had
er achieved. The patternform had grown what resembled a
twork of veins over each man's skin; tubules infiltrated the
dy, multiplying around the organs and muscles inside. Par-
les roamed through the damaged tissue, repairing cells and
equencing DNA smashed apart by the X-ray barrage. The
mputing power required to control the operation in each
rson was phenomenal; patternform had grown the process-
g nodules as well. They hung under each bed like leaf-
een wasp nests, their root tendrils connecting them to the
rallel vein network.

Lawrence looked through a couple of the doors as he and
enise walked through the section. The men were all sleep-
g.

"They look peaceful," he said.

"So did you," she told him.

The converted cabins were split with the *Norvelle* cre members who had received more physical injuries when t dragon's egg struck. Their starship's axis had snapped fro the egg impact, with two of the life support wheels bei flung off into space as the broken sections tumbled aw from One. Over a hundred crew had ejected in lifeboa Those remaining in the two intact wheels had waited for t *Koribu* to rendezvous, then transferred over for the sho journey to the *Clichane*.

Now the two remaining starships were parked in One umbra as patternform strands began to creep across them.

Simon Roderick was waiting for Lawrence and Deni outside a cabin with a closed hatch. The locks disengage and he pushed it open. They followed him in.

There was a single bed inside; the SK2 lay on it, also e cased in patternform systems. His legs and hand were bei grown back. A Skin sustainer cabinet that had been broug into the cabin was smothered in a lacework of patternfor veins; they were harvesting the organic components a blood reserves for raw material to generate new tissue. Fla cid translucent sacs the shape of legs already extended fro his stumps, with glutinous fluids circulating inside.

Denise's expression tightened as she looked down at t unconscious man. "What do you intend to do with him?"

"For the immediate future, he will be excluded from t Board, and most of his executive privileges will be revoke House arrest, essentially. After that, who knows? I suspect depends on what form Earth's society chooses to follow."

"Good enough," Lawrence said. He ignored the dirty lo Denise threw him. "None of us exactly came out of this saints."

"No," Simon agreed. "But then I never claimed to be."

"How will your clone siblings react to all this?"

"The same as we did. Not that it will really matter." He ve both of them a pointed look. "The captains will make e the dragon knowledge is given to everybody when we urn. They're already making plans to transfer the memo- s directly into Earth's datapool before Z-B even notices ere's something different about their old starship. It'll be otected by this upgraded Prime, which should ensure equal cess."

"You sound as though you disapprove."

"I almost do." Simon gestured at his clone sibling. "We're chaotic race. His method would have given us a smooth nsition."

"Where's the fun in that?" Lawrence said. "Tear down the iculture, open your eyes, give people their identity back."

"Ah." Simon's eyebrows rose in modest censure. "I might ve guessed."

"How long before the *Clichane* is flightworthy?" Denise ked.

"Another fortnight," Simon said. "Quite remarkable, re- y. Fortunately there's plenty of spare mass to restructure issing components. After all, we hardly need the weapons, all that asset cargo now. Are you sure you don't want to me back with us? It will be an interesting time to live in."

"No," Denise said curtly.

Lawrence just smiled.

ne *Clichane*'s compression drive powered up, and the im- ense starship flashed out of space-time with a dizzying ist. The *Koribu* was left floating alone in the dragon's nbra. It was never going to fly again. Instead it was giv- g birth. Patternform had plaited the fuselage in a gridiron crystalline stems that suckled at the minerals and com- unds of the structure. From a distance it looked as though e starship was covered in a harlequin patchwork of gem ost; millions of slender amber, ruby and emerald facets

flashed and glinted in the haze of warm light that spille
around the edge of the alien. Wider sapphire proboscis
had penetrated tanks, siphoning out the liquids to contribu
to the semiorganic growths sprouting on opposite sides
the cargo section. As the weeks progressed, they swelle
out into chrysalids wrapped in a tight skin of diamon
strand silk.

The Arnoon dragon, too, was metamorphosing. The X
anti's payload bay doors had been opened to space. Inside th
bay, the cargo pod had split apart, exposing the dragon. Cry
tals threaded their way across the floor of the maintenan
bay and encrusted the spaceplane. Their tips meshed with th
particle structure of the dragon and began feeding it mol
cules and information.

Denise spent hours every day thinking with it. As the
was no place for it within the Aldebaran dragon civilizatio
it had decided to go with her, to become a part of whatev
society flowered from the genetic package. Dragon mem
ries were reviewed and analyzed for templates of the abiliti
they sought. They began to incorporate functions that woul
allow it to be free-flying, to sense in every spectrum,
power itself with sunlight, to absorb solid cold matter, to r
tain its original personality. Dozens of notions taking on sol
form.

After the *Clichane* left, Lawrence spent ten days underg
ing extensive patternform treatment, transforming his bo
and resetting his DNA. He emerged as his teenage self, wit
out Skin valves.

Denise looked him up and down and pursed her lips. "Ve
cute," she observed coyly.

The chrysalid cases split open and peeled back, reveali
the Ring Empire–era starships that patternform had ge
tated—streamlined silver and magenta ellipsoids, with
necklace of drive fins and forward-swept power shields ri
ing smoothly from the rear quarter. Lawrence gazed at h

h a reckless enthusiasm that matched his new adoles-
ıce.

'I guess this is good-bye," Denise said.

Ie gave her an awkward grimace. "Yeah." Then his smile
ırned. "No, it's bon voyage. The way things are shaping
it's not impossible that we'll meet again."

'All right, Lawrence, bon voyage it is." She gave him a
t kiss. "What are you going to name yours?"

'That's easy. *Fool's Errand.*"

Denise laughed. "Mine's the *Starflower.*"

'Sounds good."

The interior of the *Fool's Errand* comprised three circular
ınges with concave walls. In their neutral state the cream-
ored surfaces had the same texture as soft leather. Human-
led fittings could distend out of it as required. The lounges
o made perfect auditoriums, capable of providing a 360-
gree image that could show either sensor images or any of
i's that he'd loaded from the *Koribu*'s multimedia library.

Lawrence walked into the forward lounge, enjoying the
velty of a standard gravity field. A single luxurious chair
e up out of the center of the floor. He settled himself in it
l called up a visual sensor image. The front of the lounge
lted away, showing him the *Koribu*'s crystal gilded fuse-
ge dead ahead.

In his mind, a broad crown of the starship's system icons
rned a willing gold. He selected several, and the *Fool's Er-
nd* slowly backed out of the inert chrysalis. An idiot's grin
read over his face as the ship's elegance and power became
parent. And he alone commanded it. The *Starflower* rose
o view from the other side of the *Koribu.* He watched as
nise flew around to the drastically mutated maintenance
y. The Arnoon dragon waited at the center, elegant semi-
ganic segments closed against the main body, solar wing-
eets furled tight. Watching the *Starflower* touch it was like
eing two drops of water merging; the dragon was absorbed

through the shimmering hull, leaving only a slight bulge t
betray its presence. Then the *Starflower* moved out into cle
space beyond One's umbra. Lawrence could see the stran
forces gathering around the drive section as the pow
shields and fins shone like fragments of a blue-white star.
flung itself into the nullvoid.

"Thank you for your help," Lawrence told One.

"We will learn what you become," the dragon replie
"And we will remember you. This is what we are."

Lawrence selected his course, delving deep into some
the oldest memories the dragons possessed for the inform
tion. The engines gathered up their colossal strength and ir
pelled the *Fool's Errand* into the nullvoid.

Nebulas are among the most beautiful objects in the sk
revered by astronomers across the galaxy. Fluoresced l
stars hidden deep inside, they shade parsecs with the mc
magical patterns of shifting primordial light. Yet for all the
grandeur they are transient. The stars that provide their eth
real beauty also blow out a hard wind of ions that slow
disperses the gas and dust. Gravity, too, plays its part, ine
orably thinning out the streamers and clouds. Protostars pe
form the opposite function, their great glowing who
sucking in the spectral tides, compressing them down to
central spark.

Their lifetimes are measured in millions of years only-
negligible in galactic timescales.

Even the Ulodan Nebula, one of the thickest and darke
ever to form in the galaxy, was waning when the Ring Er
pire archaeologists came across the planet of the Mordiff. I
the time of the Decadence War it was just a zone of interste
lar space with a slightly higher than average gas densit
Even then the Mordiff sun was cold and shrunken, no mo
than a glimmering red ember. Just a memory within the ne
born dragon civilizations.

took Lawrence Newton and the *Fool's Errand* a long
: to find the cool star-husk with its single lonely planet.
eventually the beautiful starship fell from the sky and
led beside the one remaining relic of the Mordiff.

awrence put on a spacesuit and stepped down onto the
net's surface. There was no air left: it had bled away mil-
s of years ago. The sand under his boots had frozen to a
st harder than iron. But there was still light. High above
horizon, the galactic core formed a lambent white swirl
occupied nearly a tenth of the sky. It cast a sharp shadow
ind him as he walked.

'his was a landscape bleaker than any he had ever known.
:k outcrops were sharp and fractured: even the stones that
red the ground were jagged. Over millions of years, cold
drawn the very color from the land. He knew this planet
no future left. That knowledge didn't bother him; he had
ne for its past.

Ie paused on a low ridge and looked up at the terminus. It
strange, he thought, that a race so warlike and terrible
ld build a machine so much greater than themselves.

The terminus was a broad toroid, snow-white in color, its
er aperture measuring three kilometers in diameter. Five
nt buttress towers supported it a kilometer off the ground.
tcrops of rock rested against the base of each one, as if
y were waves breaking against a cliff. Neither cold nor en-
py had affected the titanic artifact; even geology had been
eated by it. The Ring Empire archaeologists weren't even
e it was made from matter in the normal sense. Nothing
: of the Mordiff remained, no ruins, no monuments. Only
terminus, their failed bid for immortality.

Iarsh turquoise light shone down out of the center, illu-
nating the frozen sand underneath. Lawrence was vaguely
appointed he couldn't actually see the wormhole, but the
e light acted like a veil across the aperture.

After a while Lawrence walked back to the *Fool's Errand.*

He sat on his seat in the forward lounge and guided the s
ship under the toroid. It rose slowly into the blue haze. F
second the sensors could see nothing; then they were in
the wormhole. A tube of pale violet light stretched away f
the starship. Ahead lay the future, billions of years stretch
out to the end of the galaxy, when even the terminus fell
the black hole. In the other direction lay the past.

Fool's Errand flew back into history.

The three sister planets were moving into their major c
junction. It was a spectacular sight as the bright cresc
lined up above the gentle rolling hills of New Arnc
Denise was sitting in the shade of a big cigni tree as t
slid together, its ginger leaves casting a broad dapple c
the grass around her. The cluster of seven-year-olds sit
on the ground sighed and cooed at the astral exhibition
they squinted really hard, they could just make out hair-
lines cutting across the distant planets. The silver thread
the world web spun out by the dragon in geostationary o
were becoming more complex as the englobing progress
Soon the whole world would be caged. These were exci
times for the children.

"Finish it," Jones Johnson whined plaintively. "Ple
Denise!"

But not as exciting as other people used to have,
thought in amusement. Little Jones was getting agitated,
face screwed up with urgency.

"Finish! Finish!" the rest of them chanted.

"All right," she said.

Lawrence Newton tried not to show any nerves as he p
sented his ticket to the departures desk in the center of
terminal building. Templeton Spaceport was a small affai
the north of the domed city, a couple of runways and f
hangars. It was built with arrivals in mind, twenty thous

a time when one of McArthur's starships decelerated into
bit. Traffic in the other direction was small and open to
rutiny.

The receptionist scanned his ticket in and smiled as her
eet screen scrolled confirmation. "Do you have any bag-
ge?" she asked.

"Er, one," Lawrence said. He was so relieved the Prime
d guarded him from his father's askpings he made a hash
lifting the case onto her scales. She leaned over and helped
m.

"Why are you going?" she asked.

"I, er, my family is sending me to university on Earth," he
ammered.

"Lucky you," she said brightly. "You can go through to the
unge now, Mr. Newton. Your flight leaves in forty minutes.
the snowplows can keep the runway clear."

"Thanks." His stomach felt curiously light.

Starflight! I'm going to have a starflight. I really am.

And that made up for absolutely everything.

He straightened his back and walked toward the lounge
or. Out of the corner of his eye he could see Vinnie Carl-
n hanging around by the main entrance. Lawrence gave
m a surreptitious thumbs-up. Vinnie winked back.

The lounge doors closed behind him. Ahead of him, the
cture window gave him a view out over the long arrow-
ad-shaped spaceplane that was due to take him up to the
lean. Lawrence hurried forward for a better look.

nnie Carlton watched the lounge doors shut. He went out
the taxi rank and climbed into the first of the little white
bbles. "Leith dome," he told the vehicle's AS.

A thin hail of ice twisted about them on the road. Vinnie
ttled back in the seat, closed his eyes and concentrated.
e flesh on his face rippled; characteristics began to shift,
s nose flattening slightly, chin protruding, eyebrows

straightening. He opened his eyes, now shaded halfway b
tween gray and green. When he looked in the taxi's mirr
to check, Lawrence Newton's young features looked ba
at him. It had been so many months he barely recogniz
himself.

Sadly he acknowledged that for all the perfect physical r
generation patternform had given him, he could never proje
the naïveté that his younger self had always possessed. A
inevitable consequence, he thought—not just from th
twenty years of experience he owned, but also the result
living beside his earlier self for so long. That proximity ha
been so much harder than he'd ever imagined. There ha
been so many times when, as Vinnie, he'd wanted to give th
love-stricken, horrifyingly ignorant teenage Lawrence a dan
good kicking.

But he hadn't. For all those months he'd been a go
friend on every level, gritting his teeth as he watched t
doomed love affair play out from a vantage point that was f
too close. Time and again, when they'd been out in a grou
he'd seen the fear hidden in Roselyn's eyes when Lawren
was looking the other way. She must have known he'd fi
out eventually. Still she carried on, just as besotted and trag
as that desperate teenage boy, clinging to her fragile hap
ness. He'd lost count of the times he'd almost surrendered
his stormy emotions and whispered to her: "It'll be all righ
Ultimately, he didn't have that right. For he didn't truly kn
yet if it would be.

The features on the fresh young face in the taxi's mirr
had sunk into a drawn, nervous expression as the weight
twenty years and two broken lives emerged as part of
genuine identity. "Now you've really come home," he sa
to it.

The elevator took an age to get up to the right flo
Lawrence remembered the anxiety he'd felt as he left Am
thi for Earth. This was worse. He took the memory chip fr

pocket and studied it curiously. The very last episode of *ght: Horizon.* And he still hadn't accessed the damn thing. ıst wouldn't have been right by himself. He flipped it like ›in and tucked it away again, smiling to himself. The ele- ›r doors opened, and his smile faded.

Iis legs were very unsteady as he walked down the corri- He stood outside the apartment door, too scared to move. *'ve invaded planets. I've been charged by a herd of :rorexes. I've seen dragons in their natural habitat. I've ked on the planet of the Mordiff. Now knock, you coward.*

Iis fingers rapped lightly on the door.

Roselyn opened it. She'd been crying.

'I'm an idiot," Lawrence blurted. "But I had a lot of time hink. A *lot* of time. And the one thing I want to do more n anything else in the universe is tell you that I love ı."

veral of the children had clasped their hands together as y gazed up at Denise. All of them had fallen silent. They re smiling contentedly.

Denise had about five seconds of peace while the story k in.

'But what did she say to him?"

'Did they get married?"

'Will they live happily ever after?"

She held up her hands in a plea for quiet. "I don't know ex- ly what happened next. Not there, or on Earth. But I'm e Lawrence and Roselyn must have spent a long and ›py time together. At least . . ."

'What?" Jones demanded, eyes bugging.

'It's just that Lawrence never wanted to get involved. But ıply knowing makes you involved. That's what the drag- s taught us: knowledge is the only true immortal. And he ew that in twenty years' time the *Clichane* was going to ar- e back on Earth with patternform and the dragons' memo-

ries. The human race would change. So there he was on
planet that was his true home, with the girl he loved. An
had the *Fool's Errand* with him. He had the knowledg
turn Amethi into a paradise and populate it with angels l
before Earth had the ability. Poor, dear old Lawrence. I w
der if he could resist the temptation."

About the Author

TER F. HAMILTON is the author of numerous short
ories and novels, including the acclaimed epic The
ght's Dawn trilogy (*The Reality Dysfunction*, *The
utronium Alchemist*, and *The Naked God*). Mr. Hamilton
es in England.

BRILLIANT, EPIC SCIENCE FICTION FROM ACCLAIMED AUTHOR
PETER F. HAMILTON

FALLEN DRAGON (0-446-52708-4)

a rebellious impulse, Lawrence joined the military arm of a rporation that he now recognizes to be ruthless and exploita-e. His only hope for escape is to earn enough money to buy his ace in a better corporation. When his platoon is sent to a dis-nt colony to quell a local resistance effort, Lawrence plans to the colony of their fabled gemstone, the Fallen Dragon, to get e money he needs. However, he soon discovers that the Fallen agon is not a gemstone at all . . .

DON'T MISS HAMILTON'S BESTSELLING AND BRILLIANT EPIC SCIENCE FICTION

The Night's Dawn Trilogy:

The Reality Dysfunction
Part 1: Emergence (0-446-60515-8)
Part 2: Expansion (0-446-60516-6)

The Neutronium Alchemist
Part 1: Consolidation (0-446-60517-4)
Part 2: Conflict (0-446-60546-8)

The Naked God
Part 1: Flight (0-446-60-897-1)
Part 2: Faith (0-446-60-518-2)

A Second Chance at Eden (0-446-60671-5)
short stories set in the same universe at The Night's Dawn Trilogy.

1120e

Warner Aspect is proud to announce an epic
new series from *New York Times* bestselling
author Kevin J. Anderson

The Saga of the Seven Suns
Book 1

HIDDEN EMPIRE
(0-446-52862-5)

With alien races, court intrigues, family dram
romance, battles, and all the pomp and pageantry
competing races in an expanding interstellar empi
Kevin J. Anderson's new epic science fiction series
a grand space adventure in the tradition of *Dune* a
Star Wars.

Enthusiastic Praise for HIDDEN EMPIRE

"Galactic fantasy at its adventurous and romant
best. . . . Intriguing characters love and battle amo
the stars. A great read!"

—Margaret Weis, author
Dragons of a Lost St

Watch for Book 2 of the series,

A FOREST OF STARS

Available from Warner Aspect in July 2003.

ACCLAIM FOR PETER F. HAMILTON AND *FALLEN DRAGON*, FINALIST FOR THE ARTHUR C. CLARKE AWARD

"A rousing, space-operative read."

—***New York Review of Science Fiction***

"Hamilton nicely develops character while he also does some fine world building that's as good as it gets....A sort of *Starship Trooper* as if written by Charles Dickens."

—***Publishers Weekly* (starred review)**

"A classic."

—***Cleveland Plain Dealer* on The Night's Dawn trilogy**

"The author's expansive vision of the future combines action and intrigue on a panoramic scale."

—***Library Journal***

"Absolutely vintage science fiction....Hamilton puts [sf] back into interstellar overdrive."

—***The Times (London)***

"Astounding...thrilling...surprising...an astute storyteller....Hamilton uses technology to excellent effect."

—***Science Fiction Age***

"Hamilton's joy in science-tethered flights of fancy is infectious—big, brash, sweeping, hyperbolic, exuberant, thunderously enjoyable."

—***Interzone***

"His imagination knows no bounds!"

—***Science Fiction Weekly***

"Shows how thought-provoking yet entertaining science fiction can be. Some of the best fiction...in years."

—***Midwest Book Review***

Also by Peter F. Hamilton

The Night's Dawn Trilogy
The Reality Dysfunction
Part 1: Emergence
The Reality Dysfunction
Part 2: Expansion
The Neutronium Alchemist
Part 1: Consolidation
The Neutronium Alchemist
Part 2: Conflict
The Naked God, Part 1: Flight
The Naked God, Part 2: Faith

A Second Chance at Eden
The Confederation Handbook: The Essential Guide to the Night's Dawn Series
Futures (with Stephen Baxter, Paul McAuley, and Ian McDonald)

Available from Warner Aspect

PETER F. HAMILTON

FALLEN DRAGON

WARNER BOOKS

An AOL Time Warner Company

This book is a work of fiction. Names, characters, places, and incidents are the product of the author's imagination or are used fictitiously. Any resemblance to actual events, locales, or persons, living or dead, is coincidental.

WARNER BOOKS EDITION

Cover design by Don Puckey
Cover illustration by Jim Burns

Warner Books, Inc.
1271 Avenue of the Americas
New York, NY 10020

Visit our Web site at
www.twbookmark.com

 An AOL Time Warner Company

Printed in the United States of America

Originally published in hardcover by Warner Books.
First Paperback Printing: March 2003

10 9 8 7 6 5 4 3 2 1